The Geometry of Curvature Homogeneous Pseudo-Riemannian Manifolds

ICP Advanced Texts in Mathematics   ISSN 1753-657X

Series Editor: Dennis Barden *(Univ. of Cambridge, UK)*

Published

Vol. 1   Recent Progress in Conformal Geometry
by Abbas Bahri & Yong Zhong Xu

Vol. 2   The Geometry of Curvature Homogeneous
Pseudo-Riemannian Manifolds
by Peter B. Gilkey

ICP Advanced Texts in Mathematics – Vol. 2

The Geometry of Curvature Homogeneous Pseudo-Riemannian Manifolds

Peter B. Gilkey
University of Oregon, USA

Imperial College Press

Published by

Imperial College Press
57 Shelton Street
Covent Garden
London WC2H 9HE

Distributed by

World Scientific Publishing Co. Pte. Ltd.

5 Toh Tuck Link, Singapore 596224

USA office: 27 Warren Street, Suite 401-402, Hackensack, NJ 07601

UK office: 57 Shelton Street, Covent Garden, London WC2H 9HE

British Library Cataloguing-in-Publication Data
A catalogue record for this book is available from the British Library.

**THE GEOMETRY OF CURVATURE HOMOGENEOUS
PSEUDO-RIEMANNIAN MANIFOLDS**
ICP Advanced Texts in Mathematics — Vol. 2

For photocopying of material in this volume, please pay a copying fee through the Copyright Clearance Center, Inc., 222 Rosewood Drive, Danvers, MA 01923, USA. In this case permission to photocopy is not required from the publisher.

ISBN-13 978-1-86094-785-8
ISBN-10 1-86094-785-9

Printed in Singapore by World Scientific Printers (S) Pte Ltd

Preface

This book arose out of a desire to investigate the relationship between certain algebraic properties of the curvature tensor and the underlying geometry of a pseudo-Riemannian manifold.

In Chapter 1, we discuss the geometry of the Riemannian curvature tensor. Basic geometrical notions are presented in Section 1.2. In Section 1.3, a passage to the algebraic context is given by introducing algebraic curvature tensors which are algebraic objects with the same symmetries as that of the Riemann curvature tensor. One says that a pseudo-Riemannian manifold is *curvature homogeneous* if the curvature tensor "looks the same at each point". This notion is made precise in Section 1.4. Section 1.5 presents some results from linear algebra and Section 1.6 provides concepts from differential geometry that will be needed subsequently. In Section 1.7, the geometry of the Jacobi operator is discussed and in Section 1.8, corresponding results for the curvature operator are given. Chapter 1 concludes in Section 1.9 with some general facts concerning the spectral geometry of the curvature tensor.

Chapter 2 deals with curvature homogeneous generalized plane wave manifolds. This is a class of manifolds that are geodesically complete, whose exponential map is surjective, and which have a number of other properties. In Section 2.2, we present the main geometrical results concerning this class of manifolds. The remainder of Chapter 2 deals with specific examples. Sections 2.3, 2.4 and 2.5 deal with manifolds of signature $(2,2)$, $(2,4)$, and (p,p), respectively. Section 2.6 treats generalized plane wave manifolds with flat factors, Section 2.7 discusses Nikčević manifolds, and Section 2.8 presents Dunn manifolds. Sections 2.9 and 2.10 deal with k-curvature homogeneous manifolds.

Chapter 3 discusses examples which are not generalized plane wave manifolds. Section 3.2 discusses Lorentz manifolds, Section 3.3 treats Walker manifolds of signature $(2, 2)$, Section 3.4 deals with questions of geodesic completeness and Ricci blowup, and Section 3.5 presents Fiedler manifolds.

Chapter 4 is more algebraic in nature. In Section 4.2, we present various topological results. In Section 4.3, we use the Nash embedding theorem to provide generators for the space of algebraic curvature tensors and algebraic covariant derivative tensors. Sections 4.4 and 4.5 treat Jordan Osserman algebraic curvature tensors and Szabó covariant derivative algebraic curvature tensors, respectively. In Section 4.6, we study questions in conformal geometry. Section 4.7 deals with Stanilov models. Section 4.8 treats complex geometry.

Chapter 5 contains a discussion of complex models which are both Osserman and complex Osserman; the classification is complete except in a few exceptional dimensions and ranks. Chapter 6 contains an introduction to Stanilov–Tsankov theory; this is a discussion of Jacobi Tsankov manifolds, skew Tsankov manifolds, Stanilov–Videv manifolds, and Jacobi Videv manifolds. Chapters 5 and 6 are joint work with M. Brozos-Vázquez.

Each chapter is divided into sections; the first section of a chapter provides an outline to the subsequent material in the chapter. Many sections are further divided into subsections. Theorems, lemmas, corollaries, and so forth are labeled by section. Equations which are cited are labeled by section; equations which are not cited are not labeled.

Much of this book reports on previous joint work with various authors. It is an honor and a privilege to acknowledge the contribution made by these colleagues: N. Blažić, N. Bokan, M. Brozos-Vázquez, J. Díaz-Ramos, C. Dunn, B. Fiedler, E. García-Río, R. Ivanova, J. V. Leahy, Z. Rakić, H. Sadofsky, U. Semmelman, U. Simon, G. Stanilov, I. Stavrov, A. Swann, L. Vanhecke, V. Videv, and T. Zhang. In addition to pleasant professional collaborations they have also enriched the personal life of the author.

The author expresses his appreciation to the Max Planck Institute in the Mathematical Sciences (Leipzig, Germany) where most of the writing and research took place. The author acknowledges with pleasure the support and encouragement of Lorraine Davis, Susana López-Ornat, Gwen Steigelman, and Arnie Zweig; without these individuals, the book would not have been written. The book is dedicated to my father and mother.

P. B. Gilkey

Contents

Chapter 1

The Geometry of the Riemann Curvature Tensor

1.1 Introduction

In Chapter 1, we present some basic results that we will be using subsequently; we also summarize some of the literature in the field. Throughout, we define a number of tensors. In the interests of brevity, we shall only give the non-zero entries up to the obvious symmetries. Here is a brief outline to Chapter 1.

In Section 1.2, we define the curvature of a connection on a vector bundle. We introduce affine manifolds and pseudo-Riemannian manifolds. We discuss scalar Weyl invariants, holonomy, and geodesics.

In Section 1.3, we pass to the algebraic context and discuss algebraic curvature tensors and algebraic covariant derivative curvature tensors. We introduce the Weyl tensor and we discuss k-curvature models and k-curvature homogeneity in both the affine and in the pseudo-Riemannian settings. There are certain canonical curvature tensors which will play a crucial role in our development. These are introduced in Section 1.3.2.

In Section 1.4, we give a brief survey of the literature concerning curvature homogeneity. In Theorem 1.4.1, we recall results of Prüfer, Tricerri, and Vanhecke (1996) relating the local scalar Weyl invariants to questions of homogeneity in the Riemannian setting. In Theorem 1.4.2, we recall a result of Singer (1960) and of Podesta and Spiro (1996) concerning k-curvature homogeneity; the appropriate generalization to the affine setting is given in in Theorem 1.4.3 which is due to Opozda (1996). Theorem 1.4.4 is due to Tricerri and Vanhecke (1986) in the Riemannian setting and to Cahen, Leroy, Parker, Tricerri, and Vanhecke (1991) in the Lorentzian setting and deals with manifolds which are 0-curvature modeled on irreducible symmetric spaces. We conclude Section 1.4 with a survey of the literature.

1

The subject is a vast one and we can provide only a very brief introduction to the field. For further details, the reader is urged to consult Boeckx, Kowalski, and Vanhecke (1996).

In Section 1.5, we summarize some needed results from linear algebra. We review the notion of Jordan normal form and discuss the spectrum of a linear operator. We show that given an arbitrary linear map T of a vector space V of dimension m, that there exists a non-degenerate inner product $\langle \cdot, \cdot \rangle$ on V with respect to which T is symmetric. We present some technical results concerning ordinary differential equations.

In Section 1.6, we summarize some elementary results from differential geometry. Let $\mathfrak{M}_1 = (V, \langle \cdot, \cdot \rangle, A_0, A_1)$ be a 1-model. Here $\langle \cdot, \cdot \rangle$ is a non-degenerate inner product on a vector space V of dimension m, A_0 is an algebraic curvature tensor on V, and A_1 is an algebraic covariant derivative curvature tensor on V. In Lemma 1.6.2, we show that $\mathfrak{M}_1$ is geometrically realizable; this fact plays an important role in the discussion of Section 4.3. We also introduce some technical facts concerning principal bundles.

Using symmetric and anti-symmetric bilinear forms, respectively, we define two families of algebraic curvature tensors that will play an important role in our investigations. In Theorem 1.6.1, we establish a result of Fiedler showing these algebraic curvature tensors span the vector space of all algebraic curvature tensors. In Section 1.6.4, we turn to complex geometry and give several equivalent conditions for the compatibility of an algebraic curvature tensor with a pseudo-Hermitian almost complex structure; this defines the notion of a *compatible complex model*. In Section 1.6.5, we introduce the pseudo-spheres and pseudo-complex projective spaces; any real or complex space form is locally isometric to one of these examples. Section 1.6.6 deals with conformal complex space forms. We conclude Section 1.6 in Section 1.6.7 with a quick review of Kähler geometry.

In Sections 1.7 and 1.8, we discuss natural operators related to the curvature tensor and to the covariant derivative of the curvature tensor. It is an interesting geometrical question to study when such an operator has constant eigenvalues or, more generally, constant Jordan normal form on the appropriate domain of definition. One often studies such questions first in the algebraic setting and then subsequently passes to the geometric setting; the second Bianchi identity then plays a crucial role in the analysis.

Here is a brief guide to some of the terminology that will be employed subsequently. We use the word "Osserman" when the underlying operator is related to the Jacobi operator and the word "Ivanov–Petrova" when the underlying operator is related to the skew-symmetric curvature operator. We shall use the words "spacelike", "timelike", "mixed", and "of type (r, s)" as modifiers reflecting the underlying domain of the operator. Thus, for example, the words "spacelike Osserman" will refer to the Jacobi operator on spacelike unit vectors and the words "timelike Ivanov–Petrova" will refer to the skew-symmetric curvature operator on timelike 2-planes. We will add the modifier "Jordan" when instead of studying the eigenvalues we are studying the Jordan normal form of the operator. The words "Tsankov" and "Videv" refer to various commutativity properties that will be discussed subsequently in Chapter 6.

We will usually work first in the algebraic context by defining the notions on a k-model. If $\mathcal{M}$ is a pseudo-Riemannian manifold, we add the modifier "pointwise" if the structures in question have a given property on $T_P(M)$ for each $P \in M$ but if the eigenvalues (or Jordan normal form) are permitted to vary with the point in question; the modifier "global" is added if in fact the eigenvalues (or Jordan normal form) do not vary with the point in question.

In Section 1.7, we discuss the Jacobi operator, the higher order Jacobi operator, and the Weyl conformal Jacobi operator. In Section 1.8, we treat the complex Jacobi operator, the skew-symmetric curvature operator, the Stanilov operator (higher order skew-symmetric curvature operator), the conformal skew-symmetric curvature operator, and the complex skew-symmetric curvature operator. We conclude our discussion by dealing with the Szabó operator.

In Section 1.9, we present various results concerning natural operators in Riemannian Geometry and survey the literature. If Θ is a natural operator of Riemannian geometry and if the natural domain $\mathcal{D}$ of Θ decomposes into disjoint sets $\mathcal{D}_1,...$ reflecting the various possible signatures, then one can complexify and use analytic continuation to see that the eigenvalues of Θ are constant on $\mathcal{D}_1$ if and only if they are constant on the remaining $\mathcal{D}_i$. Thus, for example, spacelike Osserman and timelike Osserman are equivalent conditions. We refer Theorem 1.9.1 for further details. In Section 1.9.2, we show k-Osserman manifolds are $m - k$ Osserman and Einstein; this is a useful duality result. In Section 1.9.3, we present work of Blažić concerning natural operators with bounded spectrum. Let $\mathfrak{M}_0$ be a 0-model of signature (p, q) where $p > 0$ and $q > 0$. We show that if the Jacobi operator of $\mathfrak{M}_0$ has bounded spectrum on the pseudospheres of timelike or spacelike unit vectors, then $\mathfrak{M}_0$ is Osserman; similar results hold for the other natural operators of Riemannian geometry.

We then survey the literature on the spectral geometry of the curvature tensor. The Osserman conjecture is discussed in Section 1.9.4, the higher order Jacobi operator is discussed in Section 1.9.5, the conformal Jacobi operator is discussed in Section 1.9.6, the Stanilov and Szabó operators are discussed in Section 1.9.7, the skew-symmetric curvature operator is discussed in Section 1.9.8, the conformal skew-symmetric curvature operator and the complex skew-symmetric curvature operator are discussed in Section 1.9.9. As was true for curvature homogeneity, the subject is a vast one and this section is only a very brief introduction; further information on these questions may be found in García-Río, Kupeli, and Vázquez-Lorenzo (2002) and in Gilkey (2002).

1.2 Basic Geometrical Notions

In this section, we shall define the basic concepts and notions we will be working with. Section 1.2.1 provides a brief introduction to linear algebra in the indefinite context. Section 1.2.2 deals with vector bundles, connections, and curvature. Section 1.2.3 introduces holonomy and parallel translation. Affine manifolds, geodesics, and completeness are discussed in Section 1.2.4. Section 1.2.5 is concerned with pseudo-Riemannian geometry, the Levi-Civita connection, and the Riemann curvature tensor. In Section 1.2.6, we use the metric to contract indices in pairs and construct scalar Weyl invariants.

1.2.1 *Vector spaces with symmetric inner products*

Let V be a real vector space of dimension m. Let V^* be the dual vector space and let $\mathrm{End}(V)$ be the vector space of all linear maps from V to V; setting $\eta^* \otimes \xi : v \to \eta^*(v)\xi$ identifies

$$\mathrm{End}(V) = V^* \otimes V\,.$$

Let $\mathrm{Gl}(V) \subset \mathrm{End}(V)$ be the *general linear group* on V; $\mathrm{Gl}(V)$ is the Lie-group of all invertible elements of $\mathrm{End}(V)$; the Lie algebra $\mathfrak{gl}(V)$ of $\mathrm{Gl}(V)$ is given by

$$\mathfrak{gl}(V) = \mathrm{End}(V)\,.$$

Let $\mathrm{Al}(V)$ be the group of invertible *affine linear maps* of V; $A \in \mathrm{Al}(V)$ if and only if $Ax = ax + b$ for some $a \in \mathrm{Gl}(V)$ and some $b \in V$. We define $\mathrm{Gl}(n)$ and $\mathrm{Al}(n)$ by taking $V = \mathbb{R}^n$.

Let $\langle \cdot, \cdot \rangle \in S^2(V^*)$ equip V with a symmetric bilinear inner product. We say that $\langle \cdot, \cdot \rangle$ is *non-degenerate* if $\langle \phi_1, \phi_2 \rangle = 0$ for all $\phi_2 \in V$ implies $\phi_1 = 0$. We define a linear map from V to V^* by setting $\phi_1^*(\phi_2) = \langle \phi_1, \phi_2 \rangle$. The inner product is non-degenerate if and only if this map is an isomorphism from V to V^*.

Let $\langle \cdot, \cdot \rangle$ be non-degenerate. Let $\mathrm{End}_\pm(V, \langle \cdot, \cdot \rangle)$ denote the subspaces of symmetric $(+)$ and anti-symmetric $(-)$ linear maps;

$$\mathrm{End}_\pm(V, \langle \cdot, \cdot \rangle) := \{\phi \in \mathrm{End}(V) : \langle \phi \xi_1, \xi_2 \rangle = \pm \langle \xi_1, \phi \xi_2 \rangle \ \forall \ \xi_i \in V\}.$$

Let $O(V, \langle \cdot, \cdot \rangle)$ be the associated *orthogonal group*;

$$O(V, \langle \cdot, \cdot \rangle) = \{T \in \mathrm{Gl}(V) : T^* \langle \cdot, \cdot \rangle = \langle \cdot, \cdot \rangle\}.$$

We let $\mathfrak{o}(V)$ denote the associated Lie algebra;

$$\mathfrak{o}(V) = \mathrm{End}_-(V, \langle \cdot, \cdot \rangle).$$

We say that $v \in V$ is *spacelike*, *timelike*, or *null* if $(v, v) > 0$, if $(v, v) < 0$, or if $(v, v) = 0$, respectively. Let $S^\pm(V, \langle \cdot, \cdot \rangle)$ be the *pseudo-spheres* of unit spacelike $(+)$ and timelike $(-)$ vectors:

$$S^\pm(V, \langle \cdot, \cdot \rangle) := \{v \in V : \langle v, v \rangle = \pm 1\}.$$

Similarly let $N(V, \langle \cdot, \cdot \rangle)$ be the *null cone*:

$$N(V, \langle \cdot, \cdot \rangle) := \{v \in V : \langle v, v \rangle = 0\}.$$

Let δ be the Kronecker symbol;

$$\delta_{ab} = \delta_a^b := \begin{cases} 1 \text{ if } a = b, \\ 0 \text{ if } a \neq b. \end{cases}$$

We can always find a basis $\mathcal{B} := \{e_1^-, ..., e_p^-, e_1^+, ..., e_q^+\}$ for V so that

$$\langle e_i^-, e_j^- \rangle = -\delta_{ij}, \quad \langle e_i^-, e_a^+ \rangle = 0, \quad \text{and} \quad \langle e_a^+, e_b^+ \rangle = \delta_{ab}.$$

Such a basis $\mathcal{B}$ is said to be an *orthonormal basis* for V. The pair (p, q) is called the *signature of V* and is independent of the particular orthonormal basis chosen. Note that $p + q = m$. We say that $\langle \cdot, \cdot \rangle$ has *neutral signature* if $p = q$.

We can define an inner product of signature (p, q) on $\mathbb{R}^{p+q}$ by setting

$$\langle \vec{x}, \vec{y} \rangle := -x_1 y_1 - ... - x_q y_q + x_{q+1} y_{q+1} + ... + x_{p+q} y_{p+q}. \tag{1.2.a}$$

Let $\mathbb{R}^{(p,q)}$ denote $\mathbb{R}^{p+q}$ with this particular inner product and let $O(p,q)$ be the associated orthogonal group. Choosing an orthonormal basis $\mathcal{B}$ for V defines an isometry from V to $\mathbb{R}^{(p,q)}$ that identifies the two Lie groups

$$O(V, \langle \cdot, \cdot \rangle) = O(p,q).$$

We say that $\pi \subset V$ is a *non-degenerate* subspace of signature (r,s) if the restriction of the inner product to the subspace π is non-degenerate and has signature (r,s). Let $\mathrm{Gr}_{r,s}(V)$ denote the *Grassmannian* of all such subspaces and let $\mathrm{Gr}^+_{r,s}(V)$ be the Grassmannian of all oriented such subspaces; these Grassmannians are connected and non-trivial if

$$1 \leq r + s \leq m - 1, \quad 0 \leq r \leq p, \quad \text{and} \quad 0 \leq s \leq q.$$

Such values of (r,s) will be said to be an *admissible pair*. Forgetting the orientation defines a double cover

$$\mathbb{Z}_2 \to \mathrm{Gr}^+_{r,s}(V) \to \mathrm{Gr}_{r,s}(V).$$

More generally, if we permit $\phi \in S^2(V^*)$ to be degenerate, we can let $\ker \phi := \{\xi : \phi(\xi, \eta) = 0 \ \forall \ \eta\}$. We pass to the quotient $V/\ker \phi$, to see that we can construct a basis for V so that $\phi(e_i, e_j) = 0$ for $i \neq j$ and so that $\phi(e_i, e_i) \in \{0, \pm 1\}$. If $\phi \in \Lambda^2(V^*)$, we can choose a basis $\{e_1, ..., e_r, f_1, ..., f_r, n_1, ..., n_s\}$ so the non-zero entries in ϕ are $\phi(e_i, f_i) = 1$.

1.2.2 *Vector bundles, connections, and curvature*

We shall work in the smooth context, unless otherwise noted. Thus the words "M is a manifold" are to be understood to mean "M is a smooth manifold", the words "f is a function" are to be understood to mean "f is a smooth function", and so on. Occasionally, we shall have to restrict to the real analytic context, but this will be clearly noted, as, for example, in Theorem 2.2.2.

Let M be a connected manifold of dimension m. Let $T(M)$ and $T^*(M)$ be the tangent and cotangent bundles of M, respectively. If $\xi \in C^\infty(T(M))$ is a vector field on M and if $\xi^* \in C^\infty(T^*(M))$ is a 1-form on M, then the natural pairing between cotangent and tangent vectors defines a function $\xi^*(\xi) \in C^\infty(M)$.

Let $\rho : V \to M$ be a vector bundle over M. Let V_P be the fiber of V over $P \in M$. Let V^* be the dual bundle of V and let $S^2(V^*)$ be the bundle of symmetric 2-cotensors on V. Let $\mathrm{End}(V)$ be the bundle of endomorphisms of V. Let $\Lambda(V)$ be the exterior algebra bundle.

Let $C^\infty(V)$ be the space of sections to V. If $\phi \in C^\infty(V)$, we shall sometimes use the notation ϕ_P to denote the element $\phi(P) \in V_P$. A section $g \in C^\infty(S^2(V^*))$ defines a symmetric bilinear inner product on V. We say g is *non-degenerate* if $g_P \in S^2(V_P^*)$ is non-degenerate for every $P \in M$.

Let ∇ be a *connection* on V. If $\xi \in C^\infty(T(M))$ is a tangent vector field on M and if $\phi \in C^\infty(V)$ is a section to V, then $\nabla_\xi \phi$ is once again a section to V. Let $f \in C^\infty(M)$. Then ∇ is bilinear, function linear in the first argument, and behaves like a derivation in the second argument:

$$\nabla_{\{\xi_1+\xi_2\}}\{\phi_1 + \phi_2\} = \nabla_{\xi_1}\phi_1 + \nabla_{\xi_1}\phi_2 + \nabla_{\xi_2}\phi_1 + \nabla_{\xi_2}\phi_2,$$
$$\nabla_{f\xi}\phi = f\nabla_\xi\phi, \quad \text{and} \quad \nabla_\xi\{f\phi\} = f\nabla_\xi\phi + \{\xi(f)\}\phi.$$

In particular, if $\xi_1(P) = \xi_2(P)$, then

$$\{\nabla_{\xi_1}\phi\}(P) = \{\nabla_{\xi_2}\phi\}P.$$

Covariant differentiation is a local operator. If $\mathcal{O}$ is an open subset of M, if $\xi_1|_{\mathcal{O}} = \xi_2|_{\mathcal{O}}$, and if $\phi_1|_{\mathcal{O}} = \phi_2|_{\mathcal{O}}$, then

$$\nabla_{\xi_1}\phi_1|_{\mathcal{O}} = \nabla_{\xi_2}\phi_2|_{\mathcal{O}}.$$

We extend ∇ to act naturally on the associated tensor bundles. For example, the *induced connections* on $C^\infty(V^*)$, on $C^\infty(S^2(V^*))$, on $C^\infty(\mathrm{End}(V))$, and on $C^\infty(\Lambda(V))$ are characterized, respectively, by the identities:

$$\{\nabla_\xi \phi^*\}\phi = \xi\{\phi^*(\phi)\} - \phi^*(\nabla_\xi\phi),$$
$$\{\nabla_\xi g\}(\phi_1, \phi_2) = \xi\{g(\phi_1, \phi_2)\} - g(\nabla_\xi\phi_1, \phi_2) - g(\phi_1, \nabla_\xi\phi_2),$$
$$\{\nabla_\xi E\}(\phi) = \nabla_\xi\{E\phi\} - E\{\nabla_\xi\phi\},$$
$$\nabla_\xi\{w_1 \wedge w_2\} = \{\nabla_\xi w_1\} \wedge w_2 + w_1 \wedge \{\nabla_\xi w_2\}.$$

Let $\rho_2 : V_2 \to M_2$ be a vector bundle over M_2. Let $f : M_1 \to M_2$. The *pull-back bundle* f^*V_2 is the vector bundle over M_1 defined by

$$f^*(V_2) := \{(P_1, v_2) \in M_1 \times V_2 : f(P_1) = \rho_2(v_2)\}$$

with the projection $\rho_1(P_1, v_2) := P_1$ from $f^*(V_2)$ to M_1. If $\phi_2 \in C^\infty(V_2)$, then the *pull-back section* is given by:

$$f^*(\phi_2)(P_1) := (P_1, \phi_2(f(P_2))).$$

If $g_2 \in S^2(V_2^*)$ defines a non-degenerate fiber metric on V_2, then the pull-back f^*g_2 is a non-degenerate fiber metric on V_1 which is defined by:

$$f^*(g_2)((P_1, v_2), (P_1, w_2)) = g_2(v_2, w_2) \, .$$

If ∇ is a connection on V_2, then the pull-back connection $f^*\nabla$ is characterized by the identity:

$$\{(f^*\nabla)_{\xi_1} f^*\phi\}(P_1) = (P_1, \{\nabla_{f_*\xi_1}\phi\}(f(P_1))) \, .$$

If $\xi_1, \xi_2 \in C^\infty(T(M))$, let $[\xi_1, \xi_2] := \xi_1\xi_2 - \xi_2\xi_1$ be the *Lie bracket*. Let $\mathcal{R}$ be the curvature operator:

$$\mathcal{R}(\xi_1, \xi_2) := \nabla_{\xi_1}\nabla_{\xi_2} - \nabla_{\xi_2}\nabla_{\xi_1} - \nabla_{[\xi_1, \xi_2]} \, .$$

The curvature is tensorial; $\{\mathcal{R}(\xi_1, \xi_2)\phi\}(P)$ depends only on $\xi_1(P)$, on $\xi_2(P)$, and on $\phi(P)$. If $\{e_i\}$ is a local frame for $T(M)$ and if $\{\phi_a\}$ is a local frame for V, then

$$\mathcal{R}\left(\sum_i u_i e_i, \sum_j v_i e_j\right)\left\{\sum_a w_a \phi_a\right\} = \sum_{i,j,a} u_i v_j w_a \mathcal{R}(e_i, e_j)\phi_a \, .$$

We may regard

$$\mathcal{R} \in C^\infty(T^*M \otimes T^*M \otimes V^* \otimes V) \, .$$

Let V be trivial over an open subset $\mathcal{O}$. Let $\{\phi_1, ..., \phi_s\}$ be a local frame for $V|_{\mathcal{O}}$ and let $x = (x_1, ..., x_m)$ be local coordinates on $\mathcal{O}$. Decompose

$$\nabla_{\partial_{x_i}}\phi_a = \sum_b \omega_{ia}{}^b \phi_b \, . \tag{1.2.b}$$

The *connection 1-form* ω completely describes the connection ∇ on $\mathcal{O}$. Let $\xi = \sum_i \xi^i \partial_{x_i}$ be a vector field on $\mathcal{O}$ and let $\phi = \sum_a \alpha^a \phi_a$ be a section to V on $\mathcal{O}$; here $\xi^i, \alpha^a \in C^\infty(\mathcal{O})$. One has:

$$\nabla_\xi \phi = \sum_{i,a} \xi^i \partial_{x_i}\{\alpha^a\}\phi_a + \sum_{i,a,b} \xi^i \alpha^a \omega_{ia}{}^b \phi_b \, .$$

The curvature tensor may then be described by:

$$\mathcal{R}(\partial_{x_i}, \partial_{x_j})\phi_a = \sum_b R_{ija}{}^b \phi_b$$

where

$$R_{ija}{}^b := \partial_{x_i}\omega_{ja}{}^b - \partial_{x_j}\omega_{ia}{}^b + \sum_c (\omega_{ic}{}^b \omega_{ja}{}^c - \omega_{jc}{}^b \omega_{ia}{}^c) \, .$$

If $V = TM$, one has *Christoffel symbols* for ∇:

$$\nabla_{\partial_{x_i}} \partial_{x_j} = \sum_k \Gamma_{ij}{}^k \partial_{x_k} \,. \tag{1.2.c}$$

Denote the ith covariant derivative of the curvature operator by

$$\nabla^i \mathcal{R} \in C^\infty(\otimes^{i+2} T^*(M) \otimes V^* \otimes V) \,.$$

We shall use the following notations for the evaluation:

$$(\nabla^i \mathcal{R})(\xi_1, \xi_2; \xi_3, ..., \xi_{i+2})\phi \quad \text{or} \quad \nabla_{\xi_{i+2}}...\nabla_{\xi_3}\mathcal{R}(\xi_1, \xi_2)\phi \,.$$

One has inductively that

$$\begin{aligned}
&(\nabla^i \mathcal{R})(\xi_1, ..., \xi_{i+2})\phi \\
&= \nabla_{\xi_{i+2}}\{(\nabla^{i-1}\mathcal{R})(\xi_1, ..., \xi_{i+1})\phi\} - (\nabla^{i-1}\mathcal{R})(\xi_1, ..., \xi_{i+1})\nabla_{\xi_{i+2}}\phi \\
&\quad - \sum_{j=1}^{i+1}(\nabla^{i-1}\mathcal{R})(\xi_1, ..., \nabla_{\xi_{i+2}}\xi_j, ..., \xi_{i+1})\phi \,.
\end{aligned}$$

It is immediate from the definition that if $i \geq 2$, then one has the following *commutation formula for covariant differentiation*

$$\begin{aligned}
&(\nabla^i\mathcal{R})(\xi_1, ..., \xi_{i+1}, \xi_{i+2}) - (\nabla^i\mathcal{R})(\xi_1, ..., \xi_{i+2}, \xi_{i+1}) \\
&= \mathcal{R}(\xi_{i+2}, \xi_{i+1})\{(\nabla^{i-2}\mathcal{R})(\xi_1, \xi_2, ..., \xi_i)\} \\
&\quad - \sum_{1 \leq j \leq i}(\nabla^{i-2}\mathcal{R})(\xi_1, ..., \mathcal{R}(\xi_{i+2}, \xi_{i+1})\xi_j, ..., \xi_i) \\
&\quad -\{(\nabla^{i-2}\mathcal{R})(\xi_1, ..., \xi_i)\}\mathcal{R}(\xi_{i+2}, \xi_{i+1}) \,.
\end{aligned} \tag{1.2.d}$$

Thus, for example, we have

$$\begin{aligned}
&(\nabla^2\mathcal{R})(\xi_1, \xi_2; \xi_3, \xi_4) - (\nabla^2\mathcal{R})(\xi_1, \xi_2; \xi_4, \xi_3) \\
&= \mathcal{R}(\xi_4, \xi_3)\mathcal{R}(\xi_1, \xi_2) - \mathcal{R}(\xi_1, \xi_2)\mathcal{R}(\xi_4, \xi_3) \\
&\quad - \mathcal{R}(\mathcal{R}(\xi_4, \xi_3)\xi_1, \xi_2) - \mathcal{R}(\xi_1, \mathcal{R}(\xi_4, \xi_3)\xi_2) \,.
\end{aligned}$$

Fix a non-singular symmetric inner product $g \in C^\infty(S^2(V^*))$ on the vector bundle V. We say that ∇ is *Riemannian* if $\nabla g = 0$. In terms of the connection 1-form defined in Eq. (1.2.b), this means equivalently that relative to any orthonormal frame field $\{\phi_1, ..., \phi_s\}$ that

$$\omega_{ia}{}^b + \omega_{ib}{}^a = 0 \,.$$

If ∇ is Riemannian, then $\mathcal{R}$ is skew-symmetric:

$$0 = g(\mathcal{R}(\xi_1, \xi_2)\phi_1, \phi_2) + g(\phi_1, \mathcal{R}(\xi_1, \xi_2)\phi_2) \,.$$

In the special case that $V = T(M)$, we say that ∇ is *torsion free* if

$$\nabla_{\xi_1}\xi_2 - \nabla_{\xi_2}\xi_1 = [\xi_1, \xi_2].$$

Equivalently in the notation of Eq. (1.2.c), this means that relative to a coordinate frame $\{\partial_{x_1}, ..., \partial_{x_m}\}$, we have the symmetry property:

$$\Gamma_{ij}{}^k = \Gamma_{ji}{}^k.$$

We set

$$\Gamma_{ijk} := \sum_l g_{kl}\Gamma_{ij}{}^l = g(\nabla_{\partial_{x_i}}\partial_{x_j}, \partial_{x_k}).$$

There is a unique torsion free Riemannian connection on $T(M)$ which is called the Levi-Civita connection; the Christoffel symbols of the Levi-Civita connection are given in Eq. (1.2.f).

1.2.3 *Holonomy and parallel translation*

Let ∇ be a connection on a vector bundle V. Let γ be a curve in M. We say that a section $\xi(t)$ to V along γ is *parallel* if $\nabla_{\dot\gamma(t)}\xi(t) = 0$. If $\{\phi_a\}$ is a local frame field, we may expand $\xi(t) = \sum_a \xi^a(t)\phi_a(\gamma(t))$. Expand $\gamma(t) = (\gamma^1(t), ..., \gamma^m(t))$ in a system of local coordinates. Then ξ is parallel if and only if

$$0 = \dot\xi^a(t) + \sum_{b,i} \dot\gamma^i(t)\xi^b(t)\omega_{ib}{}^a(\gamma(t)) = 0 \quad \forall a.$$

Consequently, given $\xi \in V_{\gamma(0)}(M)$, there is a unique parallel section $\xi(t)$ along γ with $\xi(0) = \xi_0$. The map $\mathfrak{P}_\gamma : \xi_0 \to \xi(1)$ is called *parallel translation*; $\mathfrak{P}_\gamma$ defines a linear isomorphism from $V_{\gamma(0)}$ to $V_{\gamma(1)}$. We say that γ is a *closed loop* at $P \in M$ if $\gamma(0) = \gamma(1) = P$. Let

$$\mathcal{H}_P := \{\mathfrak{P}_\gamma : \gamma(0) = \gamma(1) = P\} \subset \mathrm{Gl}(V_P)$$

be the *holonomy group* at P. Since M is connected, the isomorphism class of the Lie group $\mathcal{H}_P$ is independent of the point $P \in M$.

Let $g \in C^\infty(S^2(V^*))$ be a non-degenerate inner product on the vector bundle V and let ∇ be a Riemannian connection. The holonomy group is then a subgroup of the orthogonal group $O(V_P, g_P)$ since parallel translation preserves the inner product.

The holonomy is said to be *reducible* if there exists a proper invariant subspace; this means there exists $V_1 \subset V_P$ so that

$$0 \subsetneq V_1 \subsetneq V_P \quad \text{and} \quad \mathcal{H}_P V_1 = V_1\,.$$

Otherwise, the holonomy is said to be *irreducible*. The holonomy is said to be *decomposable* if there is a non-trivial invariant decomposition of V_P. This means that we can find a decomposition $V_P = V_1 \oplus V_2$ with

$$0 \subsetneq V_i \subsetneq T_P M \quad \text{and} \quad \mathcal{H}_P V_i = V_i\,.$$

Otherwise the holonomy is said to be *indecomposable*. If the connection is Riemannian with respect to a positive definite inner product, these two notions coincide. These notions are distinct for indefinite metrics, see, for example, the discussion in Section 3.2.

1.2.4 *Affine manifolds, geodesics, and completeness*

Let ∇ be a torsion free connection on $T(M)$. The pair $\mathcal{F} := (M, \nabla)$ is said to be an *affine manifold*. An affine manifold $\mathcal{F}$ is said to be reducible, irreducible, decomposable, or indecomposable when the holonomy has this property.

We say that a curve γ in M is a *geodesic* if $\dot\gamma$ is a parallel vector field along γ; this means that γ solves the *geodesic equation* $\nabla_{\dot\gamma}\dot\gamma = 0$. Let $\gamma = (\gamma^1(t), ..., \gamma^m(t))$ be a curve. Then γ is a geodesic if and only if

$$0 = \ddot\gamma^k + \sum_{ij} \dot\gamma^i \dot\gamma^j \Gamma_{ij}{}^k \quad \text{for} \quad 1 \leq k \leq m\,. \tag{1.2.e}$$

Thus the fundamental theorem of ordinary differential equations shows that given $P \in M$ and given $\xi_0 \in T_P M$, there is a unique geodesic $\gamma_{P,\xi}$ so that

$$\gamma_{P,\xi}(0) = P \quad \text{and} \quad \dot\gamma_{P,\xi}(0) = \xi_0\,.$$

Two torsion free connections ∇ and $\check\nabla$ are said to be *projectively equivalent* if their unparametrized geodesics coincide or equivalently by Eisenhart (1964), if there exists a 1-form $\hat\Theta$ so that

$$\check\nabla_v u - \nabla_v u = \hat\Theta(u)v + \hat\Theta(v)u\,.$$

If γ is a curve and if $s \in \mathbb{R}^+$, we rescale γ to define the curve γ^s by setting $\gamma_s(t) = \gamma(st)$. If γ is a geodesic, then γ^s is once again a geodesic. Since $\gamma^s(0) = P$ and $\dot\gamma^s(0) = s\dot\gamma(0)$, $\gamma_{P,\xi}(st) = \gamma_{P,s\xi}(t)$. Consequently by the fundamental theorem of ordinary differential equations, there exists an open neighborhood $\mathcal{O}$ of $O \in T_P M$ so that $\gamma_{P,\xi}(t)$ exists for $|t| \leq 1$ and for all $\xi \in \mathcal{O}$. We define the *exponential map*

$$\exp_P : \mathcal{O} \to M \quad \text{by} \quad \exp_P(\xi) := \gamma_{P,\xi}(1) \, .$$

It is then immediate that $\gamma_{P,\xi}(t) = \exp_P(t\xi)$. The map $\xi \to \exp_P(\xi)$ defines a coordinate system on M; such coordinates are called *geodesic coordinates* centered at P. Straight lines through the origin are geodesics in these coordinate systems.

We say that an affine manifold $\mathcal{F}$ is *geodesically complete* if all geodesics extend for infinite time; this means that $\exp_P$ is defined on all of $T_P M$ for any point P in M. We say that $\mathcal{F}$ is *strongly geodesically convex* if there exists a unique geodesic between any two points of M; $\mathcal{F}$ is complete and strongly geodesically convex if and only if the exponential map is a diffeomorphism from $T_P M$ to M for any $P \in M$.

Let $\rho(u,v) := \mathrm{Tr}\{y \to \mathcal{R}(y,u)v\}$ be the *Ricci tensor*. This tensor need not be symmetric for a general affine connection; ∇ is Ricci symmetric if and only if ∇ locally admits a parallel volume form, see Pinkall, Schwenk-Schellschmidt, and Simon (1994) for details. If $f \in C^\infty(M)$, then the *Hessian* $H_\nabla(f) \in C^\infty(S^2(T^*M))$ is defined by

$$H_\nabla(f)(u,v) := u(v(f)) - df(\nabla_u v) \, .$$

The Hessian is symmetric if and only if ∇ is torsion free.

One says $\mathcal{M} = (M,g)$ exhibits *Ricci blowup* if there exists a geodesic γ whose parameter range is $[0,T)$ for $T < \infty$ and where

$$\limsup_{t \to T} |\rho(\dot{\gamma}, \dot{\gamma})| = \infty \, .$$

Necessarily such a manifold is incomplete; it may not be embedded as an open subset of a complete affine manifold. Examples of locally homogeneous 3-dimensional Lorentz manifolds with Ricci blowup are given in Theorem 3.2.1.

1.2.5 *Pseudo-Riemannian manifolds*

Let $g \in C^\infty(S^2(T^*(M)))$. We say that g is a *pseudo-Riemannian metric of signature* (p,q) on M if g_P is a non-degenerate symmetric inner product of signature (p,q) on $T_P M$ for every point P of M. The pair $\mathcal{M} := (M,g)$ is said to be a *pseudo-Riemannian manifold of signature* (p,q). Note that $p + q = m$. We say that $\mathcal{M}$ is a *Riemannian manifold* if $p = 0$; we say that

M is a *Lorentzian manifold* if $p = 1$. Every manifold admits a Riemannian metric. There are topological obstructions to admitting metrics of higher signature. For example, if M is compact, then M admits a Lorentzian metric if and only if the Euler-Poincaré characteristic of M vanishes. In Theorem 4.2.3, we relate the signature of indefinite metrics on spheres to the Adams number.

The *Levi-Civita connection* on M is a connection on $T(M)$ which is characterized by the following properties:

$$\nabla_{\xi_1}\xi_2 - \nabla_{\xi_2}\xi_1 = [\xi_1, \xi_2] \qquad \text{[torsion free]}$$
$$\xi_3 g(\xi_1, \xi_2) = g(\nabla_{\xi_3}\xi_1, \xi_2) + g(\xi_1, \nabla_{\xi_3}\xi_2) \quad \text{[Riemannian]}.$$

Let $x = (x_1, ..., x_m)$ be a system of local coordinates on M. Set

$$g_{ij} := g(\partial_{x_i}, \partial_{x_j})$$

and let g^{ij} be the inverse matrix;

$$\sum_j g_{ij} g^{jk} = \delta_i^k.$$

Introduce *Christoffel symbols* $\Gamma_{ijk} := g(\nabla_{\partial_{x_i}}\partial_{x_j}, \partial_{x_k})$. We then have

$$\Gamma_{ijk} = \tfrac{1}{2}\{\partial_{x_j} g_{ik} + \partial_{x_i} g_{jk} - \partial_{x_k} g_{ij}\}. \qquad (1.2.\text{f})$$

It is then easily checked that

$$\Gamma_{ijk} = \Gamma_{jik} \qquad \text{so} \quad \nabla \text{ is torsion free,}$$
$$\Gamma_{ijk} + \Gamma_{ikj} = \partial_{x_i} g_{jk} \quad \text{so} \quad \nabla \text{ is Riemannian}.$$

One has that

$$\nabla_{\partial_{x_i}}\partial_{x_j} = \sum_k \Gamma_{ij}{}^k \partial_{x_k} \quad \text{where} \quad \Gamma_{ij}{}^k = \sum_\ell g^{\ell k}\Gamma_{ij\ell}.$$

Relative to the coordinate frame, the Lie bracket is trivial and the derivatives of the metric encode the Levi-Civita connection. With an orthonormal frame $\{e_i\}$, the reverse is true. Let $[e_i, e_j] = \sum_k C_{ijk} e_k$ describe the Lie bracket of a local orthonormal frame. We then have

$$\nabla_{e_i} e_j = \sum_k \tfrac{1}{2}\{C_{ijk} - C_{jki} + C_{kij}\} e_k.$$

The corresponding affine structure is given by taking $\mathcal{F}(\mathcal{M}) := (M, \nabla)$ and the curvature tensor $R \in \otimes^4(T^*(M))$ of $\mathcal{M}$ is defined by setting:

$$R(\xi_1, \xi_2, \xi_3, \xi_4) := g(\mathcal{R}(\xi_1, \xi_2)\xi_3, \xi_4).$$

This satisfies the usual curvature symmetries:

$$\begin{aligned}
R(\xi_1, \xi_2, \xi_3, \xi_4) &= -R(\xi_2, \xi_1, \xi_3, \xi_4) = R(\xi_3, \xi_4, \xi_1, \xi_2), \\
0 &= R(\xi_1, \xi_2, \xi_3, \xi_4) + R(\xi_2, \xi_3, \xi_1, \xi_4) + R(\xi_3, \xi_1, \xi_2, \xi_4).
\end{aligned} \tag{1.2.g}$$

The first relations are $\mathbb{Z}_2$ symmetries, the final relation is the *first Bianchi identity*. Relative to a coordinate frame,

$$\mathcal{R}(\partial_{x_i}, \partial_{x_j})\partial_{x_k} = \sum_\ell R_{ijk}{}^\ell \partial_{x_\ell} \quad \text{and} \quad R(\partial_{x_i}, \partial_{x_j}, \partial_{x_k}, \partial_{x_\ell}) = R_{ijk\ell}$$

where $R_{ijk\ell} = \sum_n g_{n\ell} R_{ijk}{}^n$ and where

$$R_{ijk}{}^\ell = \partial_{x_i}\Gamma_{jk}{}^\ell - \partial_{x_j}\Gamma_{ik}{}^\ell + \sum_n \left\{ \Gamma_{in}{}^\ell \Gamma_{jk}{}^n - \Gamma_{jn}{}^\ell \Gamma_{ik}{}^n \right\}.$$

We set

$$\ker(R) := \{ \eta \in TM : R(\eta, \xi_1, \xi_2, \xi_3) = 0 \quad \forall \quad \xi_i \in TM \}.$$

We define the covariant derivative $\nabla R \in C^\infty(\otimes^5 T^*(M))$ by setting

$$\nabla R(\xi_1, \xi_2, \xi_3, \xi_4; \xi_5) := g(\nabla_{\xi_5}\mathcal{R}(\xi_1, \xi_2)\xi_3, \xi_4).$$

This tensor has the symmetries

$$\begin{aligned}
\nabla R(\xi_1, \xi_2, \xi_3, \xi_4; \xi_5) &= -\nabla R(\xi_2, \xi_1, \xi_3, \xi_4; \xi_5) \\
&= \nabla R(\xi_3, \xi_4, \xi_1, \xi_2; \xi_5), \\
0 = \nabla R(\xi_1, \xi_2, \xi_3, \xi_4; \xi_5) &+ \nabla R(\xi_2, \xi_3, \xi_1, \xi_4; \xi_5) \\
&+ \nabla R(\xi_3, \xi_1, \xi_2, \xi_4; \xi_5), \\
0 = \nabla R(\xi_1, \xi_2, \xi_3, \xi_4; \xi_5) &+ \nabla R(\xi_1, \xi_2, \xi_4, \xi_5; \xi_3) \\
&+ \nabla R(\xi_1, \xi_2, \xi_5, \xi_3; \xi_4).
\end{aligned} \tag{1.2.h}$$

The first relations are $\mathbb{Z}_2$ symmetries, the next relation is the covariant derivative of the first Bianchi identity, and the final relation is the *second Bianchi identity*. The tensors $\nabla^i R \in C^\infty(\otimes^{4+i} T^*(M))$ for $i > 1$ are defined similarly.

Let $\mathcal{R}$ be the curvature operator. The *Jacobi operator* is given by:

$$\mathcal{J}(\xi) : \eta \to \mathcal{R}(\eta, \xi)\xi.$$

Let σ be a geodesic in $\mathcal{M}$. A vector field Y along σ is said to be a *Jacobi vector field* if the *Jacobi equation* is satisfied:

$$\ddot{Y} + J(\dot{\sigma})Y = 0\,.$$

The fundamental theorem of ordinary differential equations shows that a Jacobi vector field Y along a curve is specified by $Y(0)$ and by $\dot{Y}(0)$.

Jacobi vector fields arise in the study of geodesic sprays. We say that $T : [a, b] \times [c, d] \to M$ is a *geodesic spray* if the curves $\sigma_v : u \to T(u, v)$ are geodesics for every $v \in [c, d]$. The following is well known:

Lemma 1.2.1

(1) Let T be a geodesic spray. Then $T_*(\partial_v)$ is a Jacobi vector field along σ_v for every v.

(2) Let $Y(u)$ be a Jacobi vector field along a geodesic $\sigma : [0, T] \to M$. Then there exists $\varepsilon > 0$ and a geodesic spray $T : [0, \varepsilon] \times [0, \varepsilon] \to M$ so that $\sigma(u) = T(u, 0)$ and so that $Y(u) = T_*(\partial_v)|_{v=0}$ for $u \in [0, \varepsilon]$.

One can symmetrize the Jacobi or polarize the Jacobi operator to define

$$\mathcal{J}(x, y) : \eta \to \tfrac{1}{2}\{\mathcal{R}(\eta, x)y + \mathcal{R}(\eta, y)x\}\,.$$

Note that $\rho(x, y) = \mathrm{Tr}\{\mathcal{J}(x, y)\}$ and $\mathcal{J}(x) = \mathcal{J}(x, x)$. We say that $\mathcal{M}$ is *Einstein* if there is a constant c_1 so that $\rho = c_1 g$. We say that $\mathcal{M}$ is *k-stein* if there exist constants c_i such that

$$\mathrm{Tr}\{\mathcal{J}(\xi)^i\} = c_i g(\xi, \xi)^i \quad \text{for all} \quad \xi \in T(M) \quad \text{and} \quad 1 \leq i \leq k\,.$$

This definition is motivated by the observation that 1-stein and Einstein are equivalent notions as noted by Carpenter, Gray, and Willmore (1982).

1.2.6 *Scalar Weyl invariants*

Let $\nabla^k \mathcal{R}$ be the kth covariant derivative of the curvature operator defined by the Levi-Civita connection. Let $x := (x_1, ..., x_m)$ be local coordinates on M. Expand

$$\nabla_{\partial_{x_{j_1}}} ... \nabla_{\partial_{x_{j_l}}} \mathcal{R}(\partial_{x_{i_1}}, \partial_{x_{i_2}})\partial_{x_{i_3}} = R_{i_1 i_2 i_3}{}^{k}{}_{;j_\ell, ..., j_1} \partial_{x_k}$$

where we adopt the *Einstein convention* and sum over repeated indices. Scalar invariants of the metric can be formed by using the metric tensors g^{ij} and g_{ij} to *fully contract* all indices. For example, one may use the

Einstein convention to define the *scalar curvature* τ, the *norm of the Ricci tensor* $|\rho|^2$, and the *norm of the full curvature tensor* $|R|^2$ by:

$$\tau := g^{ij} R_{kij}{}^k ,$$
$$|\rho|^2 := g^{i_1 j_1} g^{i_2 j_2} R_{ki_1 j_1}{}^k R_{li_2 j_2}{}^l , \quad \text{and} \qquad (1.2.\mathrm{i})$$
$$|R|^2 := g^{i_1 j_1} g^{i_2 j_2} g^{i_3 j_3} g_{i_4 j_4} R_{i_1 i_2 i_3}{}^{i_4} R_{j_1 j_2 j_3}{}^{j_4} .$$

Such invariants are called *Weyl invariants*; if all possible such invariants vanish, then $\mathcal{M} = (M, g)$ is said to be *VSI* (vanishing scalar invariants).

In Chapter 2, we will discuss a number of families of pseudo-Riemannian manifolds. Many of these families will have vanishing scalar invariants; the task then is to construct other invariants to distinguish them. For example, in Theorem 2.3.2, we consider the following family of manifolds. Let $(x, y, \tilde{x}, \tilde{y})$ be coordinates on $\mathbb{R}^4$, let $f \in C^\infty(\mathbb{R})$, and let $\mathcal{M}_f := (\mathbb{R}^4, g_f)$ where g_f is the metric of neutral signature $(2, 2)$ on $\mathbb{R}^4$ given by:

$$g_f(\partial_x, \partial_x) = -2f(y), \quad g_f(\partial_x, \partial_{\tilde{x}}) = g_f(\partial_y, \partial_{\tilde{y}}) = 1 .$$

We show that these manifolds all have vanishing scalar invariants. Assume that $f^{(3)} \neq 0$. For $k \geq 2$, set:

$$\alpha_k(f) := \left\{ f^{(k+2)} \{ f^{(2)} \}^{k-1} \{ f^{(3)} \}^{-k} \right\} .$$

We shall show presently that:

(1) If there is a local isometry $\Phi : \mathcal{M}_{f_1} \to \mathcal{M}_{f_2}$ with $\Phi(P_1) = P_2$, then $\alpha_k(f_1)(P_1) = \alpha_k(f_2)(P_2)$ for all integers $k \geq 2$.
(2) If f_1 and f_2 are real analytic and if $\alpha_k(f_1)(P_1) = \alpha_k(f_2)(P_2)$ for all integers $k \geq 2$, then there is an isometry $\Phi : \mathcal{M}_{f_1} \to \mathcal{M}_{f_2}$ such that $\Phi(P_1) = P_2$.

1.3 Algebraic Curvature Tensors and Homogeneity

Section 1.3.1 deals with algebraic curvature tensors and covariant derivative algebraic curvature tensors. In Section 1.3.2, we introduce the canonical curvature tensors associated to a symmetric or an anti-symmetric bilinear form. In Section 1.3.3, we decompose the action of the orthogonal group on the space of algebraic curvature tensors to define the Weyl conformal curvature tensor. The notion of a k-curvature model $\mathfrak{M}_k$ is introduced in Section 1.3.4; this encodes information on the covariant derivatives of the curvature up to order k. In Section 1.3.5, various notions of homogeneity are presented. In Section 1.3.6 we introduce the notion of a Killing vector field; in Section 1.3.7 we discuss nilpotent curvature.

1.3.1 *Algebraic curvature tensors*

Let V be a real vector space of dimension m which is equipped with a non-singular inner product $\langle \cdot, \cdot \rangle$.

Definition 1.3.1 A 4-tensor $A \in \otimes^4 V^*$ is said to be an *algebraic curvature tensor* if it satisfies the relations of Eq. (1.2.g); this means that

(1) $A(\xi_1, \xi_2, \xi_3, \xi_4) = -A(\xi_2, \xi_1, \xi_3, \xi_4)$.
(2) $A(\xi_1, \xi_2, \xi_3, \xi_4) + A(\xi_2, \xi_3, \xi_1, \xi_4) + A(\xi_3, \xi_1, \xi_2, \xi_4) = 0$.
(3) $A(\xi_1, \xi_2, \xi_3, \xi_4) = -A(\xi_1, \xi_2, \xi_4, \xi_3)$.
(4) $A(\xi_1, \xi_2, \xi_3, \xi_4) = A(\xi_3, \xi_4, \xi_1, \xi_2)$.

Let $\{e_1, ..., e_n\}$ be a basis for V. We set $g_{ij} := \langle e_i, e_j \rangle$. If $A \in \otimes^4(V^*)$, we set $A_{ijkl} := A(e_i, e_j, e_k, e_l)$. The components g_{ij} determine the inner product; the components A_{ijkl} determines the tensor A. If A is an algebraic curvature tensor, then the *Ricci tensor* is defined by

$$\rho_A(\xi_1, \xi_2) = \sum_{jk} g^{jk} A(\xi_1, e_j, e_k, \xi_2).$$

There is a bit of redundancy in Definition 1.3.1.

Lemma 1.3.1 *Let $A \in \otimes^4 V^*$. Suppose that Properties (1) and (2) of Definition 1.3.1 are satisfied. Then the following assertions are equivalent:*

(1) $A(\xi_1, \xi_2, \xi_3, \xi_4) = -A(\xi_1, \xi_2, \xi_4, \xi_3)$.
(2) $A(\xi_1, \xi_2, \xi_3, \xi_4) = A(\xi_3, \xi_4, \xi_1, \xi_2)$.

Proof. Assume that (1) holds. Let $\xi_i \in V$. Set

$$A(\xi_1, \xi_2, \xi_3, \xi_4) = a_1, \quad A(\xi_3, \xi_4, \xi_1, \xi_2) = a_1 + \varepsilon_1,$$
$$A(\xi_1, \xi_3, \xi_2, \xi_4) = a_2, \quad A(\xi_2, \xi_4, \xi_1, \xi_3) = a_2 + \varepsilon_2,$$
$$A(\xi_2, \xi_3, \xi_1, \xi_4) = a_3, \quad A(\xi_1, \xi_4, \xi_2, \xi_3) = a_3 + \varepsilon_3.$$

We then use the first Bianchi identity to compute:

$$0 = A(\xi_1, \xi_2, \xi_3, \xi_4) + A(\xi_2, \xi_3, \xi_1, \xi_4) + A(\xi_3, \xi_1, \xi_2, \xi_4)$$
$$= a_1 - a_2 + a_3,$$
$$0 = A(\xi_1, \xi_2, \xi_4, \xi_3) + A(\xi_2, \xi_4, \xi_1, \xi_3) + A(\xi_4, \xi_1, \xi_2, \xi_3)$$
$$= -a_1 + a_2 - a_3 + \varepsilon_2 - \varepsilon_3,$$

$$0 = A(\xi_1, \xi_3, \xi_4, \xi_2) + A(\xi_3, \xi_4, \xi_1, \xi_2) + A(\xi_4, \xi_1, \xi_3, \xi_2)$$
$$= -a_2 + a_1 + a_3 + \varepsilon_1 + \varepsilon_3,$$
$$0 = (\xi_2, \xi_3, \xi_4, \xi_1) + A(\xi_3, \xi_4, \xi_2, \xi_1) + A(\xi_4, \xi_2, \xi_3, \xi_1)$$
$$= -a_3 - a_1 + a_2 - \varepsilon_1 + \varepsilon_2.$$

This yields the relations:

$$0 = \varepsilon_2 - \varepsilon_3 = \varepsilon_1 + \varepsilon_3 = -\varepsilon_1 + \varepsilon_2$$

from which it follows that $\varepsilon_1 = \varepsilon_2 = \varepsilon_3 = 0$ which establishes Assertion (2). The implication (2) $\Rightarrow$ (1) is immediate. $\qquad\square$

Definition 1.3.2   A 5-tensor $A_1 \in \otimes^5 V^*$ is said to be a *covariant derivative algebraic curvature tensor* if it satisfies the relations of Eq. (1.2.h):

(1) $A_1(\xi_1, \xi_2, \xi_3, \xi_4; \xi_5) = -A_1(\xi_2, \xi_1, \xi_3, \xi_4; \xi_5)$.
(2) $A_1(\xi_1, \xi_2, \xi_3, \xi_4; \xi_5) + A_1(\xi_2, \xi_3, \xi_1, \xi_4; \xi_5) + A_1(\xi_3, \xi_1, \xi_2, \xi_4; \xi_5) = 0$.
(3) $A_1(\xi_1, \xi_2, \xi_3, \xi_4; \xi_5) + A_1(\xi_1, \xi_2, \xi_4, \xi_5; \xi_3) + A_1(\xi_1, \xi_2, \xi_5, \xi_3; \xi_4) = 0$.
(4) $A_1(\xi_1, \xi_2, \xi_3, \xi_4; \xi_5) = -A_1(\xi_1, \xi_2, \xi_4, \xi_3; \xi_5)$.
(5) $A_1(\xi_1, \xi_2, \xi_3, \xi_4; \xi_5) = A_1(\xi_3, \xi_4, \xi_1, \xi_2; \xi_5)$.

The same argument used to prove Lemma 1.3.1 shows that (4) and (5) are equivalent in Definition 1.3.2 given that (1) and (2) hold. One sets $\mathcal{A}lg_0(V) \subset \otimes^4 V^*$ (respectively $\mathcal{A}lg_1(V) \subset \otimes^5 V^*$) to be the space of all algebraic curvature tensors (respectively of all algebraic covariant derivative tensors). We do not introduce the spaces $\mathcal{A}lg_k(V)$ for $k \geq 2$ as the symmetries are more complicated and involve lower order terms as is shown by Eq. (1.2.d).

In the study of affine geometry, one has the following notion:

Definition 1.3.3   We say that $\mathcal{A} \in \otimes^2 V^* \otimes \mathrm{End}(V) = \otimes^3 V^* \otimes V$ is an *affine curvature operator* if

(1) $\mathcal{A}(\xi_1, \xi_2)\xi_3 = -\mathcal{A}(\xi_2, \xi_1)\xi_3$.
(2) $\mathcal{A}(\xi_1, \xi_2)\xi_3 + \mathcal{A}(\xi_2, \xi_3)\xi_1 + \mathcal{A}(\xi_3, \xi_1)\xi_2 = 0$.
(3) $\mathrm{Tr}\{\mathcal{A}(\xi_1, \xi_2)\} = 0$.

In the presence of an inner product $\langle \cdot, \cdot \rangle$ on V, one can lower indices and define $A \in \otimes^4 V^*$ by the identity:

$$A(\xi_1, \xi_2, \xi_3, \xi_4) = \langle \mathcal{A}(\xi_1, \xi_2)\xi_3, \xi_4 \rangle .$$

Let $\mathbb{F}_g$ be the set of such tensors; they are characterized by the properties:

(1) $A(\xi_1, \xi_2, \xi_3, \xi_4) = -A(\xi_2, \xi_1, \xi_3, \xi_4)$.

(2) $A(\xi_1,\xi_2,\xi_3,\xi_4) + A(\xi_2,\xi_3,\xi_1,\xi_4) + A(\xi_3,\xi_1,\xi_2,\xi_4) = 0$.
(3) $\sum_{kl} g^{kl} A_{ijkl} = 0$.

Let $S_0^2(V^*, \langle\cdot,\cdot\rangle)$ be the vector space of trace free symmetric bilinear forms. If $A \in \mathbb{F}_g$ and if $S \in \Lambda^2(V^*) \otimes S_0^2(V^*, \langle\cdot,\cdot\rangle)$, let

$$\pi_s(A)_{ijkl} = \tfrac{1}{2}\{A_{ijkl} + A_{ijlk}\} \in \Lambda^2(V^*) \otimes S_0^2(V^*, \langle\cdot,\cdot\rangle),$$
$$\alpha(S)_{ijkl} := \tfrac{1}{2}\{S_{kjil} + S_{ikjl} - S_{ljik} - S_{iljk}\} \in \Lambda^2(\Lambda^2(V^*)).$$

One then has the following relationship between the spaces $\mathcal{A}\mathrm{lg}_0$ and $\mathbb{F}_g$ as representation spaces for the orthogonal group as shown by Blažić, Gilkey, Nikĕcvić, and Simon (2006):

Lemma 1.3.2 *There is a natural short exact sequence*

$$0 \to \mathcal{A}\mathrm{lg}_0 \to \mathbb{F}_g \xrightarrow{\pi_s} \Lambda^2(V^*) \otimes S_0^2(V^*, \langle\cdot,\cdot\rangle) \to 0.$$

The map $\mathrm{Id} + \alpha$ *splits the projection* π_s.

Proof. If $A \in \ker(\pi_s)$, then A satisfies:

$$A_{ijkl} = -A_{jikl}, \quad A_{ijkl} + A_{jkil} + A_{kijl} = 0, \quad A_{ijkl} = -A_{ijlk}.$$

By Lemma 1.3.1, $A_{ijkl} = A_{klij}$ and thus A is an algebraic curvature tensor. Conversely, of course, if $A \in \mathcal{A}\mathrm{lg}_0$, then $A_{ijkl} + A_{ijlk} = 0$. Thus

$$\ker(\pi_s) \cap \mathbb{F}_g = \mathcal{A}\mathrm{lg}_0.$$

If $S \in \Lambda^2(V^*) \otimes S_0^2(V^*, \langle\cdot,\cdot\rangle)$, let $T = \alpha(S)$. We have

$$T_{ijkl} = \tfrac{1}{2}\{S_{kjil} + S_{ikjl} - S_{ljik} - S_{iljk}\},$$
$$T_{jikl} = \tfrac{1}{2}\{S_{kijl} + S_{jkil} - S_{lijk} - S_{jlik}\} = -T_{jikl},$$
$$T_{klij} = \tfrac{1}{2}\{S_{ilkj} + S_{kilj} - S_{jlki} - S_{kjli}\} = -T_{ijkl}.$$

Thus $\alpha(S) \in \Lambda^2(\Lambda^2(V^*))$. In particular $\pi_s\alpha(S) = 0$ so

$$\pi_s(\mathrm{id} + \alpha) = \mathrm{Id} \quad \text{on} \quad \Lambda^2(V^*) \otimes S_0^2(V^*, \langle\cdot,\cdot\rangle).$$

To prove that $S + \alpha(S) \in \mathbb{F}_g$, we need only check that the first Bianchi identity is satisfied:

$$\begin{aligned}
&(\mathrm{id} + \alpha)S_{ijkl} + (\mathrm{id} + \alpha)S_{jkil} + (\mathrm{id} + \alpha)S_{kijl} \\
&= S_{ijkl} + S_{jkil} + S_{kijl} \\
&\quad + \tfrac{1}{2}\{S_{kjil} + S_{ikjl} + S_{jikl} + S_{ikjl} + S_{jikl} + S_{kjil}\} \\
&\quad - \tfrac{1}{2}\{S_{ljik} + S_{lkji} + S_{likj} + S_{iljk} + S_{jlki} + S_{klij}\} \\
&= 0 \, .
\end{aligned}$$

The orthogonally equivariant decomposition

$$\mathcal{F}_g = \mathcal{A}\mathrm{lg}_0 \oplus \Lambda^2(V^*) \otimes S_0^2(V^*, \langle \cdot, \cdot \rangle)$$

of the Lemma now follows. $\qquad\qquad\square$

This decomposition is not irreducible. The map α can be used to define a natural short exact sequence

$$\begin{aligned}
0 \to &\{\Lambda^2(V^*) \otimes S_0^2(V^*, \langle \cdot, \cdot \rangle)\} \cap \mathbb{F}_g \to \Lambda^2(V^*) \otimes S^2(V^*) \\
&\xrightarrow{\ \alpha\ } \Lambda_0^2(\Lambda^2(V^*)) \to 0
\end{aligned}$$

which is split by the map

$$\beta(T)_{ijkl} := \tfrac{1}{2}(T_{kjil} - T_{kijl})$$

where

$$\Lambda_0^2(\Lambda^2(V^*)) := \beta^{-1}\{\Lambda^2(V^*) \otimes S_0^2(V^*, \langle \cdot, \cdot \rangle)\} \, .$$

Thus we may decompose

$$\mathbb{F}_g = \mathcal{A}\mathrm{lg}_0 \oplus \big\{\{\Lambda^2(V^*) \otimes S_0^2(V^*, \langle \cdot, \cdot \rangle)\} \cap \mathbb{F}_g\big\} \oplus \Lambda_0^2(\Lambda^2 V^*)$$

as a representation space of the orthogonal group. This leads to the following result of Blažić, Gilkey, Nikěcvić, and Simon (2006):

Theorem 1.3.1

(1) There is an $O(V, \langle \cdot, \cdot \rangle)$ equivariant orthogonal decomposition of

$$\mathbb{F}_g \approx W_1 \oplus W_2 \oplus W_4 \oplus W_5 \oplus W_6 \oplus W_7 \oplus W_8$$

as the direct sum of irreducible $O(V, \langle \cdot, \cdot \rangle)$ modules where:

$$\dim\{W_1\} = 1,$$
$$\dim\{W_2\} = \dim\{W_5\} = \tfrac{1}{2}(m-1)(m+2),$$
$$\dim\{W_4\} = \tfrac{1}{2}m(m-1),$$
$$\dim\{W_6\} = \tfrac{1}{12}m(m+1)(m-3)(m+2),$$
$$\dim\{W_7\} = \tfrac{1}{8}(m-1)(m-2)(m+1)(m+4),$$
$$\dim\{W_8\} = \tfrac{1}{8}m(m-1)(m-3)(m+2).$$

(2) There are the following isomorphisms as $O(\langle\cdot,\cdot\rangle)$ modules:

(a) $W_1 \approx \mathbb{R}$.
(b) $W_2 \approx W_5 \approx S_0^2(V^*, \langle\cdot,\cdot\rangle)$.
(c) $W_4 \approx \Lambda^2(V^*)$.
(d) $W_6 = \{A \in \mathrm{Alg}_0 : \rho_A = 0\} = $ Weyl conformal curvature tensors.
(e) $W_8 \approx \{\Theta \in \otimes^4 V^* : \Theta_{ijkl} = -\Theta_{jikl} = -\Theta_{klij}, g^{il}\Theta_{ijkl} = 0\}$.

The module W_7 is a bit more difficult to describe explicitly and we refer to Blažić, Gilkey, Nikěcvić, and Simon (2006) for details. The notation is taken from that established by Bokan (1990); if one drops the condition that $g^{kl}A_{ijkl} = 0$, one then obtains the space of curvature tensors $\mathfrak{R}(V)$ and has

$$\mathfrak{R}(V) = \mathbb{F}_g(V) \oplus W_3 \quad \text{where} \quad W_3 \approx \Lambda^2(V^*).$$

We shall return to this class again in Section 6.3 in our discussion of Jacobi Tsankov tensors.

1.3.2 *Canonical curvature tensors*

Let $S^2(V^*)$ and $\Lambda^2(V^*)$ be the spaces of symmetric and anti-symmetric bilinear forms on V. If $\Phi_+ \in S^2(V^*)$ and if $\Phi_- \in \Lambda^2(V^*)$, define 4-tensors

$$\begin{aligned}
A_{\Phi_+}(x,y,z,w) &:= \Phi_+(x,w)\Phi_+(y,z) - \Phi_+(x,z)\Phi_+(y,w), \\
A_{\Phi_-}(x,y,z,w) &:= \Phi_-(x,w)\Phi_-(y,z) - \Phi_-(x,z)\Phi_-(y,w) \\
&\quad -2\Phi_-(x,y)\Phi_-(z,w)\,.
\end{aligned} \qquad (1.3.\mathrm{a})$$

Define a corresponding linear map $\phi \in \mathrm{End}(V)$ so that $\langle \phi x, y \rangle = \Phi(x,y)$; if $\Phi \in S^2(V^*)$, then $\phi \in \mathrm{End}_+(V, \langle\cdot,\cdot\rangle)$ is symmetric; if $\Phi \in \Lambda^2(V^*)$, then $\phi \in \mathrm{End}_-(V, \langle\cdot,\cdot\rangle)$ is skew-symmetric. The associated curvature operators are then defined by

$$\begin{aligned}
\mathcal{A}_{\Phi_+}(x,y)z &:= \langle \phi_+ y, z \rangle \phi_+ x - \langle \phi_+ x, z \rangle \phi_+ y, \\
\mathcal{A}_{\Phi_-}(x,y)z &:= \langle \phi_- y, z \rangle \phi_- x - \langle \phi_- x, z \rangle \phi_- y - 2\langle \phi_- x, y \rangle \phi_- z\,.
\end{aligned} \qquad (1.3.\mathrm{b})$$

In Lemma 1.6.3 we will show that these tensors are algebraic curvature tensors and in Theorem 1.6.1, we will show that

$$\mathcal{A}lg_0 = \mathrm{Span}_{\phi_+ \in S^2(V^*)}\{A_{\phi_+}\} = \mathrm{Span}_{\phi_- \in \Lambda^2(V^*)}\{A_{\phi_-}\};$$

this result is originally due to Fiedler (2003a). If $\phi = c\,\mathrm{Id}$ and, correspondingly, if $\Phi = c\langle\cdot,\cdot\rangle$, then A is said to have *constant sectional curvature c.* The curvature tensor of constant sectional curvature $+1$ will be denoted by

$$A_{\langle\cdot,\cdot\rangle}(x,y,z,w) := \langle x,w\rangle\langle y,z\rangle - \langle x,z\rangle\langle y,w\rangle\,.$$

Algebraic curvature tensors defined by a symmetric bilinear form appear in hypersurface theory. Let M^m be a hypersurface in $\mathbb{R}^{m+k}$. We work in the positive definite setting, but there is a natural pseudo-Riemannian generalization. Fix a point $P \in M$ and choose an orthonormal basis $\{\nu_1, ..., \nu_k\}$ for the normal plane at P. Let

$$L_i(\xi_1,\xi_2) := \langle d_{\xi_1}\xi_2, \nu_i\rangle$$

be the second fundamental form defined by ν_i at P; this is a symmetric bilinear form on $T_P M$. We then have:

$$R_P = \sum_{i=1}^{k} R_{L_i}\,.$$

The curvature tensors which arise from skew-symmetric bilinear forms also have geometric significance. Let $\langle\cdot,\cdot\rangle$ be a Riemannian inner product on V. Let $\mathcal{F} = \{J_1, ..., J_\ell\}$ be a *Clifford family* on V. Here the J_i are skew-symmetric endomorphisms of V which satisfy the *Clifford commutation relations*:

$$J_i J_j + J_j J_i = -2\delta_{ij}\,\mathrm{Id}\,.$$

The J_i form an anti-commuting family of Hermitian almost complex structures on V. Given real constants c_i, following Eq. (1.3.a), one defines:

$$A := c_0 A_{\langle\cdot,\cdot\rangle} + \sum_{i=1}^{\ell} c_i A_{J_i}\,.$$

We shall show presently in Lemma 1.6.3 that $A \in \mathcal{A}lg_0(V)$. Such a curvature tensor is said to be *given by a Clifford family.*

Let $\mathbb{CP}^n$ and $\mathbb{HP}^n$ denote complex projective space and quaternionic projective space. We endow these spaces with the *Fubini–Study.* Let J be the natural almost complex structure on $\mathbb{CP}^n$ and let $\{J_1, J_2, J_3\}$ be the natural quaternionic structure on $\mathbb{HP}^n$. We then have, see Section 1.6.5 for details,

$$R_{\mathbb{CP}^n} = R_{\mathrm{Id}} + R_J \quad \text{and} \quad R_{\mathbb{HP}^n} = R_{\mathrm{Id}} + R_{J_1} + R_{J_2} + R_{J_3}\,.$$

1.3.3 *The Weyl conformal curvature tensor*

There is a natural representation of the orthogonal group $O(V, \langle \cdot, \cdot \rangle)$ on the space of algebraic curvature tensors. This representation is not irreducible but decomposes as the direct sum of 3 irreducible representations which we can describe as follows. It is not necessary to assume the metric in question is positive definite. Let $g_{ij} := \langle e_i, e_j \rangle$ and let g^{ij} be the inverse matrix relative to some basis $\{e_i\}$ for V. Let A be an algebraic curvature tensor on V. The associated Ricci tensor ρ_A and scalar curvature τ_A are then defined by contracting indices:

$$\rho_A(x, y) := g^{ij} A(x, e_i, e_j, y) \quad \text{and} \quad \tau_A := g^{ij} \rho_A(e_i, e_j)\,.$$

One then has $O(V, \langle \cdot, \cdot \rangle)$ equivariant maps

$$\sigma_\rho : A \to \rho_A \in S^2(V^*) \quad \text{and} \quad \sigma_\tau : A \to \tau_A \in \mathbb{R}\,.$$

The space of *algebraic Weyl curvature tensors*

$$\mathcal{W}(V, \langle \cdot, \cdot \rangle) := \ker(\sigma_\rho)$$

is an irreducible representation space for $O(V, \langle \cdot, \cdot \rangle)$ and:

$$\mathcal{A}\mathrm{lg}_0(V) = \mathcal{W}(V, \langle \cdot, \cdot \rangle) \oplus S^2(V^*)$$

as an $O(V, \langle \cdot, \cdot \rangle)$ representation space. Note that $\mathcal{W}$ is the space W_6 of Theorem 1.3.1. The further decomposition of $S^2(V^*)$ as the direct sum of the trace free tensors and the scalar multiples of the identity then completes the decomposition of $\mathcal{A}\mathrm{lg}_0(V)$ as a direct sum of irreducible $O(V, \langle \cdot, \cdot \rangle)$ modules. More explicitly, one sets $S_0^2(V^*, \langle \cdot, \cdot \rangle) := \ker(\tau_A)$ and then uses τ_A to decompose

$$S^2(V^*) = S_0^2(V^*, \langle \cdot, \cdot \rangle) \oplus \mathbb{R}\,.$$

This then leads to the full decomposition

$$\mathcal{A}\mathrm{lg}_0(V) = \mathcal{W}(V, \langle \cdot, \cdot \rangle) \oplus S_0^2(V^*, \langle \cdot, \cdot \rangle) \oplus \mathbb{R}\,.$$

Orthogonal projection $\pi_{\mathcal{W}}$ from $\otimes^4 V^*$ to $\mathcal{W}(V, \langle \cdot, \cdot \rangle)$ is given by:

$$
\begin{aligned}
\pi_{\mathcal{W}}(A)(x, y, z, w) = {} & A(x, y, z, w) \\
& - \tfrac{1}{m-2}\{\rho_A(x, w)\langle y, z \rangle + \langle x, w \rangle \rho_A(y, z)\} \\
& + \tfrac{1}{m-2}\{\rho_A(x, z)\langle y, w \rangle + \langle x, z \rangle \rho_A(y, w)\} \\
& + \tfrac{1}{(m-1)(m-2)}\tau_A\{\langle x, w \rangle \langle y, z \rangle - \langle x, z \rangle \langle y, w \rangle\}.
\end{aligned}
$$

If $\mathcal{M}$ is a pseudo-Riemannian manifold, then $\pi_{\mathcal{W}} R$ is said to be the associated *Weyl conformal curvature tensor*. Let $g_1 = \alpha g_2$ be a conformally equivalent metric where $\alpha \in C^\infty(M)$ is a positive function. The Weyl conformal curvature tensor $W \in \otimes^4(T^*(M))$ simply rescales:

$$
W_{g_1}(x, y, z, w) = \alpha W_{g_2}(x, y, z, w).
$$

We may summarize the relevant facts we need as follows:

Lemma 1.3.3 $\mathcal{A}lg_0(V)$ *is an irreducible* $\mathrm{Gl}(V)$ *module. As an* $O(V, \langle \cdot, \cdot \rangle)$ *module, we may decompose* $\mathcal{A}lg_0(V) = \mathcal{W}(V, \langle \cdot, \cdot \rangle) \oplus S_0^2(V^*, \langle \cdot, \cdot \rangle) \oplus \mathbb{R}$.

1.3.4   *Models*

Suppose given $A_i \in \otimes^{4+i} V^*$ for $0 \le i \le k$. We assume $A_0 \in \mathcal{A}lg_0(V)$ and $A_1 \in \mathcal{A}lg_1(V)$ but impose no relations on A_i for $i \ge 2$ as giving the requisite symmetries would be unnecessarily complex. Let

$$
\mathfrak{M}_k := (V, \langle \cdot, \cdot \rangle, A_0, A_1, ..., A_k)
$$

be the associated *k-model*. If $k = \infty$, then the string is infinite. If $\mathfrak{M}_k^1$ (respectively $\mathfrak{M}_k^2$) is a k-model defined on a vector space V^1 (respectively on V^2), then ϕ is said to be an *isomorphism* from $\mathfrak{M}_k^1$ to $\mathfrak{M}_k^2$ if ϕ is a linear isomorphism from V^1 to V^2 so that

$$
\phi^*\{\langle \cdot, \cdot \rangle^2\} = \langle \cdot, \cdot \rangle^1 \quad \text{and} \quad \phi^*\{A_i^2\} = A_i^1 \quad \text{for} \quad 0 \le i \le k.
$$

The notion of an *affine k-model* is defined similarly. The inner product $\langle \cdot, \cdot \rangle$ plays no role and instead of considering $A_i \in \otimes^{4+i}(V^*)$, one considers

$$
\mathcal{A}_i \in \otimes^{3+i}(V^*) \otimes V = \otimes^{2+i} V^* \otimes \mathrm{End}(V)
$$

which are algebraic counterparts of $\nabla^i \mathcal{R}$. If $\mathfrak{M}_k$ is a k-model, we can define the associated affine k-model $\mathfrak{F}_k$ by using the inner product to raise indices to define $\mathcal{A}_i \in \otimes^{2+i} V^* \otimes \mathrm{End}(V)$. In the interests of notational simplicity, we set $A := A_0$ and $\mathcal{A} := \mathcal{A}_0$. In the interests of brevity, we shall sometimes say that $\mathfrak{M}$ is a *model* if it is a 0-model or that $\mathfrak{F}$ is an affine model if it is an affine 0-model.

A model $\mathfrak{M}_k$ is said to be *decomposable* if there exists a non-trivial orthogonal decomposition $V = V_1 \oplus V_2$ that induces a direct sum decomposition $A_i = A_{i,1} \oplus A_{i,2}$ for $1 \le i \le k$; if $\mathfrak{M}_k$ is not decomposable, then $\mathfrak{M}_k$ is said to be indecomposable. Similarly an affine model $\mathfrak{F}_k$ is said to be *decomposable* if there exists a non-trivial direct sum decomposition $V = V_1 \oplus V_2$ that induces a decomposition $\mathcal{A}_i = \mathcal{A}_{i,1} \oplus \mathcal{A}_{i,2}$.

If P is a point of a pseudo-Riemannian manifold $\mathcal{M}$, one defines the *associated k-model* $\mathfrak{M}_k(\mathcal{M}, P)$ by setting:

$$\mathfrak{M}_k(\mathcal{M}, P) := (T_P M, g_P, R_P, ..., \nabla^k R_P).$$

This is a purely algebraic structure which encodes the covariant derivatives of the curvature operator up to order k at P as well as describing the inner product on $T_P(M)$. Similarly if $\mathcal{F}$ is an affine manifold, one defines the *affine k-model* of $\mathfrak{F}_k(\mathcal{F}, P)$ at $P \in M$ by setting:

$$\mathfrak{F}_k(\mathcal{F}, P) := (T_P M, \mathcal{R}_P, ..., \nabla^k \mathcal{R}_P). \tag{1.3.c}$$

Many geometric properties have algebraic counterparts. For example, if $\mathfrak{F}_0$ is an affine 0-model, one may define the *Jacobi operator* $\mathcal{J}$ and the *Ricci tensor* ρ, by setting, respectively,

$$\mathcal{J}(v_1, v_2) : v_3 \to \tfrac{1}{2}\{\mathcal{A}(v_3, v_1)v_2 + \mathcal{A}(v_3, v_2)v_1\},$$
$$\mathcal{J}(v) := \mathcal{J}(v, v), \quad \text{and} \quad \rho(v_1, v_2) := \operatorname{Tr} \mathcal{J}(v_1, v_2).$$

We say that $\mathfrak{M}_0$ is *Einstein* if $\rho = c\langle \cdot, \cdot \rangle$; we say that $\mathfrak{M}_0$ is *k-stein* if there exist constants c_i so that

$$\operatorname{Tr}\{\mathcal{J}(v)^i\} = c_i \langle v, v \rangle^i \quad \text{for} \quad 1 \le i \le k.$$

Let $\mathfrak{M}_1$ be a 1-model. We will show in Section 4.3 that $\mathfrak{M}_1$ is *geometrically realizable*; this means that there exists a pseudo-Riemannian manifold $\mathcal{M}$ and that there is a point $P \in M$ so that $\mathfrak{M}_1$ is isomorphic to $\mathfrak{M}_1(\mathcal{M}, P)$. Thus the relations of Eqs. (1.2.g) and (1.2.h) are the only universal relations that R and ∇R satisfy; the relations that $\nabla^2 R$ satisfy are more complicated owing to the intertwining relation given in Eq. (1.2.d). We refer to Belger and Kowalski (1994) for further details.

Let $\mathfrak{M}$ and $\mathcal{F}$ be k-models. Let $\mathrm{Gl}(\mathfrak{M})$ and $\mathrm{Gl}(\mathfrak{F})$ be the Lie group of all isomorphisms $\mathfrak{M}$ and $\mathfrak{F}$; let $\mathfrak{gl}(\mathfrak{M})$ and $\mathfrak{gl}(\mathfrak{F})$ be the associated Lie-algebras.

If $T \in \mathrm{End}(V)$, extend T to act on the tensor algebra as a derivation. We then have:

$$\mathfrak{gl}(\mathfrak{F}) = \{T \in \mathrm{End}(V) : T \cdot \mathcal{A}_i = 0 \text{ for } 0 \le i \le k\},$$

$$\mathfrak{gl}(\mathfrak{M}) = \{T \in \mathrm{End}(V) : T \cdot \langle \cdot, \cdot \rangle = 0, T \cdot A_i = 0 \text{ for } 0 \le i \le k\}.$$

Let $\mathcal{F}$ be a family of k-models. We assume an object $\alpha(\mathfrak{M}_k)$ is associated to each $\mathfrak{M}_k \in \mathcal{F}$. We say that α is an *invariant* of the family if $\mathfrak{M}_k^1$ isomorphic to $\mathfrak{M}_k^2$ implies $\alpha(\mathfrak{M}^1) = \alpha(\mathfrak{M}^2)$. For example, $|\rho|^2$ is a scalar invariant of the family of all 0-models. We shall often omit specifying the family $\mathcal{F}$ if it is clear from the context.

1.3.5 *Various notions of homogeneity*

A pseudo-Riemannian manifold $\mathcal{M} = (M, g)$ is said to be *k-curvature homogeneous* if $\mathfrak{M}_k(\mathcal{M}, P)$ and $\mathfrak{M}_k(\mathcal{M}, Q)$ are isomorphic for any two points $P, Q \in M$. We say that $\mathfrak{M}_k$ is a k-model for $\mathcal{M}$ if $\mathfrak{M}_k$ is isomorphic to $\mathfrak{M}_k(\mathcal{M}, P)$ for any P in M. Clearly $\mathcal{M}$ is k-curvature homogeneous if and only if it admits a k-model. In the interests of brevity, we shall sometimes simply say that $\mathcal{M}$ is *curvature homogeneous* if it is 0-curvature homogeneous.

One says that $\mathcal{M}$ is *locally homogeneous* if for any two points $P, Q \in M$, there are neighborhoods U_P and U_Q of P and Q in M, respectively, and an isometry ψ from U_P to U_Q such that $\psi P = Q$. One says that $\mathcal{M}$ is *homogeneous* if $U_P = U_Q = M$ and thus ψ is globally defined.

If $\mathcal{H}$ is a homogeneous space, let $\mathfrak{M}(\mathcal{H}) := \mathfrak{M}_k(\mathcal{H}, Q)$ for any point $Q \in H$; the isomorphism class of $\mathfrak{M}(\mathcal{H})$ is independent of the point $Q \in H$. Let $\mathcal{M}$ be another pseudo-Riemannian manifold which is not necessarily locally homogeneous. We say that $\mathcal{M}$ is *k-modeled on* $\mathcal{H}$ and that $\mathfrak{M}(\mathcal{H})$ is a *k-model for* $\mathcal{M}$ if $\mathfrak{M}_k(\mathcal{H})$ and $\mathfrak{M}_k(\mathcal{M}, P)$ are isomorphic for any $P \in M$.

There are similar notions in the affine context. In Theorem 3.2.1, we use work of Gilkey and Nikčević (2005d) to exhibit a complete Lorentzian manifold which is 1-affine modeled on a homogeneous Lorentz manifold, 0-curvature modeled on an indecomposable symmetric Lorentzian manifold, and which is not 1-curvature homogeneous. Thus affine curvature homogeneity and curvature homogeneity are different notions. In Sections 2.9 and 2.10, we present examples which are k-curvature homogeneous but not $k + 1$-affine curvature homogeneous where k is arbitrarily large.

One can weaken the notion of curvature homogeneity slightly. Suppose given $A_i \in \otimes^{4+i}(V^*)$. One says that $\mathcal{M}$ is *weakly k-curvature homogeneous* if for every point $P \in M$, there is an isomorphism $\Phi : T_P M \to V$ so that $\Phi^* A_i = \nabla^i R_P$. There is no requirement that Φ preserve an inner product. In Section 2.4, examples are given which show that affine curvature homogeneity and weak curvature homogeneity are different notions; the metric is required to lower indices.

In the Riemannian setting, a homogeneous manifold is necessarily complete. This is not true in the higher signature context. In Theorem 2.3.6, we will give an example of a manifold of signature $(2, 2)$ which is not homogeneous, but which contains a proper dense open submanifold $\mathcal{O}$ so that $\mathcal{O}$ is homogeneous; necessarily $\mathcal{O}$ is incomplete.

Let $G(\mathcal{M})$ (respectively $G(\mathcal{F})$, $G(\mathfrak{M}_k)$, $G(\mathfrak{F}_k)$) be the group of isometries (respectively affine diffeomorphisms or isomorphisms) of a pseudo-Riemannian manifold $\mathcal{M}$ (respectively of an affine manifold $\mathcal{F}$, of a k-model $\mathfrak{M}_k$, or of an affine k-model $\mathfrak{F}_k$). We let $\mathfrak{g}(\cdot)$ be the associated Lie algebra. Let $G_P(\mathcal{M})$ (respectively $G_P(\mathcal{F})$) be the isotropy subgroup fixing a point P of M. If $\mathcal{M}$ or $\mathcal{F}$ are homogeneous, then there are natural identifications

$$\mathcal{M} = G(\mathcal{M})/G_P(\mathcal{M}) \quad \text{and} \quad \mathcal{F} = G(\mathcal{F})/G_P(\mathcal{F}).$$

1.3.6 *Killing vector fields*

Let $\mathcal{M}$ be a pseudo-Riemannian manifold. We say that $X \in C^\infty(T(M))$ is a *Killing vector field* on $\mathcal{M}$ if

$$g(\nabla_\xi X, \eta) + g(\nabla_\eta X, \xi) = 0 \quad \text{for all} \quad \xi, \eta \in C^\infty(T(M)).$$

If $\xi \in \mathfrak{g}(\mathcal{M})$, then necessarily ξ is a Killing vector field. The implication can be reversed in the real analytic context for generalized plane wave manifolds as we shall see shortly. In Theorem 2.3.6, we study the Lie algebra for a family of real analytic pseudo-Riemannian manifolds of signature $(2, 2)$. All are 1-curvature homogeneous and 0-modeled on the same indecomposable symmetric space. Some are symmetric, some are homogeneous, and some are inhomogeneous.

One says that $\mathcal{M}$ is a *local symmetric space* if $\nabla R = 0$; local symmetric spaces are locally homogeneous. One says that $\mathcal{M}$ is a *symmetric space* if $\nabla R = 0$ and if $\mathcal{M}$ is complete; symmetric spaces are homogeneous – see Theorem 2.2.3 for a special case of this result.

1.3.7   *Nilpotent curvature*

The eigenvalue 0 plays a distinguished role in the study of the Jacobi operator. We say that a 0-model $\mathfrak{M}_0$ is *spacelike nilpotent Osserman* if 0 is the only eigenvalue of $\mathcal{J}(v)$ or equivalently if $\mathcal{J}(v)^m = 0$ for any $v \in S^+(V, \langle \cdot, \cdot \rangle)$. The notion *timelike nilpotent Osserman* is defined similarly; we will show in Theorem 1.9.1 that these are equivalent notions. By using the appropriate operators, one can define the notions of *higher order nilpotent Osserman of type* (r, s), *nilpotent Ivanov–Petrova of type* (r, s), *nilpotent Stanilov of type* (r, s), and *nilpotent Szabó* similarly; only the value $k = (r + s)$ is relevant. In Section 3.5, we give examples of pseudo-Riemannian manifolds which are Jacobi nilpotent or Szabó nilpotent of arbitrarily high order.

1.4   Curvature Homogeneity – a Brief Literature Survey

In this section, we provide a brief review of some well known results related to curvature homogeneity that we shall need subsequently.

1.4.1   *Scalar Weyl invariants in the Riemannian setting*

As discussed in Section 1.2.6, we can form scalar invariants by contracting indices in pairs in the covariant derivatives of the curvature tensor; for example, the scalar curvature τ, the norm of the Ricci tensor $|\rho|^2$, and the norm of the full curvature tensor $|R|^2$ are given by

$$\tau := g^{ij} R_{kij}{}^k,$$
$$|\rho|^2 := g^{i_1 j_1} g^{i_2 j_2} R_{k i_1 j_1}{}^k R_{l i_2 j_2}{}^l, \quad \text{and}$$
$$|R|^2 := g^{i_1 j_1} g^{i_2 j_2} g^{i_3 j_3} g_{i_4 j_4} R_{i_1 i_2 i_3}{}^{i_4} R_{j_1 j_2 j_3}{}^{j_4}$$

where we adopt the *Einstein convention* and sum over repeated indices. All scalar invariants of the metric arise in this fashion, see, for example, the treatment in Atiyah, Bott, and Patodi (1973) or in Weyl (1946).

In the Riemannian setting, the scalar Weyl invariants determine the local geometry of the manifold. We refer to Prüfer, Tricerri, and Vanhecke (1996) for the proof of the following result:

Theorem 1.4.1   (Prüfer, Tricerri, and Vanhecke) *If all local scalar Weyl invariants up to order $\frac{1}{2}m(m - 1)$ are constant on a Riemannian* *manifold $\mathcal{M}$, then $\mathcal{M}$ is locally homogeneous and $\mathcal{M}$ is determined up to local isometry by these invariants.*

This result fails in the pseudo-Riemannian setting; Koutras and McIntosh (1996) and Pravda, Coley, and Milson (2002) have exhibited examples of non-flat manifolds all of whose scalar Weyl invariants vanish. We shall exhibit many other families of pseudo-Riemannian manifolds subsequently where this result fails. In Theorem 2.2.1, we shall show that all the scalar invariants of a generalized plane wave manifold vanish. Nevertheless, there are non-trivial scalar invariants of certain families of generalized plane wave manifolds which are not of Weyl type. We refer to Theorems 2.3.4, 2.4.1, 2.5.2, 2.7.2, 2.8.1, 2.9.3, and 2.10.2 for examples of pseudo-Riemannian manifolds which are not flat but all of whose scalar Weyl invariants vanish. The primary technical difficulty is, of course, constructing isometry invariants which are not of Weyl type.

1.4.2   *Relating curvature homogeneity and homogeneity*

It is clear that local homogeneity implies k-curvature homogeneity for any k. The following result is due to Singer (1960) in the Riemannian setting and to Podesta and Spiro (1996) in the general context:

Theorem 1.4.2   [Singer, Podesta and Spiro] *There exists an integer $k_{p,q}$ so that if $\mathcal{M}$ is a complete simply connected pseudo-Riemannian manifold of signature (p,q) which is $k_{p,q}$-curvature homogeneous, then (M,g) is homogeneous.*

The integer $k_{p,q}$ is called the *Singer number*. We will show in Section 2.9 that $k_{p,q} \geq \min(p,q)$. There is some evidence to suggest that the correct estimate is in fact $k_{p,q} = \min(p,q) + 1$. For example, in Theorem 3.2.1 (7), we give an example of a 3-dimensional Lorentzian manifold which is 1-curvature homogeneous but not curvature homogeneous.

There is a related result in the affine setting in the real analytic context due to Opozda (1997). Let $\mathfrak{gl}(\mathfrak{F}_k(\mathcal{F}, P))$ be the Lie-algebra of the k-affine curvature model at P defined in Eq. (1.3.c). One says that *the curvature sequence stabilizes at level k_0* if

$$\mathfrak{gl}(\mathfrak{F}_k(\mathcal{F}, P)) = \mathfrak{gl}(\mathfrak{F}_{k_0}(\mathcal{F}, P)) \quad \text{for every} \quad k \geq k_0.$$

Theorem 1.4.3   (Opozda) *Let $\mathcal{F}$ be a real analytic affine manifold. Assume that $\mathcal{F}$ is k-affine curvature homogeneous and that the curvature* *sequence stabilizes at level $k - 1$. Then $\mathcal{F}$ is locally affine homogeneous.*

Theorem 1.4.3 shows that if the dimension m of (M, ∇) is fixed, then there is an integer $k_{\mathcal{F}}(m)$ for which k-affine curvature homogeneity implies affine homogeneity; this number is called the *Opozda number*. The examples of Section 2.9 show $k_{\mathcal{F}}(2p + 6) \geq p + 3$ for $p \geq 0$.

1.4.3 *Manifolds modeled on symmetric spaces*

One says that a pseudo-Riemannian manifold $\mathfrak{M}$ is a *local symmetric space* if $\nabla R = 0$; a complete local symmetric space is said to be a *symmetric space*. If $\mathfrak{M}$ is a simply connected symmetric space, then the geodesic involution is an isometry and $\mathfrak{M}$ is a homogeneous space. One has the following rigidity result of Tricerri and Vanhecke (1986) in the Riemannian setting and of Cahen, Leroy, Parker, Tricerri, and Vanhecke (1991) in the Lorentzian setting:

Theorem 1.4.4

(1) **Tricerri and Vanhecke** *A Riemannian curvature homogeneous manifold which is 0-curvature modeled on an irreducible symmetric space is locally symmetric.*

(2) **Cahen et al.** *A Lorentzian curvature homogeneous manifold which is 0-curvature modeled on an irreducible symmetric space has constant sectional curvature.*

For example, any manifold of constant sectional curvature is modeled on a pseudo-sphere. Similarly manifold which is modeled on a pseudo-complex projective space is locally isometric to a pseudo-complex projective space. These are rigid geometries. We refer to Lemmas 1.6.7 and 1.6.8 for further details.

On the other hand, there exist curvature homogeneous spaces modeled on indecomposable symmetric spaces which are not even locally homogeneous. For example, in Section 2.10, we shall construct neutral signature generalized plane wave metrics on $\mathbb{R}^{4p+6}$ which are 0-modeled on an indecomposable symmetric space, which are $(p + 2)$-curvature homogeneous, and which are not $p + 3$ affine curvature homogeneous. We refer to Bueken (1997a), Bueken and Djoric (2000), Bueken and Vanhecke (1997), Cahen, Leroy, Parker, Tricerri, and Vanhecke (1991) for additional work.

1.4.4 *Historical survey*

Here is a very brief and necessarily incomplete introduction to the history of this subject. There are 2-curvature homogeneous affine manifolds which are not locally affine homogeneous, see, for example, García-Río, Kupeli, Vázquez-Abal, and Vázquez-Lorenzo (1999), Kowalski, Opozda, and Vlášek (1999), Kowalski, Opozda, and Vlášek (2000), Kowalski, Opozda, and Vlášek (2004), and Opozda (1996).

In the Riemannian setting ($p = 0$), Takagi (1974) showed that there are 0-curvature homogeneous complete non-compact manifolds which are not locally homogeneous; subsequently Ferus, Karcher, and Münzner (1981) exhibited compact examples. Many other examples are known Calvaruso, Marinosci, and Perrone (2000), Kowalski and Prüfer (1994), Kowalski, Tricerri, and Vanhecke (1992a), Kowalski, Tricerri, and Vanhecke (1992b), Tomassini (1997), Tricerri (1988), Tsukada (1988), Vanhecke (1991). There are no known Riemannian manifolds which are 1-curvature homogeneous but not locally homogeneous.

In the Lorentzian setting ($p = 1$), Cahen, Leroy, Parker, Tricerri, and Vanhecke (1991) showed that there exist curvature homogeneous manifolds which are not locally homogeneous; Bueken and Djoric (2000) and Bueken and Vanhecke (1997) showed there exist 1-curvature homogeneous Lorentzian manifolds which are not locally homogeneous. One could conjecture that a 2-curvature homogeneous Lorentzian manifold must be locally homogeneous.

The constants $k_{p,q}$ of Theorem 1.4.2 were first studied in the Riemannian setting. Singer (1960) proved that $k_{0,m} < \frac{1}{2}m(m-1)$; subsequently Yamato (1989) established the the bound $3m - 5$ and Gromov (1986) established the bound $\frac{3}{2}m - 1$. Work of Sekigawa, Suga, and Vanhecke (1992) and Sekigawa, Suga, and Vanhecke (1995) showed that any 1-curvature homogeneous complete simply connected Riemannian manifold of dimension $m < 5$ is homogeneous; thus $k_{0,2} = k_{0,3} = k_{0,4} = 1$. We refer to Boeckx, Kowalski, and Vanhecke (1996) for further details concerning k-curvature homogeneous manifolds in the Riemannian setting.

In the higher signature setting, results of Gilkey and Nikčević (2004d) can be used to show $k_{p,q} \geq \min(p,q)$. One could conjecture that in fact $k_{p,q} = \min(p,q) + 1$.

Opozda (1997) exhibited a 2-dimensional example showing that 1-affine curvature homogeneity does not imply local homogeneity; it is also known that if $\mathcal{F}$ is an analytic 2-dimensional affine manifold which is 2-affine curvature homogeneous and if ∇ has symmetric Ricci tensor, then $\mathcal{F}$ is locally homogeneous. The examples we shall construct presently also show that in general, the Opozda number is unbounded as the dimension is allowed to increase.

1.5   Results from Linear Algebra

In Section 1.5.1, results are given concerning the spectral theory of linear maps which are symmetric or anti-symmetric with respect to a Riemannian (positive definite) inner product. Spectral theory and traces are related in Section 1.5.2 and the Jordan normal form is discussed in Section 1.5.3. In Section 1.5.4 it is shown that if T is an arbitrary linear map of a vector space V, then there exists a non-degenerate inner product on V relative to which T is symmetric. In Section 1.5.5, a technical result is established which is needed subsequently in Chapter 2. It is demonstrated that any solution to the ordinary differential equation $h'' = kh'h'$, where k is constant, has the form

$$h = ae^{\lambda y} + h_0 \quad \text{or} \quad h = a(y+b)^c + h_0\,.$$

1.5.1   *Symmetric and anti-symmetric operators*

Let V be a vector space of dimension m which is equipped with a Riemannian (positive definite) inner product $\langle \cdot, \cdot \rangle$. The conjugacy class of a symmetric or of an anti-symmetric linear map is determined in this setting by the eigenvalue structure. The following is well known:

Lemma 1.5.1   *Let $\langle \cdot, \cdot \rangle \in S^2(V^*)$ be positive definite.*

(1) Let $T \in \mathrm{End}_+(V, \langle \cdot, \cdot \rangle)$ be a symmetric linear map. There exists an orthonormal basis $\{e_i\}$ for V and $\lambda_i \in \mathbb{R}$ so $Te_i = \lambda_i e_i$. If $\lambda = \max_i \lambda_i$ or if $\lambda = \min_i \lambda_i$, then $Tv = \lambda v$ if and only if $\langle v, Tv \rangle = \lambda|v|^2$.

(2) Let $T \in \mathrm{End}_-(V, \langle \cdot, \cdot \rangle)$ be a skew-symmetric linear map. There exists orthonormal basis $\{e_1^+, e_1^-, ..., e_\ell^+, e_\ell^-, f_1, ..., f_k\}$ for V and $\lambda_i \in \mathbb{R}^+$ so

$$Te_i^+ = \lambda_i e_i^-, \quad Te_i^- = -\lambda_i e_i^+, \quad \text{and} \quad Tf_j = 0\,.$$

1.5.2   *The spectrum of an operator*

Let $T \in \mathrm{End}(V)$. In general, of course, T is not diagonalizable. We say that a is an *eigenvalue* of T if $\det(T - a\,\mathrm{Id}) = 0$. We let the spectrum of T denote the collection of eigenvalues where each eigenvalue is repeated according to multiplicity. The following is a useful observation. We omit the proof and refer instead to Lemma 2.1.6 in Gilkey (1995).

Lemma 1.5.2 *Let $T_1, T_2 \in \text{End}(V)$. The following assertions are equivalent.*

(1) T_1 and T_2 have the same spectrum.
(2) $\text{Tr}(T_1^i) = \text{Tr}(T_2^i)$ for $1 \leq i \leq \dim(V)$.

1.5.3 *Jordan normal form*

Let $\mathfrak{J}(k, a)$ be the *Jordan block* of size k for a real eigenvalue $a \in \mathbb{R}$:

$$\mathfrak{J}(k, a) := \begin{pmatrix} a & 1 & 0 & \ldots & 0 & 0 \\ 0 & a & 1 & \ldots & 0 & 0 \\ \ldots & \ldots & \ldots & \ldots & \ldots & \ldots \\ 0 & 0 & 0 & \ldots & a & 1 \\ 0 & 0 & 0 & \ldots & 0 & a \end{pmatrix}. \tag{1.5.a}$$

We define a Jordan block of size $2k \times 2k$ corresponding to the pair of complex conjugate eigenvalues $\{a + \sqrt{-1}b, a - \sqrt{-1}b\}$ by first setting

$$A_{a,b} := \begin{pmatrix} a & b \\ -b & a \end{pmatrix}, \text{ and } I_2 := \begin{pmatrix} 1 & 0 \\ 0 & 1 \end{pmatrix}$$

and then setting

$$\mathfrak{J}(k, a, b) := \begin{pmatrix} A_{a,b} & I_2 & 0 & \ldots & 0 & 0 \\ 0 & A_{a,b} & I_2 & \ldots & 0 & 0 \\ \ldots & \ldots & \ldots & \ldots & \ldots & \ldots \\ 0 & 0 & 0 & \ldots & A_{a,b} & I_2 \\ 0 & 0 & 0 & \ldots & 0 & A_{a,b} \end{pmatrix}. \tag{1.5.b}$$

We refer to Adkins and Weintraub (1992) for the proof of the following:

Lemma 1.5.3 *Let T be a linear transformation of a vector space V. Relative to a suitably chosen basis for V, T decomposes as a direct sum of the Jordan blocks described in Eqs. (1.5.a) and (1.5.b). Furthermore, the unordered collection of Jordan blocks is determined by T.*

The *Jordan normal form* of T is the unordered collection of Jordan blocks described in Lemma 1.5.3. We say that two linear maps T and $\tilde{T}$ of

V are *Jordan equivalent* if any of the following three equivalent conditions are satisfied:

(1) There exist bases $\mathcal{B} = \{e_1, ..., e_m\}$ and $\tilde{\mathcal{B}} = \{\tilde{e}_1, ..., \tilde{e}_m\}$ for V so that the matrix representation of T with respect to the basis $\mathcal{B}$ is equal to the matrix representation of $\tilde{T}$ with respect to the basis $\tilde{\mathcal{B}}$.
(2) There exists an isomorphism ψ of V so $T = \psi\tilde{T}\psi^{-1}$; this means that T and $\tilde{T}$ are *conjugate*.
(3) The Jordan normal forms of T and $\tilde{T}$ are equal.

1.5.4 *Self-adjoint maps in the higher signature setting*

Let $\langle\cdot,\cdot\rangle \in S^2(V^*)$ have signature (p,q). In the positive definite setting $(p = 0)$, the Jordan normal form of a symmetric linear map is determined by the eigenvalue structure since T is diagonalizable. Similarly, if T is skew-symmetric, then the eigenvalue structure is determined since T can be written as the direct sum of blocks of the form

$$\begin{pmatrix} 0 & b \\ -b & 0 \end{pmatrix} \quad \text{and} \quad (0)\,.$$

However, the eigenvalue structure does not determine the Jordan normal form of a symmetric map or of a skew-symmetric map in the higher signature setting $(p > 0$ and $q > 0)$. In fact, there is no obstruction to an operator being symmetric in the higher signature setting:

Lemma 1.5.4 *Let $T \in \mathrm{End}(V)$. There exists a non-degenerate inner product $\langle\cdot,\cdot\rangle \in S^2(V^*)$ so that T is symmetric with respect to $\langle\cdot,\cdot\rangle$.*

Proof. By Lemma 1.5.3, T can be decomposed as the sum of Jordan blocks. Consequently, it suffices to prove Lemma 1.5.4 for the special cases $T = \mathfrak{J}(k,a)$ and $T = \mathfrak{J}(k,a,b)$ described in Eqs. (1.5.a) and (1.5.b).

Let $\{e_1, ..., e_k\}$ be the standard basis for $\mathbb{R}^k$. The Jordan block $\mathfrak{J}(k,a)$ defines the linear transformation:

$$\mathfrak{J}(k,a)e_i := \begin{cases} ae_i + e_{i-1} & \text{if } i > 1, \\ ae_i & \text{if } i = 1\,. \end{cases}$$

We define a non-degenerate inner product $\langle\cdot,\cdot\rangle$ on $\mathbb{R}^k$ by setting:

$$\langle e_i, e_j\rangle := \delta_{i+j,k+1}\,.$$

Since $\mathfrak{J}(k,a) = a \cdot \mathrm{Id} + \mathfrak{J}(k,0)$, and since Id is symmetric with respect to any inner product, we may take $a = 0$. We have

$$\langle \mathfrak{J}(k,0)e_i, e_j \rangle = \delta_{i-1+j,k+1} \, .$$

As this is symmetric in the indices i and j, the desired result follows.

To study the Jordan block $\mathfrak{J}(k,a,b)$, let $\{e_1, f_1, ..., e_k, f_k\}$ be the usual basis for $\mathbb{R}^{2k}$. Then

$$\mathfrak{J}(k,a,b)e_i = \begin{cases} ae_i - bf_i + e_{i-1} & \text{if } i > 1, \\ ae_i - bf_i, & \text{if } i = 1, \end{cases}$$

$$\mathfrak{J}(k,a,b)f_i = \begin{cases} be_i + af_i + f_{i-1} & \text{if } i > 1, \\ be_i + af_i, & \text{if } i = 1. \end{cases}$$

We define the inner product

$$\langle e_i, e_j \rangle = \delta_{i+j,k+1},$$
$$\langle e_i, f_j \rangle = 0 \quad \text{for all} \quad i,j,$$
$$\langle f_i, f_j \rangle = -\delta_{i+j,k+1} \, .$$

We may decompose $\mathfrak{J}(k,a,b) = \mathfrak{J}(k,0,0) + B$ where $Be_i = ae_i - bf_i$ and $Bf_i = be_i + af_j$. The proof that $\mathfrak{J}(k,0,0)$ is symmetric is the same as that given above to show $\mathfrak{J}(k,0)$ is symmetric and is therefore omitted; the bases $\{e_1, ..., e_k\}$ and $\{f_1, ..., f_k\}$ do not interact. To show that B is symmetric, we compute:

$$\langle Be_i, e_j \rangle = a\delta_{i+j,k+1}, \quad \langle Bf_i, f_j \rangle = -a\delta_{i+j,k+1},$$
$$\langle Be_i, f_j \rangle = b\delta_{i+j,k+1}, \quad \langle e_i, Bf_j \rangle = b\delta_{i+j,k+1} \, .$$

This establishes the desired relations. $\qquad\square$

1.5.5 *Technical results concerning differential equations*

Lemma 1.5.5 *Let $\mathcal{O}$ be a connected open subset of $\mathbb{R}$ and let $h \in C^\infty(\mathcal{O})$. Assume that $h' \neq 0$ and that $hh''(h')^{-2}$ is constant. Then either one has that $h(y) = ae^{\lambda y}$ or one has that $h(y) = a(y+b)^c$.*

Proof. We have the equation $h''h = kh'h'$. Thus

$$\int \tfrac{h''}{h'} = k \int \tfrac{h'}{h} \quad \text{so} \quad \ln(h') = k\ln(h) + \beta \quad \text{so}$$
$$h' = e^\beta h^k \quad \text{so} \quad \int \tfrac{h'}{h^k} = e^\beta y + \gamma \, .$$

If $k = 1$, this implies $\ln(h) = e^\beta y + \gamma$ or equivalently $h = e^\gamma e^{e^\beta y}$ which leads to an exponential solution. If $k \neq 1$, then $h^{1-k} = (1 - k)(e^\beta y + \gamma)$; this leads to a solution involving powers of a translate of y. $\qquad\square$

We shall need the following two results in Section 3.2 to study the geodesic structure on certain 3-dimensional Lorentz manifolds.

Lemma 1.5.6   *Let $h : \mathbb{R} \to (-\infty, 0)$ be smooth. Let $[0, T)$ be the maximal domain of the solution y to the ordinary differential equation $y'' = h(y)$ where $y(0) = y_0$ and $y'(0) = y_0'$. If $T < \infty$, then*

$$\lim_{t \to T} y(t) = \lim_{t \to T} y'(t) = -\infty \quad and \quad \limsup_{y \to T} \left| \frac{h(y(t))}{y(t)} \right| = \infty \,.$$

Proof.   Since $y'' < 0$, y' is monotonically decreasing and y is bounded from above on $[0, T)$. Suppose first that y is bounded from below on $[0, T)$. This implies that y'' is bounded and hence y' is bounded as well on $[0, T)$. Let

$$y_1 = \lim_{t \to T} \inf y(t) \quad and \quad y_1' = \lim_{t \to T} y'(t) \,.$$

The fundamental theorem of ordinary differential equations shows that there exists $\kappa > 0$ so that if $|z_1 - y_1| < \kappa$ and if $|z_1' - y_1'| < \kappa$ then there exists a solution z to the equation $z'' = h(z)$ with initial conditions $z(s) = z_1$ and $z'(s) = z_1'$ which is valid on the interval $[s, s + \kappa)$. We choose

$$s \in (T - \tfrac{1}{2}\kappa, T) \quad \text{so that} \quad |y(s) - y_1| < \kappa \quad and \quad |y'(s) - y_1'| < \kappa \,.$$

Let $z'' = h(z)$ be defined on $[s, s + \kappa)$ with $z(s) = y(s)$ and $z'(s) = y'(s)$. Then z extends y to the region $[0, T + \tfrac{1}{2}\kappa)$ which contradicts the assumption that $[0, T)$ was a maximal domain.

Thus y is not bounded from below on $[0, T)$ so $\lim_{t \to T} y'(t) = -\infty$. Consequently, y is monotonically decreasing for t close to T so one has as well that $\lim_{t \to T} y(t) = -\infty$. Suppose

$$\limsup_{t \to T} \frac{h(y(t))}{y(t)} < \infty \,.$$

This means that there exists $C < \infty$ so $|h(y(t))| \le C|y(t)|$ on $[t_0, T)$. We then have

$$\{\ln|y(t)|\}'' = \left\{\frac{y'(t)}{y(t)}\right\}' = \frac{y''(t)}{y(t)} - \left\{\frac{y'(t)}{y(t)}\right\}^2$$
$$= \frac{h(y(t))}{y(t)} - \left\{\frac{y'(t)}{y(t)}\right\}^2 \le C.$$

This implies $\ln|y(t)|$ is bounded from above and hence $|y(t)|$ is bounded from above on $[t_0, T)$ which is false. This contradiction shows $\limsup_{t \to T} \frac{h(y(t))}{y(t)} = \infty$. $\qquad\square$

We shall also need the following result:

Lemma 1.5.7 *Let $h : \mathbb{R} \to (-\infty, 0)$ be smooth.*

(1) Let $\alpha > 0$. Let $\{t_n\}_{n \ge 1}$ be a sequence of real numbers with $t_1 = 1$ and with $t_{n+1} - t_n \ge n^\alpha$ for $n \ge 1$. Then $t_n \ge \frac{n^{1+\alpha}}{(1+\alpha)2^{1+\alpha}}$.

(2) Let $\epsilon > 0$ and $\delta > 0$. Suppose that $h(y) < -\epsilon|y|^{1+\delta}$ for $y \le -1$. Let $[0, T)$ be the maximal domain of definition for the solution y to the ordinary differential equation $y'' = h(y)$ with initial conditions given by $y(0) = -1$ and $y'(0) = -1$. Then $T < \infty$ and $\lim_{t \to T} y(t) = -\infty$.

Proof. We prove Assertion (1) by induction on n; it holds trivially for $n = 1$. We take $n \ge 2$ and use the comparison test to compute:

$$t_n > t_n - t_1 = \sum_{k=2}^{n} \left\{ t_k - t_{k-1} \right\} \ge \int_1^n (x-1)^\alpha dx$$
$$= \frac{(n-1)^{1+\alpha}}{1+\alpha} = \frac{n^{1+\alpha}}{(1+\alpha)(1+\frac{1}{n-1})^{1+\alpha}} \ge \frac{n^{1+\alpha}}{(1+\alpha)2^{1+\alpha}}.$$

To prove Assertion (2), we suppose first $T = \infty$ and argue for a contradiction. Choose $\tau \ge 1$ so that

$$\tau\varepsilon \ge 2^{1+\delta/2}(1 + \delta/2).$$

With our initial conditions, $y'' < 0$ so y' is monotonically decreasing and $y' \le -1$. This implies y decreases monotonically. Let $\Delta_n = \tau \cdot n^{-1-\delta/2}$. Let $s_1 = 0$ and let $s_{n+1} = s_n + \Delta_n$ for $n \ge 2$. As $\delta > 0$,

$$S := \lim_{n \to \infty} s_n = \sum_{n=1}^{\infty} \tau n^{-1-\delta/2} < \infty.$$

Consider the following statements:

$(1)_n$ $y'(s_n) \leq -n^{1+\delta/2}$.

$(2)_n$ $y(s_n) \leq -n$.

$(3)_n$ $y'(s_{n+1}) - y'(s_n) \leq -2^{1+\delta/2}(1 + \frac{1}{2}\delta)n^{\delta/2}$.

We establish these statements by induction on n. Statements $(1)_n$ and $(2)_n$ hold $n = 1$ by the choice of our initial conditions. Since y and y' decrease monontonically, we may estimate

$$y''(s) \leq -\epsilon|y(s)|^{1+\delta} \leq -\epsilon|y(s_n)|^{1+\delta} \leq -\epsilon n^{1+\delta} \text{ for } s \in [s_n, s_{n+1}],$$
$$y'(s_{n+1}) - y'(s_n) \leq -\Delta_n \epsilon n^{1+\delta} = -\tau n^{-1-\delta/2}\epsilon n^{1+\delta}$$
$$\leq -2^{1+\delta/2}(1 + \delta/2)n^{\delta/2}.$$

Statements $(1)_n$ and $(2)_n$ are thus seen to imply Statement $(3)_n$.

Statements $(3)_k$ for $1 \leq k \leq n$ together with Assertion (1) imply Statement $(1)_{n+1}$. Finally, we use Statement $(1)_n$ together with Statement $(2)_n$ to establish Statement $(2)_{n+1}$ by computing:

$$y'(s) \leq y'(s_n) \leq -n^{1+\delta/2} \quad \text{for} \quad s \geq s_n,$$
$$y(s_{n+1}) \leq y(s_n) + \Delta_n y'(s_n) \leq -n - \tau n^{-1-\delta/2}n^{1+\delta/2} \leq -n - 1.$$

This establishes the truth of all the 3 statements. Thus, $\lim_{s \to S} y(s) = -\infty$. This contradicts the assumption that $T = \infty$.

This shows that y must be defined on a maximal domain $[0, T)$ for $T < \infty$; we use Lemma 1.5.6 to see $\lim_{t \to T} y(t) = -\infty$. $\qquad\square$

1.6   Results from Differential Geometry

In this section, we summarize some results from Differential Geometry that we shall need. In Section 1.6.1, we discuss principle bundles. In Section 1.6.2, we show any 1-model is geometrically realizable. In Section 1.6.3, we give generating sets for the space of algebraic curvature tensors in terms of the canonical curvature tensors defined in Section 1.3.2. We also show that if $A_{\Phi_1} = A_{\Phi_2}$ where the $\Phi_i \in S^2(V^*)$ have rank at least 3, then $\Phi_1 = \Phi_2$. In Section 1.6.4, we turn to complex geometry and give several equivalent conditions for the compatibility of an algebraic curvature tensor with a pseudo-Hermitian almost complex structure. Section 1.6.5 deals with space forms and complex space forms, Section 1.6.6 deals with conformal complex space forms, and Section 1.6.7 is concerned with Kähler geometry.

1.6.1 *Principle bundles*

We say that $\pi : E \to B$ is a *fiber bundle* with fiber F if there exists an open cover $\mathcal{O}_\alpha$ of B and fiber preserving diffeomorphisms Φ_α from $\pi^{-1}(\mathcal{O}_\alpha)$ to $\mathcal{O}_\alpha \times F$. Let $\pi : P \to B$ be a fiber bundle whose fiber is a Lie group G. We say that P is a *principle bundle* and write

$$G \to P \to B$$

if G acts freely on P from the right and if $B = P/G$. Equivalently, this means that the transition functions of P are given by *left* multiplication by the group G. Let S^n denote the usual round sphere, let $\mathbb{CP}^n$ denote complex projective space, let $\mathrm{Gr}_k(n)$ denote the Grassmannian of k-planes in $\mathbb{R}^n$, and let $\mathrm{Gr}_k^+(n)$ denote the oriented Grassmannian of k-planes in $\mathbb{R}^n$. As examples, we have the following principle bundles:

$$U(1) \to U(n+1) \to \mathbb{CP}^n,$$
$$SO(n) \to SO(n+1) \to S^n,$$
$$O(k) \times O(n-k) \to O(n) \to \mathrm{Gr}_k(n),$$
$$SO(k) \times SO(n-k) \to SO(n) \to \mathrm{Gr}_k^+(n).$$

The following is well known:

Lemma 1.6.1 *Let a Lie group G act on a space X from the left. If $x \in X$, let $G \cdot x$ be the orbit and let $G_x = \{g \in G : gx = x\}$ be the isotropy subgroup.*

(1) We have a principle bundle $G_x \to G \to G \cdot x$.

(2) $\dim\{G\} = \dim\{G_x\} + \dim\{G \cdot x\}$.

1.6.2 *Geometric realizability*

Although the following is well-known, see for example Belger and Kowalski (1994) where a more general result is established, we shall give the proof to keep the development as self-contained as possible and to establish notation needed subsequently.

Let $\{x_1, ..., x_m\}$ be local coordinates on a pseudo-Riemannian manifold $\mathcal{M}$. Let

$$\partial_i := \partial_{x_i} := \frac{\partial}{\partial x_i}.$$

Denote the components of the metric g, of the curvature tensor R, and of the covariant derivative of the curvature tensor ∇R with respect to the coordinate frame by:

$$g_{ij} := g(\partial_i, \partial_j),$$
$$R_{ijkl} := R(\partial_i, \partial_j, \partial_k, \partial_l),$$
$$R_{ijkl;n} := R(\partial_i, \partial_j, \partial_k, \partial_l; \partial_n).$$

Lemma 1.6.2

(1) Let $\{x_1, ..., x_m\}$ be local coordinates on a pseudo-Riemannian manifold $\mathcal{M}$. If $\{\partial_i g_{jk}\}(P) = 0$, then:

(a) $R_{ijkl}(P) = \frac{1}{2}\{\partial_i\partial_k g_{jl} + \partial_j\partial_l g_{ik} - \partial_i\partial_l g_{jk} - \partial_j\partial_k g_{il}\}(P)$.
(b) $R_{ijkl;n}(P) = \frac{1}{2}\{\partial_i\partial_k\partial_n g_{jl} + \partial_j\partial_l\partial_n g_{ik} - \partial_i\partial_l\partial_n g_{jk} - \partial_j\partial_k\partial_n g_{il}\}(P)$.

(2) Let $\mathfrak{M}_1$ be a 1-model. There exists a point P of a pseudo-Riemannian manifold $\mathcal{M}$ so that $\mathfrak{M}_1$ is isomorphic to $\mathfrak{M}_1(\mathcal{M}, P)$.

Proof. Let P be a point of a pseudo-Riemannian manifold $\mathcal{M}$. Let x be a system of local coordinates on M; we may assume without loss of generality that P corresponds to the origin of the coordinate system. Suppose that the 1 jets of the metric vanish at the origin. We establish Assertion (1) by computing:

$$\Gamma_{ijk} := g(\nabla_{\partial_i}\partial_j, \partial_k) = \tfrac{1}{2}(\partial_i g_{jk} + \partial_j g_{ik} - \partial_k g_{ij}) = O(|x|),$$
$$R_{ijkl} = \partial_i\Gamma_{jkl} - \partial_j\Gamma_{ikl} + O(|x|^2),$$
$$R_{ijkl;n} = \partial_n R_{ijkl} + O(|x|).$$

Let $\mathfrak{M} = (V, \langle\cdot,\cdot\rangle, A_0, A_1)$ be a 1-model. To prove Assertion (2), choose an orthonormal basis $\{e_1, ..., e_m\}$ for V so that $\langle e_i, e_j \rangle = \pm\delta_{ij}$. Use this orthonormal basis to identify $V = \mathbb{R}^m$. Let A_{ijkl} and $A_{1,ijkl;n}$ denote the components of A and of A_1, respectively, relative to this orthonormal basis. Define

$$g_{ik} := \langle e_i, e_k \rangle - \tfrac{1}{3}\sum_{jl} A_{ijlk}x_j x_l - \tfrac{1}{6}\sum_{jln} A_{1,ijlk;n}x_j x_l x_n.$$

Clearly $g_{ik} = g_{ki}$. As $g_{ij}(0) = \langle e_i, e_j \rangle$, g is non-degenerate at the origin and hence is non-degenerate on some neighborhood of the origin. Since the 1 jets of the metric vanish at 0, we may apply Assertion (1) to compute that:

$$R_{ijkl}(0) = \tfrac{1}{2}\{\partial_i\partial_k g_{jl} + \partial_j\partial_l g_{ik} - \partial_i\partial_l g_{jk} - \partial_j\partial_k g_{il}\}(0)$$
$$= \tfrac{1}{6}\{-A_{jikl} - A_{jkil} - A_{ijlk} - A_{iljk} + A_{jilk} + A_{jlik} + A_{ijkl} + A_{ikjl}\}$$
$$= \tfrac{1}{6}\{4A_{ijkl} - 2A_{iljk} - 2A_{iklj}\} = A_{ijkl}\,.$$

We complete the proof of Assertion (2) by computing:

$$R_{ijkl;n}(0) = \tfrac{1}{2}\{\partial_i\partial_k\partial_n g_{jl} + \partial_j\partial_l\partial_n g_{ik} - \partial_i\partial_l\partial_n g_{jk} - \partial_j\partial_k\partial_n g_{il}\}(0)$$
$$= \tfrac{1}{12}\{-A_{jikl;n} - A_{jkil;n} - A_{jnkl;i} - A_{jknl;i} - A_{jinl;k} - A_{jnil;k}$$
$$-A_{ijlk;n} - A_{iljk;n} - A_{inlk;j} - A_{ilnk;j} - A_{ijnk;l} - A_{injk;l}$$
$$+A_{jilk;n} + A_{jlik;n} + A_{jnlk;i} + A_{jlnk;i} + A_{jink;l} + A_{jnik;l}$$
$$+A_{ijkl;n} + A_{ikjl;n} + A_{inkl;j} + A_{iknl;j} + A_{ijnl;k} + A_{injl;k}\}$$
$$= \tfrac{1}{12}\{(4A_{ijkl;n} - 2A_{jkil;n} + 2A_{jlik;n}) + (-2A_{jnkl;i} - 2A_{inlk;j})$$
$$+(-2A_{jinl;k} - 2A_{ijnk;l}) + (-A_{ilnk;j} - A_{jnil;k})$$
$$+(-A_{injk;l} - A_{jknl;i}) + (A_{jlnk;i} + A_{injl;k}) + (A_{jnik;l} + A_{iknl;j})\}$$
$$= \tfrac{1}{12}\{6A_{ijkl;n} + 2A_{ijkl;n} + 2A_{ijkl;n} + A_{ilkj;n} + A_{jkli;n} - A_{jlki;n} - A_{iklj;n}\}$$
$$= \tfrac{1}{12}\{10A_{ijkl;n} + 2A_{ilkj;n} + 2A_{ikjl;n}\} = \tfrac{1}{12}\{10A_{ijkl;n} - 2A_{ijlk;n}\}$$
$$= A_{ijkl;n}\,. \qquad \square$$

1.6.3 *The canonical algebraic curvature tensors*

Let $S^2(V^*)$ and $\Lambda^2(V^*)$ be the spaces of symmetric and anti-symmetric bilinear forms on V, respectively. If $\Phi_+ \in S^2(V^*)$ and if $\Phi_- \in \Lambda^2(V^*)$, adopt the notation of Eq. (1.3.a) to define the canonical 4-tensors

$$A_{\Phi_+}(x,y,z,w) := \Phi_+(x,w)\Phi_+(y,z) - \Phi_+(x,z)\Phi_+(y,w),$$
$$A_{\Phi_-}(x,y,z,w) := \Phi_-(x,w)\Phi_-(y,z) - \Phi_-(x,z)\Phi_-(y,w)$$
$$-2\Phi_-(x,y)\Phi_-(z,w)\,.$$

Let $\mathfrak{M}_0^w := (V, A_\Phi)$ be the weak 0-model defined by Φ. If $\langle\cdot,\cdot\rangle$ is a non-degenerate inner product on V, let $\mathfrak{M}_0 := (V, \langle\cdot,\cdot\rangle, A_\Phi)$ be the associated 0-model. Let

$$\ker(A_\Phi) := \{\eta \in V : A_\Phi(\eta, \xi_1, \xi_2, \xi_3) = 0 \ \forall \ \xi_i \in V\}\,.$$

Lemma 1.6.3 *Let $\Phi \in S^2(V^*)$ or $\Phi \in \Lambda^2(V^*)$. Assume* $\mathrm{Rank}\{\Phi\} \geq 2$.

(1) $A_\Phi \in \mathcal{A}lg_0$.

(2) If there is a decomposition $V = V^1 \oplus V^2$ with $A_\Phi = A^1 \oplus A^2$, then either $V^1 \subset \ker\Phi$ or $V^2 \subset \ker\Phi$.

(3) If Φ is non-degenerate, then $\mathfrak{M}_0^w$ is indecomposable.

(4) If $\ker\Phi$ is totally isotropic, then $\mathfrak{M}_0$ is indecomposable.

(5) $\ker(A_\Phi) = \ker\Phi$.

Proof.   We must ensure that the identities of Eq. (1.2.g) are satisfied to establish Assertion (1). Let $A = A_\Phi$ where $\Phi \in S^2(V)$ or $\Phi \in \Lambda^2(V)$. The curvature symmetries

$$A(x, y, z, w) = -A(y, x, z, w) = A(z, w, x, y)$$

are then immediate by inspection. Only the first Bianchi identity needs to be established. If $\Phi \in S^2(V^*)$, then:

$$\begin{aligned}
A(x, &y, z, w) + A(y, z, x, w) + A(z, x, y, w) \\
&= \Phi(x, w)\Phi(y, z) - \Phi(x, z)\Phi(y, w) \\
&\quad + \Phi(y, w)\Phi(z, x) - \Phi(y, x)\Phi(z, w) \\
&\quad + \Phi(z, w)\Phi(x, y) - \Phi(z, y)\Phi(x, w) = 0 \,.
\end{aligned}$$

Similarly, if $\Phi \in \Lambda^2(V^*)$, then:

$$\begin{aligned}
A(x, &y, z, w) + A(y, z, x, w) + A(z, x, y, w) \\
&= \Phi(x, w)\Phi(y, z) - \Phi(x, z)\Phi(y, w) - 2\Phi(x, y)\Phi(z, w) \\
&\quad + \Phi(y, w)\Phi(z, x) - \Phi(y, x)\Phi(z, w) - 2\Phi(y, z)\Phi(x, w) \\
&\quad + \Phi(z, w)\Phi(x, y) - \Phi(z, y)\Phi(x, w) - 2\Phi(z, x)\Phi(y, w) = 0 \,.
\end{aligned}$$

This completes the proof of Assertion (1).

We follow the discussion in Dunn (2006) to prove Assertion (2). Assume there exists a non-trivial direct sum decomposition $V = V^1 \oplus V^2$ with

$$A_\Phi = A_1 \oplus A_2, \quad V^1 \not\subset \ker \Phi, \quad V^2 \not\subset \ker \Phi \,.$$

If $v \in V$, we expand $v = v^1 + v^2$ for $v^i \in V^i$. We argue for a contradiction.

Assume that Φ is symmetric. Choose $v_1^1 \in V^1$ so that $v_1^1 \notin \ker \Phi$. Suppose first $\Phi(v_1^1, v_1^1) \neq 0$. Since $\operatorname{Rank}(\Phi) \geq 2$, we can choose $v_2 \in V$ so $\Phi(v_1^1, v_2) = 0$ and $\Phi(v_2, v_2) \neq 0$. Decompose $v_2 = v_2^1 + v_2^2$ for $v_2^i \in V^i$. Then:

$$\begin{aligned}
0 \neq \Phi(v_1^1, v_1^1)\Phi(v_2, v_2) &= \Phi(v_1^1, v_1^1)\Phi(v_2, v_2) - \Phi(v_1^1, v_2)\Phi(v_1^1, v_2) \\
&= A_\Phi(v_1^1, v_2, v_2, v_1^1) = A_\Phi(v_1^1, v_2^1, v_2^1, v_1^1) \\
&= \Phi(v_1^1, v_1^1)\Phi(v_2^1, v_2^1) - \Phi(v_1^1, v_2^1)\Phi(v_1^1, v_2^1) \,.
\end{aligned}$$

This shows $\pi := \operatorname{Span}\{v_1^1, v_2^1\}$ is 2-dimensional and $\Phi|_\pi$ is non-degenerate. If, on the other hand, we have $\Phi(v_1^1, v_1^1) = 0$, since $v_1^1 \notin \ker \Phi$, we can choose v_2 so $\Phi(v_1^1, v_2) \neq 0$. Consequently:

$$
\begin{aligned}
0 \neq -\Phi(v_1^1, v_2)\Phi(v_1^1, v_2) &= \Phi(v_1^1, v_1^1)\Phi(v_2, v_2) - \Phi(v_1^1, v_2)\Phi(v_1^1, v_2) \\
&= A_\Phi(v_1^1, v_2, v_2, v_1^1) = A_\Phi(v_1^1, v_2^1, v_2^1, v_1^1) \\
&= \Phi(v_1^1, v_1^1)\Phi(v_2^1, v_2^1) - \Phi(v_1^1, v_2^1)\Phi(v_1^1, v_2^1) \,.
\end{aligned}
$$

Thus once again $\Phi|_\pi$ is non-degenerate. Thus we may conclude:

$$
\mathrm{Rank}(\Phi|_{V_1}) \geq 2 \quad \text{and} \quad \mathrm{Rank}(\Phi|_{V_2}) \geq 2 \,.
$$

Since $\mathrm{Rank}\{\Phi|_{V_1}\} \neq 0$, we may find $v_1^1 \in V_1$ so $\Phi(v_1^1, v_1^1) \neq 0$. Since $\mathrm{Rank}\{\Phi|_{V_2}\} \geq 2$, we can choose $v_2^2 \in V_2$ and $v_2^2 \notin \ker \Phi$ so $\Phi(v_1^1, v_2^2) = 0$. Since $v_2^2 \notin \ker \Phi$, we can choose v_3 so $\Phi(v_2^2, v_3) \neq 0$. We then have

$$
\begin{aligned}
A_\Phi(v_1^1, v_2^2, v_3, v_1^1) &= \Phi(v_1^1, v_1^1)\Phi(v_2^2, v_3) - \Phi(v_1^1, v_2^2)\Phi(v_1^1, v_3) \\
&= \Phi(v_1^1, v_1^1)\Phi(v_2^2, v_3) \neq 0 \,.
\end{aligned}
$$

This contradiction establishes Assertion (2) if Φ is symmetric.

We now prove Assertion (2) if Φ is skew-symmetric. One has:

$$
\begin{aligned}
&A_\Phi(x, y, y, x) \\
&= \Phi(x, x)\Phi(y, y) - \Phi(x, y)\Phi(y, x) - 2\Phi(x, y)\Phi(y, x) \\
&= 3\Phi(x, y)^2 \,.
\end{aligned}
$$

Choose $v_1^1 \in V_1$ with $v_1^1 \notin \ker \Phi$. Choose v_2 so $\Phi(v_1^1, v_2) \neq 0$. Decompose $v_2 = v_2^1 + v_2^2$ for $v_2^i \in V^i$. This implies

$$
3\Phi(v_1^1, v_2^1) = A_\Phi(v_1^1, v_2^1, v_2^1, v_1^1) = A_\Phi(v_1^1, v_2, v_2, v_1^1) = 3\Phi(v_1^1, v_2)^2 \neq 0 \,.
$$

Consequently we may conclude:

$$
\mathrm{Rank}(\Phi|_{V_1}) \geq 2 \quad \text{and} \quad \mathrm{Rank}(\Phi|_{V_2}) \geq 2 \,.
$$

If there exists $v_2^2 \in V^2$ so $\Phi(v_1^1, v_2^2) \neq 0$, then $A_\Phi(v_1^1, v_2^2, v_2^2, v_1^1) \neq 0$ which is false. Thus

$$
V^1 \perp_\Phi V^2 \,.
$$

Choose $v^i \in V^i$ and $w^i \in V^i$ so that $\Phi(v^i, w^j) = \delta^{ij}$. Let $v = v^1 + v^2$, $w = w^1 + w^2$. Then:

$$
\begin{aligned}
A_\Phi(v, w, w, v) &= 3\Phi(v, w)^2 = 3\{\Phi(v^1 + v^2, w^1 + w^2)\}^2 \\
&= 3\{1 + 1\}^2 = 12 \\
&= A_\Phi(v^1, w^1, w^1, v^1) + A_\Phi(v^2, w^2, w^2, v^2) \\
&= 3\Phi(v^1, w^1)^2 + 3\Phi(v^2, w^2)^2 = 6\,.
\end{aligned}
$$

This contradiction establishes Assertion (2) if Φ is anti-symmetric.

If Φ is non-degenerate, then $\ker \Phi = \{0\}$ and hence $V_i \not\subset \ker \Phi$. Thus Assertion (3) follows immediately from Assertion (2).

Suppose $\ker \Phi$ is totally isotropic with respect to an inner product $\langle \cdot, \cdot \rangle$. If $V = V_1 \oplus V_2$ is an orthogonal direct sum decomposition with respect to an inner product $\langle \cdot, \cdot \rangle$, then V_1 and V_2 are not totally isotropic with respect to $\langle \cdot, \cdot \rangle$. Thus $V_i \not\subset \ker \Phi$ and Assertion (4) follows from Assertion (2).

Clearly if $\xi \in \ker \Phi$, then $\xi \in \ker(A_\Phi)$. Conversely, suppose $v_1 \notin \ker \Phi$. The arguments given to prove Assertion (2) show that there exists v_2 so $A_\Phi(v_1, v_2, v_2, v_1) \neq 0$ and thus $v_1 \notin \ker(A_\Phi)$; Assertion (5) follows. $\square$

We generalize Lemma 1.6.3 (5) as follows. If $A_0 \in \mathcal{A}lg_0(V)$, let

$$
\ker(A_0) := \{\eta \in V : A_0(\eta, \xi_1, \xi_2, \xi_3) = 0 \ \forall\, \xi_i \in V\}\,.
$$

Let π be the natural projection from V to $\bar{V} := V/\ker(A_0)$. The algebraic curvature tensor A_0 descends to $\bar{V}$ to define an algebraic curvature tensor $\bar{A}_0 \in \mathcal{A}lg_0(\bar{V})$ so that $\pi^* \bar{A}_0 = A_0$.

Lemma 1.6.4   *Let* $\mathfrak{M}_0 := (V, \langle \cdot, \cdot \rangle, A_0)$. *If* $\bar{\mathfrak{M}}_0^w := (\bar{V}, \bar{A}_0)$ *is indecomposable and if* $\ker(A_0)$ *is totally isotropic, then* $\mathfrak{M}_0$ *is indecomposable.*

Proof.   We suppose, to the contrary, that there exists a non-trivial orthogonal direct sum decomposition $V = V^1 \oplus V^2$ which decomposes $A_0 = A_0^1 \oplus A_0^2$. We argue for a contradiction.

Suppose there exists $v^i \in V^i$ with $0 \neq \pi(v^1) = \pi(v^2)$. Because $0 \neq \pi(v^1)$, $v^1 \notin \ker(A_0)$. Consequently, we can choose $\xi_1, \xi_2, \xi_3 \in V$ so $A_0(v^1, \xi_1, \xi_2, \xi_3) \neq 0$. Decompose $\xi_i = \xi_i^1 + \xi_i^2$ for $\xi_i^1 \in V^1$ and $\xi_i^2 \in V^2$. Since $A_0 = A_0^1 \oplus A_0^2$ and since $v^1 \in V^1$,

$$
0 \neq A_0(v^1, \xi_1, \xi_2, \xi_3) = A_0(v^1, \xi_1^1, \xi_2^1, \xi_3^1)\,.
$$

On the other hand, since $\pi(v^1) = \pi(v^2)$, we have $v^1 - v^2 \in \ker(A_0)$ so

$$
0 \neq A_0(v^1, \xi_1^1, \xi_2^1, \xi_3^1) = A_0(v^2, \xi_1^1, \xi_2^1, \xi_3^1)\,.
$$

This contradicts the decomposition $A_0 = A_0^1 \oplus A_0^2$ and shows

$$\pi(V^1) \cap \pi(V^2) = \{0\}\,.$$

Since $\pi(V^1) + \pi(V^2) = \pi(V) = \bar{V}$, we have a direct sum decomposition

$$\bar{V} = \pi(V^1) \oplus \pi(V^2) \quad \text{and} \quad \bar{A}_0 = \bar{A}_0^1 \oplus \bar{A}_0^2\,.$$

Since $(\bar{V}, \bar{A}_0)$ is assumed to be indecomposable, one of the two summands is trivial; without loss of generality we assume the notation is chosen so $\pi(V_2) = \{0\}$. This implies $V^2 \subset \ker(A_0)$ and hence V^2 is totally isotropic. This is false as the decomposition $V = V^1 \oplus V^2$ is assumed to be an orthogonal direct sum decomposition with respect to $\langle \cdot, \cdot \rangle$. $\qquad \square$

The following is a useful observation we shall need Section 2.5:

Lemma 1.6.5

(1) Let $\Phi_i \in S^2(V^*)$. If $\operatorname{Rank}(\Phi_1) \geq 3$ and if $A_{\Phi_1} = A_{\Phi_2}$, then $\Phi_1 = \pm\Phi_2$.
(2) Let $\Phi_i \in \Lambda^2(V^*)$. If $A_{\Phi_1} = A_{\Phi_2}$, then $\Phi_1 = \pm\Phi_2$.

Proof. Suppose that $\Phi_1, \Phi_2 \in S^2(V^*)$ with $\operatorname{Rank}(\Phi_1) \geq 3$. Choose a basis $\{e_1, ..., e_r\}$ for V so that

$$\Phi_1(e_i, e_j) = 0 \quad \text{and} \quad \Phi_1(e_i, e_i) = \varepsilon_i \quad \text{where} \quad \varepsilon_i \in \{0, \pm 1\}\,.$$

Since $\operatorname{Rank}(\Phi_1) \geq 3$, we can assume $\varepsilon_i \neq 0$ for $i = 1, 2, 3$. By replacing Φ_1 by $-\Phi_1$ if necessary, we may assume that $\varepsilon_1 = 1$ and $\varepsilon_2 = 1$. Let $\pi := \operatorname{Span}\{e_1, e_2\}$. By diagonalizing the quadratic form $\Phi_2|_\pi$ with respect to the positive definite quadratic form $\Phi_1|_\pi$, we can further normalize the choice of $\{e_1, e_2\}$ so that:

$$\Phi_2(e_1, e_1) = \varrho_1, \quad \Phi_2(e_1, e_2) = 0, \quad \Phi_2(e_2, e_2) = \varrho_2\,.$$

Let $A_i := A_{\Phi_i}$. We have

$$\begin{aligned}
1 &= \Phi_1(e_1, e_1)\Phi_1(e_2, e_2) - \Phi_1(e_1, e_2)^2 = A_1(e_1, e_2, e_2, e_1) \\
&= A_2(e_1, e_2, e_2, e_1) = \Phi_2(e_1, e_1)\Phi_2(e_2, e_2) - \Phi_2(e_1, e_2)^2 = \varrho_1 \varrho_2\,.
\end{aligned}$$

Thus by replacing Φ_2 by $-\Phi_2$ if necessary, we may assume that $\varrho_1 > 0$ and $\varrho_2 > 0$. If $k \geq 3$, then

$$\begin{aligned}
0 &= \Phi_1(e_1, e_1)\Phi_1(e_2, e_k) - \Phi_1(e_1, e_2)\Phi_1(e_1, e_k) = A_1(e_1, e_2, e_k, e_1) \\
&= A_2(e_1, e_2, e_k, e_1) = \Phi_2(e_1, e_1)\Phi_2(e_2, e_k) - \Phi_2(e_1, e_2)\Phi_2(e_1, e_k) \\
&= \Phi_2(e_2, e_k)\,.
\end{aligned}$$

Thus $\Phi_2(e_2, e_k) = 0$ for $k \geq 3$. Similarly $\Phi_2(e_1, e_k) = 0$ for $k \geq 3$. Thus

$$\varepsilon_k = \Phi_1(e_1, e_1)\Phi_1(e_k, e_k) - \Phi_1(e_1, e_k)^2 = A_1(e_1, e_k, e_k, e_1)$$
$$= A_2(e_1, e_k, e_k, e_1) = \Phi_2(e_1, e_1)\Phi_2(e_k, e_k) - \Phi_2(e_1, e_k)^2$$
$$= \varrho_1 \varrho_k,$$
$$\varepsilon_k = \Phi_1(e_2, e_2)\Phi_1(e_k, e_k) - \Phi_1(e_2, e_k)^2 = A_1(e_2, e_k, e_k, e_2)$$
$$= A_2(e_2, e_k, e_k, e_2) = \Phi_2(e_2, e_2)\Phi_2(e_k, e_k) - \Phi_2(e_2, e_k)^2$$
$$= \varrho_2 \varrho_k \,.$$

Setting $k = 3$ then yields $\varrho_1 = \varrho_2$ and thus $\varrho_1^2 = 1$. Since $\varrho_1 > 0$, we have $\varrho_1 = \varrho_2 = +1$. We can now conclude that $\varrho_k = \varepsilon_k$ for all $k \geq 3$. Consequently $\Phi_1 = \Phi_2$; this proves Assertion (1).

Suppose that $\Phi_1, \Phi_2 \in \Lambda^2(V^*)$. Choose a basis

$$\{e_1, ..., e_s, f_1, ..., f_s, n_1, ..., n_t\}$$

for V so that the non-zero components of Φ_1 are $\Phi_1(e_i, f_i) = 1$. We have

$$A_{\Phi_i}(x, y, y, x) = 3\Phi_i(x, y)^2 \,.$$

Consequently, the non-zero components of Φ_2 are $\Phi_2(e_i, f_i) = \varepsilon_i = \pm 1$. Let $A_i := A_{\Phi_i}$. If $i < j$, then

$$12 = 3\Phi_1(e_i + e_j, f_i + f_j)^2 = A_{\Phi_1}(e_i + e_j, f_i + f_j, f_i + f_j, e_i + e_j)$$
$$= A_{\Phi_2}(e_i + e_j, f_i + f_j, f_i + f_j, e_i + e_j) = 3\Phi_1(e_i + e_j, f_i + f_j)^2$$
$$= 3(\varepsilon_i + \varepsilon_j)^2 \,.$$

Thus either $\varepsilon_i = \varepsilon_j = +1$ for all i, j and $\Phi_1 = \Phi_2$ or $\varepsilon_i = \varepsilon_j = -1$ for all i, j and $\Phi_1 = -\Phi_2$. $\qquad\qquad\square$

Remark 1.6.1   Lemma 1.6.5 (1) can fail if $\mathrm{Rank}(\Phi_1) < 3$. Suppose that $\dim(V) = 2$ and that Φ_i are any two non-degenerate symmetric bilinear forms on V. If $\{v_1, v_2\}$ is a basis for V, then $A_{\Phi_i}(v_1, v_2, v_2, v_1) = c_i$ for some non-zero constants c_i and thus A_{Φ_1} and A_{Φ_2} are multiples. It does not, however, follow that Φ_1 and Φ_2 are multiples.

We can now establish a basic result in the field. It was originally proved by Fiedler (2003a) using Young diagrams; subsequently a direct proof was given in Gilkey (2002). We give a third proof here; a still different proof that $\mathcal{Alg}_0(V) = \mathrm{Span}_{\Phi \in S^2(V^*)}\{A_\Phi\}$ will follow from the discussion in Section 4.3 which uses the Embedding Theorem of Nash (1956).

Theorem 1.6.1 [Fiedler]

$$\mathcal{A}\mathrm{lg}_0(V) = \mathrm{Span}_{\Phi \in S^2(V^*)}\{A_\Phi\} = \mathrm{Span}_{\Phi \in \Lambda^2(V^*)}\{A_\Phi\}\,.$$

Proof. To simplify the discussion, we introduce as temporary notation:

$$\mathcal{C}^+ := \mathrm{Span}_{\Phi \in S^2(V^*)}\{A_\Phi\} \quad \text{and} \quad \mathcal{C}^- := \mathrm{Span}_{\Phi \in \Lambda^2(V^*)}\{A_\Phi\}\,.$$

These are clearly subspaces which are invariant under the action of $\mathrm{Gl}(V)$. By Lemma 1.3.3, $\mathcal{A}\mathrm{lg}_0(V)$ is an irreducible $\mathrm{Gl}(V)$ module; in particular, there are no proper invariant subspaces. The Lemma now follows since $\mathcal{C}^\pm \neq \{0\}$. $\qquad\qquad\square$

Remark 1.6.2 The proof in fact establishes a slightly stronger result. One can restrict the generating elements Φ to those which have rank 2 since these also generate non-trivial $\mathrm{Gl}(V)$ modules.

Fiedler also gave generators for $\mathcal{A}\mathrm{lg}_1(V)$. If

$$\Psi \in S^2(V) \quad \text{and} \quad \Psi_1 \in S^3(V),$$

define $A_{1,\Psi,\Psi_1} \in \mathcal{A}\mathrm{lg}_1(V)$ by:

$$\begin{aligned}
A_{1,\Psi,\Psi_1}(x,y,z,w;v) := \;&\Psi_1(x,w,v)\Psi(y,z) + \Psi(x,w)\Psi_1(y,z,v) \\
&-\Psi_1(x,z,v)\Psi(y,w) - \Psi(x,z)\Psi_1(y,w,v)\,.
\end{aligned} \tag{1.6.a}$$

If one thinks of Ψ_1 as the symmetrized covariant derivative of Ψ, then A_{1,Ψ,Ψ_1} can be regarded, at least formally speaking, as the covariant derivative of A_Ψ. Fiedler (2003b) used group representation theory to show:

Theorem 1.6.2 **(Fiedler)** $\mathcal{A}\mathrm{lg}_1(V) = \mathrm{Span}_{\Psi \in S^2(V), \Psi_1 \in S^3(V)}\{A_{1,\Psi,\Psi_1}\}$.

In Section 4.3, we will give a proof of Theorem 1.6.2 which is based on the Embedding Theorem of Nash (1956).

1.6.4 *Complex geometry*

Let $\mathfrak{M}_0 = (V, \langle\cdot,\cdot\rangle, A)$ be a 0-model. An isometry $J \in O(V, \langle\cdot,\cdot\rangle)$ is said to be a *pseudo-Hermitian almost complex structure* on V if additionally $J^2 = -\,\mathrm{id}$. We use J to define a complex structure on V by setting

$$(a + b\sqrt{-1})v := av + bJv \quad \text{for} \quad v \in V \quad \text{and} \quad a + b\sqrt{-1} \in \mathbb{C}\,.$$

Thus a linear transformation T of V is said to be *complex linear* if T commutes with J. A subspace π of V is said to be a *complex subspace* if $J\pi = \pi$. If $\dim_{\mathbb{R}} \pi = 2$ and if π is complex, then π is said to be a complex line. We set $\mathbb{CP}^{\pm}(V, \langle \cdot, \cdot \rangle, J)$ to be the projective spaces of all complex spacelike $(+)$ or complex timelike $(-)$ lines in V; there are no complex mixed lines. We shall also sometimes simply denote these spaces by $\mathbb{CP}^{\pm}(V)$. Let

$$\pi_x := \mathrm{Span}\{x, Jx\} \quad \text{for} \quad x \in S^{\pm}(V, \langle \cdot, \cdot \rangle)$$

be the associated complex line. The map $x \to \pi_x$ defines the *Hopf fibrations*

$$S^1 \to S^{\pm}(V, \langle \cdot, \cdot \rangle) \to \mathbb{CP}^{\pm}(V, \langle \cdot, \cdot \rangle, J) \,. \tag{1.6.b}$$

The following operators are independent of the particular unit vector x which was chosen and depend only on the underlying complex plane:

$$\mathcal{J}(\pi_x) := \mathcal{J}(x) + \mathcal{J}(Jx) \quad \text{and} \quad \mathcal{A}(\pi_x) := \mathcal{A}(x, Jx) \,.$$

We say that $\mathfrak{M} = (V, \langle \cdot, \cdot \rangle, J, A)$ is a *complex 0-model* if $(V, \langle \cdot, \cdot \rangle, A)$ is a 0-model and if J is a pseudo-Hermitian almost complex structure on V. The following Lemma shows several different compatibility conditions between J and A are equivalent in this setting:

Lemma 1.6.6   *Let $\mathfrak{M} = (V, \langle \cdot, \cdot \rangle, J, A)$ be a complex 0-model. The following assertions are equivalent; if any (and hence all) are satisfied, we say A and J are compatible:*

(1) $J^* A = A$.
(2) $\mathcal{J}(\pi)$ is complex linear for every π in $\mathbb{CP}^{\pm}(V, \langle \cdot, \cdot \rangle, J)$.
(3) $\mathcal{A}(\pi)$ is complex linear for every π in $\mathbb{CP}^{\pm}(V, \langle \cdot, \cdot \rangle, J)$.

Proof.   Suppose $J^* A = A$. Then for all x, y, z, we have that:

$$A(y, x, x, z) + A(y, Jx, Jx, z) = A(Jy, x, x, Jz) + A(Jy, Jx, Jx, Jz) \,.$$

Replacing z by Jz yields

$$A(y, x, x, Jz) + A(y, Jx, Jx, Jz) = -A(Jy, x, x, z) - A(Jy, Jx, Jx, z) \,.$$

This implies that

$$\langle \mathcal{J}(\pi_x) y, Jz \rangle = -\langle \mathcal{J}(\pi_x) Jy, z \rangle \,.$$

This shows that $J\mathcal{J}(\pi_x) = \mathcal{J}(\pi_x) J$ as desired. Thus Assertion (1) implies Assertion (2). Similarly, we may compute that for all x, y, z we have:

$$\langle J\mathcal{A}(\pi_x) y, z \rangle = -\langle \mathcal{A}(\pi_x) y, Jz \rangle = -A(x, Jx, y, Jz)$$
$$= -A(Jx, JJx, Jy, JJz) = A(x, Jx, Jy, z) = \langle \mathcal{A}(\pi_x) Jy, z \rangle \,.$$

Thus $J\mathcal{A}(\pi_x) = \mathcal{A}(\pi_x)J$ so Assertion (1) also implies Assertion (3).

We use an argument shown to us by Brozos-Vázquez to prove that Assertion (2) implies Assertion (1). If $\mathcal{J}(\pi_x)$ is complex then

$$J\{\mathcal{J}(x) + \mathcal{J}(Jx)\} = \{\mathcal{J}(x) + \mathcal{J}(Jx)\}J\,.$$

Suppose this identity holds for all $x \in S^{\pm}(V, \langle\cdot,\cdot\rangle)$. We can rescale to see this holds for all non-degenerate x. Since the set of non-degenerate vectors is dense in V, this holds for all $x \in V$. Consequently, after moving J across the inner product, we see that for all x, y, z,

$$-\langle(\mathcal{J}(x) + \mathcal{J}(Jx))y, Jz\rangle = \langle(\mathcal{J}(x) + \mathcal{J}(Jx))Jy, z\rangle$$

which implies that

$$-A(y, x, x, Jz) - A(y, Jx, Jx, Jz) = A(Jy, x, x, z) + A(Jy, Jx, Jx, z)\,.$$

Polarizing this identity and replacing z by Jz yields

$$
\begin{aligned}
A(y, x, w, z) &+ A(y, w, x, z) + A(y, Jx, Jw, z) + A(y, Jw, Jx, z)\\
&= A(Jy, x, w, Jz) + A(Jy, w, x, Jz) + A(Jy, Jx, Jw, Jz)\\
&\quad + A(Jy, Jw, Jx, Jz).
\end{aligned}
\tag{1.6.c}
$$

Interchanging arguments $1 \leftrightarrow 2$ and $3 \leftrightarrow 4$ then yields:

$$
\begin{aligned}
A(x, y, z, w) &+ A(w, y, z, x) + A(Jx, y, z, Jw) + A(Jw, y, z, Jx)\\
&= A(x, Jy, Jz, w) + A(w, Jy, Jz, x) + A(Jx, Jy, Jz, Jw)\\
&\quad + A(Jw, Jy, Jz, Jx).
\end{aligned}
\tag{1.6.d}
$$

If we change $x \leftrightarrow y$ and $z \leftrightarrow w$ in Eq. (1.6.d) we get

$$
\begin{aligned}
A(y, x, w, z) &+ A(z, x, w, y) + A(Jy, x, w, Jz) + A(Jz, x, w, Jy)\\
&= A(y, Jx, Jw, z) + A(z, Jx, Jw, y) + A(Jy, Jx, Jw, Jz)\\
&\quad + A(Jz, Jx, Jw, Jy).
\end{aligned}
\tag{1.6.e}
$$

Adding (1.6.c) and (1.6.e) and simplifying yields:

$$
\begin{aligned}
A(y, x, w, z) &+ A(y, w, x, z)\\
&= A(Jy, Jx, Jw, Jz) + A(Jy, Jw, Jx, Jz)\,.
\end{aligned}
\tag{1.6.f}
$$

In Eq. (1.6.f) we change $y \to x$, $x \to w$ and $w \to y$. This yields:

$$
\begin{aligned}
A(x, w, y, z) &+ A(x, y, w, z)\\
&= A(Jx, Jw, Jy, Jz) + A(Jx, Jy, Jw, Jz)\,.
\end{aligned}
\tag{1.6.g}
$$

We add $2(1.6.\text{f})$ and $(1.6.\text{g})$ to see

$$A(y,x,w,z) + 2A(y,w,x,z) + A(x,w,y,z)$$
$$= A(Jy,Jx,Jw,Jz) + 2A(Jy,Jw,Jx,Jz) + A(Jx,Jw,Jy,Jz)\,. \qquad (1.6.\text{h})$$

By the first Bianchi identity,

$$A(y,x,w,z) + A(x,w,y,z) = A(y,w,x,z),$$
$$A(Jy,Jx,Jw,Jz) + A(Jx,Jw,Jy,Jz) = A(Jy,Jw,Jx,Jz)\,. \qquad (1.6.\text{i})$$

We use Eqs. $(1.6.\text{h})$ and $(1.6.\text{i})$ to see that

$$3A(y,w,x,z) = 3A(Jy,Jw,Jx,Jz)$$

for all x,y,z,w. Thus Assertion (2) implies Assertion (1).

Finally, we show Assertion (3) implies Assertion (1). We use an argument that was shown to us by Salamon. Suppose $\mathcal{A}(x,Jx)J = J\mathcal{A}(x,Jx)$ for every $x \in S^{\pm}(V,\langle \cdot,\cdot\rangle)$. We can rescale to see this holds for every non-degenerate x and hence by continuity for every x in V. Thus for all x,z,w, we have

$$J\mathcal{A}(x,Jx) = \mathcal{A}(x,Jx)J,$$
$$\Rightarrow \langle JA(x,Jx)z,w\rangle - \langle A(x,Jx)Jz,w\rangle = 0,$$
$$\Rightarrow A(x,Jx,z,Jw) + A(x,Jx,Jz,w) = 0\,.$$

Polarizing yields an identity for all x,y,z,w:

$$0 = A(y,Jx,z,Jw) + A(x,Jy,z,Jw) + A(y,Jx,Jz,w)$$
$$+ A(x,Jy,Jz,w)\,.$$

Interchange the first two arguments in the first and third terms to see

$$0 = -A(Jx,y,z,Jw) + A(x,Jy,z,Jw) - A(Jx,y,Jz,w)$$
$$+ A(x,Jy,Jz,w)\,.$$

Replace (x,w) by (Jx,Jw) to show:

$$0 = -A(x,y,z,w) - A(Jx,Jy,z,w) + A(x,y,Jz,Jw)$$
$$+ A(Jx,Jy,Jz,Jw)\,. \qquad (1.6.\text{j})$$

Interchange the first two arguments with the final two arguments:

$$0 = -A(z,w,x,y) - A(z,w,Jx,Jy) + A(Jz,Jw,x,y)$$
$$+ A(Jz,Jw,Jx,Jy)\,.$$

Change notation to interchange x and z and y and w to see:

$$0 = -A(x,y,z,w) - A(x,y,Jz,Jw) + A(Jx,Jy,z,w) \\ + A(Jx,Jy,Jz,Jw)\,. \tag{1.6.k}$$

We add Eqs. (1.6.j) and (1.6.k) to conclude

$$-A(x,y,z,w) + A(Jx,Jy,Jz,Jw) = 0$$

and complete the proof that Assertion (3) implies Assertion (1). $\qquad\square$

1.6.5 *Rank 1-symmetric spaces*

Let $\mathfrak{M}_0 = (V, \langle \cdot, \cdot \rangle, A)$ be a 0-model. If $\{e_1, e_2\}$ is an orthonormal basis for a non-degenerate 2-plane π, then the *sectional curvature* of π is defined by

$$\kappa(\pi) := A(e_1, e_2, e_2, e_1)\,.$$

Definition 1.6.1 One says that a model $\mathfrak{M}_0 = (V, \langle \cdot, \cdot \rangle, A)$ has *constant sectional curvature* c if $\kappa(\pi) = c$ on every spacelike and timelike 2-plane and if $\kappa(\pi) = -c$ on every mixed 2-plane. One says that a pseudo-Riemannian manifold $\mathcal{M}$ is a *space form* if $\mathfrak{M}_0(\mathcal{M}, P)$ has constant sectional curvature c at every point $P \in M$; the constant is allowed, in principle, to vary with the point P.

Remark 1.6.3 Let $A_{\langle \cdot, \cdot \rangle}$ be the canonical curvature tensor of Eq. (1.3.a):

$$A_{\langle \cdot, \cdot \rangle}(x, y, z, w) := \langle x, w \rangle \langle y, z \rangle - \langle x, z \rangle \langle y, w \rangle\,.$$

$\mathfrak{M}_0$ has constant sectional curvature c if and only if $A = cA_{\langle \cdot, \cdot \rangle}$.

Let $\mathbb{R}^{(p,q)}$ denote $\mathbb{R}^{p+q}$ with the canonical inner product of signature (p, q) given in Eq. (1.2.a). We consider the pseudospheres:

$$S_+^{(p,q)} := \{\xi \in \mathbb{R}^{(p+1,q)} : \langle \xi, \xi \rangle = +1\},$$
$$S_-^{(p,q)} := \{\xi \in \mathbb{R}^{(p,q+1)} : \langle \xi, \xi \rangle = -1\}\,.$$

We note that $S_-^{(0,q)}$ has two components; each component has constant sectional curvature -1 and is isometric to hyperbolic space.

The geometry is very rigid in this setting. The following is well known; see, for example, Lemmas 1.14.2 and 2.6.1 of Gilkey (2002).

Lemma 1.6.7 *Adopt the notation established above.*

(1) The manifolds $S_\pm^{(p,q)}$ are spaceforms of signature (p,q) with constant sectional curvature ± 1.

(2) If $\mathcal{M}$ is a space form of signature (p,q) with constant sectional curvature ± 1 of dimension $m \geq 3$, then $\mathcal{M}$ is locally isometric to $S_\pm^{(p,q)}$.

Definition 1.6.2 A model $\mathfrak{M}_0 = (V, \langle \cdot, \cdot \rangle, A)$ is said to be a *complex space form* if there exists a Hermitian almost complex structure J on $(V, \langle \cdot, \cdot \rangle)$ so $A = c_0 A_{\langle \cdot, \cdot \rangle} + c_1 A_J$. We say that a pseudo-Riemannian manifold $\mathcal{M}$ is a *complex space form* if $\mathfrak{M}_0(\mathcal{M}, P)$ is a complex space form at each point $P \in M$; the constants c_0 and c_1 being (in principle) allowed to vary with the point in question.

Let J be the standard Hermitian almost complex structure on $\mathbb{R}^{(2r,2s)}$. We have the Hopf fibrations of Eq. (1.6.b):

$$S^1 \to S^-(\mathbb{R}^{(2p+2,2q)}) \xrightarrow{h} \mathbb{CP}_-^{(2p,2q)},$$
$$S^1 \to S^+(\mathbb{R}^{(2p,2q+2)}) \xrightarrow{h} \mathbb{CP}_+^{(2p,2q)}.$$

The metrics on the horizontal distributions $\ker(h_*)$ have signature $(2p, 2q)$ and are invariant under the action of S^1. They induce, therefore, metrics called the *Fubini–Study metrics* on the associated projective spaces. The following is well known; see, for example, Lemma 1.15.1 and Lemma 3.6.4 of Gilkey (2002):

Lemma 1.6.8

(1) Let $2p + 2q \geq 4$. The manifold $\mathbb{CP}_\pm^{(2p,2q)}$ is a complex space form of signature (p,q) and the curvature is given by $R = \pm\{R_{\mathrm{id}} + R_J\}$.

(2) Let (M) be a contractible complex space form of signature $(2p, 2q)$ with $2p + 2q \geq 6$. Then

    (a) $c_0(P) = c_1(P) = c$ is constant.
    (b) J can be chosen to vary smoothly with P.
    (c) $\nabla J = 0$ and $\nabla R = 0$.
    (d) If $c = \pm 1$, then $\mathcal{M}$ is locally isometric to $\mathbb{CP}_\pm^{(p,q)}$.

The 4-dimensional geometries are exceptional; see, for example, the discussion in Olszak (1989).

In the Riemannian setting, Lemmas 1.6.7 and 1.6.8 illustrate Theorem 1.4.4; any Riemannian manifold whose curvature tensor is modeled on that of $S_\pm^m$ or on $\mathbb{CP}_\pm^{(0,2q)}$ is locally isomorphic to $S_\pm^m$ or $\mathbb{CP}_\pm^{(0,2q)}$; these manifolds have rigid geometries.

We can also discuss quaternionic generalizations. Let

$$\mathbb{H} := \mathrm{Span}_{\mathbb{R}}\{1, i, j, k\}$$

be the quaternions where

$$ij = -ji = k \quad \text{and} \quad i^2 = j^2 = k^2 = -1\,.$$

Give $\mathbb{H}^k = \mathbb{R}^{4k}$ the canonical quaternion structure. Let S^3 be the unit quaternions. Quaternionic multiplication defines a Hopf fibration

$$S^3 \to S^-(\mathbb{R}^{(4q+4,4p)}) \to \mathbb{HP}_-^{(4q,4p)}$$
$$S^3 \to S^+(\mathbb{R}^{(4q,4p+4)}) \to \mathbb{HP}_+^{(4q,4p)}\,.$$

The metrics on the horizontal distributions have signature $(4q, 4p)$; they are S^3 invariant and induce metrics called the *Fubini–Study* metrics on the associated projective spaces.

Lemma 1.6.9 *The manifolds* $\mathbb{HP}_\pm^{(4p,4q)}$ *are pseudo-Riemannian manifolds of signature* $(4p, 4q)$. *If* $P \in \mathbb{HP}_\pm^{(4q,4p)}$, *then there is a Clifford family* $\mathcal{F}(P) = \{I, J, K\}$ *on the tangent space so that curvature tensor is given by*

$$R_P = \pm\{R_{\mathrm{id}} + R_I + R_J + R_K\}\,.$$

The rank 1-symmetric spaces are classified. The manifolds

$$\{S_\pm^{(0,q)}, \mathbb{CP}_\pm^{(0,2q)}, \mathbb{HP}_\pm^{(0,4q)}\}$$

are Riemannian *rank* 1*-symmetric spaces* and together with the Cayley plane and its negative curvature dual, comprise the complete list of all the Riemannian rank 1-symmetric spaces. These spaces play a central role in the Osserman conjecture as we shall see in Section 1.9.4.

1.6.6 *Conformal complex space forms*

We work in the Riemannian context. Let Φ be a Hermitian almost complex structure on TM; necessarily $m = 2n$ is even. We say that (M, g) is a *complex space form* if $R = \lambda_0 R_0 + \lambda_1 R_\Phi$ for smooth functions λ_0 and λ_1 where $\lambda_1 \neq 0$. If (M, g) is a complex space form and if $m \geq 6$, then one can show that $\lambda_0 = \lambda_1$ and that λ_0 is constant. By rescaling the metric, we may assume $\lambda_0 = \pm 1$. If $\lambda_0 = 1$, then (M, g) is locally isometric to complex projective space with the Fubini–Study metric; if $\lambda_0 = -1$, then (M, g) is locally isometric to the negative curvature dual. We refer to Tricerri and Vanhecke (1981) for further details. Let W be the Weyl conformal curvature tensor discussed in Section 1.3.3. One says that (M, g) is a *conformal complex space form* if $W = \lambda_0 R_0 + \lambda_1 R_\Phi$ for some Hermitian almost complex structure on TM where λ_0 and λ_1 are smooth functions on M with $\lambda_1 \neq 0$. We refer to Blažić and Gilkey (2004) for the proof of the following result:

Theorem 1.6.3   *Let (M, g) be a conformal complex space form so that $m \geq 8$. Then (M, g) is locally conformally equivalent to either complex projective space with the Fubini–Study metric or to the negative curvature dual.*

1.6.7 *Kähler geometry*

Let $\mathcal{M} = (M, g)$ be a pseudo-Riemannian manifold with a Hermitian almost complex structure J; J is an isometry of TM with $J^2 = -\,\mathrm{id}$. One says that $(\mathcal{M}, J)$ is *almost Kähler* if $\nabla J = 0$ and that $(\mathcal{M}, J)$ is Kähler if additionally the almost complex structure in question is integrable; this means that there exist coordinates $z_j = x_j + \sqrt{-1} y_j$ so that

$$J\partial_{x_j} = \partial_{y_j} \quad \text{and} \quad J\partial_{y_j} = -\partial_{x_j}\,.$$

Newlander and Nirenberg (1957) provide necessary and sufficient conditions that an almost complex structure giving rise to a complex structure.

Any holomorphic submanifold of a Kähler manifold is Kähler. As the Fubini–Study metric on $\mathbb{CP}^k$ is Kähler, any algebraic variety is Kähler. Kähler geometry provides a useful family of compatible examples:

Lemma 1.6.10   *If $\mathcal{M}$ is an almost Kähler manifold, then J and R are compatible.*

Proof.   Since $\nabla J = 0$, we have $\nabla_x J = J\nabla_x$ so $J\mathcal{R}(x, y) = \mathcal{R}(x, y)J$ for all x, y. Thus as a special case $J\mathcal{R}(x, Jx) = \mathcal{R}(x, Jx)J$ and hence J and R are compatible by Lemma 1.6.6. $\qquad\square$

1.7 The Geometry of the Jacobi Operator

Let $\mathfrak{M}_0 = (V, \langle \cdot, \cdot \rangle, A)$ be a 0-model and let $\mathcal{A}$ be the associated curvature operator;

$$\langle \mathcal{A}(u, v)w, x \rangle = A(u, v, w, x)\,.$$

In Section 1.7, we shall discuss natural operators which are related to the Jacobi operator. Section 1.7.1 deals with the classical Jacobi operator, Section 1.7.2 treats the higher order Jacobi operator, Section 1.7.3 considers the conformal Jacobi operator, and Section 1.7.4 is concerned with the complex Jacobi operator.

1.7.1 *The Jacobi operator*

The *Jacobi operator* $\mathcal{J}(v)$ is the linear map of V defined by:

$$\mathcal{J}(v) : w \to \mathcal{A}(w, v)v \,.$$

The curvature identities show that $\mathcal{J}$ is symmetric since one has:

$$\langle \mathcal{J}(v)w, z \rangle = A(w, v, v, z) = A(z, v, v, w) = \langle \mathcal{J}(v)z, w \rangle \,.$$

We polarize $\mathcal{J}$ and define

$$\mathcal{J}(v_1, v_2) : w \to \tfrac{1}{2}\{\mathcal{A}(w, v_1)v_2 + \mathcal{A}(w, v_2)v_1\} \,.$$

One then has $\mathcal{J}(v) = \mathcal{J}(v, v)$. We note that the *Ricci tensor* ρ is given by

$$\rho(v_1, v_2) = \mathrm{Tr}\{\mathcal{J}(v_1, v_2)\} \,.$$

The Jacobi operator determines the curvature:

Lemma 1.7.1 *Let $\mathfrak{M}_0$ be a 0-model. If $\mathcal{J} = 0$, then $A = 0$.*

Proof. Suppose that $\mathcal{J} = 0$. Then $A(y, x, x, z) = 0$ for all x, y, z. Polarizing in x then yields $0 = A(y, x, v, z) + A(y, v, x, z)$. Consequently,

$$\begin{aligned}
0 &= A(y, x, v, z) + A(y, v, z, x) + A(y, z, x, v) \\
&= A(y, x, v, z) - A(y, v, x, z) + A(y, z, x, v) \\
&= A(y, x, v, z) + A(y, x, v, z) - A(y, x, z, v) \\
&= 3A(y, x, v, z) \,.
\end{aligned}$$

This shows $A = 0$ as desired. $\qquad\square$

One then has the following

Corollary 1.7.1 *Let $\mathfrak{M}^i := (V, \langle \cdot, \cdot \rangle, A^i)$ be 0-models for $i = 1, 2$. If $\mathcal{J}_{\mathfrak{M}^1} = \mathcal{J}_{\mathfrak{M}^2}$, then $A^1 = A^2$.*

One says that $\mathfrak{M}_0$ is *spacelike Osserman* (respectively *timelike Osserman*) if the eigenvalues of $\mathcal{J}$ are constant on the pseudosphere $S^+(V, \langle \cdot, \cdot \rangle)$ (respectively on $S^-(V, \langle \cdot, \cdot \rangle)$) of unit spacelike (respectively timelike) vectors in V; the multiplicities are then necessarily constant as well. We will show in Theorem 1.9.1 that these are equivalent notions so one simply speaks of an Osserman algebraic curvature tensor in this setting. One says that $\mathfrak{M}_0$ is *spacelike Jordan Osserman* if the Jordan normal form of $\mathcal{J}$ is constant on $S^+(V, \langle \cdot, \cdot \rangle)$. The notion *timelike Jordan Osserman* is defined similarly; spacelike Jordan Osserman and timelike Jordan Osserman are different notions as we shall see presently. It is clear that spacelike (respectively timelike) Jordan Osserman implies spacelike (respectively timelike) Osserman. The designation "Osserman" is used owing to the seminal paper of Osserman (1990).

In the Riemannian setting ($p = 0$), the eigenvalue structure determines the Jordan normal form. This is not, however, the case in higher signatures. If $p \geq 2$ and $q \geq 2$, there are examples where $\mathfrak{M}_0$ is spacelike Osserman but not spacelike Jordan Osserman. There are also examples where $\mathfrak{M}_0$ is spacelike Jordan Osserman but not timelike Jordan Osserman. We refer to the discussion in Theorems 1.7.1, 1.7.2, 2.6.1, and 2.7.3 for further details.

Let $\mathcal{F} := \{J_1, ..., J_\ell\}$ be a *Clifford family* on V. The J_i are skew-adjoint endomorphisms of V satisfying the Clifford commutation relations:

$$J_i J_j + J_j J_i = -2\delta_{ij} \, \mathrm{id} \, .$$

Following the notation established in Eq. (1.3.a), one defines

$$A_{\langle \cdot, \cdot \rangle}(x, y, z, w) := \langle x, w \rangle \langle y, z \rangle - \langle x, z \rangle \langle y, w \rangle,$$
$$A_{J_i}(x, y, z, w) := \langle J_i x, w \rangle \langle J_i y, z \rangle - \langle J_i x, z \rangle \langle J_i y, w \rangle$$
$$-2 \langle J_i x, y \rangle \langle J_i z, w \rangle \, .$$

Theorem 1.7.1   *Let $\mathcal{F}$ be a Clifford family on V. Let $\mathfrak{M} := (V, \langle \cdot, \cdot \rangle, A)$ where $A = c_0 A_{\langle \cdot, \cdot \rangle} + \sum_i c_i A_{J_i}$. Then $\mathfrak{M}$ is spacelike and timelike Jordan Osserman.*

Proof.   Let $x \in S^+(V, \langle \cdot, \cdot \rangle)$. The spectral resolution of $\mathcal{J}(x)$ is given by

$$\mathcal{J}(x)y = \begin{cases} 0 & \text{if} \quad y \in \mathrm{Span}\{x\}, \\ (c_0 + 3c_i)y & \text{if} \quad y \in \mathrm{Span}\{J_i x\}, \\ c_0 y & \text{if} \quad y \in \mathrm{Span}\{x, J_1 x, ... J_\ell x\}^\perp \, . \end{cases}$$

This shows that $\mathfrak{M}$ is spacelike Jordan Osserman. Similarly one can show that A is timelike Jordan Osserman. $\qquad\Box$

Let $\mathcal{M}$ be a pseudo-Riemannian manifold of signature (p, q). We say that $\mathcal{M}$ is *pointwise* spacelike Osserman, *pointwise* timelike Osserman, *pointwise* spacelike Jordan Osserman, or *pointwise* timelike Jordan Osserman if the 0-model $\mathfrak{M}_0(\mathcal{M}, P)$ has this property for every point $P \in M$; the eigenvalues or Jordan normal form being permitted to vary with the point in question. We say that $\mathcal{M}$ is *globally* spacelike Osserman, *globally* timelike Osserman, *globally* spacelike Jordan Osserman, or *globally* timelike Jordan Osserman if the structures in question do not in fact vary with P.

One says that $\mathcal{M}$ is a 2-*point homogeneous space* if the isometries of $\mathcal{M}$ act transitively on the pseudo-sphere bundles $S^{\pm}(\mathcal{M})$. It is clear that if $\mathcal{M}$ is a 2-point homogeneous space, then $\mathcal{M}$ is spacelike and timelike Jordan Osserman. Thus, in particular the standard round sphere S^m and complex projective space $\mathbb{CP}^k$ are Jordan Osserman. We refer to the discussion in Section 1.9.4 for other examples.

If $\mathcal{M}$ is a 2-dimensional Riemannian manifold, then $\mathcal{M}$ is pointwise Osserman; it is globally Osserman if and only if $\mathcal{M}$ has constant sectional curvature. Furthermore, there exist pointwise Osserman 4-dimensional Riemannian manifolds which are not globally Osserman; see, for example, the discussion in Gilkey, Swann, and Vanhecke (1995). We shall survey some of the relevant results in this area in Section 1.9.4.

1.7.2 *The higher order Jacobi operator*

Recall that a pair (r, s) is said to be *admissible* if one has

$$1 \leq r + s \leq m - 1, \quad 0 \leq r \leq p, \quad \text{and} \quad 0 \leq s \leq q.$$

Equivalently, this means that the Grassmannian $\mathrm{Gr}_{r,s}(V, \langle \cdot, \cdot \rangle)$ is a connected manifold of positive dimension.

Let $\{e_i\}$ be a basis for a non-degenerate linear subspace π of signature (r, s). Denote the components of the inner product by $g_{ij} := \langle e_i, e_j \rangle$; let g^{ij} be the inverse matrix. Stanilov and Videv (1998) defined a *higher order Jacobi operator*; this is the symmetric linear map given by:

$$\mathcal{J}(\pi) := \sum_{i,j} g^{ij} \mathcal{J}(e_i, e_j).$$

It is independent of the basis chosen for π. If $\{e_1^-, ..., e_r^-, e_1^+, ..., e_s^+\}$ is an orthonormal basis for π, one then has in particular that

$$\mathcal{J}(\pi) := \mathcal{J}(e_1^+) + ... + \mathcal{J}(e_s^+) - \mathcal{J}(e_1^-) - ... - \mathcal{J}(e_r^-).$$

If $r = 0$ so $\pi \in \mathrm{Gr}_{0,s}(V, \langle \cdot, \cdot \rangle)$ is spacelike, then the unit sphere $S(\pi)$ is compact and there is a universal constant $c_{\mathcal{J}}(s)$ so that

$$\mathcal{J}(\sigma) = c_{\mathcal{J}}(s) \int_{v \in S(\pi)} \mathcal{J}(v) dv.$$

Consequently, the higher order Jacobi operator may be regarded as an average Jacobi operator in this special case. It was introduced first in the Riemannian setting by Stanilov and Videv (1992).

One says that a 0-model $\mathfrak{M}_0 = (V, \langle \cdot, \cdot \rangle, A)$ is *Osserman of type* (r, s) if the eigenvalues of $\mathcal{J}(\cdot)$ are constant on the Grassmannian $\mathrm{Gr}_{r,s}(V, \langle \cdot, \cdot \rangle)$. It is immediate that $\mathfrak{M}_0$ is Osserman of type $(1, 0)$ (respectively $(0, 1)$) if and only if $\mathfrak{M}_0$ is spacelike (respectively timelike) Osserman. In fact, only the value $k := r + s$ is relevant. We will show in Theorem 1.9.1 that if $\mathfrak{M}_0$ is Osserman of type (r, s) for any admissible (r, s) with $r + s = k$, then $\mathfrak{M}_0$ is Osserman of type $(\bar{r}, \bar{s})$ for every admissible $(\bar{r}, \bar{s})$ with $\bar{r} + \bar{s} = k$; thus one speaks of k-Osserman in this setting.

One adds the words "Jordan" if additionally the Jordan normal form is constant. The following examples show that there exist admissible pairs (r, s) and $(\bar{r}, \bar{s})$ with $r + s = \bar{r} + \bar{s}$ so that there are models which are Jordan Osserman of type (r, s) but which are not Jordan Osserman of type $(\bar{r}, \bar{s})$. We postpone until Chapter 2 a discussion of similar geometric examples.

Example 1.7.1   Let $\{e_1^-, ..., e_p^-, e_1^+, ..., e_q^+\}$ be an orthonormal basis for a vector space V of signature (p, q). Set

$$\Phi_a e_i^{\pm} = \begin{cases} \pm(e_i^+ + e_i^-) & \text{if } i \le a, \\ 0 & \text{if } i > a, \end{cases}$$

$$A_a(x, y)z := \langle \Phi_a y, z \rangle \Phi_a x - \langle \Phi_a x, z \rangle \Phi_a y.$$

The map Φ_a is self-adjoint and A_a is the associated canonical curvature operator. We refer to Theorem 3.3.2 of Gilkey (2002) for the proof of the following result of Stavrov (2003a):

Theorem 1.7.2   *Adopt the notation established above.*

(1) If $1 \le k \le m - 1$, then A_a is k-Osserman.

(2) Let $2 \le r \le p$ and $2 \le s \le q$. Then A_a is Jordan Osserman of type

    (a) $(1, 0)$ or $(p - 1, q)$ if and only if $p = a$.
    (b) $(0, 1)$ or $(p, q - 1)$ if and only if $q = a$.
    (c) $(r, 0)$ or $(p - r, q)$ if and only if $p - a + 2 \le r$.
    (d) $(0, s)$ or $(p, q - s)$ if and only if $q - a + 2 \le s$.

(3) If $1 \leq r \leq p-1$ and if $1 \leq s \leq q-1$, then A_a is not Jordan Osserman of type (r,s).

Example 1.7.2 Let $\{e_1^-, ..., e_p^-, e_1^+, ..., e_q^+\}$ be an orthonormal basis for V. Choose $a \geq 1$ so $2a \leq \min(p,q)$. Define a skew-adjoint linear map Φ_a of V and associated canonical curvature tensor A_a by setting:

$$\Phi_a e_k^\pm := \begin{cases} \pm(e_{2i}^- + e_{2i}^+) & \text{if} \quad k = 2i-1 \leq 2a, \\ \mp(e_{2i-1}^- + e_{2i-1}^+) & \text{if} \quad k = 2i \leq 2a, \\ 0 & \text{if} \quad k > 2a, \end{cases}$$

$$A_a(x,y,z,w) := \langle \Phi_a x, w \rangle \langle \Phi_a y, z \rangle - \langle \Phi_a x, z \rangle \langle \Phi_a y, w \rangle$$
$$- 2 \langle \Phi_a x, y \rangle \langle \Phi_a z, w \rangle.$$

We can interchange the roles of spacelike and timelike vectors by changing the sign of the inner product. Thus we may always assume that $p \leq q$. We refer to Gilkey and Ivanova (2002a) for the proof of the following result:

Theorem 1.7.3 *Adopt the notation established above. If $p \leq q$, then:*

(1) A_a is k-Osserman for $1 \leq k \leq \dim V - 1$.

(2) Suppose that $2a < p$. Then A_a is Jordan Osserman of type $(p,0)$ and $(0,q)$; A_a is not Jordan Osserman of type (r,s) otherwise.

(3) Suppose that $2a = p < q$. Then A_a is Jordan Osserman of type $(r,0)$ and of type (r,q) for any $1 \leq r \leq p-1$; A_a is not Jordan Osserman otherwise.

(4) Suppose that $2a = p = q$. Then A_a is Jordan Osserman of type $(r,0)$, of type (r,q), of type $(0,s)$, and of type (p,s) for $1 \leq r \leq p-1$ and $1 \leq s \leq q-1$; A_a is not Jordan Osserman otherwise.

We shall present additional results concerning the higher order Jacobi operator in Section 1.9.5. We also refer to Theorems 2.5.1, 2.6.1, and 2.7.3 for additional examples.

1.7.3 *The conformal Jacobi operator*

Let P be a point of a pseudo-Riemannian manifold (M,g). Let

$$\mathcal{W}_P := (T_P M, g_P, W_P)$$

where $W_P := \pi_{\mathcal{W}} R_P$ is the associated Weyl conformal curvature tensor defined in Section 1.3.3. We say that (M,g) is *conformally spacelike Osserman* (respectively *conformally timelike Osserman*) if $\mathcal{W}_P$ is spacelike

Osserman (respectively timelike Osserman) for every point P of M. One adds the modifier "Jordan" if instead the Jordan normal form is constant. The eigenvalue structure, or Jordan normal form, is permitted to vary with the point P of M; the technical distinction between "global" and "pointwise" plays no role in this setting. We will show in Section 1.9.6 that this is a conformally invariant condition.

1.7.4   *The complex Jacobi operator*

Let $\langle \cdot, \cdot \rangle$ be a non-degenerate inner product of signature (p, q) on a vector space V. We say that $\mathfrak{M} := (V, \langle \cdot, \cdot \rangle, J, A)$ is a *complex 0-model* if J is a Hermitian almost complex structure and if $A \in \mathcal{A}\mathrm{lg}_0(V)$; such structures exist if and only if p and q are even. If J and A are compatible and if the eigenvalues of $J(\pi)$ are constant on $\mathbb{CP}^+(V, \langle \cdot, \cdot \rangle, J)$ (respectively on $\mathbb{CP}^-(V, \langle \cdot, \cdot \rangle, J)$), then we say that $\mathfrak{M}$ is *complex spacelike Osserman* (respectively *complex timelike Osserman*) if $q > 0$ (respectively $p > 0$). These are equivalent notions if $p > 0$ and if $q > 0$ so we shall simply speak of $\mathfrak{M}$ being *complex Osserman*. The notions *complex spacelike Jordan Osserman* and *complex timelike Jordan Osserman* are defined similarly.

Let $\mathcal{M} := (M, g, J)$ be an almost complex Hermitian manifold. Here g is a pseudo-Riemannian manifold of signature (p, q) and J is a Hermitian almost complex structure on $T(M)$. We assume $J^*R = R$ as a compatibility condition. We say that $\mathcal{M}$ is *pointwise* complex spacelike Osserman if this property holds for $\mathfrak{M}(\mathcal{M}, P)$ for every $P \in M$; we say that $\mathcal{M}$ is *globally* complex spacelike Osserman if the eigenvalues do not vary with P. Other notions are defined similarly. We refer to Theorems 5.1.1 and 5.1.3 for examples of Riemannian complex Osserman 0-models and manifolds. In this section we content ourselves by showing the analogue of Lemma 1.7.1 fails in this context; we refer to Brozos-Vázquez, García-Río, and Gilkey (2006) for a further discussion of this question:

Lemma 1.7.2   *Let V be a Riemannian vector space of dimension $m \equiv 0$ mod 4. Then there exists a non-zero algebraic curvature tensor A on V and a Hermitian almost complex structure J on V so that J and A are compatible and so that the Riemannian complex 0-model $\mathfrak{M} = (V, \langle \cdot, \cdot \rangle, J, A)$ satisfies $\mathcal{J}(\pi) = 0$ for all $\pi \in \mathbb{CP}(V, \langle \cdot, \cdot \rangle, J)$. If $m \geq 8$, $\mathfrak{M}$ can be chosen so that $\mathfrak{M}$ is not Osserman.*

Proof.   Choose an isometry to identify $V = \mathbb{R}^m$ with the canonical positive definite Euclidean inner product $\langle \cdot, \cdot \rangle$. If K is a skew-symmetric linear map of $\mathbb{R}^m$, let A_K be the canonical algebraic curvature tensor of Eq. (1.3.b):

$$A_K(x, y)z = \langle Ky, z \rangle Kx - \langle Kx, z \rangle Ky - 2\langle Kx, y \rangle Kz \,.$$

It is then immediate that

$$\mathcal{J}_K(x)y = 3\langle y, Kx \rangle Kx \,.$$

Since $m \equiv 0 \bmod 4$, we can choose an isometry to identify V with the quaternions $\mathbb{H}^{\bar{m}}$ where $m = 4\bar{m}$. Let $J = i$ and $K = j$ define skew-adjoint endomorphisms of V with $J^2 = K^2 = -\operatorname{Id}$ and $JK + KJ = 0$. Let $A := A_K - A_{JK}$. We then have

$$\mathcal{J}_A(x)y = 3\langle y, Kx \rangle Kx - 3\langle y, JKx \rangle JKx \,.$$

Since $\{x, Jx, Kx\}$ is an orthonormal set, this shows that $\mathcal{J}_A(x)Kx = 3Kx$ and thus A is not the zero algebraic curvature operator. One also has that

$$
\begin{aligned}
\mathcal{J}_A(\pi_x)y &= 3\langle y, Kx \rangle Kx + 3\langle y, KJx \rangle KJx \\
&\quad - 3\langle y, JKx \rangle JKx - 3\langle yJKJx \rangle JKJx \\
&= 3\langle y, Kx \rangle Kx + 3\langle y, KJx \rangle KJx \\
&\quad - 3\langle y, KJx \rangle KJx - 3\langle yKx \rangle K \\
&= 0 \,.
\end{aligned}
$$

This shows $\mathcal{J}_A(\pi_x) = 0$. Consequently, by Lemma 1.6.6, A and J are compatible. This completes the proof if $m = 4$.

Suppose $\bar{m} \geq 2$. Take a non-trivial decomposition $\mathbb{H}^{\bar{m}} = \mathbb{H}_+ \oplus \mathbb{H}_-$. Set $J_1 = i$, $J_2 = j$, and $J_3 = \pm k$ on $\mathbb{H}_\pm$; if $x_\pm \in S(\mathbb{H}_\pm)$, then $J_1 J_2 J_3 x_\pm = \pm x_\pm$. Define:

$$A := A_{J_2} - A_{J_1 J_2} - A_{J_3} + A_{J_1 J_3} \,.$$

The same calculations as those given above show that $\mathfrak{M}$ is complex Osserman and that $\mathcal{J}(\pi) = 0$ for any $\pi \in \mathbb{CP}(V, \langle \cdot, \cdot \rangle, J)$. We have $J_1 J_2 x_+ = J_3 x_+$ and $J_1 J_3 x_+ = -J_2 x_+$. Consequently

$$
\mathcal{J}_A(x_+)y = \begin{cases}
(3 - 3)y & \text{if } y \in \operatorname{Span}\{J_2 x_+\}, \\
(3 - 3)y & \text{if } y \in \operatorname{Span}\{J_3 x_+\}, \\
0 & \text{if } y \perp \operatorname{Span}\{J_2 x_+, J_3 x_+\} \,.
\end{cases}
$$

This shows $\mathcal{J}_A(x_+) = 0$. On the other hand, if we take $x = (x_+ + x_-)/\sqrt{2}$, we have

$$J_2 x = (J_2 x_+ + J_2 x_-)/\sqrt{2}, \qquad J_3 x = (J_3 x_+ + J_3 x_-)/\sqrt{2},$$
$$J_1 J_2 x = (J_3 x_+ - J_3 x_-)/\sqrt{2}, \quad J_2 J_3 x = (-J_2 x_+ + J_2 x_-)/\sqrt{2}$$

forms an orthonormal set. Thus

$$\mathcal{J}_A(x)y = \begin{cases} 3y & \text{if } y \in \mathrm{Span}\{J_2 x, J_3 x\}, \\ -3y & \text{if } y \in \mathrm{Span}\{J_1 J_2 x, J_1 J_3 x\}. \end{cases}$$

This show that $\mathfrak{M}$ is not Osserman. $\qquad\qquad\square$

1.8   The Geometry of the Curvature Operator

In this section, we continue our discussion of natural operators related to the curvature tensor and focus on the skew-symmetric curvature operator. The geometry of the classic skew-symmetric operator is discussed in Section 1.8.1, the geometry of the conformal skew-symmetric curvature operator is presented in Section 1.8.2, the geometry of the Stanilov or higher order skew-symmetric curvature operator is studied in Section 1.8.3, and the geometry of the complex skew-symmetric curvature operator is related in Section 1.8.4. Throughout this section, let $\mathfrak{M}_0 := (V, \langle \cdot, \cdot \rangle, A)$ be a 0-model.

1.8.1   *The skew-symmetric curvature operator*

If $\{e_1, e_2\}$ is an oriented orthonormal basis for an oriented non-degenerate 2-plane π, the *skew-symmetric curvature operator* $\mathcal{A}(\pi)$ is defined by:

$$\mathcal{A}(\pi) : x \to A(e_1, e_2)x.$$

Let $\tilde{e}_1 = a_{11}e_1 + a_{12}e_2$ and $\tilde{e}_2 = a_{21}e_1 + a_{22}e_2$ be another orthonormal basis. Then

$$\mathcal{A}(\tilde{e}_1, \tilde{e}_2) = (a_{11}a_{22} - a_{12}a_{21})\mathcal{A}(e_1, e_2).$$

Since $\det(a) = +1$, this is independent of the particular orthonormal basis chosen. Furthermore, $\mathcal{A}(\pi)$ is skew-symmetric since

$$\langle \mathcal{A}(\pi)v, w \rangle = A(e_1, e_2, v, w) = -A(e_1, e_2, w, v) = -\langle \mathcal{A}(\pi)w, v \rangle.$$

Let $-\pi$ denote π with the reversed orientation; as $\{e_1, -e_2\}$ is an oriented orthonormal basis for $-\pi$,

$$\mathcal{A}(-\pi) = -\mathcal{A}(\pi).$$

One says that a 0-model $\mathfrak{M}_0$ is *spacelike* (respectively *timelike* or *mixed*) *Ivanov–Petrova* if the eigenvalues of the associated curvature operator $\mathcal{A}$ are constant on the Grassmannian of oriented spacelike (respectively timelike or mixed) 2-planes. The word "Jordan" is added if additionally the Jordan normal form is constant. The designation "Ivanov–Petrova" is used owing to the seminal paper by Ivanov and Petrova (1998).

We follow the notation established in Eq. (1.3.a) to construct examples.

Theorem 1.8.1 *Let* $\mathfrak{M}_0 := (V, \langle \cdot, \cdot \rangle, A_\phi)$ *where* ϕ *is self-adjoint.*

(1) If $\phi^2 = \pm \operatorname{Id}$, *then* $\mathfrak{M}_0$ *is spacelike, timelike, and mixed Ivanov–Petrova.*

(2) If $\phi^2 = 0$ *and if* $\ker \phi$ *contains no spacelike vectors, then* $\mathfrak{M}_0$ *is spacelike Ivanov–Petrova.*

We note that if $\phi^2 = \operatorname{Id}$ and if ϕ is self-adjoint, then ϕ is an isometry of $(V, \langle \cdot, \cdot \rangle)$; if $\phi^2 = -\operatorname{Id}$ and if ϕ is self-adjoint, then ϕ is a para-isometry of $(V, \langle \cdot, \cdot \rangle)$; this means that $\langle \phi u, \phi v \rangle = -\langle u, v \rangle$. Note that para-isometries exist if and only if $p = q$.

Proof. Suppose that ϕ is self-adjoint and that $\phi^2 = \varepsilon \operatorname{id}$ where $\varepsilon = \pm 1$. Let $\{x, y\}$ be an oriented orthonormal basis for a non-degenerate 2-plane π. We have

$$\mathcal{A}_\phi(x, y)z = \langle \phi y, z \rangle \phi x - \langle \phi x, z \rangle \phi y.$$

If $\{x, y\}$ is spacelike, then:

$$\mathcal{A}(x, y)\phi x = -\varepsilon \phi y, \quad \mathcal{A}(x, y)\phi y = \varepsilon \phi x,$$
$$\mathcal{A}(x, y)z = 0 \text{ if } y \perp \operatorname{Span}\{\phi x, \phi y\}.$$

Thus if π is spacelike, $\mathcal{A}(\pi)$ is an almost complex structure on $\phi\pi$ and vanishes on $\phi\pi^\perp$. Consequently A is spacelike Jordan Ivanov–Petrova. Similarly, if $\{x, y\}$ is timelike, then

$$\mathcal{A}(x, y)\phi x = \varepsilon \phi y, \quad \mathcal{A}(x, y)\phi y = -\varepsilon \phi x,$$
$$\mathcal{A}(x, y)z = 0 \text{ if } y \perp \operatorname{Span}\{\phi x, \phi y\}.$$

This shows that $\mathfrak{M}_0$ is timelike Jordan Ivanov–Petrova. Finally, if x is spacelike and y is timelike, then

$$\mathcal{A}(x,y)\phi x = -\varepsilon\phi y, \quad \mathcal{A}(x,y)\phi y = -\varepsilon\phi x,$$
$$\mathcal{A}(x,y)z = 0 \text{ if } y \perp \mathrm{Span}\{\phi x, \phi y\}.$$

Again, the Jordan normal form is determined so $\mathfrak{M}_0$ is mixed Jordan Ivanov–Petrova. This completes the proof of Assertion (1).

If $\phi^2 = 0$, then $\mathcal{A}(x,y)^2 = 0$. Thus the Jordan normal form is determined by $\mathrm{Rank}(\mathcal{A})$. Since

$$\mathrm{Range}(\mathcal{A}(x,y)) \subset \mathrm{Span}\{\phi x, \phi y\},$$

$\mathrm{Rank}(\mathcal{A}) \leq 2$. Suppose that $\ker\phi$ contains no spacelike vectors. Let $\{x,y\}$ be an orthonormal basis for a spacelike 2-plane π. Since $ax + by$ is again spacelike and non-zero for $(a,b) \neq (0,0)$, $\phi(ax+by) \neq 0$ and thus $\{\phi x, \phi y\}$ is a linearly independent set. Thus we can find z_1 and z_2 so

$$\langle z_1, \phi x\rangle = 1, \quad \langle z_1, \phi y\rangle = 0, \quad \langle z_2, \phi x\rangle = 0, \quad \langle z_2, \phi y\rangle = 1.$$

This shows $\mathcal{A}(x,y)z_2 = \phi x$ and $\mathcal{A}(x,y)z_1 = \phi y$ and thus $\mathrm{Rank}(\mathcal{A}(x,y)) = 2$. Assertion (2) now follows. $\qquad\square$

There are 4-dimensional examples that play an important role. Let $\{e_1, e_2, e_3, e_4\}$ be an orthonormal basis for $\mathbb{R}^{(0,4)}$. Let

$$\begin{aligned}
A^{(0,4)}_{1212} &= -1, & A^{(0,4)}_{1234} &= 2, & A^{(0,4)}_{1313} &= 2, & A^{(0,4)}_{1324} &= 1, \\
A^{(0,4)}_{1414} &= 2, & A^{(0,4)}_{1423} &= -1, & A^{(0,4)}_{2323} &= 2, & A^{(0,4)}_{2314} &= -1, \\
A^{(0,4)}_{2424} &= 2, & A^{(0,4)}_{2413} &= 1, & A^{(0,4)}_{3434} &= -1, & A^{(0,4)}_{3412} &= 2.
\end{aligned} \tag{1.8.a}$$

Similarly, let $\{e_1^-, e_2^-, e_3^+, e_4^+\}$ be an orthonormal basis for $\mathbb{R}^{(2,2)}$. Define

$$\begin{aligned}
A^{(2,2)}_{1212} &= -1, & A^{(2,2)}_{1234} &= -2, & A^{(2,2)}_{1313} &= -2, & A^{(2,2)}_{1324} &= -1, \\
A^{(2,2)}_{1414} &= -2, & A^{(2,2)}_{1423} &= 1, & A^{(2,2)}_{2323} &= -2, & A^{(2,2)}_{2314} &= 1, \\
A^{(2,2)}_{2424} &= -2, & A^{(2,2)}_{2413} &= -1, & A^{(2,2)}_{3434} &= -1, & A^{(2,2)}_{3412} &= -2.
\end{aligned} \tag{1.8.b}$$

The tensor of Eq. (1.8.b) may be defined by complexifying the tensor of Eq. (1.8.a) and setting

$$e_1^- := \sqrt{-1}e_1, \quad e_2^- := \sqrt{-1}e_2, \quad e_3^+ := e_3, \quad e_4^+ := e_4.$$

The following result in the Riemannian setting is established by Ivanov and Petrova (1998); we refer to the discussion in Gilkey and Semmelmann (2000) for the neutral signature result:

Theorem 1.8.2 $(\mathbb{R}^{(0,4)}, A^{(0,4)})$ *and* $(\mathbb{R}^{(2,2)}, A^{(2,2)})$ *are Ivanov–Petrova.*

The following family was introduced by Gilkey and Nikčević (2004a). In Section 2.7, we shall construct pseudo-Riemannian manifolds modeled on this example. For $s \geq 2$, let $\{U_1, ..., U_s, T_1, ..., T_s, V_1, ..., V_s\}$ be a basis for $\mathbb{R}^{3s}$. Let

$$\mathfrak{M}_0 := (\mathbb{R}^{3s}, \langle \cdot, \cdot \rangle, A) \quad \text{where}$$
$$\langle U_i, V_i \rangle = \langle V_i, U_i \rangle = 1, \quad \langle T_i, T_i \rangle = -1, \quad \text{and}$$
$$A(U_i, U_j, U_j, T_i) = 1 \quad \text{for} \quad i \neq j .$$

Theorem 1.8.3

(1) $\mathfrak{M}_0$ is spacelike Jordan Ivanov–Petrova.
(2) $\mathfrak{M}_0$ is timelike Ivanov–Petrova.
(3) $\mathfrak{M}_0$ is not timelike Jordan Ivanov–Petrova.

One says that a pseudo-Riemannian manifold $\mathcal{M} = (M, g)$ is *pointwise spacelike* (respectively *timelike* or *mixed*) *Ivanov–Petrova* if $\mathfrak{M}_0(\mathcal{M}, P)$ has this structure for any $P \in M$; the structure in principle being permitted to vary with P. One replaces the word "pointwise" by the word "globally" if the structures in question is in fact independent of P. Manifolds of constant sectional curvature are globally Jordan Ivanov–Petrova; we refer to the discussion in Section 1.9.8 for other examples and for a survey of the literature in this area.

In Theorem 2.5.1, we give examples of manifolds which are spacelike and timelike Jordan Ivanov–Petrova but which are not mixed Jordan Ivanov–Petrova.

1.8.2 *The conformal skew-symmetric curvature operator*

Let W be the conformal Weyl tensor of a pseudo-Riemannian manifold $\mathcal{M}$. We say that $\mathcal{M}$ is *conformally spacelike* (respectively *timelike*) *Jordan Ivanov–Petrova* if W_P is spacelike (respectively timelike) Jordan Ivanov–Petrova for every point P of M. The Jordan normal form is permitted to vary with the point P of M; the technical distinction between "global" and "pointwise" plays no role in this setting. In Section 1.9.9, we will show that this is a conformal notion.

1.8.3 *The Stanilov operator*

A higher order generalization of the skew-symmetric curvature operator has been introduced by Stanilov (2000); see also Stanilov (2004). Suppose that $\mathfrak{M}_0 := (V, \langle \cdot, \cdot \rangle, A)$ is a 0-model. Let $\{e_1, ..., e_\nu\}$ be a basis for a non-degenerate ν-plane π where $\nu \geq 3$. The *Stanilov operator* is defined by setting:

$$\Theta(\pi) := \sum_{i,j,k,l} g^{ik} g^{jl} \mathcal{A}(e_i, e_j) \mathcal{A}(e_k, e_l) \,.$$

This self-adjoint operator is independent of the basis chosen for π. If π is spacelike, then Θ can be regarded as the average square skew-symmetric curvature operator because there exists a universal constant $c_\Theta(k)$ so:

$$\Theta(\pi) = c_\Theta(k) \int_{\sigma \in \mathrm{Gr}_2^+(\pi)} \mathcal{A}(\sigma)^2 d\sigma \,.$$

It is necessary to square $\mathcal{A}$ to obtain a non-zero average since $\mathcal{A}(\cdot)$ changes sign if the orientation of π is reversed; thus

$$0 = \int_{\sigma \in \mathrm{Gr}_2^+(\pi)} \mathcal{A}(\sigma) d\sigma \,.$$

If the eigenvalues of Θ are constant on $\mathrm{Gr}_{r,s}(V, \langle \cdot, \cdot \rangle)$, then $\mathfrak{M}_0$ is said to be *Stanilov of type* (r, s). As with the higher order Jacobi operator, only the value $r + s$ is relevant by Theorem 1.9.1; $\mathfrak{M}_0$ is k-*Stanilov* if $\mathfrak{M}_0$ is Stanilov of type (r, s) for any (and hence for all) admissible (r, s) with $r + s = k$.

One says that a pseudo-Riemannian manifold $\mathcal{M} = (M, g)$ is pointwise Stanilov of type (r, s) if $\mathfrak{M}_0(\mathcal{M}, P)$ is Stanilov of type (r, s) for all points P of M. The word "Jordan" is added if the Jordan normal form is constant. The word "pointwise" is replaced by the word "globally" if the structures do not depend on P. In Section 4.7, we will establish the basic results concerning these manifolds. Examples will be given in Theorems 2.5.1 and 2.7.3.

1.8.4 *The complex skew-symmetric curvature operator*

We say that a complex 0-model $\mathfrak{M} := (V, \langle \cdot, \cdot \rangle, J, A)$ is *complex skew-symmetric curvature operator* if J and A are compatible (as described in Lemma 1.6.6) and if the eigenvalues of $\mathcal{A}(\cdot)$ are constant on $\mathbb{CP}^+(V, \langle \cdot, \cdot \rangle, J)$.

The notions of *timelike complex Ivanov–Petrova*, or *spacelike complex Jordan Ivanov–Petrova*, or *timelike complex Jordan Ivanov–Petrova* are defined similarly. Since there are no non-degenerate complex lines of signature $(1,1)$, the notion "mixed" does not appear in this setting.

There are several families of examples. Let ϕ be a linear transformation of V. If $\phi^* = \pm\phi$, then we have a canonical algebraic curvature tensor associated to ϕ by Eq. (1.3.a):

$$
\mathcal{A}_\phi(x,y)z := \begin{cases} \langle \phi y, z\rangle \phi x - \langle \phi x, z\rangle \phi y & \text{if } \phi = \phi^*, \\ \langle \phi y, z\rangle \phi x - \langle \phi x, z\rangle \phi y - 2\langle \phi x, y\rangle \phi z & \text{if } \phi = -\phi^*. \end{cases}
$$

Fix a Hermitian almost complex structure on V.

Definition 1.8.1 Let J be a Hermitian almost structure on $(V, \langle\cdot,\cdot\rangle)$. Let ϕ be a linear map of V to V.

(1) ϕ is said to be *J-admissible* if

    (a) Either $\phi = \phi^*$ and $\phi J = \pm J\phi$ or $\phi = -\phi^*$ and $J\phi = -\phi J$.
    (b) Either $\phi^2 = \mathrm{Id}$, or $\phi^2 = -\mathrm{Id}$, or $\phi^2 = 0$ and $\ker \phi = \mathrm{Range}\,\phi$.

(2) A pair $\{\phi_1, \phi_2\}$ is said to be *J-admissible* if

    (a) Both ϕ_1 and ϕ_2 are J-admissible.
    (b) We have $\phi_1 J = J\phi_1$, $\phi_2 J = -J\phi_2$, and $\phi_1^* \phi_2 + \phi_2^* \phi_1 = 0$.
    (c) If $\phi_1^2 = \phi_2^2 = 0$ and if π is any non-degenerate complex line, then we have that $\phi_1 \pi \cap \phi_2 \pi = \{0\}$.

We refer to Gilkey and Ivanova (2001a) for the proof of the following:

Theorem 1.8.4 *Let* $\mathfrak{M} := (V, \langle\cdot,\cdot\rangle, J, A)$ *be a complex 0-model.*

(1) If ϕ is J-admissible and if $A = cA_\phi$, then $\mathfrak{M}$ is complex spacelike and timelike Jordan Ivanov–Petrova.

(2) If $\{\phi_1, \phi_2\}$ is J-admissible and if $A = c_1 A_{\phi_1} + c_2 A_{\phi_2}$, then $\mathfrak{M}$ is complex spacelike and timelike Jordan Ivanov–Petrova.

There is also another family of examples one can consider involving Clifford module structures. We refer to Theorem 2.11.5 of Gilkey (2002) for the proof of the following result:

Theorem 1.8.5 *Let* $\mathfrak{M} := (V, \langle\cdot,\cdot\rangle, J, A)$ *be a complex 0-model.*

(1) If $A = c_0 A_{\langle\cdot,\cdot\rangle} + c_1 A_J$, then $\mathfrak{M}$ is complex spacelike and timelike Jordan Ivanov–Petrova.

(2) Let ϕ be skew-adjoint. Assume that $\phi^2 = \pm\,\mathrm{id}$ and that $\phi J = -J\phi$. If $A = c_0 A_{\langle\cdot,\cdot\rangle} + c_1 A_J + c_2 A_\phi + c_3 A_{J\phi}$, then $\mathfrak{M}$ is complex spacelike and timelike Jordan Ivanov–Petrova.

The following Theorem follows from Theorem 1.8.4. It provides examples in the context of Riemannian geometry:

Theorem 1.8.6   *Let (M, g) be a Riemannian manifold.*

(1) If (M, g) has constant sectional curvature c and if J is any Hermitian almost complex structure on M, then (M, g, J) is complex Ivanov–Petrova.

(2) Suppose that (M, g) is a complex space form and that J is the canonical almost complex structure. Then (M, g, J) is complex Ivanov–Petrova.

1.8.5   *The Szabó operator*

Let $\mathfrak{M}_1 = (V, \langle\cdot,\cdot\rangle, A, A_1)$ be a 1-model; here $\langle\cdot,\cdot\rangle$ is a non-degenerate inner product on V, $A \in \mathcal{Alg}_0(V)$ and $A_1 \in \mathcal{Alg}_1(V)$. Let $\mathcal{A}_1$ be the associated covariant derivative curvature operator; $\mathcal{A}_1$ is characterized by the identity:

$$\langle \mathcal{A}_1(v_1, v_2; w)v_3, v_4 \rangle = A_1(v_1, v_2, v_3, v_4; w)\,.$$

In analogy to the Jacobi operator, one defines the *Szabó operator* by:

$$\mathcal{S}(v) : w \to \mathcal{A}_1(w, v; v)v\,.$$

This is a symmetric linear operator. One says that $\mathfrak{M}_1$ is *spacelike* (respectively *timelike*) *Szabó* if the eigenvalues of $\mathcal{S}$ are constant on $S^+(V, \langle\cdot,\cdot\rangle)$ (respectively on $S^-(V, \langle\cdot,\cdot\rangle)$). These are equivalent notions as we shall see subsequently so one just speaks of Szabó 1-models.

One says that a pseudo-Riemannian manifold $\mathcal{M}$ is *pointwise spacelike* (respectively *timelike*) *Szabó* if the 1-model $\mathfrak{M}_1(\mathcal{M}, P)$ has this property for all P. The modifiers "Jordan" and "globally" have the same meaning as that employed in previous sections. The designation "Szabó" is used owing to the seminal paper by Szabó (1991). There are no known Jordan Szabó algebraic curvature tensors other than $A_1 = 0$. It is known that any Riemannian or Lorentzian Szabó tensor is necessarily zero. There are non-trivial Szabó algebraic curvature tensors in the higher signature setting $(p > 1, q > 1)$ that we will discuss presently.

The following is a useful remark that generalizes Lemma 1.7.1; it shows that S encodes all the information that A_1 does.

Lemma 1.8.1   *Let $\mathfrak{M}_1$ be a 1-model. If $S = 0$, then $A_1 = 0$.*

Proof.   Suppose $S = 0$. We then have for all x, y, w that

$$A_1(x, y, y, w; y) = 0. \tag{1.8.c}$$

We polarize Eq. (1.8.c). Set $y(t) := y + tx$, expand in powers of t, and set the term which is linear in t to zero to see:

$$\begin{aligned}
0 &= A_1(x, x, y, w; y) + A_1(x, y, x, w; y) + A_1(x, y, y, w; x) \\
&= A_1(x, y, x, w; y) - A_1(x, y, w, x; y) - A_1(x, y, x, y; w) \\
&= -2A_1(x, y, w, x; y) + A_1(x, y, y, x; w).
\end{aligned} \tag{1.8.d}$$

Setting $w = x$ in Eq. (1.8.c) yields $A_1(x, y, y, x; y) = 0$. Polarization in y then yields

$$0 = 2A_1(x, y, w, x; y) + A_1(x, y, y, x; w). \tag{1.8.e}$$

We add Eqs. (1.8.d) and (1.8.e) to see that $0 = A_1(y, x, x, y; w)$. At this stage, the argument follows that given to prove Lemma 1.7.1. We polarize in x and in y to see $0 = A_1(y, x, v, z; w) + A_1(y, v, x, z; w)$. We use the curvature symmetries to see:

$$\begin{aligned}
0 &= A_1(y, x, v, z; w) + A_1(y, v, z, x; w) + A_1(y, z, x, v; w) \\
&= A_1(y, x, v, z; w) - A_1(y, v, x, z; w) + A_1(y, z, x, v; w) \\
&= A_1(y, x, v, z; w) + A_1(y, x, v, z; w) - A_1(y, x, z, v; w) \\
&= 3A_1(y, x, v, z; w).
\end{aligned}$$

This shows A_1 vanishes identically as desired. $\square$

In Section 1.9.7, we review some of the literature concerning this operator.

1.9   Spectral Geometry of the Curvature Tensor

In Section 1.9.1, we will show that spacelike Osserman and timelike Osserman are equivalent concepts; similarly spacelike Ivanov–Petrova and timelike Ivanova–Petrova are equivalent and so forth; a theorem of this kind first being proved by [García-Río, Kupeli, Vázquez-Abal, and Vázquez-Lorenzo (1999)] for the Jacobi operator. In Section 1.9.2, we establish a fundamental duality result for the higher order Jacobi operator. In Section 1.9.3, we present work of Blažić concerning natural operators with bounded spectrum.

In studying the geometry of the curvature tensor, one supposes that the eigenvalues, or more generally the Jordan normal form, of a natural operator in Riemannian geometry is constant on a natural domain of definition and then seeks to understand the geometric consequences which follow. This can give, for example, characterizations of 2-point homogeneous spaces and of local symmetric spaces. There is often a purely algebraic content to the investigation where one classifies such structures on a k-model. One then investigates the relevant integrability conditions that arise in geometry using the first and second Bianchi identities. We give a brief review of some of the results in this area; we shall concentrate on the Riemannian ($p = 0$) and Lorentzian ($p = 1$) settings and postpone a discussion of the higher signature setting for the moment. We refer to García-Río, Kupeli, and Vázquez-Lorenzo (2002) and to Gilkey (2002) for a more complete treatment. Section 1.9.4 deals with the Jacobi operator, Section 1.9.5 deals with the higher order Jacobi operator, Section 1.9.6 deals with the conformal and complex Jacobi operators, Section 1.9.7 deals with the Stanilov and Szabó operators, Section 1.9.8 deals with the skew-symmetric curvature operator, Section 1.9.9 deals with the conformal skew-symmetric curvature operator and with the complex skew-symmetric curvature operator.

1.9.1   *Analytic continuation*

The following result is fundamental in the study of the spectral geometry of the Riemann curvature tensor. Let $\mathfrak{M}_0$ be a 0-model of signature (p, q). Recall that (r, s) is said to be an *admissible pair* if $1 \leq r + s \leq m - 1$, $0 \leq r \leq p$, and $0 \leq s \leq q$.

Theorem 1.9.1

(1) The following assertions are equivalent and if either holds, then a 0-model $\mathfrak{M}_0$ is said to be Osserman:

   (a) $q > 0$ and $\mathfrak{M}_0$ is spacelike Osserman.
   (b) $p > 0$ and $\mathfrak{M}_0$ is timelike Osserman.

(2) Let $2 \leq k \leq m - 2$. The following assertions are equivalent and if either

holds, then a 0-model $\mathfrak{M}_0$ is said to be k-Osserman:

(a) $\exists\ (r,s)$ admissible with $r + s = k$ so $\mathfrak{M}_0$ is Osserman of type (r,s).
(b) $\mathfrak{M}_0$ is Osserman of type (r,s) $\forall$ admissible pairs with $r + s = k$.

(3) The following assertions are equivalent and if either holds, then a complex 0-model $\mathfrak{M}_0 = (V, \langle \cdot, \cdot \rangle, J, A)$ is said to be complex Osserman:

(a) $q > 0$ and $\mathfrak{M}_0$ is complex spacelike Osserman.
(b) $p > 0$ and $\mathfrak{M}_0$ is complex timelike Osserman.

(4) The following assertions are equivalent and if any holds, then a 0-model $\mathfrak{M}_0$ is said to be Ivanov–Petrova.

(a) $p \geq 2$ and $\mathfrak{M}_0$ is timelike Ivanov–Petrova.
(b) $p \geq 1$ and $q \geq 1$ and $\mathfrak{M}_0$ is mixed Ivanov–Petrova.
(c) $q \geq 2$ and $\mathfrak{M}_0$ is spacelike Ivanov–Petrova.

(5) The following assertions are equivalent and if either holds, then a 0-model $\mathfrak{M}_0$ is said to be k-Stanilov:

(a) $\exists\ (r,s)$ admissible with $r + s = k$ so $\mathfrak{M}_0$ is Stanilov of type (r,s).
(b) $\mathfrak{M}_0$ is Stanilov of type (r,s) $\forall$ admissible pairs with $r + s = k$.

(6) The following assertions are equivalent and if either holds, then a complex 0-model $\mathfrak{M}_0 = (V, \langle \cdot, \cdot \rangle, J, A)$ is said to be complex Ivanov–Petrova:

(a) $q > 0$ and $\mathfrak{M}_0$ is complex spacelike Ivanov–Petrova.
(b) $p > 0$ and $\mathfrak{M}_0$ is complex timelike Ivanov–Petrova.

(7) The following assertions are equivalent and if either holds, then a 1-model $\mathfrak{M}_1$ is said to be Szabó:

(a) $q > 0$ and $\mathfrak{M}_1$ is spacelike Szabó.
(b) $p > 0$ and $\mathfrak{M}_1$ is timelike Szabó.

Proof. Suppose $p > 0$ and $q > 0$. Assume that $\mathfrak{M}_0$ is spacelike Osserman. We shall show that $\mathfrak{M}_0$ is timelike Osserman. The proof of the reverse implication is similar. This will establish Assertion (1).

As $\mathfrak{M}_0$ is spacelike Osserman, we may apply Lemma 1.5.2 to see there are universal constants c_i so $\operatorname{Tr}\{\mathcal{J}(x)^i\} = c_i$ for $x \in S^+(V, \langle \cdot, \cdot \rangle)$. Set

$$f_i(\xi) := \operatorname{Tr}\{\mathcal{J}(\xi)^i\} - c_i(\xi, \xi)^i.$$

Since we have included the appropriate scaling factor, it follows that $f_i(\xi)$ vanishes on the open subset of spacelike vectors of V. On the other hand, $f_i(\xi)$ is polynomial in the components of ξ relative to any basis for V. Since $f_i(\xi)$ vanishes on a non-empty open subset of V, it now follows that f_i is constant on all of V. Thus $\mathrm{Tr}\{\mathcal{J}(x)^i\} = (-1)^i c_i$ on $S^-(V, \langle\cdot,\cdot\rangle)$. Another application of Lemma 1.5.2 then implies that $\mathfrak{M}_0$ is timelike Osserman.

To prove Assertion (2), we complexify. Let $V_\mathbb{C} := V \otimes_\mathbb{R} \mathbb{C}$. Fix k. Extend $\langle\cdot,\cdot\rangle$ and A to be complex multilinear. Set

$$V^k := V \times ... \times V, \qquad \mathcal{O} := \{\vec{v} \in V^k : \det\{\langle v_i, v_j\rangle \neq 0\},$$

$$V_\mathbb{C}^k := V_\mathbb{C} \times ... \times V_\mathbb{C}, \quad \mathcal{O}_\mathbb{C} := \{\vec{v} \in V_\mathbb{C}^k : \det\{\langle v_i, v_j\rangle \neq 0\},$$

$$\mathcal{O}_{\bar{r},\bar{s}} := \{\vec{v} \in V^k : \mathrm{Span}\{v_i\} \in \mathrm{Gr}_{(\bar{r},\bar{s})}(V)\}\,.$$

We note that $\mathcal{O}_\mathbb{C}$ is a connected open subset of $V_\mathbb{C}^k$ with real points

$$\mathcal{O}_\mathbb{C} \cap V^k = \mathcal{O} = \cup_{\bar{r}+\bar{s}=k}\mathcal{O}_{\bar{r},\bar{s}}\,.$$

If $\vec{v} \in \mathcal{O}_\mathbb{C}$, then $\pi_{\vec{v}} := \mathrm{Span}_\mathbb{C}\{v_1, ..., v_k\}$ is a non-degenerate complex k-dimensional subspace of V. Let

$$\mathcal{J}(\pi_{\vec{v}}) = \sum_{i,j} g^{ij} \mathcal{J}(v_i, v_j) \quad \text{where} \quad g_{ij} := \langle v_i, v_j\rangle\,.$$

This is independent of the particular basis for π which is chosen. By assumption, the eigenvalues of $\mathcal{J}(\vec{v})$ are constant on $\mathcal{O}_{r,s}$. Consequently, there are constants c_i so that

$$\mathrm{Tr}\{\mathcal{J}(\pi_{\vec{v}})^i\} = c_i \quad \text{if} \quad \vec{v} \in \mathcal{O}_{r,s}\,.$$

Thus by the *Identity Theorem*, $\mathrm{Tr}\{\mathcal{J}(\pi_{\vec{v}})^i\} = c_i$ for $\vec{v} \in \mathcal{O}_\mathbb{C}$. We restrict to see $\mathrm{Tr}\{\mathcal{J}(\pi_{\vec{v}})^i\} = c_i$ for $\vec{v} \in \mathcal{O}_{\bar{r},\bar{s}}$ and hence $\mathfrak{M}_0$ is Osserman of type $(\bar{r}, \bar{s})$. This proves Assertion (2); the proof of Assertions (4), (5), (6), and (7) is similar and is therefore omitted. $\square$

In Theorems 2.5.1 and 2.6.1, we exhibit examples which are Jordan Osserman of certain but not all types for a given $r + s = k$. Consequently Theorem 1.9.1 fails if we replace the words "Osserman" by "Jordan Osserman". Similarly, we have examples which are spacelike Jordan Ivanov–Petrova but not timelike Jordan Ivanov–Petrova and timelike Jordan Ivanov–Petrova but not spacelike Jordan Ivanov–Petrova.

1.9.2 *Duality*

There is a basic duality result which is used in the study of k-Osserman manifolds. Let

$$\mathcal{J}(\xi, \eta)y := \tfrac{1}{2}\{\mathcal{R}(y, \xi)\eta + \mathcal{R}(y, \eta)\xi\}$$

be *polarized* or *symmetrized Jacobi operator*. If ρ is the Ricci tensor, then

$$\rho(\xi, \eta) := \mathrm{Tr}\{\mathcal{J}(\xi, \eta)\}\,.$$

Theorem 1.9.2 *Fix $1 \leq k \leq m - 1$. Let $\mathfrak{M}_0$ be a 0-model of signature (p, q) which is k-Osserman. Then*

(1) $\mathfrak{M}_0$ is Einstein.

(2) $\mathfrak{M}_0$ is $m - k$ Osserman.

(3) If $\mathfrak{M}_0$ is Jordan Osserman of type (r, s), then $\mathfrak{M}_0$ is Jordan Osserman of type $(p - r, q - s)$.

Proof. We follow the discussion in Gilkey, Stanilov, and Videv (1998). We adopt the notation established in the proof of Assertion (2) of Theorem 1.9.1. Let $\{v_1, ..., v_k\}$ be a $\mathbb{C}$-basis for a linear subspace σ of $V_{\mathbb{C}}$ which has complex dimension k. We assume $\vec{v} := (v_1, ..., v_k) \in \mathcal{O}_{\mathbb{C}}$ and set

$$\mathcal{J}(\sigma) := \sum_{ij} g^{ij}\, \mathcal{J}(v_i, v_j)\,.$$

This is independent of the particular basis chosen. We introduce the complex null cone

$$\mathcal{N}_{\mathbb{C}} := \{v \in V_{\mathbb{C}} : \langle v, v \rangle = 0\};$$

this is a nowhere dense closed subset of $V_{\mathbb{C}}$. Let $x \in \mathcal{N}_{\mathbb{C}}$. Choose y so $\langle x, y \rangle \neq 0$; since $\mathcal{N}_{\mathbb{C}}$ is nowhere dense, we may assume that $y \notin \mathcal{N}_{\mathbb{C}}$ and normalize y so $\langle y, y \rangle = 1$. Let

$$\tau := \mathrm{Span}_{\mathbb{C}}\{x, y\}\,.$$

This is a non-degenerate 2-plane since

$$\det \begin{pmatrix} \langle x, x \rangle & \langle x, y \rangle \\ \langle x, y \rangle & \langle y, y \rangle \end{pmatrix} = -\langle x, y \rangle^2 \neq 0\,.$$

Let $W := \tau^{\perp}$; this is a non-degenerate subspace of complex dimension $m - 2$. Since $k - 1 \leq m - 2$, there is a non-degenerate subspace σ of complex dimension $k - 1$ contained in W. Let

$$\xi_t := x + ty \quad \text{and} \quad g(t) := \langle \xi_t, \xi_t \rangle = t^2 + 2t\langle x, y \rangle\,.$$

Let $t_0 := 0$ and $t_1 := -\frac{1}{2}\langle x, y \rangle$ be the zeros of this quadratic polynomial. We consider the complex k-plane $\pi(t) := \sigma \oplus \mathrm{Span}\{\xi_t\}$; $\pi(t)$ is non-degenerate for $t \neq t_i$.

The eigenvalues of $\mathcal{J}(\pi(t))$ are constant if $0 < t < |t_1|$. Let

$$c_1 := \mathrm{Tr}(\mathcal{J}(\pi(t))) .$$

We then have

$$\begin{aligned}
g(t)c_1 &= g(t)\,\mathrm{Tr}\{[\mathcal{J}(\sigma) + g(t)^{-1}\mathcal{J}(\xi_t)]\} \\
&= \mathrm{Tr}\{[g(t)\mathcal{J}(\sigma) + \mathcal{J}(\xi_t)]\} \quad \text{for} \quad 0 < t < |t_1| .
\end{aligned}$$

We take the limit as $t \to 0$ to see

$$\mathrm{Tr}(\mathcal{J}(x)) = 0 \quad \text{if} \quad x \in \mathcal{N}_{\mathbb{C}} . \tag{1.9.a}$$

Let $\xi_1, \xi_2 \in V_{\mathbb{C}}$ with

$$\langle \xi_1, \xi_1 \rangle = \langle \xi_2, \xi_2 \rangle = 1 \quad \text{and} \quad \langle \xi_1, \xi_2 \rangle = 0 .$$

Let $x_{\pm} := \xi_1 \pm \sqrt{-1}\xi_2 \in \mathcal{N}_{\mathbb{C}}$. We expand

$$\mathcal{J}(x_{\pm}) = \mathcal{J}(\xi_1) - \mathcal{J}(\xi_2) \pm \sqrt{-1}\mathcal{J}(\xi_1, \xi_2) .$$

We apply Eq. (1.9.a) to see:

$$0 = \mathrm{Tr}(\mathcal{J}(x_{\pm})) = \mathrm{Tr}(\mathcal{J}(\xi_1)) - \mathrm{Tr}(\mathcal{J}(\xi_2)) \pm \mathrm{Tr}(\sqrt{-1}\mathcal{J}(\xi_1, \xi_2)) . \tag{1.9.b}$$

One may use Eq. (1.9.b) to see that

$$\rho(\xi_1, \xi_1) = \rho(\xi_2, \xi_2) = c \quad \text{and} \quad \rho(\xi_1, \xi_2) = 0 .$$

Consequently $\rho(\xi, \xi) = c|\xi|^2$ for any $\xi \in V_{\mathbb{C}}$ and hence A is Einstein.

We restrict to the real setting. Let $\{e_1^-, ..., e_q^-, e_1^+, ..., e_p^+\}$ be an orthonormal basis for V. We have

$$\begin{aligned}
c\langle \xi, \eta \rangle &= \mathrm{Tr}\{\mathcal{J}(\xi, \eta)\} \\
&= -A(e_1^-, \xi, \eta, e_1^-) - ... - A(e_p^-, \xi, \eta, e_p^-) \\
&\quad + A(e_1^+, \xi, \eta, e_1^+) + ... + A(e_q^+, \xi, \eta, e_q^+) \\
&= \langle \{-\mathcal{J}(e_1^-) - ... - \mathcal{J}(e_p^-) + \mathcal{J}(e_1^+) + ... + \mathcal{J}(e_1^+)\}\xi, \eta \rangle .
\end{aligned}$$

This shows that

$$-\mathcal{J}(e_1^-) - ... - \mathcal{J}(e_p^-) + \mathcal{J}(e_1^+) + ... + \mathcal{J}(e_q^+) = c\,\mathrm{id} . \tag{1.9.c}$$

Let τ be a non-degenerate linear subspace of signature (r, s) and let $\tau^{\perp}$ be the corresponding non-degenerate linear subspace of signature $(p-r, q-s)$.

It follows from Eq. (1.9.c) that

$$\mathcal{J}(\tau) + \mathcal{J}(\tau^{\perp}) = c\,\mathrm{id}\ .$$

The remaining assertions now follow. $\qquad\square$

1.9.3 *Bounded spectrum*

The higher signature setting $(p > 0, q > 0)$ is significantly different from the Riemannian setting in at least one respect as the following results will show. We shall follow Blažić (2005) throughout.

If $T \in \mathrm{End}(V)$, let $\mathrm{Spec}\{T\}$ be the complex eigenvalues of T where each eigenvalue is repeated according to multiplicity. One says that a 0-model $\mathfrak{M}_0$ has *bounded spacelike Jacobi spectrum* if there exists a constant K so

$$\lambda \in \mathrm{Spec}\{\mathcal{J}(x)\} \Rightarrow |\lambda| \le K \quad \forall x \in S^{+}(V, \langle \cdot, \cdot \rangle)\,.$$

The notion of bounded timelike Jacobi spectrum is defined similarly. The other natural curvature operators give rise to similar notions where the domain is specified appropriately. Since the domains in question are compact if V is Riemannian, all operators necessarily have bounded spectrum in this setting.

If $\mathfrak{M}_0$ is Osserman, then necessarily $\mathfrak{M}_0$ has bounded spacelike and timelike Jacobi spectrum. The converse holds in the higher signature context. We refer to Blažić (2005) for the proof of the following result:

Theorem 1.9.3 **[Blažić]** *Let $\mathfrak{M}_0 := (V, \langle \cdot, \cdot \rangle, A)$ be a 0-model of signature (p, q) where $p \ge 1$ and $q \ge 1$.*

(1) If $\mathfrak{M}_0$ has bounded spacelike Jacobi spectrum, then $\mathfrak{M}_0$ is Osserman.
(2) If $\mathfrak{M}_0$ has bounded timelike Jacobi spectrum, then $\mathfrak{M}_0$ is Osserman.

Proof. We suppose $\mathfrak{M}_0$ has bounded spacelike Jacobi spectrum; the time-like case is similar. Let $\mathcal{T}_i(x) := \mathrm{Tr}\{[\mathcal{J}(x)]^i\}$. Since $\mathcal{T}_i(x)$ is the sum of the ith powers of the eigenvalues, the functions $\mathcal{T}_i$ are uniformly bounded on $S^{+}(V, \langle \cdot, \cdot \rangle)$. Let $\{e_1^-, ..., e_p^-, e_1^+, ..., e_q^+\}$ be an orthonormal basis for V. Let

$$\mathcal{O} := \{x \in S^{+}(V, \langle \cdot, \cdot \rangle) : \langle e_1^+, x \rangle > 1\};$$

$\mathcal{O}$ is a non-empty open subset of $S^{+}(V, \langle \cdot, \cdot \rangle)$ since $q \ge 1$. If $x \in \mathcal{O}$, we may choose $\theta_0 \in \mathbb{R}$ so that $\cosh \theta_0 = \langle e_1^+, x \rangle$ and express

$$x = \cosh \theta_0 e_1^+ + \sinh \theta_0 v^-$$

where

$$v_- := \frac{1}{\sinh\theta_0}\left\{ -\sum_i \langle x, e_i^-\rangle e_i^- + \sum_{j>1} \langle x, e_j^+\rangle e_j^+ \right\} \in S^-(e_1^\perp, \langle\cdot,\cdot\rangle).$$

Let $v_\theta := \cosh\theta e_1^+ + \sinh\theta v_- \in S^+(V, \langle\cdot,\cdot\rangle)$. There exists n_0 such that

$$\mathcal{T}_i(v_\theta) = \sum_{-n_0 \le n \le n_0} a_n(e_1^+, v_-)e^{n\theta}.$$

By assumption, $\mathcal{T}_i(v_\theta)$ is a bounded function of $\theta \in \mathbb{R}$. Consequently, $a_n(e_1^+, v_-)$ vanishes for $n \ne 0$ so $\mathcal{T}_i(v_\theta) = \mathcal{T}_i(v_+)$ is constant. Thus

$$\mathcal{T}_i(x) = \mathcal{T}_i(e_1^+) \quad \text{for} \quad x \in \mathcal{O}.$$

As $\mathcal{T}_i$ is real analytic and constant on a non-empty open set, $\mathcal{T}_i$ is constant on the component of $S^+(V, \langle\cdot,\cdot\rangle)$ that contains e_1^+; as $\mathcal{J}(x) = \mathcal{J}(-x)$, $\mathcal{T}_i$ is constant on $S^+(V, \langle\cdot,\cdot\rangle)$. This implies that $\mathcal{J}(v)$ has constant spectrum on $S^+(V, \langle\cdot,\cdot\rangle)$ and hence $\mathfrak{M}_0$ is Osserman as desired. $\qquad\square$

The proof given above extends without change to establish a similar result for the Szabó operator, for the complex Jacobi operator, and for the complex skew-symmetric curvature operator since the appropriate domain of definition can be regarded as being the pseudo-projective spaces

$$\mathbb{RP}^\pm(V, \langle\cdot,\cdot\rangle) := S^\pm(V, \langle\cdot,\cdot\rangle)/\mathbb{Z}_2.$$

Before discussing the other operators, we must establish a technical result that deals with Grassmannians. Let $\{e^-, e^+\}$ be an orthonormal basis for a 2-plane of signature $(1,1)$. We define a *hyperbolic boost* $T = T(e^-, e^+, \theta)$ by setting:

$$Ty = \begin{cases} \cosh\theta e^- + \sinh\theta e^+ & \text{if } y = e^-, \\ \cosh\theta e^+ + \sinh\theta e^- & \text{if } y = e^+, \\ y & \text{if } y \perp \mathrm{Span}\{e^-, e^+\}. \end{cases}$$

Let G_h be the closed Lie subgroup of $O(V, \langle\cdot,\cdot\rangle)$ which generated by these hyperbolic boosts.

Lemma 1.9.1   *Let $p \ge 1$ and let $q \ge 1$.*

(1) G_h is the connected component of the identity in $O(V, \langle\cdot,\cdot\rangle)$.
(2) G_h acts transitively on $\mathrm{Gr}_{r,s}^+(V, \langle\cdot,\cdot\rangle)$ and $\mathrm{Gr}_{r,s}(V, \langle\cdot,\cdot\rangle)$.

Proof. Let $\{e_1^-, ..., e_p^-, e_1^+, ..., e_q^+\}$ be an orthonormal basis for V. Let

$$T_{ij}^{--}y := \langle y, e_i^- \rangle e_j^- - \langle y, e_j^- \rangle e_i^-,$$
$$T_{ia}^{-+}y := \langle y, e_i^- \rangle e_a^+ - \langle y, e_a^+ \rangle e_i^-,$$
$$T_{ab}^{++}y := \langle y, e_a^+ \rangle e_b^+ - \langle y, e_b^+ \rangle e_a^+.$$

The collection of elements

$$\{T_{ij}^{--}\}_{i<j} \cup \{T_{ia}\}_{ia} \cup \{T_{ab}\}_{a<b}$$

is a basis for the Lie algebra of $O(V, \langle \cdot, \cdot \rangle)$. To complete the proof of Assertion (1), we must only show that these endomorphisms belong to the Lie algebra $\mathfrak{g}_h$ of G_h.

Let $T(\theta) = T(e_i^-, e_a^+, \theta)$. Then $T'(\theta) = -T_{ia}^{-+}$ so $T_{ia}^{-+} \in \mathfrak{g}_h$. Since

$$[T_{i1}^{-+}, T_{j1}^{-+}] = T_{ij}^{--} \quad \text{and} \quad [T_{1a}^{-+}, T_{1b}^{-+}] = T_{ab}^{++},$$

the remaining generators belong to $\mathfrak{g}_h$ as well. This proves Assertion (1); Assertions (2) and (3) follow from Assertion (1). $\qquad\square$

We can now prove:

Theorem 1.9.4 **[Blažić]** *Let* $\mathfrak{M}_0 := (V, \langle \cdot, \cdot \rangle)$ *be a 0-model of signature* (p, q) *where* $p \geq 1$, $q \geq 1$, *and* $p + q \geq 3$.

(1) Let $p \geq 2$. *If* $\mathrm{Spec}\{\mathcal{R}\}$ *is bounded on* $\mathrm{Gr}_{2,0}^+(V)$, $\mathfrak{M}_0$ *is Ivanov–Petrova.*
(2) If $\mathrm{Spec}\{\mathcal{R}\}$ *is bounded on* $\mathrm{Gr}_{1,1}^+(V)$, $\mathfrak{M}_0$ *is Ivanov–Petrova.*
(3) Let $q \geq 2$. *If* $\mathrm{Spec}\{\mathcal{R}\}$ *is bounded on* $\mathrm{Gr}_{0,2}^+(V)$, $\mathfrak{M}_0$ *is Ivanov–Petrova.*

Proof. Let $T = T(e^-, e^+, \theta)$ be a hyperbolic boost. If π is a non-degenerate 2-plane, let $\pi(\theta) = T(\theta)\pi$. Then $\mathrm{Tr}\{[\mathcal{R}(\pi(\theta))]^i\}$ is a Laurent polynomial in $\{e^\theta, e^{-\theta}\}$. If the spectrum is bounded, this polynomial is bounded. The argument given to prove Theorem 1.9.3 then shows the polynomial is constant and hence $\mathrm{Spec}\{\mathcal{R}(\pi(\theta))\}$ is independent of θ. Thus the spectrum of $\mathcal{R}$ is constant on the orbits of the hyperbolic boosts and hence, by continuity, on the orbits of the closed Lie group G_h that these boosts generate. Since G_h acts transitively on the Grassmannians in question, the desired result follows. $\qquad\square$

This proof extends without change to establish a similar result for the higher order Jacobi operator and for the Stanilov operator. The conformal Jacobi operator and the conformal skew-symmetric curvature operator can be treated similarly.

1.9.4 *The Jacobi operator*

We continue the discussion of Section 1.7.1. Let $\mathfrak{M}_0$ be a 0-model. The Jacobi operator $\mathcal{J}$ is defined by

$$\mathcal{J}(x) : y \to \mathcal{R}(y, x)x \, .$$

One says that a 0-model $\mathfrak{M}_0$ is *Osserman* if the eigenvalues of $\mathcal{J}(\cdot)$ are constant on the pseudo-spheres $S^{\pm}(V, \langle \cdot, \cdot \rangle)$. One says that a pseudo-Riemannian manifold $\mathcal{M}$ is *pointwise Osserman* if $\mathfrak{M}_0(\mathcal{M}, P)$ is Osserman for every point $P \in M$; one says that $\mathcal{M}$ is *globally Osserman* if the eigenvalue structure does not vary with P.

A complete simply connected Riemannian manifold $\mathcal{M}$ is a *2-point homogeneous space* if the group of isometries $G(\mathcal{M})$ acts transitively on the unit sphere bundle $S(T(M, g))$ or, equivalently, if given two pairs of points (P_1, Q_1) and (P_2, Q_2) with equal Riemannian distances, then there exists an isometry ϕ of M with $\phi P_1 = P_2$ and $\phi Q_1 = Q_2$. It is known that $\mathcal{M}$ is a 2-point homogeneous space if and only if either $\mathcal{M}$ is flat or $\mathcal{M}$ is a rank 1 symmetric space – see Section 1.6.5 for details.

There is a corresponding local classification. Impose no global constraints on $\mathcal{M}$. One says that $\mathcal{M}$ is a *local 2-point homogeneous space* if the pseudo-group of local isometries of $\mathcal{M}$ acts transitively on $S(M, g)$. Again the local geometry is very rigid in this setting; every point of M has an open neighborhood which is either flat or which is isometric to an open subset of a rank 1 symmetric space.

If a Riemannian manifold $\mathcal{M}$ is a local 2-point homogeneous space, then necessarily the eigenvalues of the Jacobi operator $\mathcal{J}(\cdot)$ are constant on $S(M, g)$ and hence M is Osserman. Osserman (1990) wondered if the converse held; this question has been called the Osserman conjecture by subsequent authors. This conjecture has been established if $m \neq 16$ following work of Chi (1988), Nikolayevsky (2003b), Nikolayevsky (2004), and Nikolayevsky (2005); there are also some partial results in dimension 16. The proof consists of 2 parts. First one classifies the Osserman algebraic curvature tensors. Then one applies the Bianchi identities to complete the geometric classification.

We introduce some additional pieces of notation. Let $\mathcal{F} := \{J_1, ..., J_\ell\}$ be a Clifford family. This means that the J_i are Hermitian almost complex structures on $(V, \langle \cdot, \cdot \rangle)$ which satisfy the Clifford commutation relations

$$J_i J_j + J_j J_i = -2\delta_{ij} \, .$$

We use Eq. (1.3.a) to define an associated algebraic curvature tensor

$$A := c_0 A_{\langle \cdot, \cdot \rangle} + \sum_i c_i A_{J_i} \,. \tag{1.9.d}$$

Let $\mathfrak{M}_0 = (V, \langle \cdot, \cdot \rangle, A)$ be a Riemannian 0-model. Let $x \in S(V, \langle \cdot, \cdot \rangle)$. Since $\mathcal{J}(x)x = 0$ and since $\mathcal{J}(x)$ is self-adjoint, $\mathcal{J}(x)$ preserves $x^\perp$. The *reduced Jacobi operator* is given by setting:

$$\tilde{\mathcal{J}}(x) := \mathcal{J}(x)|_{x^\perp} \quad \text{for} \quad x \in S(V, \langle \cdot, \cdot \rangle) \,.$$

Chi (1988) used topological methods to study the eigenvalue structure of an Osserman manifold and thereby prove:

Theorem 1.9.5 [Chi]

(1) Let $\mathfrak{M} = (V, \langle \cdot, \cdot \rangle, A)$ be a Riemannian Osserman 0-model. Then:

 (a) If $\tilde{\mathcal{J}}$ has only one eigenvalue, then $A = c A_{\langle \cdot, \cdot \rangle}$ and $\mathfrak{M}$ has constant sectional curvature.

 (b) If $\tilde{\mathcal{J}}$ has two distinct eigenvalues and if one of the eigenvalues has multiplicity 1, then $A = c_0 A_{\langle \cdot, \cdot \rangle} + c_1 A_J$ where J is a Hermitian almost complex structure on V.

(2) Let $\mathcal{M} = (M, g)$ be a globally Osserman Riemannian manifold. Then:

 (a) If $\tilde{\mathcal{J}}$ has 1 eigenvalue, then $\mathcal{M}$ has constant sectional curvature.

 (b) If $\tilde{\mathcal{J}}$ has two distinct eigenvalues and if one of the eigenvalues has multiplicity 1, then $\mathcal{M}$ is locally isometric to a rank 1 symmetric space. In particular, the eigenvalues are in a ratio of 1 to 4.

When Theorem 1.9.5 is combined with work of Nikolayevsky (2003b), Nikolayevsky (2004), and Nikolayevsky (2005), one has an almost complete answer to the question Osserman raised by solving the Osserman conjecture.

Theorem 1.9.6 [Chi–Nikolayevsky] *Let $\mathfrak{M}_0 = (V, \langle \cdot, \cdot \rangle, A)$ be a Riemannian 0-model and let $\mathcal{M}$ be a Riemannian manifold of dimension m.*

(1) If $\mathfrak{M}_0$ is Osserman and if $m \neq 16$, then A is given by a Clifford module structure as described in Eq. (1.9.d).

(2) If $\mathcal{M}$ is pointwise Osserman and if $m \neq 2, 4, 16$, then $\mathcal{M}$ is either flat or locally isometric to a rank 1 symmetric space.

(3) If $m = 2, 4$ and if $\mathcal{M}$ is globally Osserman, then $\mathcal{M}$ is either flat or locally isometric to a rank 1 symmetric space.

It is known that Assertion (1) fails in dimension 16; the algebraic curvature tensor defined by the Cayley plane is an Osserman algebraic curvature tensor which is *not* given by a Clifford module structure as noted by Nikolayevsky (2003a). There are also some partial results available in dimension $m = 16$ which are due to Nikolayevsky (2003a) but the situation is still not clear in that exceptional dimension.

In the Lorentzian setting, the Osserman conjecture been settled both in the algebraic and in the geometric settings by work of Blažić, Bokan, and Gilkey (1997a) and of García-Río, Kupeli, and Vázquez-Abal (1997):

Theorem 1.9.7 [Blažić, Bokan and Gilkey; García-Río, Kupeli and Vázquez-Abal]

(1) If $\mathfrak{M}_0$ is a Lorentzian Osserman 0-model, then $\mathfrak{M}_0$ has constant sectional curvature.

(2) If $\mathcal{M}$ is a Lorentzian Osserman manifold, then $\mathcal{M}$ has constant sectional curvature.

The picture is very different when $p \geq 2$ and $q \geq 2$. In Theorem 2.3.2, we exhibit complete manifolds found by Dunn, Gilkey, and Nikčević (2005) of signature $(2, 2)$ which are spacelike and timelike Jordan Osserman but which are not locally homogeneous. In Theorem 3.3.1, we present results of Díaz-Ramos, García-Río, and Vázquez-Lorenzo (2006) giving manifolds of signature $(2, 2)$ which are spacelike and timelike Jordan Osserman on a nontrivial open dense subset but which are not spacelike and timelike Jordan Osserman on the entire manifold. The Jacobi operator is not nilpotent for these examples. The Jordan normal form of a spacelike Jordan Osserman algebraic curvature can be arbitrarily complicated in the neutral signature setting, see Theorem 4.4.1 for details.

However, the situation is very different if $p < q$, i.e. if the spacelike directions in a certain sense dominate the timelike directions since by Theorem 4.4.2 if A is a spacelike Jordan Osserman algebraic curvature tensor on a vector space V of signature (p, q), where $p < q$, then $\mathcal{J}(x)$ is diagonalizable for any $x \in S^+(V, \langle \cdot, \cdot \rangle)$. We shall investigate the neutral signature setting and the setting when $q < p$ presently.

We refer to Blažić, Bokan, Gilkey, and Rakić (1997b), Blažić, Bokan, and Rakić (1997), Blažić, Bokan, and Rakić (1998), Blažić, Bokan, and Rakić (2000), Blažić, Bokan, and Rakić (2001a), Bonome, Castro, and García-Río (2001), Bonome, Castro, García-Río, Hervella, and Vázquez-Lorenzo (1998), Díaz-Ramos, García-Río, and Vázquez-Lorenzo (2006),

Dotti and Druetta (1999a), Dotti and Druetta (1999b), Dotti and Druetta (2000), Fiedler and Gilkey (2003), García-Río and D. Kupeli (1996), García-Río, Kupeli, Vázquez-Abal, and Vázquez-Lorenzo (1999), García-Río, Vázquez-Abal, and Vázquez-Lorenzo (1998), Gilkey and Ivanova (2001b), Gilkey and Ivanova (2002a), Gilkey and Ivanova (2002b), Gilkey, Ivanova, and Zhang (2002), Gilkey, Ivanova, and Zhang (2003), Gilkey and Nikčević (2004a), Gilkey and Nikčević (2004b), Gilkey, Swann, and Vanhecke (1995), Osserman (1990), Rakic (1997), Rakic (1999), Stanilov and Videv (1995), and Stanilov and Videv (1998) for additional work in this area. In particular, we refer to García-Río, Kupeli, and Vázquez-Lorenzo (2002) for a more exhaustive discussion of the subject.

1.9.5 *The higher order Jacobi operator*

Let $\mathfrak{M}_0$ be a 0-model. We recall the notation established in Section 1.7.2. Let $\{v_i\}$ be a basis for a non-degenerate k-plane π. Let $g_{ij} := \langle v_i, v_j \rangle$ for $1 \le i, j \le k$ and let g^{ij} denote the inverse matrix. The *higher order Jacobi operator*, introduced by Stanilov and Videv (1992) in the Riemannian setting, is defined quite generally by

$$\mathcal{J}(\pi) : y \to \sum_{i,j} g^{ij} \mathcal{A}(y, v_i) v_j \, .$$

One says that $\mathfrak{M}_0$ is *k-Osserman* if the eigenvalues of $\mathcal{J}$ are constant on the Grassmannian of non-degenerate k-planes. One says that a pseudo-Riemannian manifold $\mathcal{M}$ is *pointwise k-Osserman* if $\mathfrak{M}(\mathcal{M}, P)$ is k-Osserman for every $P \in M$; one says that $\mathcal{M}$ is *globally k-Osserman* if the eigenvalue structure is independent of P.

The classification is complete in the Lorentzian and the Riemannian settings. The case $k = 1$ follows from Theorems 1.9.6 and 1.9.7; the case $k = m - 1$ follows from $k = 1$ using the duality result of Theorem 1.9.1. Thus we may suppose $2 \le k \le m - 2$ and $m \ge 4$. We adopt the notation of Eq. (1.9.d). We refer to Gilkey (2001b) for the proof in the Riemannian setting and to Gilkey and Stavrov (2002) for the proof in the Lorentzian setting of the following result:

Theorem 1.9.8 [**Gilkey, Gilkey–Stavrov**] *Let* $2 \le k \le m - 2$.

(1) Let $\mathfrak{M}_0$ be a Riemannian k-Osserman 0-model.

    (a) If m is odd, then $A = cA_{\langle \cdot, \cdot \rangle}$.

(b) If m is even, then either $A = cA_{\langle \cdot, \cdot \rangle}$ or $A = cA_J$ where J is a Hermitian almost complex structure on $\mathfrak{M}_0$.

(2) *Let $\mathfrak{M}_0$ be a Lorentzian k-Osserman 0-model. Then $\mathfrak{M}_0$ has constant sectional curvature.*

(3) *Let $\mathcal{M}$ be either a Riemannian or a Lorentzian pointwise k-Osserman manifold. Then $\mathcal{M}$ has constant sectional curvature.*

Again, the situation is quite different in the higher signature setting as the examples described in Theorems 1.7.2 and 1.7.3 show. We refer to Gilkey (1992), Gilkey and Ivanova (2002a), Gilkey and Ivanova (2002b), Gilkey, Stanilov, and Videv (1998), Stanilov (1992), Stanilov and Videv (1992), and Stavrov (2003a) for additional details.

1.9.6   *The conformal and complex Jacobi operators*

We adopt the notation of Section 1.7.3. Recall that two pseudo-Riemannian metrics g_1 and g_2 on a manifold M are said to be *conformally equivalent* if there is a positive scaling function $\alpha \in C^\infty(M)$ so that $g_1 = \alpha g_2$. We let $[g]$ be the set of all pseudo-Riemannian metrics on M which are conformally equivalent to g.

Theorem 1.9.9   *Let $g_1 \in [g_2]$. Let $\mathcal{M}_i := (M, g_i)$.*

(1) *If $\mathcal{M}_1$ is conformally Osserman, then $\mathcal{M}_2$ is conformally Osserman.*

(2) *If $\mathcal{M}_1$ is conformally spacelike (respectively timelike) Jordan Osserman, then $\mathcal{M}_2$ is conformally spacelike (respectively timelike) Jordan Osserman.*

Proof.   As $g_1 = \alpha g_2$, the conformal Weyl tensors rescale:

$$W_{g_1} = \alpha W_{g_2} \, .$$

Let $x \in S^+(M, g_2)$. Let

$$\tilde{x} := \frac{1}{\sqrt{\alpha(P)}} x$$

be the corresponding g_1 spacelike or timelike unit vector. Let π be a space-like 2-plane.

$$J_{W_{g_1}}(\tilde{x}) = \frac{1}{\alpha(P)} J_{W_{g_2}}(x) \quad \text{and} \quad W_{g_1}(\pi) = \frac{1}{\alpha(P)} W_{g_2}(\pi) \, .$$

The Lemma now follows as the Jordan normal forms rescale. $\square$

Additional results in this area may be found in Blažić and Gilkey (2004), in Blažić and Gilkey (2005), and in Blažić, Gilkey, Nikĕcvić, and Simon (2005a).

1.9.7 *The Stanilov and the Szabó operators*

We continue the discussion of Section 1.8. The Stanilov operator has not been studied to any great extent in the literature. It was first introduced by Stanilov (2000); we also refer to later work of Stanilov (2004) and of Gilkey, Nikčević, and Videv (2004).

Let $\mathfrak{M}_1$ be a 1-model. The Szabó operator, defined in Section 1.8.5, is a generalization of the Jacobi operator defined by

$$\mathcal{S}(x) : y \to \mathcal{A}_1(y, x; x)x \,.$$

One says that $\mathfrak{M}_1$ is *Szabó* if the eigenvalues of $\mathcal{S}$ are constant on the pseudo-spheres of V. One says that a pseudo-Riemannian manifold $\mathcal{M}$ is *pointwise Szabó* if $\mathfrak{M}_1(\mathcal{M}, P)$ is Szabó for every point P of M; one says that $\mathcal{M}$ is *globally Szabó* if the structures are independent of P. The following result is due to Szabó (1991) in the Riemannian setting and to Gilkey and Stavrov (2002) in the Lorentzian setting; we postpone the proof until Section 4.5:

Theorem 1.9.10 [Szabó, Gilkey–Stavrov]

(1) Let $\mathfrak{M}_1$ be a Szabó 1-model which is either Riemannian or Lorentzian. Then $A_1 = 0$.

(2) If $\mathcal{M}$ be either a Riemannian or a Lorentzian pointwise Szabó manifold. Then $\mathcal{M}$ is locally symmetric.

There are no known 1-models $\mathfrak{M}_1$ with $A_1 \neq 0$ which are spacelike Jordan Szabó. It has been shown by Gilkey and Stavrov (2002) that if A_1 is a spacelike Jordan Szabó algebraic covariant derivative curvature tensor on a vector space of signature (p, q), where $q \equiv 1 \bmod 2$ and $p < q$ or where $q \equiv 2 \bmod 4$ and $p < q - 1$, then $A_1 = 0$. This algebraic result yields an elementary proof of the geometrical fact that any pointwise totally isotropic pseudo-Riemannian manifold with such a signature (p, q) is locally symmetric. The general question of finding non-trivial spacelike Jordan Szabó covariant algebraic curvature tensors, or conversely showing none exists, remains open.

1.9.8 *The skew-symmetric curvature operator*

Let $\mathfrak{M}_0 = (V, \langle \cdot, \cdot \rangle, A)$ be a 0-model. If $\{e_1, e_2\}$ is an orthonormal basis for a non-degenerate oriented 2-plane, one defines

$$\mathcal{A}(\pi) := \mathcal{A}(e_1, e_2).$$

More generally, if $\{v_1, v_2\}$ is an oriented basis for π, one may set

$$\mathcal{A}(\pi) := |\det(g_{ij})|^{-1/2} \mathcal{A}(v_1, v_2).$$

One says $\mathfrak{M}_0$ is *Ivanov–Petrova* if the eigenvalues of $\mathcal{A}$ are constant on the Grassmannian of non-degenerate 2-planes. One says that a pseudo-Riemannian manifold $\mathcal{M}$ is *Ivanov–Petrova* if $\mathfrak{M}_0(\mathcal{M}, P)$ is Ivanov–Petrova for all $P \in M$.

The algebraic classification is complete in the Riemannian setting. We refer to Gilkey, Leahy, and Sadofsky (1999), Gilkey (1999a), Ivanov and Petrova (1998), and Nikolayevsky (2004c) for the proof of the following result:

Theorem 1.9.11   *Let $\mathfrak{M}_0$ be a Riemannian Ivanov–Petrova 0-model with* $\dim(V) = m \geq 4$. *Then either there exists a self-adjoint isometry ϕ of V with $\phi^2 = \mathrm{id}$ so that $A = cA_\phi$, or $m = 4$ and A is isomorphic to a multiple of the algebraic curvature tensor described in Eq. (1.8.a).*

Manifolds of constant sectional curvature are Ivanov–Petrova; they correspond to taking $\phi = \mathrm{id}$. There are, however, other non-trivial geometric examples.

Definition 1.9.1   Let (S_K, g_K) be a pseudo-Riemannian manifold of signature $(\bar{p}, \bar{q})$ and constant sectional curvature K; (S_K, g_K) is determined up to local isometry by the parameters (p, q, K) and has the local geometry of the pseudo-spheres described in Lemma 1.6.7. Let $\varepsilon = \pm 1$ and let

$$f(t) = \varepsilon K t^2 + At + B \quad \text{where} \quad A^2 - 4\varepsilon KB \neq 0.$$

Choose a connected open interval $I \subset \mathbb{R}$ where $f(t) \neq 0$. Let

$$M := I \times S_K \quad \text{and} \quad g_M := \varepsilon dt^2 + f(t)g_{S_K}.$$

If S_K is Riemannian, if $f(t) > 0$, and if $\varepsilon = +1$, then g_M is Riemannian. The arguments of Ivanov and Petrova (1998) in the Riemannian setting extend immediately to the pseudo-Riemannian setting to show (M, g_M) is Ivanov–Petrova. We refer to Gilkey (2002) for further details.

The classification of Ivanov–Petrova manifolds is complete in the Riemannian setting:

Theorem 1.9.12 *Let M be a Riemannian Ivanov–Petrova manifold of dimension $m \geq 4$. Then either M has constant sectional curvature or M is locally isomorphic to one of the manifolds given in Definition 1.9.1.*

Theorems 1.9.11 and 1.9.12 were established in dimension $m = 4$ by Ivanov and Petrova. Subsequently, Gilkey, Leahy, and Sadofsky dealt with the cases $m \neq 4, 7, 8$. The case $m = 8$ was handled separately by Gilkey and Nikolayevsky completed the proof by treating the case $m = 7$.

In the indefinite setting, the Jordan normal form plays a crucial role. One says that $\mathfrak{M}_0$ is *spacelike Jordan Ivanov–Petrova* if the Jordan normal form of $\mathcal{A}$ is constant on the Grassmannian of spacelike 2-planes. In this situation, let $\mathrm{Rank}(\mathcal{A})$ be the rank of $\mathcal{A}(\pi)$ for any (and hence all) spacelike 2-plane. One has the following generalizations of Theorems 1.9.11 and 1.9.12:

Theorem 1.9.13 *Let $\mathfrak{M}_0$ be a 0-model of signature (p, q) where $q \geq 5$. The following conditions are equivalent:*

(1) $\mathfrak{M}_0$ is spacelike rank 2 Jordan Ivanov–Petrova.
(2) There exists a non-zero constant C and a self-adjoint map ϕ of V so that $R = CR_\phi$ where either ϕ is an isometry of V or ϕ is a paraisometry of V or $\phi^2 = 0$ and $\ker(\phi)$ contains no spacelike vectors.

Theorem 1.9.14 *Let M be a spacelike rank 2 Jordan Ivanov–Petrova manifold of signature (p, q) where $q \geq 5$. Assume that the curvature operator is not nilpotent for at least one point of M. Then either M has constant sectional curvature or M is locally isometric to one of the manifolds given in Definition 1.9.1.*

Remark 1.9.1 Theorem 1.9.14 does not complete the classification of the spacelike rank 2 Ivanov–Petrova manifolds; what remain to be considered are those where $\mathcal{R}$ is nilpotent. In Section 2.3, we present work of Dunn, Gilkey, and Nikčević (2005) giving a family of manifolds of signature $(2, 2)$ which are nilpotent timelike and spacelike Ivanov–Petrova but which are not mixed Ivanov–Petrova. In Section 2.7, we exhibit manifolds which are spacelike Ivanov–Petrova of rank 4 and which are not timelike Ivanov–Petrova.

Theorems 1.9.13 and 1.9.14 focus attention on the rank 2-spacelike Jordan Ivanov–Petrova manifolds. Assertion (1) in the following theorem is due to Gilkey (1999a) (in dimension $m = 8$), to Gilkey, Leahy, and Sadofsky (1999) (in dimensions $m \neq 7, 8$), and to Nikolayevsky (2004c) (in dimension 7); Assertions (2) and (3) are due to Zhang (2002); Assertion (4) is due to Stavrov (2003a).

Theorem 1.9.15   *Let $\mathfrak{M}_0 = (V, \langle \cdot, \cdot \rangle, A)$ is a 0-model of signature (p, q) such that $\mathcal{R}_A$ has constant spacelike rank $r > 0$.*

(1) Let $p = 0$, let $q \geq 5$. Then $r = 2$.

(2) Let $p = 1$ and let $q \geq 9$. Then $r = 2$.

(3) Let $p = 2$ and let $q \geq 10$. Assume neither q nor $q + 2$ are powers of 2. Then $r = 2$.

(4) Let $p \leq \frac{1}{4}q - 6$. Assume that $\{q, q + 1, ..., q + p\}$ does not contain a power of 2. Then $r = 2$.

We shall show in Theorem 2.7.3 that there exist algebraic curvature tensors which are spacelike Jordan Ivanov–Petrova rank 4; here, in contrast to the setting of Theorem 1.9.15, one has $q >> p$. We refer to Gilkey and Ivanova (2001a), Gilkey and Zhang (2002a), Gilkey and Ivanova (2002b), Ivanova (1996a), Ivanova (1998a), Ivanova and Stanilov (1995), Stanilov (2000), Stavrov (2004b), and Zhang (2000) for additional related work.

1.9.9   *The conformal skew-symmetric curvature operator*

Adopt the notation established in Section 1.8.2. The proof of Theorem 1.9.9 generalizes immediately to yield:

Theorem 1.9.16   *Let $g_1 \in [g_2]$. Let $\mathcal{M}_i := (M, g_i)$.*

(1) If $\mathcal{M}_1$ is conformally Ivanov–Petrova, then $\mathcal{M}_2$ is conformally Ivanov–Petrova.

(2) If $\mathcal{M}_1$ is conformally spacelike (respectively timelike) Jordan Ivanov–Petrova, then $\mathcal{M}_2$ is conformally spacelike (respectively timelike) Jordan Ivanov–Petrova.

Apart from the examples discussed in Section 1.8.4, not much is known about complex Ivanov–Petrova manifolds. Theorem 4.2.5 does give information about the eigenvalue structure of a Riemannian Ivanov–Petrova manifold as we shall discuss in Chapter 3. We refer to Gilkey and Ivanova (2001a) for a few additional results.

Chapter 2

Curvature Homogeneous Generalized Plane Wave Manifolds

2.1 Introduction

Chapter 2 is devoted to the study of geometric properties of various families of examples which arise as generalized plane wave manifolds.

The basic geometric properties of generalized plane wave manifolds are introduced in Section 2.2. In Theorem 2.2.1, we show that such manifolds are geodesically complete, Ivanov–Petrova, Osserman, Stanilov, Szabó, and Ricci flat. They also have vanishing scalar invariants and have nilpotent holonomy.

All the Weyl invariants of generalized plane wave manifolds vanish so a fundamental task is the construction of scalar invariants which are not of Weyl type. In Theorem 2.2.2, we construct global isometries between analytic generalized plane wave manifolds whose ∞-models are isomorphic; this result will play an important role in Sections 2.3, 2.9, and 2.10 when we construct a complete set of isometry invariants for certain families of analytic generalized plane wave manifolds. In Theorem 2.2.3, we present a result which will be used to study symmetric spaces in this family.

In the rest of Chapter 2, we study specific families of generalized plane wave manifolds which have useful geometric properties. In Section 2.3, we examine manifolds of signature $(2, 2)$ and follow the treatment in Dunn, Gilkey, and Nikčević (2005) to generalize results of Derdzinski (2000); another family of manifolds of signature $(2, 2)$ which are not generalized plane wave manifolds will be discussed subsequently in Section 3.3. Let $\{x, y, \tilde{x}, \tilde{y}\}$ be coordinates on $\mathbb{R}^4$ and let $f \in C^\infty(\mathbb{R})$. Let $\mathcal{M} := (\mathbb{R}^4, g)$ be the pseudo-Riemannian manifold of neutral signature $(2, 2)$ where

$$g(\partial_x, \partial_x) := -2f(y), \quad g(\partial_x, \partial_{\tilde{x}}) = g(\partial_y, \partial_{\tilde{y}}) := 1 \,.$$

The pseudo-Riemannian manifold $\mathcal{M}$ is a generalized plane wave manifold; if f is real analytic, then $\mathcal{M}$ can be realized as a hypersurface in $\mathbb{R}^{(2,3)}$. If $f'' > 0$, $\mathcal{M}$ is modeled on an irreducible symmetric space. If in addition $f''' > 0$, $\mathcal{M}$ is 1-curvature homogeneous, 1-modeled on a homogeneous space, and 2-curvature homogeneous if and only if $\mathcal{M}$ is homogeneous. The dimension of the group of isometries of such a manifold is determined. Let

$$\alpha_k(f, P) := \left\{ f^{(k+2)} \{f^{(2)}\}^{k-1} \{f^{(3)}\}^{-k} \right\}(P) \quad \text{for} \quad k \geq 2.$$

The α_k form a complete system of isometry invariants if f is real analytic.

Section 2.4 is concerned with the family of generalized plane wave manifolds of signature $(2, 4)$ which were introduced in Gilkey and Nikčević (2005a). Let $f \in C^\infty(\mathbb{R})$. Let $\{x, \tilde{x}, y, \tilde{y}, z_1, z_2\}$ be coordinates on $\mathbb{R}^6$. Let $\mathcal{M} := (\mathbb{R}^6, g)$ where

$$g(\partial_x, \partial_{\tilde{x}}) = g(\partial_y, \partial_{\tilde{y}}) = g(\partial_{z_1}, \partial_{z_1}) = g(\partial_{z_2}, \partial_{z_2}) := 1,$$
$$g(\partial_x, \partial_x) := -2(yz_1 + f(y)z_2).$$

If $f^{(2)} > 0$, then $\mathcal{M}$ is 0-curvature homogeneous, $\mathcal{M}$ is weakly 1-curvature homogeneous, and $\mathcal{M}$ is not 1-affine curvature homogeneous.

Results of Dunn, Gilkey, and Nikčević (2005), of Dunn and Gilkey (2005), of Gilkey, Ivanova, and Zhang (2002), of Gilkey, Ivanova, and Zhang (2003), of Gilkey and Zhang (2002b), and of Stavrov (2003a) are presented in Section 2.5. Let $\{x_1, ..., x_p, \tilde{x}_1, ..., \tilde{x}_p\}$ be coordinates on $\mathbb{R}^{2p}$. Let $f \in C^\infty(\mathbb{R}^p)$ and let $\mathcal{M} := (\mathbb{R}^{2p}, g)$ where g is the neutral signature (p, p) metric

$$g(\partial_{x_i}, \partial_{x_j}) := \partial_{x_i} f \cdot \partial_{x_j} f \quad \text{and} \quad g(\partial_{x_i}, \partial_{\tilde{x}_j}) := \delta_{ij}.$$

Let $H_{ij} := \partial_{x_i} \partial_{x_j} f$ be the Hessian. We show that $\mathcal{M}$ can be realized as a hypersurface in $\mathbb{R}^{(p,p+1)}$ with second fundamental form H. If H is non-degenerate and if $p = 2$, then $\mathcal{M}$ is Jordan Osserman. If $p \geq 3$ and H is positive or negative definite, then $\mathcal{M}$ is Jordan Osserman. If $p \geq 3$ and H is indefinite, then $\mathcal{M}$ is neither spacelike nor timelike Jordan Osserman. The manifold $\mathcal{M}$ is both spacelike and timelike Jordan Ivanov–Petrova; it is not mixed Jordan Ivanov–Petrova. If H is positive definite, set

$$\alpha := \left| H^{i_1 j_1} H^{i_2 j_2} H^{i_3 j_3} H^{i_4 j_4} H^{i_5 j_5} \nabla R_{i_1 i_2 i_3 i_4; i_5} \nabla R_{j_1 j_2 j_3 j_4; j_5} \right|$$

where we adopt the Einstein convention and sum over repeated indices ranging from 1 to p; only the ∂_{x_i} indices enter. We will show that α is a local isometry invariant of the 1-model. The manifold $\mathcal{M}$ is 0-curvature homogeneous; it is not 1-curvature homogeneous for generic values of f. Except in the special case that $\mathcal{M}$ is locally symmetric, these manifolds are neither spacelike nor timelike Jordan Szabó. We give the values of (r, s) for which $\mathcal{M}$ is Jordan Osserman of type (r, s) for certain defining functions f. In Section 2.6, we continue the discussion where we add in a flat factor. We discuss the values of (r, s) for which $\mathcal{M} \times \mathbb{R}^{(a,b)}$ is higher order Jordan Osserman of type (r, s).

In Section 2.7, we discuss Nikčević manifolds; these were first introduced in Gilkey and Nikčević (2004a). The crucial point is that the timelike directions dominate. Fix $s \geq 2$. Let $(\vec{u}, \vec{t}, \vec{v})$ be coordinates on $\mathbb{R}^3$ where

$$\vec{u} := (u_1, ..., u_s), \quad \vec{t} := (t_1, ..., t_s), \quad \vec{v} := (v_1, ..., v_s).$$

Previous examples have involved just a single warping function. We now assume given a family of functions $f_i \in C^\infty(\mathbb{R})$ for $1 \leq i \leq s$. We define a pseudo-Riemannian manifold $\mathcal{M} := (\mathbb{R}^{3s}, g)$ of signature $(2s, s)$ by setting:

$$g(\partial_i^u, \partial_i^u) := -2\sum_i \{f_i(u_i) + u_i t_i\},$$

$$g(\partial_i^t, \partial_i^t) := -1, \quad \text{and} \quad g(\partial_i^u, \partial_i^v) := 1.$$

The manifold $\mathcal{M}$ is 0-curvature homogeneous, indecomposable, spacelike Jordan Osserman, and not timelike Jordan Osserman. It is skew Tsankov if $s = 2$. We study the values of (r, s) so $\mathcal{M}$ is higher order Jordan Osserman. We construct isometry invariants of $\mathcal{M}$ and show that $\mathcal{M}$ is not generically 1-curvature homogeneous. The manifold $\mathcal{M}$ is spacelike Jordan Ivanov–Petrova of rank 4, it is not timelike Jordan Ivanov–Petrova, it is k-spacelike Jordan Stanilov for $2 \leq k \leq s$; $\mathcal{M}$ is k-timelike Jordan Stanilov if and only if $k = 2s$.

Dunn manifolds are discussed in Section 2.8. Let

$$\{u_0, u_1, ..., u_s, v_0, ..., v_s, t_1, ..., t_s\}$$

be coordinates on $\mathbb{R}^{3s+2}$, let $f_i \in C^\infty(\mathbb{R})$, and let ε_i for $1 \leq i \leq s$ be a choice of signs. Let $\mathcal{M} := (\mathbb{R}^{3s+2}, g)$ where

$$g(\partial_{u_0}, \partial_{u_i}) := 2f_i(u_i)t_i, \quad g(\partial_{u_i}, \partial_{u_i}) := -2u_0 t_i,$$

$$g(\partial_{u_i}, \partial_{v_i}) := 1, \quad \quad g(\partial_{t_i}, \partial_{t_i}) := \varepsilon_i.$$

These are generalized plane wave manifolds of quite general signatures which are 0-curvature homogeneous, 1-curvature homogeneous for certain but not all f_i, and not 2-curvature homogeneous. We exhibit local invariants which are not of Weyl type and which have a very different flavor from those studied previously in Chapter 2.

In Sections 2.9 and 2.10 we take up the task of creating generalized plane wave manifolds which 0-modeled on a symmetric space, $p+2$-modeled on a homogeneous space, and not $p+3$ affine-curvature homogeneous. In Section 2.9, we present examples of signature $(p+3, p+3)$ described in Gilkey and Nikčević (2004d); the symmetric space in question is not indecomposable. In Section 2.10, we give examples of $(3 + 2p, 3 + 2p)$ constructed in Gilkey and Nikčević (2005b) exhibiting many of the same phenomena but where the underlying symmetric space is indecomposable. We also derive some results about the isometry groups of this second family of examples.

By Theorem 1.4.2, there exists an integer $k_{p,q}$ so that if $\mathcal{M}$ is a complete simply connected pseudo-Riemannian manifold of signature (p, q) which is $k_{p,q}$ curvature homogeneous, then $\mathcal{M}$ is homogeneous. Let $r = \min(p, q)$. In Section 2.3, we will show that if $r = 2$, then $k_{p,q} \geq r$ by exhibiting a family of manifolds of signature (p, q) which are 1-curvature homogeneous but not 2-curvature homogeneous. Results from Section 2.9 show that if $r \geq 3$, then $k_{p,q} \geq r$; a family of manifolds of signature (p, q) which are $r - 1$ curvature homogeneous but not r-curvature homogeneous is exhibited. We conjecture that the correct lower bound is in fact $\min(p, q) + 1$; it is known, for example, that $k_{0,q} \geq 1$ and $k_{1,q} \geq 2$.

2.2   Generalized Plane Wave Manifolds

All the manifolds we will discuss in Chapter 2 fit into the following very general framework.

Definition 2.2.1   Let $x = (x_1, ..., x_m)$ be the usual coordinates on $\mathbb{R}^m$. We say that a pseudo-Riemannian manifold $\mathcal{M} := (\mathbb{R}^m, g)$ is a *generalized plane wave manifold* if

$$\nabla_{\partial_{x_i}} \partial_{x_j} = \sum_{k > \max(i,j)} \Gamma_{ij}{}^k(x_1, ..., x_{k-1}) \partial_{x_k} . \qquad (2.2.a)$$

Such a system of coordinates will be said to *give $\mathcal{M}$ a generalized plane wave structure*.

In this section, we shall derive the basic properties of this family; the remainder of Chapter 2 is devoted to the study of specific examples.

Here are the three main results of this section:

Theorem 2.2.1 *Let M be a generalized plane wave manifold. Then:*

(1) M is complete and strongly geodesically convex. The exponential map $\exp_P$ is a diffeomorphism from $T_P M$ to M for any point $P \in M$.

(2) $\nabla_{\partial_{x_{j_1}}} ... \nabla_{\partial_{x_{j_\nu}}} R(\partial_{x_{i_1}}, \partial_{x_{i_2}}) \partial_{x_{i_3}}$

$\qquad = \sum_{k > \max(i_1, i_2, i_3, j_1, ... j_\nu)} R_{i_1 i_2 i_3}{}^k{}_{;j_1 ... j_\nu}(x_1, ..., x_{k-1}) \partial_{x_k}.$

(3) M is nilpotent Osserman, nilpotent Ivanov–Petrova, nilpotent Szabó, nilpotent k-Osserman and nilpotent k-Stanilov for any k, Ricci flat, and Einstein.

(4) All the scalar Weyl invariants of M vanish.

(5) If e is the parallel vector field along a curve γ with $e(0) = \partial_{x_i}$, then $e(t) - \partial_{x_i} \in \mathrm{Span}_{j > i}\{\partial_{x_j}\}$. Thus the holonomy group $\mathcal{H}_P(M)$ is contained in the group of lower triangular matrices.

(6) If Y is the Jacobi vector field along a geodesic γ with $Y(0) = 0$ and $\dot{Y}(0) = \partial_{x_i}$, then $Y(t) - t\partial_{x_i} \in \mathrm{Span}_{j > i}\{\partial_{x_j}\}$.

The ∞-model encodes complete information about generalized plane wave manifolds in the real analytic category. We have the following:

Theorem 2.2.2 *Let $\mathcal{M}_1 = (M_1, g_1)$ and $\mathcal{M}_2 = (M_2, g_2)$ be real analytic generalized plane wave manifolds. Assume there exists an isomorphism Φ between $\mathfrak{M}_\infty(\mathcal{M}_1, P_1)$ and $\mathfrak{M}_\infty(\mathcal{M}_2, P_2)$. Then an isometry from $\mathcal{M}_1$ to $\mathcal{M}_2$ may be constructed by setting*

$$\phi := \exp_{P_2, \mathcal{M}_2} \circ \Phi \circ \exp_{P_1, \mathcal{M}_1}^{-1}.$$

Proof. It has been noted by Belger and Kowalski (1994) about analytic pseudo-Riemannian metrics that work of E. Cartan shows that the "metric g is uniquely determined, up to local isometry, by the tensors R, ∇R, ..., $\nabla^k R$, ... at one point."; see also Gray (1973) for related work. Theorem 2.2.2 now follows from Theorem 2.2.1. $\qquad\square$

The following observation is a special case which follows from more general results of E. Cartan.

Theorem 2.2.3 *Let M be a generalized plane wave manifold. Assume that $\nabla R = 0$. Then the geodesic symmetry $S_P : Q \to \exp_P\{-\exp_P^{-1} Q\}$ is an isometry. Furthermore, M is a homogeneous space.*

Theorem 2.2.1 will be central to our discussion of generalized plane wave manifolds throughout Chapter 2. In Sections 2.2.1 through 2.2.6, we establish Theorem 2.2.1. In Section 2.2.1, we study the geodesics of generalized plane wave manifolds. In Section 2.2.2, we study the curvature tensor. In Section 2.2.3, we examine the associated spectral geometry of the curvature tensor. In Section 2.2.4, we show these manifolds have vanishing scalar invariants. In Section 2.2.5, parallel translation is investigated and in Section 2.2.6, the structure of the Jacobi vector fields is determined.

The isometries of a generalized plane wave manifold often have a partial affine structure. We examine this behaviour in Section 2.2.7. These results will be used subsequently in Section 2.5, in Section 2.7, in Section 2.8, and in Section 3.5. We conclude our discussion in Section 2.2.8 by giving the proof of Theorem 2.2.3.

Throughout this section, we adopt the convention that the empty sum is 0. Thus, for example, $\sum_{i,j<k}$ is 0 if $k = 1$.

2.2.1 *The geodesic structure*

We begin the proof of Theorem 2.2.1 by examining the geodesic structure of a generalized plane wave manifold. We note that it is rather rare in Riemannian geometry that the equations for geodesics can be solved in explicit form in global coordinates. However Eq. (2.2.a) permits the use of a recursive formalism. Let $\gamma(t) = (\gamma^1(t), ..., \gamma^m(t))$ be a curve in $\mathbb{R}^m$. Then γ is a geodesic if and only if the geodesic equation of Eq. (1.2.e) is satisfied. In the present context, that means that for all k one has that:

$$\ddot{\gamma}^k(t) + \sum_{i,j<k} \dot{\gamma}^i(t)\dot{\gamma}^j(t)\Gamma_{ij}{}^k(\gamma^1, ..., \gamma^{k-1})(t) = 0 \,.$$

To solve this system of equations, we define $\gamma(t; \vec{\gamma}_0, \vec{\gamma}_1)$ by:

$$\gamma^1(t) := \gamma_0^1 + \gamma_1^1 t,$$

and inductively for $k > 1$, by setting:

$$\gamma^k(t) := \gamma_0^k + \gamma_1^k t - \int_0^t \int_0^s \sum_{i,j<k} \dot{\gamma}^i(r)\dot{\gamma}^j(r)\Gamma_{ij}{}^k(\gamma^1, ..., \gamma^{k-1})(r)drds \,.$$

Then $\gamma(0; \gamma_0, \gamma_1) = \gamma_0$ while $\dot{\gamma}(0; \gamma_0, \gamma_1) = \gamma_1$. Thus every geodesic arises in this way so all geodesics extend for infinite time. Furthermore, given $P, Q \in \mathbb{R}^m$, there is a unique geodesic $\gamma = \gamma_{P,Q}$ so that $\gamma(0) = P$ and $\gamma(1) = Q$ where

$$\gamma_0^k = P^k, \quad \gamma_1^1 = Q_1 - P_1$$

and for $k > 1$,

$$\gamma_1^k = Q^k - P^k + \int_0^1 \int_0^s \sum_{i,j<k} \dot\gamma^i(r)\dot\gamma^j(r)\Gamma_{ij}{}^k(\gamma^1,...,\gamma^{k-1})(r)\,dr\,ds\,.$$

This establishes Assertion (1) of Theorem 2.2.1.

2.2.2 *The curvature tensor*

We adopt the Einstein convention and sum over repeated indices to expand

$$\begin{aligned}
R_{ijk}{}^l &= \partial_{x_i}\Gamma_{jk}{}^l(x_1,...,x_{l-1}) - \partial_{x_j}\Gamma_{ik}{}^l(x_1,...,x_{l-1}) \\
&\quad + \Gamma_{in}{}^l(x_1,...,x_{l-1})\Gamma_{jk}{}^n(x_1,...,x_{n-1}) \\
&\quad - \Gamma_{jn}{}^l(x_1,...,x_{l-1})\Gamma_{ik}{}^n(x_1,...,x_{n-1})\,.
\end{aligned}$$

As we can restrict the quadratic sums to $n < l$,

$$R_{ijk}{}^l = R_{ijk}{}^l(x_1,...,x_{l-1})\,.$$

Suppose $l \le k$. Then $\Gamma_{jk}{}^l = \Gamma_{ik}{}^l = 0$. Furthermore for either of the quadratic terms to be non-zero, there must exist an index n with $k < n$ and $n < l$. This is not possible if $l \le k$. Thus $R_{ijk}{}^l = 0$ if $l \le k$.

Suppose $l \le i$. Then

$$\partial_{x_i}\Gamma_{jk}{}^l(x_1,...,x_{l-1}) = 0 \quad \text{and} \quad \partial_{x_j}\Gamma_{ik}{}^l = \partial_{x_j}0 = 0\,.$$

We have $\Gamma_{in}{}^l = 0$. For the other quadratic term to be non-zero, there must exist an index n so $i < n$ and $n < l$. This is not possible if $l \le i$. This shows $R_{ijk}{}^l = 0$ if $l \le i$; similarly $R_{ijk}{}^l = 0$ if $l \le j$.

This establishes Assertion (2) of Theorem 2.2.1 if $\nu = 0$ by establishing those assertions that relate to the undifferentiated curvature tensor R. To study ∇R, we expand

$$R_{ijk}{}^n{}_{;l} = \partial_l R_{ijk}{}^n(x_1,...,x_{n-1}) \tag{2.2.b}$$

$$- \sum_r R_{rjk}{}^n(x_1,...,x_{n-1})\Gamma_{li}{}^r(x_1,...,x_{r-1}) \tag{2.2.c}$$

$$- \sum_r R_{irk}{}^n(x_1,...,x_{n-1})\Gamma_{lj}{}^r(x_1,...,x_{r-1}) \tag{2.2.d}$$

$$- \sum_r R_{ijr}{}^n(x_1,...,x_{n-1})\Gamma_{lk}{}^r(x_1,...,x_{r-1}) \tag{2.2.e}$$

$$- \sum_r R_{ijk}{}^r(x_1,...,x_{r-1})\Gamma_{lr}{}^n(x_1,...,x_{n-1})\,. \tag{2.2.f}$$

We have $r < n$ in Eqs. (2.2.c)-(2.2.f). Consequently

$$R_{ijk}{}^{n}{}_{;l} = R_{ijk}{}^{n}{}_{;l}(x_1, ..., x_{n-1}).$$

To show $R_{ijk}{}^{n}{}_{;l} = 0$ if $n \leq \max(i, j, k, l)$, we note that

(1) $\partial_l R_{ijk}{}^{n}(x_1, ..., x_{n-1}) = 0$ if $n \leq \max(i, j, k, l)$ in Eq. (2.2.b);
(2) $n > \max(r, j, k)$ and $r > \max(i, l)$ so $n > \max(i, j, k, l)$ in Eq. (2.2.c);
(3) $n > \max(i, r, k)$ and $r > \max(l, j)$ so $n > \max(i, j, k, l)$ in Eq. (2.2.d);
(4) $n > \max(i, j, r)$ and $r > \max(k, l)$ so $n > \max(i, j, k, l)$ in Eq. (2.2.e);
(5) $n > \max(l, r)$ and $r > \max(i, j, k)$ so $n > \max(i, j, k, l)$ in Eq. (2.2.f).

This establishes Assertion (2) of Theorem 2.2.1 if $\nu = 1$ so we are dealing with ∇R. The argument is the same for higher values of ν and is therefore omitted.

2.2.3 *The geometry of the curvature tensor*

Let $\mathcal{J}$ be the Jacobi operator defined in Section 1.7.1, let $\mathcal{R}$ be the skew-symmetric curvature operator defined in Section 1.8.1, and let $\mathcal{S}$ be the Szabó operator defined in Section 1.8.5. By Assertion (2) of Theorem 2.2.1,

$$\mathcal{J}(\xi)\partial_{x_i} \subset \mathrm{Span}_{k>i}\{\partial_{x_k}\}, \quad \mathcal{S}(\xi)\partial_{x_i} \subset \mathrm{Span}_{k>i}\{\partial_{x_k}\},$$
$$\mathcal{R}(\pi)\partial_{x_i} \subset \mathrm{Span}_{k>i}\{\partial_{x_k}\}.$$

Thus $\mathcal{J}$, $\mathcal{R}$, and $\mathcal{S}$ are nilpotent so $\mathcal{M}$ is nilpotent Osserman, nilpotent Ivanov–Petrova, and nilpotent Szabó. A similar argument shows $\mathcal{M}$ is nilpotent k-Osserman and nilpotent k-Stanilov for any k. Furthermore, because $\mathcal{J}(\xi)$ is nilpotent, $\rho(\xi, \xi) = \mathrm{Tr}(\mathcal{J}(\xi)) = 0$. This implies $\rho = 0$ so $\mathcal{M}$ is Ricci flat and hence Einstein. This completes the proof of Assertion (3) of Theorem 2.2.1.

2.2.4 *Local scalar invariants*

Let Θ be a Weyl monomial which is formed by contracting upper and lower indices in pairs in the variables $\{g^{ij}, g_{ij}, R_{i_1 i_2 i_3}{}^{i_4}{}_{;j_1...}\}$. The single upper index in R plays a distinguished role. We choose a representation for Θ so the number of g_{ij} variables is minimal; for example, we can eliminate the $g_{i_3 i_4}$ variable in Eq. (1.2.i) by expressing:

$$|R|^2 = g^{i_1 j_1} g^{i_2 j_2} R_{i_1 i_2 k}{}^{l} R_{j_2 j_1 l}{}^{k}.$$

Suppose there is a g_{ij} variable in this minimal representation; this means that

$$\Theta = g_{ij} R_{u_1 u_2 u_3}{}^i{}_{;\ldots} R_{v_1 v_2 v_3}{}^j{}_{;\ldots\ldots}.$$

Suppose further that $g^{u_1 w_1}$ appears in Θ; consequently

$$\Theta = g_{ij} g^{u_1 w_1} R_{u_1 u_2 u_3}{}^i{}_{;\ldots} R_{v_1 v_2 v_3}{}^j{}_{;\ldots\ldots}.$$

We could then raise and lower an index to express

$$\Theta = R^{w_1}{}_{u_2 u_3 j;\ldots} R_{v_1 v_2 v_3}{}^j{}_{;\ldots\ldots} = R_{j u_3 u_2}{}^{w_1}{}_{;\ldots} R_{v_1 v_2 v_3}{}^j{}_{;\ldots\ldots}$$

which has one less $g_{..}$ variable. This contradicts the assumed minimality. Thus u_1 must be contracted against an upper index; a similar argument shows that u_2, u_3, v_1, v_2, and v_3 are contracted against an upper index as well. Consequently

$$\Theta = g_{ij} R_{u_1 u_2 u_3}{}^i{}_{;\ldots} R_{v_1 v_2 v_3}{}^j{}_{;\ldots} R_{w_1 w_2 w_3}{}^{u_1}{}_{;\ldots\ldots}.$$

Suppose w_1 is not contracted against an upper index. We then have

$$\Theta = g_{ij} g^{w_1 x_1} R_{u_1 u_2 u_3}{}^i{}_{;\ldots} R_{v_1 v_2 v_3}{}^j{}_{;\ldots} R_{w_1 w_2 w_3}{}^{u_1}{}_{;\ldots\ldots}$$

where we use the curvature symmetries, the covariant derivative of the second Bianchi identity of Eq. (1.2.h) and, if necessary, commute covariant derivatives using Eq. (1.2.d) to ensure that the index w_1 appears in the position indicated. Thus

$$\begin{aligned}
\Theta &= R_{u_1 u_2 u_3 j;\ldots} R_{v_1 v_2 v_3}{}^j{}_{;\ldots} R^{x_1}{}_{w_2 w_3}{}^{u_1}{}_{;\ldots\ldots} \\
&= g^{u_1 y_1} R_{u_1 u_2 u_3 j;\ldots} R_{v_1 v_2 v_3}{}^j{}_{;\ldots} R^{x_1}{}_{w_2 w_3 y_1;\ldots\ldots} \\
&= R^{y_1}{}_{u_2 u_3 j;\ldots} R_{v_1 v_2 v_3}{}^j{}_{;\ldots} R^{x_1}{}_{w_2 w_3 y_1;\ldots} \\
&= R_{j u_3 u_2}{}^{y_1}{}_{;\ldots} R_{v_1 v_2 v_3}{}^j{}_{;\ldots} R_{w_2 w_3 y_1}{}^{x_1}{}_{;\ldots}
\end{aligned}$$

which has one less g_{ij} variable. Thus w_1 is contracted against an upper index so

$$\Theta = g_{ij} R_{u_1 u_2 u_3}{}^i{}_{;\ldots} R_{v_1 v_2 v_3}{}^j{}_{;\ldots} R_{w_1 w_2 w_3}{}^{u_1}{}_{;\ldots} R_{x_1 x_2 x_3}{}^{w_1}{}_{;\ldots\ldots}.$$

We continue in this fashion to build a monomial of infinite length. This is not possible. Thus we can always find a representation for Θ which contains no g_{ij} variables in the summation.

We suppose the evaluation of Θ is non-zero and argue for a contradiction. To simplify the notation, group all the lower indices together. By considering the pairing of upper and lower indices, we see that we can expand Θ in cycles:

$$\Theta = R_{...i_r...}{}^{i_1} R_{...i_1...}{}^{i_2} ... R_{...i_{r-1}...}{}^{i_r} ... .$$

By Theorem 2.2.1 (2), $R_{...j...}{}^{l} = 0$ if $l \leq j$. Thus the sum runs over indices where $i_r < i_1 < i_2 < ... < i_r$. As this is the empty sum, we see that $\Theta = 0$ as desired. Thus all the scalar Weyl invariants of a generalized plane wave manifold vanish which establishes Assertion (4) of Theorem 2.2.1.

2.2.5 *Parallel vector fields and holonomy*

Let $\gamma = (\gamma^1, ..., \gamma^m)$ be a curve in $\mathbb{R}^m$. Let $X = \sum_j a^j(t)\partial_{x_j}$ be a vector field along γ; X is parallel along γ if and only if for every ℓ we have

$$0 = \dot{a}^\ell(t) + \sum_{j,k<\ell} \Gamma_{jk}{}^\ell(\gamma(t))\dot{\gamma}^j(t)a^k(t) .$$

To solve these equations with $X(0) = \partial_{x_i}$, we take $a^\ell = 0$ for $\ell < i$ and we let $a^i = 1$. For $\ell > i$, we define a^ℓ inductively by setting:

$$a^\ell(t) = - \sum_{j,k<\ell} \int_0^t \Gamma_{jk}{}^\ell(\gamma(s))\dot{\gamma}^j(s)a^k(s)ds .$$

This determines the parallel vector fields on $\mathcal{M}$ and shows the holonomy is lower triangular. This establishes Assertion (5) of Theorem 2.2.1.

2.2.6 *Jacobi vector fields*

Let γ be a geodesic; $X = \sum_j a^j(t)\partial_{x_j}$ is a Jacobi vector field along γ if and only if for all ℓ one has that

$$0 = \ddot{a}^\ell + \sum_{j,k<\ell} \partial_t\left\{\Gamma_{jk}{}^\ell\dot{\gamma}^j a^k\right\} + \sum_{jkuv<\ell} \Gamma_{jk}{}^u\Gamma_{vu}{}^k\dot{\gamma}^j a^k\dot{\gamma}^v$$
$$+ \sum_{j,k,v} R_{jkv}{}^\ell\dot{\gamma}^j a^j\dot{\gamma}^v .$$

This can be solved recursively with the initial conditions $X(0) = 0$ and $\dot{X}(0) = \partial_{x_i}$ by taking $a^k = 0$ for $k < i$, $a^i = t$, and for $k > i$

$$a^k(t) = -\int_0^t \int_0^s \left\{ \sum_{j,k<\ell} \partial_t \left\{ \Gamma_{jk}{}^\ell \dot{\gamma}^j a^k \right\} + \sum_{jkuv<\ell} \Gamma_{jk}{}^u \Gamma_{vu}{}^k \dot{\gamma}^j a^k \dot{\gamma}^v \right.$$
$$\left. + \sum_{j,k,v} R_{jkv}{}^\ell \dot{\gamma}^j a^j \dot{\gamma}^v \right\} dr\, ds \,.$$

This completes the proof of Theorem 2.2.1.

2.2.7 *Isometries*

Let $\mathrm{Gl}(n) \subset \mathrm{Al}(n)$ be the general linear group and the affine linear group on $\mathbb{R}^n$. We follow the discussion in Dunn, Gilkey, and Nikčević (2005) to establish the following result.

Lemma 2.2.1 *Let $(x_1, ..., x_a, y_1, ..., y_b)$ be coordinates on $\mathbb{R}^m$ which give $\mathcal{M}$ a generalized plane wave structure. Let*

$$\mathcal{X} := \mathrm{Span}_{1 \le i \le a}\{\partial_{x_i}\} \qquad and \qquad \mathcal{Y} := \mathrm{Span}_{1 \le \mu \le b}\{\partial_{y_\mu}\} \,.$$

Assume $\nabla_{\partial_{x_i}}\partial_{x_j} \in \mathcal{Y}$, $\nabla_{\partial_{x_i}}\partial_{y_\mu} = \nabla_{\partial_{y_\mu}}\partial_{x_i} = \nabla_{\partial_{y_\mu}}\partial_{y_\nu} = 0$. If ϕ is an isometry of $\mathcal{M}$ with $\phi_* \mathcal{Y} = \mathcal{Y}$, then there exists $A_1 \in \mathrm{Al}(a)$ and a smooth map $A_2 : \mathbb{R}^a \to \mathrm{Al}(b)$ so that

$$\phi(\vec{x}, \vec{y}) = (A_1 \vec{x}, A_2(\vec{x})\vec{y}) \,.$$

Proof. We apply the discussion of Section 2.2.1 concerning the geodesics in generalized plane wave manifolds. Decompose $\phi(\vec{x}, 0) = (\phi_1(\vec{x}), \phi_2(\vec{x}))$. Let

$$\Phi(\vec{x}) := \phi_*(\vec{x}, 0) : T_{(\vec{x},0)}(\mathbb{R}^m) \to T_{(\phi_1(\vec{x}),\phi_2(\vec{x}))}(\mathbb{R}^m) \,.$$

Since $\Phi(\vec{x})\mathcal{Y} = \mathcal{Y}$,

$$\phi_*(\partial_{x_i}) = \sum_{j=1}^a \Phi_{11;ij}\partial_{x_j} + \sum_{\mu=1}^b \Phi_{21;i\mu}\partial_{y_\mu}, \qquad \phi_*(\partial_{y_\mu}) = \sum_{\nu=1}^b \Phi_{22;\mu\nu}\partial_{y_\nu} \,.$$

As $\nabla_{\partial_{y_i}}\partial_{y_j} = 0$, $\gamma(t) := (\vec{x}, t\vec{y})$ is a geodesic. Then $\gamma_1 := \phi(\gamma(t))$ is a geodesic with $\gamma_1(0) = (\phi_1(\vec{x}), \phi_2(\vec{x}))$ and $\dot{\gamma}_1(0) = (0, \Phi_{22}(\vec{x})\vec{y})$. Thus

$$\gamma_1(t) = (\phi_1(\vec{x}), \phi_2(\vec{x}) + t\Phi_{22}(\vec{x})\vec{y}) \,.$$

We set $t = 1$ to conclude:

$$\phi(\vec{x}, \vec{y}) = (\phi_1(\vec{x}), \phi_2(\vec{x}) + \Phi_{22}(\vec{x})\vec{y}) \, .$$

We let $\star$ indicate terms which are not of interest. Let $\gamma(t) := (t\vec{x}, \star(t))$ be a geodesic starting at $(0,0)$. Then $\gamma_1(t) := \phi(\gamma(t))$ is a geodesic with $\gamma_1(0) = (\phi_1(0), \star)$ and $\dot{\gamma}_1(0) = (\Phi_{11}(0)\vec{x}, \star)$. Consequently

$$\gamma_1(t) = (\phi_1(0) + t\Phi_{11}(0)\vec{x}, \star) \, .$$

Again setting $t = 1$ yields $\phi(\vec{x}, 0) = (\phi_1(0) + \Phi_{11}(0)\vec{x}, \phi_2(x))$. Consequently

$$\phi(\vec{x}, \vec{y}) = (\phi_1(0) + \Phi_{11}(0)\vec{x}, \phi_2(\vec{x}) + \Phi_{22}(\vec{x})\vec{y})$$

has the desired form. $\qquad\qquad\square$

Remark 2.2.1   In proving this Lemma, we only needed the fact that ϕ was geodesic preserving which is implied by the somewhat weaker assumption that ϕ was an affine morphism; this implies $\phi^*\nabla = \nabla$.

Corollary 2.2.1   *Let $(x_1, ..., x_a, y_1, ..., y_b)$ be coordinates on $\mathbb{R}^m$ which give $\mathcal{M}$ a generalized plane wave structure. Let*

$$\mathcal{X} := \mathrm{Span}_{1 \leq i \leq a}\{\partial_{x_i}\} \qquad and \qquad \mathcal{Y} := \mathrm{Span}_{1 \leq \mu \leq b}\{\partial_{y_\mu}\} \, .$$

Assume $\nabla_{\partial_{x_i}}\partial_{x_j} \in \mathcal{Y}$, $\nabla_{\partial_{x_i}}\partial_{y_\mu} = \nabla_{\partial_{y_\mu}}\partial_{x_i} = 0$, and $\nabla_{\partial_{y_\mu}}\partial_{y_\nu} = 0$. Assume that $\mathcal{Y} = \ker(\mathcal{R})$. Then:

(1) Any Killing vector field on $\mathcal{M}$ has the form:

$$X = \sum_{i=1}^{a}\left(\xi_i + \sum_{j=1}^{a} A_{ij}x_j\right)\partial_{x_i} + \sum_{\mu}^{b}\left(\tilde{\xi}_\mu(x) + \sum_{\nu=1}^{b}\tilde{A}_{\mu\nu}(x)y_\nu\right)\partial_{y_\mu} \, .$$

(2) If $a = b$, if $g(\partial_{x_i}, \partial_{y_\mu}) = \delta_{i\mu}$, and if $g(\partial_{y_\mu}, \partial_{y_\nu}) = 0$, then $\tilde{A}_{ij}(x) = -A_{ji}$ is constant.

Proof.   Since $\ker(\mathcal{R}) = \mathrm{Span}\{\partial_{y_i}\}$, $\phi_* \mathrm{Span}\{\partial_{y_i}\} = \mathrm{Span}\{\partial_{y_i}\}$ for any isometry ϕ. We apply Lemma 2.2.1 to the 1-parameter flows generated by a killing vector field X. Differentiating the affine transformations given in Lemma 2.2.1 then establishes Assertion (1). To prove Assertion (2), we apply the Killing equation

$$g(\nabla_\xi X, \eta) + g(\nabla_\eta X, \xi)$$

for any ξ, η. We compute:

$$\nabla_{\partial_{x_j}} X = \sum_{j=1}^{a} A_{ij} \partial_{x_i} + \mathcal{Y}, \quad \nabla_{\partial_{y_\nu}} = \sum_{\nu=1}^{b} \tilde{A}_{\mu\nu} \partial_{y_\nu},$$

$$g(\nabla_{\partial_{x_j}} X, \partial_{y_\nu}) + g(\partial_{x_j}, \nabla_{\partial_{y_\nu}} X) = A_{j\nu} + \tilde{A}_{\nu j}(x).$$

The desired result now follows. $\qquad\square$

We can also discuss more general filtrations. We restrict ourselves to a 3-step filtration in the interests of simplicity. Let

$$(x_1, ..., x_a, y_1, ..., y_b, z_1, ..., z_c)$$

be coordinates on $\mathbb{R}^m$ giving $\mathcal{M}$ a generalized plane wave structure. As the proof of the following result is completely analogous to the proof given of Lemma 2.2.1, we shall omit the details:

Lemma 2.2.2 *Let* $\mathcal{X} := \mathrm{Span}\{\partial_{x_i}\}$, *let* $\mathcal{Y} := \mathrm{Span}\{\partial_{y_\mu}\}$, *and let* $\mathcal{Z} := \mathrm{Span}\{\partial_{z_\sigma}\}$. *Assume that* $\nabla_{\partial_{x_i}} \partial_{x_j} \in \mathcal{Y} + \mathcal{Z}$, *that* $\nabla_{\partial_{x_i}} \partial_{y_\mu} \in \mathcal{Z}$, *that* $\nabla_{\partial_{y_\mu}} \partial_{y_\nu} \in \mathcal{Z}$, *and that* $\nabla_{\partial_{x_i}} \partial_{z_\sigma} = \nabla_{\partial_{y_\mu}} \partial_{z_\sigma} = \nabla_{\partial_{z_\sigma}} \partial_{z_\tau} = 0$. *Let* ϕ *be an isometry of* $\mathcal{M}$ *with* $\phi_* \mathcal{Z} = \mathcal{Z}$ *and with* $\phi_*\{\mathcal{Y} + \mathcal{Z}\} = \mathcal{Y} + \mathcal{Z}$. *Then there exists* $A_1 \in \mathrm{Al}(a)$, *there exists a smooth map* $A_2 : \mathbb{R}^a \to \mathrm{Al}(b)$, *and there exists a smooth map* $A_3 : \mathbb{R}^{a+b} \to \mathrm{Al}(c)$ *so that*

$$\phi(\vec{x}, \vec{y}, \vec{z}) = (A_1 \vec{x}, A_2(\vec{x})\vec{y}, A_3(\vec{x}, \vec{y})\vec{z}).$$

2.2.8 *Symmetric spaces*

Proof of Theorem 2.2.3. As $\mathcal{M}$ is a generalized plane wave manifold, $\exp_P$ is a diffeomorphism from $T_P M$ to M so the geodesic symmetry $\mathcal{S}_P$ is globally defined. Fix a geodesic $\sigma(t) := \exp_P(t\xi)$ in $\mathcal{M}$. Let

$$T(t, s; \eta) := \exp_P(t(\xi + s\eta))$$

define a geodesic spray on $\mathcal{M}$. Let $Y(t; \eta) := T_*(\partial_s)$ be the corresponding Jacobi vector field along σ. The vector field Y satisfies the Jacobi equation

$$\ddot{Y} + \mathcal{J}(\sigma)Y = 0 \quad \text{with} \quad Y(0) = 0 \quad \text{and} \quad \dot{Y}(0) = \eta \qquad (2.2.\mathrm{g})$$

along σ as discussed in Section 1.2.5; it is completely determined by Eq. (2.2.g). Since

$$\mathcal{S}_P T(t, s; \eta) = \exp_P(-t(\xi + s\eta)),$$

$\mathcal{S}_*Y$ again arises from a geodesic spray and hence is a Jacobi vector field along σ. Taking into account the reversed orientation of σ, we see that $\mathcal{S}_*Y(0) = 0$ and $\dot{\mathcal{S}}_*Y(0) = \eta$. Thus by the fundamental theorem of ordinary differential equations,

$$\mathcal{S}_*Y = Y\,. \tag{2.2.h}$$

Choose Y_i so $Y_i(0) = 0$ and $\dot{Y}_i(0) = \partial_{x_i}$. By Theorem 2.2.1,

$$Y_i(t) - t\partial_{x_i} \in \mathrm{Span}_{j>i}\{\partial_{x_j}\}\,.$$

In particular, $\{Y_i(t)\}$ is a frame for $T_{\sigma(t)}M$ provided that $t \neq 0$. It is clear that $\mathcal{S}_*$ is an isometry of $T_P M$. Thus we can take $t \neq 0$. Thus by Eq. (2.2.h), to prove $\mathcal{S}$ is an isometry, it suffices to verify that

$$g(Y_i(t), Y_j(t)) = g(Y_i(-t), Y_j(-t)) \quad \text{for} \quad 1 \leq i, j \leq m\,. \tag{2.2.i}$$

Choose a parallel frame $\{e_i\}$ along σ with $e_i(0) = \partial_{x_i}$;

$$e_i(t) - \partial_{x_i} \in \mathrm{Span}_{j>i}\{\partial_{x_j}\}\,.$$

Because $\partial_t R_{ijkl}(t) = \nabla R(e_i, e_j, e_k, e_l; \dot{\sigma}) = 0$, $R(e_i, \dot{\sigma})\dot{\sigma} = \sum_{i<j} c_{ij} e_j$ for suitably chosen constants c_{ij}. Decompose

$$Y_i(t) = \sum_j a_{ij}(t) e_j(t)\,.$$

Relative to a parallel frame, the Jacobi equation takes the form:

$$\ddot{a}_{ij}(t) + \sum_{i<j, k<j} c_{kj} a_{ik}(t) \quad \text{for all } i, j\,.$$

If $Y_i(t)$ solves this equation with given initial conditions, then $-Y_i(-t)$ again solves the equation with the same initial conditions. Consequently $a_{ij}(-t) = -a_{ij}(t)$. Since

$$g(Y_i(t), Y_j(t)) = \sum_{kl} a_{ik} a_{j\ell} g(e_k, e_\ell)$$

and since $g(e_k, e_\ell)$ is independent of the parameter t, Eq. (2.2.i) holds; this shows the geodesic involution is an isometry.

Let $P, Q \in M$. We suppose $P \neq Q$. Since $\exp_P$ is a diffeomorphism from $T_P M$ to M, we can choose a geodesic σ so $\sigma(0) = P$ and $\sigma(1) = Q$. Let $S = \sigma(\frac{1}{2})$. Then the geodesic involution centered at S interchanges P and Q and is an isometry. Thus the pseudo-Riemannian manifold $\mathcal{M}$ is a homogeneous space. $\qquad\square$

2.3 Manifolds of Signature $(2, 2)$

In this section, we consider the following family of 4-dimension neutral signature generalized plane wave manifolds; they are also Fiedler manifolds as will be discussed in Section 3.5.

Definition 2.3.1 Let $(x, y, \tilde{x}, \tilde{y})$ be coordinates on $\mathbb{R}^4$, let $f \in C^\infty(\mathbb{R})$, and let $\mathcal{M} = \mathcal{M}_f := (\mathbb{R}^4, g)$ where $g = g_f$ is the metric of neutral signature $(2, 2)$ on $\mathbb{R}^4$ given by:

$$g(\partial_x, \partial_x) := -2f(y), \quad g(\partial_x, \partial_{\tilde{x}}) := 1, \quad g(\partial_y, \partial_{\tilde{y}}) := 1.$$

Lemma 2.3.1 *The manifold $\mathcal{M}$ of Definition 2.3.1 is a generalized plane wave manifold. The only non-zero component of $\nabla^k R$ is*

$$\nabla^k R(\partial_x, \partial_y, \partial_y, \partial_x; \partial_y, ..., \partial_y) = f^{(k+2)}.$$

Proof. The possibly non-zero Christoffel symbols are:

$$g(\nabla_{\partial_x}\partial_x, \partial_y) = f', \quad g(\nabla_{\partial_x}\partial_y, \partial_x) = g(\nabla_{\partial_y}\partial_x, \partial_x) = -f'.$$

Consequently, the non-zero covariant derivatives are:

$$\nabla_{\partial_x}\partial_x = f'\partial_{\tilde{y}} \quad \text{and} \quad \nabla_{\partial_x}\partial_y = \nabla_{\partial_y}\partial_x = -f'\partial_{\tilde{x}}. \tag{2.3.a}$$

Since $f = f(y)$, this has the proper triangular form relative to the ordering of the variables $\{x, y, \tilde{x}, \tilde{y}\}$ and thus $\mathcal{M}$ is a generalized plane wave manifold. The quadratic terms play no role in computing either the curvature tensor or the covariant derivatives of the curvature tensor; thus $\nabla^k R$ has the desired form. $\square$

Derdzinski (2000) studied these manifolds showing:

Theorem 2.3.1 **[Derdzinski]**

(1) If $f^{(3)} \neq 0$, then $\alpha_2 := f^{(4)}f^{(2)}(f^{(3)})^{-2}$ is an isometry invariant.
(2) $\mathcal{M}$ is symmetric if and only if $f^{(3)} = 0$.
(3) $\mathcal{M}$ is 0-curvature homogeneous if and only if $f^{(2)} = 0$ identically or if $f^{(2)}$ never vanishes.

This shows there exist neutral signature Ricci-flat 4-dimensional pseudo-Riemannian manifolds which are curvature homogeneous but which are not locally homogeneous. The main results of this section are the following Theorems summarizing results of Dunn, Gilkey, and Nikčević (2005) which extend the result of Derdzinski cited above.

Theorem 2.3.2 *Let M be as in Definition 2.3.1. If f is real analytic, then M is isometric to a hypersurface in $\mathbb{R}^{(2,3)}$.*

If $f^{(2)}$ vanishes identically, then f is flat. To ensure that M is curvature homogeneous and non-flat, we suppose therefore $f^{(2)}$ never vanishes. We shall assume $f^{(2)} > 0$ as the case $f^{(2)} < 0$ is similar.

Theorem 2.3.3 *Let M be as in Definition 2.3.1 where $f^{(2)} > 0$.*

(1) M is 0-modeled on the indecomposable symmetric space M_{y^2}.
(2) M is spacelike and timelike Jordan Osserman Ivanov–Petrova.
(3) M is not mixed Jordan Ivanov–Petrova.

If $f^{(3)}$ vanishes identically, then M is a symmetric space. To construct 1-curvature homogeneous spaces which are not symmetric spaces, we suppose $f^{(3)}$ never vanishes; the sign plays no role. Recall that the k-model and the k-affine model are given, respectively, by

$$\mathfrak{M}_k(M, P) := (T_P M, g_P, R_P, ..., \nabla^k R_P),$$
$$\mathfrak{F}_k(M, P) := (T_P M, \mathcal{R}_P, ..., \nabla^k \mathcal{R}_P).$$

Clearly $\mathfrak{F}_k$ is determined by $\mathfrak{M}_k$.

Theorem 2.3.4 *Let M be as in Definition 2.3.1 where $f^{(2)} > 0$ and $f^{(3)} \neq 0$. Set $\alpha_k(f) := f^{(k+2)}\{f^{(2)}\}^{k-1}\{f^{(3)}\}^{-k}$ for $k \geq 2$.*

(1) M is 1-modeled on the homogeneous space M_{ev}.
(2) If f_1 and f_2 are real analytic and if $\alpha_k(f_1)(P_1) = \alpha_k(f_2)(P_2)$ for all $k \geq 2$, then there is an isometry $\Phi : M_{f_1} \to M_{f_2}$ with $\Phi(P_1) = P_2$.
(3) If $\mathfrak{F}_k(M_{f_1}, P_1) \approx \mathfrak{F}_k(M_{f_2}, P_2)$, then $\alpha_k(f_1)(P_1) = \alpha_k(f_2)(P_2)$.

The manifolds M where $f^{(2)}$ is always positive and where $f^{(3)}$ is always non-zero form an interesting family. The invariants α_k are affine invariants which capture the underlying isometry type of the manifold, at least for real analytic f. Furthermore, 2-affine curvature homogeneity implies homogeneity. We note that in Sections 2.9 and 2.10 we will construct higher dimensional examples which have similar properties.

Theorem 2.3.5 *Let M be as in Definition 2.3.1. If M is 2-affine curvature homogeneous, then M is homogeneous.*

The methods used in Section 2.9 would in fact show that α is an invariant of the affine model and that 2-affine homogeneity implies homogeneity; we omit details in the interests of brevity.

We now study the isometry groups and Killing vector fields:

Theorem 2.3.6 *Let M be as in Definition 2.3.1 where f is real analytic. Let $\mathfrak{g}$ be the Lie algebra of Killing vector fields on M.*

(1) Let $f^{(2)} = 0$. Then M is flat and $\dim\{\mathfrak{g}\} = 10$.

(2) Let $f^{(2)} = ay^2$ for $a \neq 0$. Then M is an indecomposable symmetric space and $\dim\{\mathfrak{g}\} = 8$.

(3) Let $f^{(2)} = ae^{by}$ for $a, b \in \mathbb{R}$ for $a \neq 0$ and $b \neq 0$. Then M is a homogeneous space and $\dim\{\mathfrak{g}\} = 6$.

(4) Let $f^{(2)} = a(y + b)^n$ for $a, b \in \mathbb{R}$, $a \neq 0$, and $n = 3, 4, 5, \ldots$. Then M is not a homogeneous space. We have $\dim\{\mathfrak{g}\} = 6$. The submanifold $\{x : y + b > 0\}$ is a homogeneous space which is not complete.

(5) If $f^{(2)}$ is other than in (1)-(4), then M is not a homogeneous space and $\dim\{\mathfrak{g}\} = 5$.

By Theorem 1.4.2, there exists an integer $k_{p,q}$, called the *Singer number*, so that if M is a complete simply connected pseudo-Riemannian manifold of signature (p, q) which is $k_{p,q}$-curvature homogeneous, then M is homogeneous. We have

Theorem 2.3.7 *If $\min(p, q) = 2$, then $k_{p,q} \geq 2$.*

We shall return to this subject again in Section 2.9 when we study higher signature examples.

The remainder of this Section is devoted to the proof of these results. In Section 2.3.1, we prove Theorem 2.3.2, in Section 2.3.2, we prove Theorem 2.3.3, and in Section 2.3.3, we prove Theorems 2.3.4 and 2.3.5. In Section 2.3.4, we establish Theorem 2.3.6. We also give the Killing vector fields quite explicitly for these manifolds. We conclude in Section 2.3.5 by establishing Theorem 2.3.7; the primary technical difficulty here is to pass from signature $(2, 2)$ to signatures $(2, 2 + s)$ and $(2 + s, 2)$.

Remark 2.3.1 We will return to these examples in Section 3.5. It will follow from the discussion there that if $f'' \neq 0$, then M is Jacobi nilpotent of order 1. Furthermore, if $f''' \neq 0$, then M is Szabó nilpotent of order 1.

2.3.1 *Immersions as hypersurfaces in flat space*

Let M be as in Definition 2.3.1 where f is real analytic. We wish to show that M is isometric to a hypersurface in $\mathbb{R}^{(2,3)}$.

Definition 2.3.2   Let $\{e_1, e_2, \tilde{e}_1, \tilde{e}_2, \breve{e}\}$ be a basis for $\mathbb{R}^5$. Define an inner product on $\mathbb{R}^5$ of signature $(2,3)$ whose non-zero components are:

$$\langle e_1, \tilde{e}_1 \rangle = \langle e_2, \tilde{e}_2 \rangle = \langle \breve{e}, \breve{e} \rangle := 1 \,.$$

Define an embedding of $\mathbb{R}^4$ in $\mathbb{R}^{(2,3)}$ by setting:

$$\Psi(x, y, \tilde{x}, \tilde{y}) := xe_1 + ye_2 + \tilde{x}\tilde{e}_1 + \tilde{y}\tilde{e}_2 + \{\tfrac{1}{2}x^2 + f(y)\}\breve{e} \,.$$

Let $\tilde{\mathcal{M}} = \tilde{\mathcal{M}}_f := (\mathbb{R}^4, \tilde{g} := \Psi^*\langle \cdot, \cdot \rangle)$ be the associated hypersurface where:

$$\tilde{g}(\partial_x, \partial_{\tilde{x}}) = \tilde{g}(\partial_y, \partial_{\tilde{y}}) = 1, \quad \tilde{g}(\partial_x, \partial_x) = x^2,$$
$$\tilde{g}(\partial_x, \partial_y) = xf'(y), \quad \tilde{g}(\partial_y, \partial_y) = f'(y)f'(y) \,.$$

Theorem 2.3.2 will follow from the following slightly stronger result:

Lemma 2.3.2   *Let f be real analytic and let $P \in \mathbb{R}^4$. Let $\mathcal{M}$ be as in Definition 2.3.1 and let $\tilde{\mathcal{M}}$ be as in Definition 2.3.2. Let $P \in \mathbb{R}^4$.*

(1) $\tilde{\mathcal{M}}$ is a generalized plane wave manifold.
(2) $\mathfrak{M}_\infty(\mathcal{M}, P)$ is isomorphic to $\mathfrak{M}_\infty(\tilde{\mathcal{M}}, P)$.
(3) There is an isometry Φ from $\mathcal{M}$ to $\tilde{\mathcal{M}}$ with $\Phi(P) = P$.

Proof.   Let $x_1 := x$, $x_2 := y$, $\tilde{x}_1 := \tilde{x}$, and $\tilde{x}_2 := \tilde{y}$. Let $\tilde{g}_{ij} := \tilde{g}(\partial_{x_i}, \partial_{x_j})$. The non-zero Christoffel symbols are given by:

$$\tilde{g}(\nabla_{\partial_{x_i}}\partial_{x_j}, \partial_{x_k}) = \tfrac{1}{2}\{\partial_{x_i}\tilde{g}_{jk} + \partial_{x_j}\tilde{g}_{ik} - \partial_{x_k}\tilde{g}_{ij}\} \,.$$

Consequently

$$\nabla_{\partial_{x_i}}\partial_{x_j} = \sum_k \tfrac{1}{2}\{\partial_{x_i}\tilde{g}_{jk} + \partial_{x_j}\tilde{g}_{ik} - \partial_{x_k}\tilde{g}_{ij}\}\partial_{\tilde{x}_k},$$
$$\nabla_{\partial_{x_i}}\partial_{\tilde{x}_j} = \nabla_{\partial_{\tilde{x}_j}}\partial_{x_i} = \nabla_{\partial_{\tilde{x}_i}}\partial_{\tilde{x}_j} = 0 \,.$$

Since $\tilde{g}_{ij} = \tilde{g}_{ij}(x_1, x_2)$, this has the proper triangular form with respect to the coordinate ordering $\{x_1, x_2, \tilde{x}_1, \tilde{x}_2\}$ and thus $\tilde{M}$ is a generalized plane wave manifold. This establishes Assertion (1).

Let $\{X, Y, \tilde{X}, \tilde{Y}\}$ be a basis for $\mathbb{R}^4$. Fix $P \in \mathbb{R}^4$. Let $\mathfrak{M}_\infty(P)$ be the ∞-model where

$$\begin{aligned}
\langle X, \tilde{X} \rangle &= \langle Y, \tilde{Y} \rangle := 1, \\
A_k(P)(X, Y, Y, X; Y, ..., Y) &:= f^{(k+2)}(P) \,.
\end{aligned} \tag{2.3.b}$$

We show $\mathfrak{M}_\infty(\mathcal{M}, P)$ is isomorphic to $\mathfrak{M}_\infty(P)$ as follows. Set

$$X := \partial_x + f\partial_{\tilde{x}}, \quad Y := \partial_y, \quad \tilde{X} := \partial_{\tilde{x}}, \quad \tilde{Y} := \partial_{\tilde{y}} \,.$$

This normalizes the metric appropriately without changing the curvature tensor; the relations of Eq. (2.3.b) then follow from Lemma 2.3.1.

Next we show that $\mathfrak{M}_\infty(\tilde{\mathcal{M}}, P)$ is isomorphic to $\mathfrak{M}_\infty(P)$; this will complete the proof of Assertion (2) of Lemma 2.3.2. One verifies easily that the normal of the embedding is given by

$$\nu = -x\partial_{\tilde{x}} - f'\partial_{\tilde{y}} + \breve{e}\,.$$

Since the second fundamental form $\tilde{L}(\xi_1, \xi_2)$ is given by $\tilde{L}(\xi_1, \xi_2) = \xi_1 \xi_2 \Psi \cdot \nu$,

$$\tilde{L}(\partial_x, \partial_x) = 1, \quad \tilde{L}(\partial_y, \partial_y) = f^{(2)}(y), \quad \tilde{L}(\partial_x, \partial_y) = 0\,.$$

Consequently, the curvature tensor $\tilde{R}$ of $\tilde{\mathcal{M}}$ is given by

$$\tilde{R}(\partial_x, \partial_y, \partial_y, \partial_x) = \tilde{L}(\partial_x, \partial_x)\tilde{L}(\partial_y, \partial_y) - \tilde{L}(\partial_x, \partial_y)^2 = f^{(2)}\,.$$

As the Christoffel symbols play no role in the computation of $\nabla^\nu \tilde{R}$,

$$\nabla^\nu \tilde{R}(\partial_x, \partial_y, \partial_y, \partial_x; \partial_y, ..., \partial_y) = \partial_y^\nu \tilde{R}(\partial_x, \partial_y, \partial_y, \partial_x) = f^{(2+\nu)}$$

for $\nu > 0$. To normalize the metric to be hyperbolic, we set

$$X_1 := \partial_x - \tfrac{1}{2}\tilde{g}(\partial_x, \partial_x)\partial_{\tilde{x}} - \tfrac{1}{2}\tilde{g}(\partial_x, \partial_y)\partial_{\tilde{y}}, \quad \tilde{X}_1 := \partial_{\tilde{x}},$$
$$Y_1 := \partial_y - \tfrac{1}{2}\tilde{g}(\partial_y, \partial_x)\partial_{\tilde{x}} - \tfrac{1}{2}\tilde{g}(\partial_y, \partial_y)\partial_{\tilde{y}}, \quad \tilde{Y}_1 := \partial_{\tilde{y}}\,.$$

This does not change the curvature tensor; the relations of Eq. (2.3.b) relating to the curvature tensor then follow from the computations performed above to show $\mathfrak{M}_\infty(\tilde{\mathcal{M}}, P) = \mathfrak{M}_\infty(P)$. This establishes Assertion (2).

Since $\mathcal{M}$ and $\tilde{\mathcal{M}}$ are analytic generalized plane wave manifolds with $\mathfrak{M}_\infty(\mathcal{M}, P)$ isomorphic to $\mathfrak{M}_\infty(\tilde{\mathcal{M}}, P)$, Assertion (3) follows from Theorem 2.2.2. $\qquad\qquad\square$

2.3.2 *Spectral properties of the curvature tensor*

Let $\mathcal{M}$ be as in Definition 2.3.1. We suppose that $f^{(2)} > 0$. By Lemma 2.3.1, $\mathcal{M}_{y^2}$ is a symmetric space. Let $\{X, Y, \tilde{X}, \tilde{Y}\}$ be a basis for $\mathbb{R}^4$. Let $\mathfrak{M}_0 := (\mathbb{R}^4, \langle \cdot, \cdot \rangle, A)$ where

$$\langle X, \tilde{X} \rangle = \langle Y, \tilde{Y} \rangle = 1 \quad \text{and} \quad A(X, Y, Y, X) = 1\,.$$

To see that $\mathfrak{M}_0$ is a 0-curvature model for $\mathcal{M}$, we set

$$X := \partial_x + f\partial_{\tilde{x}}, \qquad Y := (f^{(2)})^{-1/2}\partial_y,$$
$$\tilde{X} := \partial_{\tilde{x}}, \qquad\qquad \tilde{Y} := (f^{(2)})^{1/2}\partial_{\tilde{y}}\,.$$

In particular, $\mathfrak{M}_0$ is a 0-model for the symmetric space $\mathcal{M}_{y^2}$. We complete the proof of Assertion (1) of Theorem 2.3.3 by showing $\mathfrak{M}_0$ is indecomposable. We suppose the contrary and argue for a contradiction. Assume there exists a non-trivial orthogonal direct sum decomposition $\mathbb{R}^4 = V_1 \oplus V_2$ which induces a decomposition of A. Decompose $\xi = \xi_1 + \xi_2$ where $\xi_i \in V_i$. We have

$$1 = A(X, Y, Y, X) = A(X_1, Y_1, Y_1, X_1) + A(X_2, Y_2, Y_2, X_2).$$

Thus we can choose the notation so $A(X_1, Y_1, Y_1, X_1) \neq 0$. Expand

$$X_1 = a_{11}X + a_{12}Y + a_{13}\tilde{X} + a_{14}\tilde{Y},$$
$$Y_1 = a_{21}X + a_{22}Y + a_{23}\tilde{X} + a_{24}\tilde{Y}.$$

We have $0 \neq A(X_1, Y_1, Y_1, X_1) = (a_{11}a_{22} - a_{12}a_{21})^2$. If $\xi \in V_2$, then

$$0 = A(X_1, Y_1, Y, \xi) = (a_{11}a_{22} - a_{12}a_{21})A(X, Y, Y, \xi).$$
$$0 = A(X_1, Y_1, X, \xi) = (a_{11}a_{22} - a_{12}a_{21})A(X, Y, X, \xi).$$

This implies $A(X, Y, Y, \xi) = A(X, Y, X, \xi) = 0$ so $\xi \in \mathrm{Span}\{\tilde{X}, \tilde{Y}\}$. Consequently, V_2 is totally isotropic. Since the decomposition $\mathbb{R}^4 = V_1 \oplus V_2$ is an orthogonal direct sum decomposition, this is impossible. Consequently $\mathfrak{M}_0$ is indecomposable; this establishes Assertion (1) of Theorem 2.3.3.

We have shown $\mathfrak{M}_0$ is a 0-model for $\mathcal{M}$. Consequently to show $\mathcal{M}$ is spacelike and timelike Jordan Osserman Ivanov–Petrova, it suffices to prove the corresponding assertion for $\mathfrak{M}_0$. We have

$$\mathcal{R}(X, Y)X = -\tilde{Y} \quad \text{and} \quad \mathcal{R}(X, Y)Y = \tilde{X}.$$

Let $\xi = aX + bY + \tilde{a}\tilde{X} + \tilde{b}\tilde{Y} \in \mathbb{R}^4$. Assume ξ is not null. This implies $(a, b) \neq (0, 0)$. One has $\mathcal{J}(\xi)\tilde{X} = \mathcal{J}(\xi)\tilde{Y} = 0$. Furthermore:

$$\mathcal{J}(\xi)(aX + bY) = 0,$$
$$\mathcal{J}(\xi)(-bX + aY) = (a^2 + b^2)(-b\tilde{X} + a\tilde{Y}).$$

Thus $\mathcal{J}(\xi)^2 = 0$ and $\mathrm{rank}\{\mathcal{J}(\xi)\} = 1$. This shows that the Jordan normal form of $\mathcal{J}(\cdot)$ is constant on the set of non-null vectors and hence $\mathfrak{M}_0$ is spacelike and timelike Jordan Osserman.

Let $\{e_1, e_2\}$ be an oriented orthonormal basis for an oriented 2-plane π which contains no non-zero null vectors. Expand

$$e_1 = a_1X + b_1Y + \tilde{a}_1\tilde{X} + \tilde{b}_1\tilde{Y},$$
$$e_2 = a_2X + b_2Y + \tilde{a}_2\tilde{X} + \tilde{b}_2\tilde{Y}.$$

If $a_1 b_2 - a_2 b_1 = 0$, then a linear combination of e_1 and e_2 belongs to $\mathrm{Span}\{\tilde{X}, \tilde{Y}\}$. Since π contains no null vectors, this is false. Consequently $a_1 b_2 - a_2 b_1 \neq 0$. One has that

$$\mathcal{R}(\pi) = (a_1 b_2 - a_2 b_1)\mathcal{R}(X, Y), \quad \mathcal{R}(\pi)\tilde{X} = 0, \quad \mathcal{R}(\pi)\tilde{Y} = 0,$$
$$\mathcal{R}(\pi)X = -(a_1 b_2 - a_2 b_1)\tilde{Y}, \quad \mathcal{R}(\pi)Y = (a_1 b_2 - a_2 b_1)\tilde{X}.$$

Consequently since $a_1 b_2 - a_2 b_1 \neq 0$, $\mathcal{R}(\pi)^2 = 0$ and $\mathrm{rank}\{\mathcal{R}(\pi)\} = 2$. These two relations determine the Jordan normal form of the skew-symmetric curvature operator. This shows $\mathfrak{M}_0$ is spacelike and timelike Jordan Ivanov–Petrova.

Let $\pi_1 := \mathrm{Span}\{X + \tilde{X}, Y - \tilde{Y}\}$ and $\pi_2 := \mathrm{Span}\{X, \tilde{X}\}$. Then π_1 and π_2 are both planes of signature $(1, 1)$. However, $\mathcal{R}(\pi_1) \neq 0$ while $\mathcal{R}(\pi_2) = 0$. Thus $\mathfrak{M}_0$ is not mixed Jordan Ivanov–Petrova. This completes the proof of Theorem 2.3.3.

2.3.3   *A complete system of invariants*

Let $\mathcal{M} = (\mathbb{R}^4, g)$ be as in Definition 2.3.1. Let $P \in \mathbb{R}^4$. If $f^{(3)}(P) \neq 0$, set:

$$\alpha_k(f, P) := \left\{ f^{(k+2)} \{f^{(2)}\}^{k-1} \{f^{(3)}\}^{-k} \right\}(P) \quad \text{for} \quad k \geq 2.$$

We clear the previous notation. Introduce the models

$$\begin{aligned}
\mathfrak{M}_0 &:= (\mathbb{R}^4, \langle \cdot, \cdot \rangle, A), \\
\mathfrak{M}_1 &:= (\mathbb{R}^4, \langle \cdot, \cdot \rangle, A, A_1), \\
\mathfrak{M}_\infty(f, P) &:= (\mathbb{R}^4, \langle \cdot, \cdot \rangle, A, A_1, A_2(f, P), ..., A_k(f, P), ...)
\end{aligned}$$
$$(2.3.c)$$

where the non-zero components of these tensors are described by

$$\begin{aligned}
&\langle X, \tilde{X} \rangle = \langle Y, \tilde{Y} \rangle = 1, \\
&A(X, Y, Y, X) = 1, \\
&A_1(X, Y, Y, X; Y) = 1, \\
&A_k(f, P)(X, Y, Y, X; Y, ..., Y) = \alpha_k(f, P) \quad \text{for} \quad k \geq 2.
\end{aligned}$$

Lemma 2.3.3   *Let $\mathcal{M}$ be as in Definition 2.3.1. Assume that $f^{(2)}(P) > 0$ and that $f^{(3)}(P) \neq 0$. Then $\mathfrak{M}_\infty(f, P)$ is isomorphic to $\mathfrak{M}_\infty(\mathcal{M}, P)$.*

Proof.   Let $\varepsilon_1 := \{f^{(2)}\}^{-3/2} f^{(3)}$ and let $\varepsilon_2 := f^{(2)} \{f^{(3)}\}^{-1}$. Set:

$$X_1 := \varepsilon_1\{\partial_x + f\partial_{\tilde{x}}\}, \quad Y_1 := \varepsilon_2\partial_y, \quad \tilde{X}_1 := \varepsilon_1^{-1}\partial_{\tilde{x}}, \quad \tilde{Y}_1 := \varepsilon_2^{-1}\partial_{\tilde{y}}.$$

This is a hyperbolic basis for $\mathbb{R}^4$ with

$$\nabla^\nu R(X_1, Y_1, Y_1, X_1; Y_1, ..., Y_1) = \varepsilon_1^2 \varepsilon_2^{\nu+2} f^{(\nu+2)}.$$

In particular $A(X_1, Y_1, Y_1, X_1) = 1$ and $A_1(P)(X_1, Y_1, Y_1, X_1; Y_1) = 1$. Furthermore for $k \geq 2$ one has that:

$$\nabla^k R(X_1, Y_1, Y_1, X_1; Y_1, ..., Y_1) = \{f^{(2)}\}^{k-1}\{f^{(3)}\}^{-k} f^{(2+k)} = \alpha_k(f).$$

This establishes the desired isomorphism. $\qquad\qquad\square$

Assertion (1) of Theorem 2.3.4 follows from Lemma 2.3.4 and from Theorem 2.2.2. Since $\alpha(e^y, P) = 1$, Theorem 2.3.4 (1) implies $\mathcal{M}_{e^\nu}$ is a homogeneous space. Restricting to the 1-model shows $\mathfrak{M}_1(\mathcal{M}, P)$ is isomorphic to $\mathfrak{M}_1$ and hence $\mathcal{M}$ is 1-modeled on $\mathcal{M}_{y^2}$. This establishes Assertion (3) of Theorem 2.3.4.

We show that $\alpha_k(f, P)$ is an invariant of the affine k-model and establish Assertion (3) of Theorem 2.3.4 as follows. Since $f^{(3)}(P) \neq 0$, we may choose X and Y in $T_P\mathbb{R}^4$ so that $\nabla_Y R(X, Y)Y \neq 0$; for example, we could set $X = \partial_x$ and $Y = \partial_y$. Choose a linear functional ϕ so $\phi\{\nabla_Y R(X, Y)Y\} \neq 0$. We expand

$$X = a_1\partial_x + a_2\partial_y + \tilde{a}_1\partial_{\tilde{x}} + \tilde{a}_2\partial_{\tilde{y}},$$
$$Y = b_1\partial_x + b_2\partial_y + \tilde{b}_1\partial_{\tilde{x}} + \tilde{b}_2\partial_{\tilde{y}}.$$

We compute:

$$\phi\{\nabla^k R(X, Y; Y, ..., Y)Y\}$$
$$= (a_1 b_2 - a_2 b_1)b_2^k \{b_2\phi(\partial_{\tilde{x}}) - b_1\phi(\partial_{\tilde{y}})\} f^{(2+k)}.$$

In particular, taking $k = 1$ shows that

$$(a_1 b_2 - a_2 b_1)b_2 \{b_2\phi(\partial_{\tilde{x}}) - b_1\phi(\partial_{\tilde{y}})\} \neq 0.$$

We now compute:

$$\{\phi(\nabla^k \mathcal{R}(X, Y; Y, ..., Y)Y)\}\{\phi(\mathcal{R}(X, Y)Y)\}^{k-1}\{\phi(\nabla\mathcal{R}(X, Y; Y)Y)\}^{-k}$$
$$= f^{(2+k)}\{f^{(2)}\}^{k-1}\{f^{(3)}\}^{-k} \cdot b_2^{k-k}$$
$$\cdot \{(a_1 b_2 - a_2 b_1)(b_2\phi(\partial_{\tilde{x}}) - b_1\phi(\partial_{\tilde{y}}))\}^{1+(k-1)-k}$$
$$= \alpha_k(f).$$

This is independent of the choice of $\{X, Y, \phi\}$ and hence is an affine invariant. This completes the proof of Theorem 2.3.4. $\qquad\square$

Theorem 2.3.5 follows from the following somewhat more general result:

Lemma 2.3.4 *Let M be as in Definition 2.3.1. The following assertions are equivalent:*

(1) There exist $a, b \in \mathbb{R}$ so that $f^{(2)}(y) = ae^{by}$.
(2) M is homogeneous.
(3) M is 2-curvature homogeneous.
(4) M is 2-affine curvature homogeneous.

Proof. Suppose that $f^{(2)}(y) = ae^{by}$. If $a = 0$ or if $b = 0$, then $\nabla R = 0$ so M is a generalized plane wave manifold which is a symmetric space and hence homogeneous by Theorem 2.2.3. Thus we may assume $a \neq 0$ and $b \neq 0$ and hence $f^{(2)} \neq 0$ and $f^{(3)} \neq 0$. Since $\alpha_k(f)$ is constant for $k \geq 2$, Theorem 2.3.4 implies M is homogeneous. Thus $(1) \Rightarrow (2)$; the implications $(2) \Rightarrow (3)$ and $(3) \Rightarrow (4)$ are trivial.

Suppose finally that M is 2-affine curvature homogeneous. If $\nabla R = 0$, then $f^{(2)}$ is constant and we may take $a = f^{(2)}$ and $b = 0$. Thus we may assume $\nabla R \neq 0$ and hence $f^{(3)} \neq 0$. We suppose $f^{(2)} > 0$; the case $f^{(2)} < 0$ is entirely analogous. We have shown that α_2 is an affine invariant. Thus α_2 is constant. Lemma 1.5.5 then implies $f^{(2)} = ae^{by}$ or $f^{(2)} = a(y + b)^c$. This latter case is ruled out since such a function vanishes when $y = -b$. Such functions will, however, play an important role in Section 2.3.4. Thus $f^{(2)}$ has the form given in Assertion (1). $\qquad\square$

2.3.4 *Isometries*

Let M be given by Definition 2.3.1 where f is analytic. Let $G(M)$ be the Lie group of isometries of M and let $\mathfrak{g}$ be the associated Lie algebra. Any local isometry gives rise to isomorphisms of the ∞-model and hence, since the metric is analytic, to a global isometry by Theorem 2.2.2. Consequently, we may identify $\mathfrak{g}$ with the Lie algebra of Killing vector fields on M. We begin by studying the symmetry groups $G(\mathfrak{M}_0)$ and $G(\mathfrak{M}_1)$.

Lemma 2.3.5 *Let $\mathfrak{M}_i$ be given by Eq. (2.3.c) and let $G(\mathfrak{M}_i)$ be the associated isomorphism groups. Then:*

$$G(\mathfrak{M}_0) = \left\{ \begin{pmatrix} \alpha & \gamma(\alpha^{-1})^t \\ 0 & (\alpha^{-1})^t \end{pmatrix} : \alpha, \gamma \in M_2(\mathbb{R}), \ \det \alpha = \pm 1, \ \gamma + \gamma^t = 0 \right\},$$

$$G(\mathfrak{M}_1) = \left\{ \begin{pmatrix} \alpha & \gamma(\alpha^{-1})^t \\ 0 & (\alpha^{-1})^t \end{pmatrix} : \alpha, \gamma \in M_2(\mathbb{R}), \ \alpha = \begin{pmatrix} \pm 1 & 0 \\ a_{21} & 1 \end{pmatrix}, \ \gamma + \gamma^t = 0 \right\}.$$

Consequently $\dim\{G(\mathfrak{M}_0)\} = 4$ *and* $\dim\{G(\mathfrak{M}_1)\} = 2$.

Proof.   If $\Theta \in G(\mathfrak{M}_0)$, let

$$\Theta X = a_{11} X + a_{12} Y + a_{13} \tilde{X} + a_{14} \tilde{Y},$$
$$\Theta Y = a_{21} X + a_{22} Y + a_{23} \tilde{X} + a_{24} \tilde{Y},$$
$$\Theta \tilde{X} = a_{31} X + a_{32} Y + a_{33} \tilde{X} + a_{34} \tilde{Y},$$
$$\Theta \tilde{Y} = a_{41} X + a_{42} Y + a_{43} \tilde{X} + a_{44} \tilde{Y}.$$

We set

$$\Theta = \begin{pmatrix} \alpha_1 & \alpha_2 \\ \alpha_3 & \alpha_4 \end{pmatrix}$$

where

$$\alpha_1 := \begin{pmatrix} a_{11} & a_{12} \\ a_{21} & a_{22} \end{pmatrix}, \quad \alpha_2 := \begin{pmatrix} a_{13} & a_{14} \\ a_{23} & a_{24} \end{pmatrix}$$
$$\alpha_3 := \begin{pmatrix} a_{31} & a_{32} \\ a_{41} & a_{42} \end{pmatrix}, \quad \alpha_4 := \begin{pmatrix} a_{33} & a_{34} \\ a_{43} & a_{44} \end{pmatrix}.$$

As $R(\Theta X, \Theta Y, \Theta Y, \Theta X) = 1$, $(a_{11} a_{22} - a_{12} a_{21})^2 = 1$ and $\det(\alpha_1)^2 = 1$. As

$$0 = R(\Theta X, \Theta Y, \Theta X, \Theta \tilde{X}) = R(\Theta X, \Theta Y, \Theta Y, \Theta \tilde{X})$$
$$= R(\Theta X, \Theta Y, \Theta X, \Theta \tilde{Y}) = R(\Theta X, \Theta Y, \Theta Y, \Theta \tilde{Y}),$$

we have that $a_{31} = a_{32} = a_{41} = a_{42} = 0$ and thus $\alpha_3 = 0$. As

$$g(\Theta X, \Theta \tilde{X}) = 1, \ g(\Theta X, \Theta \tilde{Y}) = 0$$
$$g(\Theta Y, \Theta \tilde{X}) = 0, \ g(\Theta Y, \Theta \tilde{Y}) = 1,$$

we have the relations

$$1 = a_{11} a_{33} + a_{12} a_{34}, \ 0 = a_{11} a_{43} + a_{12} a_{44},$$
$$0 = a_{21} a_{33} + a_{22} a_{34}, \ 1 = a_{21} a_{43} + a_{22} a_{44}$$

which can be rewritten in matrix form:

$$\mathrm{Id} = \begin{pmatrix} a_{11} & a_{12} \\ a_{21} & a_{22} \end{pmatrix} \begin{pmatrix} a_{33} & a_{43} \\ a_{34} & a_{44} \end{pmatrix} = \alpha_1 \alpha_4^t.$$

The relations $g(\Theta X, \Theta X) = g(\Theta X, \Theta Y) = g(\Theta Y, \Theta Y) = 0$ yield

$$0 = a_{11} a_{13} + a_{12} a_{14},$$
$$0 = a_{11} a_{23} + a_{12} a_{24} + a_{13} a_{21} + a_{14} a_{22},$$
$$0 = a_{21} a_{23} + a_{22} a_{24}$$

which can be rewritten in matrix form as

$$0 = \begin{pmatrix} a_{11} & a_{12} \\ a_{21} & a_{22} \end{pmatrix} \begin{pmatrix} a_{13} & a_{23} \\ a_{14} & a_{24} \end{pmatrix} + \begin{pmatrix} a_{13} & a_{14} \\ a_{23} & a_{24} \end{pmatrix} \begin{pmatrix} a_{11} & a_{21} \\ a_{12} & a_{22} \end{pmatrix} = \alpha_1 \alpha_2^t + \alpha_2 \alpha_1^t \,.$$

Setting $\alpha = \alpha_1$ and $\alpha_2 = \gamma(\alpha^{-1})^t$ yields $\gamma + \gamma^t = 0$ which establishes the first implication of Assertion (1). Conversely, the implications are all reversible and thus Θ satisfies these relations implies $\Theta \in G(\mathfrak{M}_0)$.

Suppose $\Theta \in G(\mathfrak{M}_1) \subset G(\mathfrak{M}_0)$. We impose the additional relations:

$$\nabla R(\Theta X, \Theta Y, \Theta Y, \Theta X; \Theta Y) = 1,$$
$$\nabla R(\Theta X, \Theta Y, \Theta Y, \Theta X; \Theta X) = 0\,.$$

This implies $a_{22} = 1$ and $a_{12} = 0$. Since $(a_{11}a_{22} - a_{12}a_{21})^2 = 1$, we have $a_{11} = \pm 1$ and α has the desired form. The fact that $\dim\{G(\mathfrak{M}_0)\} = 4$ and $\dim\{G(\mathfrak{M}_1)\} = 2$ is now immediate. $\qquad\qquad\square$

We continue our discussion:

Lemma 2.3.6 *Let $\mathcal{M}$ be as in Definition 2.3.1. Assume $f^{(2)} > 0$. Then X is a Killing vector field on $\mathcal{M}$ if and only if*

$$X = (\xi_1 + A_{11}x + A_{12}y)\partial_x + (\xi_2 + A_{21}x + A_{22}y)\partial_y$$
$$+ (\tilde{\xi}_1(x,y) - A_{11}\tilde{x} - A_{21}\tilde{y})\partial_{\tilde{x}} + (\tilde{\xi}_2(x) - A_{12}\tilde{x} - A_{22}\tilde{y})\partial_{\tilde{y}}$$

where

$$0 = -2fA_{11} - \partial_y f \cdot (\xi_2 + A_{21}x + A_{22}y) - \partial_x \tilde{\xi}_1,$$
$$0 = -2fA_{12} - \partial_x \tilde{\xi}_2 - \partial_y \tilde{\xi}_1\,.$$

Proof. Since $f^{(2)} > 0$, $\ker(R) = \mathrm{Span}\{\partial_{\tilde{x}}, \partial_{\tilde{y}}\}$. Thus we may apply Corollary 2.2.1 to express

$$X = \alpha\partial_x + \beta\partial_y + \tilde{\alpha}\partial_{\tilde{x}} + \tilde{\beta}\partial_{\tilde{y}}$$

where

$$\alpha = \xi_1 + A_{11}x + A_{12}y, \qquad \beta = \xi_2 + A_{21}x + A_{22}y,$$
$$\tilde{\alpha} = \tilde{\xi}_1(x,y) - A_{11}\tilde{x} - A_{21}\tilde{y}, \quad \tilde{\beta} = \tilde{\xi}_2(x,y) - A_{12}\tilde{x} - A_{22}\tilde{y}\,.$$

We apply Eq. (2.3.a) to see:

$$\nabla_{\partial_x} X = (\partial_x \alpha)\partial_x + (\partial_x \beta)\partial_y + (-\partial_y f \cdot \beta + \partial_x \tilde{\alpha})\partial_{\tilde{x}} + (\partial_y f \cdot \alpha + \partial_x \tilde{\beta})\partial_{\tilde{y}}$$
$$\nabla_{\partial_y} X = (\partial_y \alpha)\partial_x + (\partial_y \beta)\partial_y + (-\partial_y f \cdot \alpha + \partial_y \tilde{\alpha})\partial_{\tilde{x}} \qquad\qquad + (\partial_y \tilde{\beta})\partial_{\tilde{y}}$$
$$\nabla_{\partial_{\tilde{x}}} X = (\partial_{\tilde{x}} \alpha)\partial_x + (\partial_{\tilde{x}} \beta)\partial_y \qquad\qquad + (\partial_{\tilde{x}} \tilde{\alpha})\partial_{\tilde{x}} \qquad\qquad + (\partial_{\tilde{x}} \tilde{\beta})\partial_{\tilde{y}}$$
$$\nabla_{\partial_{\tilde{y}}} X = (\partial_{\tilde{y}} \alpha)\partial_x + (\partial_{\tilde{y}} \beta)\partial_y \qquad\qquad + (\partial_{\tilde{y}} \tilde{\alpha})\partial_{\tilde{x}} \qquad\qquad + (\partial_{\tilde{y}} \tilde{\beta})\partial_{\tilde{y}} \,.$$

Given the form of X, the Killing Equation

$$g(\nabla_\xi X, \eta) + g(\nabla_\eta X, \xi) = 0 \quad \forall \xi, \eta,$$

which was discussed in Section 1.3.6, is equivalent to the relations:

$$0 = -2f\partial_x \alpha - \partial_y f \cdot \beta + \partial_x \tilde{\alpha},$$
$$0 = -2f\partial_y \alpha + \partial_x \tilde{\beta} + \partial_y \tilde{\alpha},$$
$$0 = \partial_y \tilde{\beta}\,.$$

The relation $\partial_y \tilde{\beta} = 0$ implies $\tilde{\xi}_2 = \tilde{\xi}_2(x)$. The Lemma now follows. $\qquad\square$

Any manifold of the form given in Definition 2.3.1 admits 5 killing vector fields. Choose a primitive F so that $\partial_y F = f$ and so that $F(0) = 0$. Define vector fields by setting

$$\begin{aligned}
X_1 &:= \partial_x, & X_2 &:= \partial_{\tilde{x}}, \\
X_3 &:= \partial_{\tilde{y}}, & X_4 &:= -y\partial_{\tilde{x}} + x\partial_{\tilde{y}}, & \text{(2.3.d)}\\
X_5 &:= y\partial_x + 2F\partial_{\tilde{x}} - \tilde{x}\partial_{\tilde{y}}\,.
\end{aligned}$$

The corresponding flows are given by:

$$\begin{aligned}
\Phi_1^t(x, y, \tilde{x}, \tilde{y}) &:= (x + t, y, \tilde{x}, \tilde{y}), \\
\Phi_2^t(x, y, \tilde{x}, \tilde{y}) &:= (x, y, \tilde{x} + t, \tilde{y}), \\
\Phi_3^t(x, y, \tilde{x}, \tilde{y}) &:= (x, y, \tilde{x}, \tilde{y} + t), & \text{(2.3.e)}\\
\Phi_4^t(x, y, \tilde{x}, \tilde{y}) &:= (x, y, \tilde{x} - ty, \tilde{y} + tx), \\
\Phi_5^t(x, y, \tilde{x}, \tilde{y}) &:= (x + ty, y, \tilde{x} + 2F(y)t, \tilde{y} + t\tilde{x} + F(y)t^2)\,.
\end{aligned}$$

Let $\mathcal{L} := \mathrm{Span}_{\mathbb{R}}\{X_1, X_2, X_3, X_4, X_5\}$.

Lemma 2.3.7 *Let $\mathcal{M}$ be as in Definition 2.3.1. Then $\mathcal{L}$ is a 5-dimensional Lie algebra of Killing vector fields.*

Proof. We apply Lemma 2.3.6.

(1) X_1 is defined by $\xi_1 = 1$; the remaining parameters are 0.
(2) X_2 is defined by $\tilde{\xi}_1 = 1$; the remaining parameters are 0.

(3) X_3 is defined by $\tilde{\xi}_2 = 1$; the remaining parameters are 0.

(4) X_4 is defined by $\tilde{\xi}_1 = -y$ and $\tilde{\xi}_2 = x$; the remaining parameters are 0.

(5) X_5 is defined by $A_{12} = 1$ and $\tilde{\xi}_1 = 2F$; the remaining parameters are 0.

It is then immediate that the X_i satisfy the conditions of Lemma 2.3.6 and hence are Killing vector fields. We verify that $\mathrm{Span}_{\mathbb{R}}\{X_1, X_2, X_3, X_4, X_5\}$ is a Lie algebra by computing:

$$
\begin{aligned}
&[X_1, X_2] = 0, \qquad [X_1, X_3] = 0, \quad [X_1, X_4] = X_3, \\
&[X_1, X_5] = 0, \qquad [X_2, X_3] = 0, \quad [X_2, X_4] = 0, \\
&[X_2, X_5] = -X_3, \quad [X_3, X_4] = 0, \quad [X_3, X_5] = 0, \\
&[X_4, X_5] = 0.
\end{aligned}
$$

The desired result now follows. $\qquad\qquad\qquad\qquad\qquad\qquad\qquad$ $\square$

Proof of Theorem 2.3.6. Since $R(\partial_x, \partial_y, \partial_y, \partial_x) = f^{(2)}$, $\mathcal{M}$ is flat if and only if $f^{(2)} = 0$. If $\mathcal{M}$ is flat, then the isometry group acts transitively; the isotropy group at a point is $O(2,2)$ which has dimension 6. Thus $\dim\{\mathfrak{g}_0\} = 10$. This proves Assertion (1) of Theorem 2.3.6.

The manifold $\mathcal{M}$ is a symmetric space if and only if $f^{(3)}$ vanishes identically or, equivalently, if $f^{(2)} = a$. If $a = 0$, then $\mathcal{M}$ is flat so we suppose $a \neq 0$. Since $\mathcal{M}$ is a symmetric space, Theorem 2.2.3 yields $\mathcal{M}$ is a homogeneous space; thus the isometry group $G(\mathcal{M})$ acts transitively. Any isomorphism of the 0-model extends to an isomorphism of the ∞-model and hence to an isometry of $\mathcal{M}$ by Theorem 2.2.2. Thus the isotropy subgroup of $G(\mathcal{M})$ is isomorphic to $G(\mathfrak{M}_0)$ and hence has dimension 4 by Lemma 2.3.5. Thus $\dim\{G(\mathcal{M})\} = 4 + 4 = 8$. This completes the proof of Assertion (2) of Theorem 2.3.6. If $f = y^2$ for the sake of completeness we note without proof that 3 additional Killing vectors may be obtained by taking:

$$
\begin{aligned}
X_6 &:= \partial_y + 2xy\partial_{\tilde{x}} - x^2\partial_{\tilde{y}}, \\
X_7 &:= x\partial_y + (yx^2 - \tilde{y})\partial_{\tilde{x}} - \tfrac{1}{3}x^3\partial_{\tilde{y}}, \\
X_8 &:= x\partial_x - y\partial_y - \tilde{x}\partial_{\tilde{x}} + \tilde{y}\partial_{\tilde{y}}.
\end{aligned}
$$

The vector fields $X_1 = \partial_x$, $X_2 = \partial_{\tilde{x}}$, and $X_3 = \partial_{\tilde{y}}$ of Eq. (2.3.d) generate commuting isometric flows described in Eq. (2.3.e) which provides a 3-dimensional translation group of isometries; the orbits of this translation group are the hyperplanes $y = c$. Suppose that $\mathcal{M}$ is homogeneous but not symmetric. Equivalently, by Lemma 2.3.4, this means $f^{(2)} = ae^{by}$ where

$a \neq 0$ and $b \neq 0$. We suppose $a > 0$ as the case $a < 0$ is similar. The 1-model is given by $\mathfrak{M}_1$ of Eq. (2.3.c). Any isomorphism of the 1-model has the form given in Lemma 2.3.5 and hence extends to an isomorphism of the ∞-model. Thus by Theorem 2.2.2, the isotropy group of $G(\mathcal{M})$ at a point P can be identified with $G(\mathfrak{M}_1(\mathcal{M}, P)) = G(\mathfrak{M}_1)$ and has dimension 2. Since $\mathcal{M}$ is a homogeneous space, $\dim\{\mathfrak{g}\} = 2 + 4 = 6$. We note that the extra Killing vector field and symmetry are given by:

$$X_6 := -\tfrac{b}{2}x\partial_x + \partial_y + \tfrac{b}{2}\tilde{x}\partial_{\tilde{x}},$$
$$\Phi_6^t(x, y, \tilde{x}, \tilde{y}) := (e^{-bt/2}x, y + t, e^{bt/2}\tilde{x}, \tilde{y}).$$

This completes the proof of Assertion (3) of Theorem 2.3.6.

Suppose $f = a(y + b)^n$ for $a \in \mathbb{R}$ and $n = 3, 4, 5, \dots$. We suppose $b = 0$ to simplify the discussion. Let $\mathcal{O}$ be the connected open halfspace $y > 0$. The invariant α_k is constant on $\mathcal{O}$. Thus by Theorem 2.3.4, $G(\mathcal{M})$ acts transitively on $\mathcal{O}$ so $\mathcal{O}$ is a homogeneous space; $\mathcal{O}$ is, of course, incomplete. The additional translational symmetry on $\mathcal{O}$ is provided by the vector field

$$X_6 := x\partial_x - \tfrac{2}{n}y\partial_y - \tilde{x}\partial_{\tilde{x}} + \tfrac{2}{n}\tilde{y}\partial_{\tilde{y}},$$
$$\Phi_6^t : (x, y, \tilde{x}, \tilde{y}) \to (e^t x, e^{-2t/n}y, e^{-t}\tilde{x}, e^{2t/n}\tilde{y}).$$

At a point $P \in \mathcal{O}$, the 1-model is defined by Eq. (2.3.c); any symmetry of this model extends to a global symmetry of $\mathcal{M}$. Thus $\dim\{\mathfrak{g}\} = 4 + 2 = 6$.

Finally suppose f does not have the forms discussed above. Chooses P so $f^{(2)}(P) \neq 0$ and $f^{(3)}(P) \neq 0$. Since $f \neq ae^{by}$ and $f \neq a(y + b)^n$, α_2 is non-constant in a neighborhood of P and hence the orbits of the isometry group are 3-dimensional at P. Since the isotropy group preserves $\mathfrak{M}_1$, the isotropy group has dimension at most 2. Thus $\dim\{\mathfrak{g}\} \leq 5$; equality holds by Lemma 2.3.7. This completes the proof of Theorem 2.3.6. If n is even, then the half spaces $y < -b$ and $y > -b$ are isometric; if n is odd, then these two half spaces are not isometric since the sign of the curvature tensor will differ. $\qquad\square$

2.3.5 *Estimating $k_{p,q}$ if $\min(p, q) = 2$*

In this section, we complete the proof of Theorem 2.3.7 to obtain a lower bound on the Singer number in certain cases. Suppose given (p, q) with $2 = \min(p, q)$. We suppose that $p = 2$ and $2 \leq q$. We must exhibit a generalized plane wave manifold $\mathcal{N}$ which is 1-curvature homogeneous but not 2-curvature homogeneous. Let $\mathcal{M}$ be as in Definition 2.3.1 where

$f^{(2)} > 0$ and $f^{(3)} > 0$. We let $\mathcal{N} := \mathcal{M} \times \mathbb{R}^{(0,q-2)}$. By Theorem 2.3.4, $\mathcal{M}$ is 1-curvature homogeneous. It now follows that the isometric product $\mathcal{N}$ also is 1-curvature homogeneous. Exactly the same argument as that given to prove Theorem 2.3.4 shows that

$$\alpha_2 := \frac{f^{(4)} f^{(2)}}{f^{(3)} f^{(3)}}$$

is an invariant of the affine 2-model; the additional flat factors play no role. Choosing $f = e^x + e^{2x}$ then yields an example of signature $(2, q)$ which is 1-curvature homogeneous, where α_2 is non-constant, and thus where $\mathcal{N}$ is not 2-affine homogeneous (and hence not 2-curvature homogeneous) as desired.

2.4 Manifolds of Signature $(2, 4)$

We consider the following family of Fiedler manifolds which was introduced in Gilkey and Nikčević (2005a):

Definition 2.4.1 Let $f \in C^\infty(\mathbb{R})$ and let $(x, y, z_1, z_2, \tilde{y}, \tilde{x})$ be coordinates on $\mathbb{R}^6$. Set $\mathcal{M} := (\mathbb{R}^6, g)$ where:

$$g(\partial_x, \partial_{\tilde{x}}) = g(\partial_y, \partial_{\tilde{y}}) = g(\partial_{z_i}, \partial_{z_i}) := 1,$$
$$g(\partial_x, \partial_x) := -2(yz_1 + f(y)z_2).$$

Lemma 2.4.1 *The manifold $\mathcal{M}$ of Definition 2.4.1 is a generalized plane wave manifold of signature $(2, 4)$ with:*

(1) $R(\partial_{z_1}, \partial_x, \partial_x, \partial_y) = 1$.
(2) $R(\partial_{z_2}, \partial_x, \partial_x, \partial_y) = f'$.
(3) $R(\partial_y, \partial_x, \partial_x, \partial_y) = f'' z_2$.
(4) $\nabla R(\partial_x, \partial_y, \partial_y, \partial_x; \partial_{z_2}) = \nabla R(\partial_x, \partial_y, \partial_{z_2}, \partial_x; \partial_y) = f''$.
(5) $\nabla R(\partial_x, \partial_y, \partial_y, \partial_x; \partial_y) = f''' z_2$.

Proof. We set $F = yz_1 + f(y)z_2 = f_1(y)z_1 + f_2(y)z_2$. The possibly non-zero Christoffel symbols are given by:

$$g(\nabla_{\partial_x} \partial_x, \partial_y) = f_1' z_1 + f_2' z_2,$$
$$g(\nabla_{\partial_y} \partial_x, \partial_x) = g(\nabla_{\partial_x} \partial_y, \partial_x) = -f_1' z_1 - f_2' z_2,$$
$$g(\nabla_{\partial_x} \partial_x, \partial_{z_i}) = f_i,$$
$$g(\nabla_{\partial_{z_i}} \partial_x, \partial_x) = g(\nabla_{\partial_x} \partial_{z_i}, \partial_x) = -f_i.$$

Consequently the possibly non-zero covariant derivatives are:

$$\nabla_{\partial_x}\partial_x = \{f_1' z_1 + f_2' z_2\}\partial_{\tilde{y}} + f_1\partial_{z_1} + f_2\partial_{z_2},$$
$$\nabla_{\partial_y}\partial_x = \nabla_{\partial_x}\partial_y = -\{f_1' z_1 + f_2' z_2\}\partial_{\tilde{x}},$$
$$\nabla_{\partial_{z_i}}\partial_x = \nabla_{\partial_x}\partial_{z_i} = -f_i\partial_{\tilde{x}}.$$

This has the proper triangular form relative to the ordering of the variables $(x, y, z_1, z_2, \tilde{y}, \tilde{x})$ and thus $\mathcal{M}$ is a generalized plane wave manifold. The quadratic terms in the Christoffel symbols play no role in computing R; the remaining assertions of the Lemma are immediate. $\qquad\square$

Theorem 2.4.1   *Let $\mathcal{M}$ be as in Definition 2.4.1. Assume that $f'' > 0$.*

(1) $\mathcal{M}$ is 0-curvature homogeneous.
(2) $\mathcal{M}_f$ is weakly 1-curvature homogeneous.
(3) $\alpha(f, P) := |f'(P)|$ is an invariant of the affine 1-model $\mathfrak{F}_1(\mathcal{M}, P)$.
(4) $\mathcal{M}_f$ is not 1-affine curvature homogeneous.

This shows that weak 1-curvature homogeneity and affine 1-curvature homogeneity are different notions.

Proof.   Assume that $f'' \neq 0$. We apply Lemma 2.4.1. Because $R(\partial_x, \partial_y, \partial_{z_1}, \partial_x) = 1$ and because $\nabla R(\partial_x, \partial_y, \partial_y, \partial_x; \partial_{z_2}) = f'' \neq 0$, we may choose $\{\varepsilon_1, \ \varepsilon_1, \ c_1, \ c_2, \ c_3\}$ so that we have the relations:

(1) $R(\partial_x, \partial_y, \partial_y, \partial_x) - 2\varepsilon_1 R(\partial_x, \partial_y, \partial_{z_1}, \partial_x) - 2\varepsilon_2 R(\partial_x, \partial_y, \partial_{z_2}, \partial_x) = 0,$
(2) $\nabla R(\partial_x, \partial_y, \partial_y, \partial_x; \partial_y) - 3\varepsilon_2 \nabla R(\partial_x, \partial_y, \partial_y, \partial_x; \partial_{z_2}) = 0,$
(3) $c_3^2(1 + (f')^2) = 1,$
(4) $c_3(1 + (f')^2)c_1^2 c_2 = 1,$
(5) $c_3 c_1^2 c_2^2 f'' = 1.$

We show that $\mathcal{M}$ is 0-curvature homogeneous and establish Assertion (2) by giving a canonical form to the 0-model. Consider the basis:

$$\begin{aligned}
X &:= c_1\{\partial_x - \tfrac{1}{2}g(\partial_x, \partial_x)\partial_{\tilde{x}}\}, & \tilde{X} &:= c_1^{-1}\partial_{\tilde{x}}, \\
Y &:= c_2\{\partial_y - \varepsilon_1\partial_{z_1} - \varepsilon_2\partial_{z_2} - \tfrac{1}{2}(\varepsilon_1^2 + \varepsilon_2^2)\partial_{\tilde{y}}\}, & \tilde{Y} &:= c_2^{-1}\partial_{\tilde{y}}, \\
Z_1 &:= c_3\{\partial_{z_1} + f'\partial_{z_2} + (\varepsilon_1 + f'\varepsilon_2)\partial_{\tilde{y}}\}, \\
Z_2 &:= c_3\{\partial_{z_2} - f'\partial_{z_1} + (\varepsilon_2 - f'\varepsilon_1)\partial_{\tilde{y}}\}.
\end{aligned} \qquad (2.4.\mathrm{a})$$

The possibly non-zero entries in g, R and ∇R are:

$$g(X, \tilde{X}) = g(Y, \tilde{Y}) = 1,$$
$$g(Z_1, Z_1) = g(Z_2, Z_2) = 1 \qquad \text{see (3) above,}$$
$$R(X, Y, Y, X) = 0 \qquad \text{see (1) above,}$$
$$R(X, Y, Z_1, X) = 1 \qquad \text{see (4) above,}$$
$$R(X, Y, Z_2, X) = 0,$$
$$\nabla R(X, Y, Y, X; Z_1) = \nabla R(X, Y, Z_1, X; Y) = f' \qquad \text{see (5) above,}$$
$$\nabla R(X, Y, Y, X; Y) = 0 \qquad \text{see (2) above,}$$
$$\nabla R(X, Y, Y, X; Z_2) = \nabla R(X, Y, Z_2, X; Y) = 1 \qquad \text{see (5) above.}$$

Let $Z_1^* := Z_1 - f' Z_2$. The non-zero entries in R and in ∇R are:

$$R(X, Y, Z_1^*, X) = 1, \quad \text{and}$$
$$\nabla R(X, Y, Y, X; Z_2) = \nabla R(X, Y, Z_2, X; Y) = 1 \, .$$

This gives a canonical form for the weak 1-model $(T_P M, R_P, \nabla R_P)$ and establishes Assertion (2).

We now show α is an affine invariant by showing that α is determined by the affine 1-model $\mathfrak{F}_1(\mathcal{M}, P)$. We have:

$$\mathcal{R}(X, Y)Z_1 = \tilde{X}, \qquad \mathcal{R}(X, Y)X = -Z_1,$$
$$\mathcal{R}(X, Z_1)Y = \tilde{X}, \qquad \mathcal{R}(X, Z_1)X = -\tilde{Y},$$
$$\nabla_{Z_1} \mathcal{R}(X, Y)Y = f' \tilde{X}, \qquad \nabla_{Z_2} \mathcal{R}(X, Y)Y = \tilde{X},$$
$$\nabla_{Z_1} \mathcal{R}(X, Y)X = -f' \tilde{Y}, \qquad \nabla_{Z_2} \mathcal{R}(X, Y)X = -\tilde{Y},$$
$$\nabla_Y \mathcal{R}(X, Y)Z_1 = f' \tilde{X}, \qquad \nabla_Y \mathcal{R}(X, Y)Z_2 = \tilde{X},$$
$$\nabla_Y \mathcal{R}(X, Z_1)Y = f' \tilde{X}, \qquad \nabla_Y \mathcal{R}(X, Z_2)Y = \tilde{X},$$
$$\nabla_Y \mathcal{R}(X, Z_1)X = -f' \tilde{Y}, \qquad \nabla_Y \mathcal{R}(X, Z_2)X = -\tilde{Y},$$
$$\nabla_Y \mathcal{R}(X, Y)X = -f' Z_1 - Z_2 \, .$$

We define the following subspaces:

$$W_1 := \text{Range}(\mathcal{R}) = \text{Span}\{\mathcal{R}(\xi_1, \xi_2)\xi_3 : \xi_i \in \mathbb{R}^6\},$$
$$W_2 := \text{Range}(\nabla \mathcal{R}) = \text{Span}\{\nabla_{\xi_1} \mathcal{R}(\xi_2, \xi_3)\xi_4 : \xi_i \in \mathbb{R}^6\},$$
$$W_3 := \text{Span}\{\mathcal{R}(\xi_1, \mathcal{R}(\xi_2, \xi_3)\xi_4)\xi_5 : \xi_i \in \mathbb{R}^6\},$$
$$W_4 := \ker(\mathcal{R}) = \{\eta \in \mathbb{R}^6 : \mathcal{R}(\xi_1, \xi_2)\eta = 0 \ \forall \ \xi_i \in \mathbb{R}^6\},$$
$$W_5 := \ker(\nabla \mathcal{R}) = \{\eta \in \mathbb{R}^6 : \nabla_{\xi_1} \mathcal{R}(\xi_2, \xi_3)\eta = 0 \ \forall \ \xi_i \in \mathbb{R}^6\} \, .$$

We wish to show that:

(1) $W_1 = \text{Span}\{\tilde{X}, \tilde{Y}, Z_1\}$,

(2) $W_2 = \text{Span}\{\tilde{X}, \tilde{Y}, f' Z_1 + Z_2\}$,

(3) $W_3 = \mathrm{Span}\{\tilde{X}, \tilde{Y}\}$,
(4) $W_4 = \mathrm{Span}\{\tilde{X}, \tilde{Y}, Z_2\}$,
(5) $W_5 = \mathrm{Span}\{\tilde{X}, \tilde{Y}, Z_1 - f'Z_2\}$.

Items (1) and (2) are immediate. We compute

$$\mathcal{R}(X, \mathcal{R}(X, Y)X)X = \mathcal{R}(X, -Z_1)X = \tilde{Y},$$
$$\mathcal{R}(X, \mathcal{R}(X, Y)X)Y = \mathcal{R}(X, -Z_1)Y = -\tilde{X}, \quad \text{so} \quad \mathrm{Span}\{\tilde{X}, \tilde{Y}\} \subset W_3.$$

We establish Item (3) by establishing the reverse inclusion:

$$\mathcal{R}(\xi_1, \mathcal{R}(\xi_2, \xi_3)\xi_4)\xi_5 = \mathcal{R}(\xi_1, aZ_1 + b\tilde{X} + c\tilde{Y})\xi_5$$
$$= \mathcal{R}(dX, aZ_1)\xi_5 \in \mathrm{Span}\{\tilde{X}, \tilde{Y}\}.$$

Since $\mathrm{Span}\{\tilde{X}, \tilde{Y}, Z_2\} \subset W_4$, we must show the reverse inclusion to establish (4). Let $\eta = aX + bY + cZ_1 + dZ_2 + e\tilde{X} + f\tilde{Y} \in W_4$. One has

$$0 = \mathcal{R}(X, Y)\eta = -aZ_1 + c\tilde{X},$$
$$0 = \mathcal{R}(X, Z_1)\eta = -a\tilde{Y} + b\tilde{X}.$$

Thus $a = b = c = 0$ and (4) follows.

It is clear $W_5 \subset \mathrm{Span}\{\tilde{X}, \tilde{Y}, Z_1 - f'Z_2\}$. We must establish the reverse inclusion. Let $\eta = aX + bY + cZ_1 + dZ_2 + e\tilde{X} + f\tilde{Y} \in W_5$. Then:

$$0 = \nabla_{Z_2}\mathcal{R}(X, Y)\eta = -a\tilde{Y} + b\tilde{X},$$
$$0 = \nabla_Y\mathcal{R}(X, Y)\eta = a(-f'Z_1 - Z_2) + (cf' + d)\tilde{X}.$$

Thus $a = b = 0$ and $d = -cf'$; this establishes Item (5).

We can now establish Assertion (3) of Theorem 2.4.1. Suppose there is an isomorphism ϕ from $\mathfrak{F}_1(\mathcal{M}_{f_1}, P_1)$ to $\mathfrak{F}_1(\mathcal{M}_{f_2}, P_2)$. We have $\phi(W_i(f_1, P_1)) = W_i(f_2, P_2)$ for $1 \leq i \leq 5$. We work modulo W_3. There exist non-zero constants a_i so

(1) $\phi(Z_1) = a_1 Z_1 \bmod W_3$ since $\phi W_1(f_1, P_1) = W_1(f_2, P_2)$.
(2) $\phi(f_1'Z_1 + Z_2) = a_2(f_2'Z_1 + Z_2) \bmod W_3$ since $\phi W_2(f_1, P_1) = W_2(f_2, P_2)$.
(3) $\phi(Z_2) = a_3 Z_2 \bmod W_3$ since $\phi W_4(f_1, P_1) = W_4(f_2, P_2)$.
(4) $\phi(Z_1 - f_1'Z_2) = a_4(Z_1 - f_2'Z_2) \bmod W_3$ since $\phi W_5(f_1, P_1) = W_5(f_2, P_2)$.

This yields

$$a_1 f_1' Z_1 + a_3 Z_2 = a_2 f_2' Z_1 + a_2 Z_2,$$
$$a_1 Z_1 - a_3 f_1' Z_2 = a_4 Z_1 - a_4 f_2' Z_2.$$

Thus $a_1 = a_4$ and $a_3 = a_2$ so $a_1 f_1' = a_2 f_2'$ and $a_2 f_1' = a_1 f_2'$. Consequently,

$$a_1 a_2 f_1' f_1' = a_2 a_1 f_2' f_2'.$$

Since the coefficients a_i are non-zero, Assertion (3) follows.

Finally, if $\mathcal{M}_f$ is 1-affine curvature homogeneous, then necessarily $\alpha(f)$ is constant or, equivalently, $(f')^2 = c$ for some constant c. This contradicts the assumption $f'' \neq 0$. $\qquad\square$

2.5 Plane Wave Hypersurfaces of Neutral Signature (p, p)

In this section, we shall present results of Dunn, Gilkey, and Nikčević (2005), of Dunn and Gilkey (2005), of Gilkey, Ivanova, and Zhang (2002), of Gilkey, Ivanova, and Zhang (2003), of Gilkey and Zhang (2002b), and of Stavrov (2003a) concerning a family of manifolds of neutral signature (p, p). We begin by giving a very general construction that includes the manifolds of Definitions 2.3.1 and 2.3.2 as special cases:

Definition 2.5.1 Introduce coordinates $\{x_1, ..., x_p, \tilde{x}_1, ..., \tilde{x}_p\}$ on $\mathbb{R}^{2p}$. Let indices i, j, k range from 1 through p. Let $\psi_{ij} = \psi_{ji}$ be a symmetric 2-tensor field where $\psi_{ij} = \psi_{ij}(x_1, ..., x_p)$ only depends on the first p coordinates. Let $\mathcal{M} := (\mathbb{R}^{2p}, g)$ be the pseudo-Riemannian metric g of signature (p, p) where

$$g(\partial_{x_i}, \partial_{x_j}) := \psi_{ij}(x_1, ..., x_p) \quad \text{and} \quad g(\partial_{x_i}, \partial_{\tilde{x}_j}) := \delta_{ij}.$$

Introduce the distributions $\mathcal{X} := \mathrm{Span}\{\partial_{x_i}\}$ and $\tilde{\mathcal{X}} := \mathrm{Span}\{\partial_{\tilde{x}_i}\}$. Let

$$R_{ijkl;n_1...n_\nu} := \nabla^\nu R(\partial_{x_i}, \partial_{x_j}, \partial_{x_k}, \partial_{x_l}; \partial_{x_{n_1}}, ..., \partial_{x_{n_\nu}}).$$

Let $\psi_{ij/k} := \partial_{x_k}\psi_{ij}$ and let $\psi_{ij/kl} := \partial_{x_k}\partial_{x_l}\psi_{ij}$.

We begin our study with the following result:

Lemma 2.5.1 *The manifold $\mathcal{M}$ of Definition 2.5.1 is a generalized plane wave manifold of neutral signature (p, p).*

(1) The non-zero components of R and of $\nabla^\nu R$ for $\nu > 0$ are given by:

$$R(\partial_{x_i}, \partial_{x_j}, \partial_{x_k}, \partial_{x_l}) = -\tfrac{1}{2}(\psi_{il/jk} + \psi_{jk/il} - \psi_{ik/jl} - \psi_{jl/ik}),$$

$$\nabla^\nu R(\partial_{x_i}, \partial_{x_j}, \partial_{x_k}, \partial_{x_l}; \partial_{x_{n_1}}...\partial_{x_{n_\nu}}) = \partial_{x_{n_1}}...\partial_{x_{n_\nu}} R(\partial_{x_i}, \partial_{x_j}, \partial_{x_k}, \partial_{x_l}).$$

(2) If $\xi_i \in T(\mathbb{R}^{2p})$, then $\mathcal{R}(\xi_1, \xi_2)\xi_3 \in \tilde{\mathcal{X}}$; $\mathcal{R}(\xi_1, \xi_2)\xi_3 = 0$ if any $\xi_i \in \tilde{\mathcal{X}}$.

Proof. By Eq. (1.2.f), the non-zero Christoffel symbols are:

$$g(\nabla_{\partial_{x_i}}\partial_{x_j}, \partial_{x_k}) = \tfrac{1}{2}(\psi_{ik/j} + \psi_{jk/i} - \psi_{ij/k}),$$

$$\nabla_{\partial_{x_i}}\partial_{x_j} = \tfrac{1}{2}\sum_k (\psi_{ik/j} + \psi_{jk/i} - \psi_{ij/k})\partial_{\tilde{x}_k}.$$

As ψ only depends on the first p coordinates $(x_1, ..., x_p)$, ∇ has the proper triangular form relative to the ordering of the variables $(x_1, ..., x_p, \tilde{x}_1, ..., \tilde{x}_p)$. Thus $\mathcal{M}$ is a generalized plane wave manifold. Furthermore,

$$R(\partial_{x_i}, \partial_{x_j}, \partial_{x_k}, \partial_{x_l}) = g((\nabla_{\partial_{x_i}}\nabla_{\partial_{x_j}} - \nabla_{\partial_{x_j}}\nabla_{\partial_{x_i}})\partial_{x_k}, \partial_{x_l})$$

$$= \tfrac{1}{2}\sum_n g(\nabla_{\partial_{x_i}}\{\psi_{jn/k} + \psi_{kn/j} - \psi_{jk/n}\}\partial_{\tilde{x}_n}, \partial_{x_l})$$

$$- \tfrac{1}{2}\sum_n g(\nabla_{\partial_{x_j}}\{\psi_{in/k} + \psi_{kn/i} - \psi_{ik/n}\}\partial_{\tilde{x}_n}, \partial_{x_l})$$

$$= \tfrac{1}{2}\{\partial_{x_i}(\psi_{jl/k} + \psi_{kl/j} - \psi_{jk/l}) - \partial_{x_j}(\psi_{il/k} + \psi_{kl/i} - \psi_{ik/l})\}$$

$$= \tfrac{1}{2}(\psi_{jl/ki} + \psi_{kl/ji} - \psi_{jk/il} - \psi_{il/jk} - \psi_{kl/ij} + \psi_{ik/jl}).$$

This determines R; $\nabla^\nu R$ is determined for $\nu > 0$ inductively – the Christoffel symbols play no role in the computation. We raise indices to determine $\mathcal{R}$ and complete the proof of the Lemma.   □

Remark 2.5.1 If we assume that $\tilde{\mathcal{X}} = \ker(R)$, then Lemma 2.2.1 and Corollary 2.2.1 can be used to get information about the isometries and Killing vector fields of $\mathcal{M}$.

The following subfamily of examples generalize the manifolds of Definition 2.3.2; they arise as immersed hypersurface in flatspace.

Definition 2.5.2 Let $\{e_1, ..., e_p, \tilde{e}_1, ..., \tilde{e}_p, \breve{e}\}$ be a basis for $\mathbb{R}^{2p+1}$. Define an inner product on $\mathbb{R}^{2p+1}$ of signature $(p, p+1)$ whose non-zero components are:

$$\langle e_i, \tilde{e}_j \rangle := \delta_{ij}, \quad \langle \breve{e}, \breve{e} \rangle := 1.$$

Let $f \in C^\infty(\mathbb{R}^p)$. Define an embedding $\Psi : \mathbb{R}^{2p} \to \mathbb{R}^{2p+1}$ by setting:

$$\Psi(x_1, ..., x_p, \tilde{x}_1, ..., \tilde{x}_p) := \sum_{i=1}^p \left\{ x_i e_i + \tilde{x}_i \tilde{e}_i \right\} + f(x_1, ..., x_p)\breve{e}.$$

Let $g := \Psi^*\langle \cdot, \cdot \rangle$ be the induced metric on $\mathbb{R}^{2p}$;

$$g(\partial_{x_i}, \partial_{x_i}) = \partial_{x_i} f \cdot \partial_{x_j} f \quad \text{and} \quad g(\partial_{x_i}, \partial_{\tilde{x}_j}) = \delta_{ij}.$$

Let $\mathcal{M} := (\mathbb{R}^{2p}, g)$ be the associated generalized plane wave manifold of signature (p,p); this corresponds to taking

$$\psi_{ij} = \partial_{x_i} f \cdot \partial_{x_j} f.$$

Let $H_{ij} := \partial_{x_i} \partial_{x_j} f$ be the Hessian of f; H is a symmetric bilinear form where $1 \leq i, j \leq p$. Let L be the second fundamental form of the embedding.

The tensor H is defined on $\mathcal{X}$ while L is defined on $\mathcal{X} \oplus \tilde{\mathcal{X}} = T\mathbb{R}^{2p}$. We show that L is supported on $\mathcal{X}$ and that the restriction of L to $\mathcal{X}$ is given by H in the following result:

Lemma 2.5.2 *The manifold $\mathcal{M}$ of Definition (2.5.2) is a generalized plane wave manifold of signature (p,p).*

(1) $L(\partial_{x_i}, \partial_{x_j}) = H_{ij}$ and $L(\partial_{x_i}, \partial_{\tilde{x}_j}) = L(\partial_{\tilde{x}_i}, \partial_{\tilde{x}_j}) = 0$.
(2) $R(\partial_{x_i}, \partial_{x_j}, \partial_{x_k}, \partial_{x_l}) = H_{il}H_{jk} - H_{ik}H_{jl}$.
(3) $\nabla^\nu R(\partial_{x_i}, \partial_{x_j}, \partial_{x_k}, \partial_{x_l}; \partial_{x_{n_1}}, ..., \partial_{x_{n_\nu}}) = \partial_{x_{n_1}}...\partial_{x_{n_\nu}} R(\partial_{x_i}, \partial_{x_j}, \partial_{x_k}, \partial_{x_l})$.

Proof. The unit normal to the hypersurface is

$$\nu := -\partial_{x_1} f \tilde{e}_1 - ... - \partial_{x_p} f \tilde{e}_p + \check{e}.$$

This determines the second fundamental form and thereby the curvature tensor. We use induction and the observation that the Christoffel symbols play no role in the computation of $\nabla^k R$ for this example to determine $\nabla^\nu R$ for $\nu > 0$; Assertion (2) can also, of course, be derived directly from Lemma 2.5.1. $\qquad\square$

The following result summarizes some of the spectral properties of the curvature tensor for these manifolds. We suppose that (r, s) is admissible; in this context this means $0 \leq r \leq p$, $0 \leq s \leq p$, and $1 \leq r + s \leq 2p - 1$.

Theorem 2.5.1 *Let $\mathcal{M}$ be as in Definition 2.5.2 with $\det(H) \neq 0$.*

(1) *If $p = 2$, or if $p \geq 3$ and if H is definite, then $\mathcal{M}$ is spacelike and timelike Jordan Osserman.*
(2) *If $p \geq 3$ and if H is indefinite, then $\mathcal{M}$ is neither spacelike Jordan Osserman nor timelike Jordan Osserman.*
(3) *Let H be definite. Then $\mathcal{M}$ is Jordan Osserman of type (a, b) if and only if $(a, b) = (0, r)$, $(a, b) = (p, p - r)$, $(a, b) = (r, 0)$, or $(a, b) = (p - r, p)$ where $0 < r \leq p$.*
(4) *$\mathcal{M}$ is spacelike and timelike Jordan Ivanov–Petrova.*
(5) *$\mathcal{M}$ is not mixed Jordan Ivanov–Petrova.*

(6) If $2 \le k \le p$, then M is k-spacelike and k-timelike Jordan Stanilov.

In Assertion (3), we may think of the values (r, s) as belonging to a square since $0 \le r \le p$ and $0 \le s \le p$; M is Jordan Osserman of type (r, s) if and only if (r, s) is admissible and if (r, s) lies on the boundary of the square.

Let W be the Weyl conformal tensor. Since M is Ricci flat, $W = R$. Thus we have the immediate consequence of Theorem 2.5.1:

Corollary 2.5.1 *Let M be as in Definition 2.5.2 with $\det(H) \neq 0$.*

(1) If $p = 2$, or if $p \ge 3$ and if H is definite, then M is conformally spacelike and conformally timelike Jordan Osserman.

(2) If $p \ge 3$ and if H is indefinite, then M is neither conformally spacelike Jordan Osserman nor conformally timelike Jordan Osserman.

(3) Let H be definite. Then M is conformally Jordan Osserman of type (a, b) if and only if $(a, b) = (0, r)$, $(a, b) = (p, p - r)$, $(a, b) = (r, 0)$, or $(a, b) = (p - r, p)$ where $0 < r \le p$.

(4) M is conformally spacelike and conformally timelike Jordan Ivanov–Petrova.

(5) M is not conformally mixed Jordan Ivanov–Petrova.

(6) If $2 \le k \le p$, then M is conformally k-spacelike and k-timelike Jordan Stanilov.

If H is invertible, set

$$\alpha := \left| \sum_{I,J} H^{i_1 j_1} H^{i_2 j_2} H^{i_3 j_3} H^{i_4 j_4} H^{i_5 j_5} \nabla R_{i_1 i_2 i_3 i_4 ; i_5} \nabla R_{j_1 j_2 j_3 j_4 ; j_5} \right| .$$

Let $M_{r,s}$ be defined by the function

$$f_{r,s} := -x_1^2 - \dots - x_r^2 + x_{r+1}^2 + \dots + x_p^2 .$$

Some curvature homogeneity properties of this family of manifolds are described in the following result:

Theorem 2.5.2 *Let M be as in Definition 2.5.2. Assume that H is non-degenerate of signature (r, s). Let $p \ge 3$.*

(1) M is 0-modeled on the indecomposable symmetric space $M_{r,s}$.

(2) If ϕ is an isomorphism of $\mathfrak{M}_0(M, P)$, then $\phi^* L = \pm L$.

(3) α is an invariant of $\mathfrak{M}_1(M, P)$.

(4) M is not 1-curvature homogeneous for generic f.

Here is a brief guide to the remainder of our discussion of these manifolds. In Section 2.5.1, we will establish Theorem 2.5.1 and in Section 2.5.2, we will establish Theorem 2.5.2.

2.5.1 *Spectral properties of the curvature tensor*

Let $\rho_{\mathcal{X}}$ and $\rho_{\tilde{\mathcal{X}}}$ be the projections of $T_P M$ on the distributions $\mathcal{X}$ and $\tilde{\mathcal{X}}$:

$$\rho_{\mathcal{X}}\partial_{x_i} = \partial_{x_i}, \quad \rho_{\mathcal{X}}\partial_{\tilde{x}_i} = 0, \quad \rho_{\tilde{\mathcal{X}}}\partial_{x_i} = 0, \quad \rho_{\tilde{\mathcal{X}}}\partial_{\tilde{x}_i} = \partial_{\tilde{x}_i}.$$

Both R and L vanish if any entry is from $\tilde{\mathcal{X}}$. By Lemma 2.5.1, the Jacobi operator and the skew-symmetric curvature operator are 2-step nilpotent:

$$\mathcal{J}(\cdot) : \mathcal{X} \to \tilde{\mathcal{X}} \to 0 \quad \text{and} \quad \mathcal{R}(\cdot) : \mathcal{X} \to \tilde{\mathcal{X}} \to 0. \tag{2.5.a}$$

We begin our study with the following result which is motivated by work of Stavrov (2003a).

Lemma 2.5.3 *Let $\mathcal{M}$ be as in Definition 2.5.2. Assume that H is non-degenerate. Let Z be a tangent vector. Let $X := \rho_{\mathcal{X}} Z \in \mathcal{X}$.*

(1) If $X = 0$, then $\mathcal{J}(Z) = 0$.
(2) If $X \neq 0$ and if $H(X, X) = 0$, then $\mathrm{Rank}\{\mathcal{J}(Z)\} = 1$.
(3) If $X \neq 0$ and if $H(X, X) \neq 0$, then $\mathrm{Rank}\{\mathcal{J}(Z)\} = p - 1$.

Proof. Let L be the second fundamental form. We have

$$g(\mathcal{J}(Z)Y, W) = L(Z, Z)L(Y, W) - L(Z, Y)L(YZ, W).$$

Furthermore, $L = 0$ on $\tilde{\mathcal{X}}$ and $H = L|_{\mathcal{X}}$. By Lemma 2.5.1, $\mathcal{J}(Z) = \mathcal{J}(X)$. Thus if $X = 0$, $\mathcal{J}(Z) = 0$. This proves Assertion (1). We have $\mathcal{J}(X)\partial_{\tilde{x}_i} = 0$ and $\mathcal{J}(X)X = 0$. Thus

$$\mathrm{Rank}(\mathcal{J}(X)) \leq p - 1.$$

Suppose that $H(X, X) \neq 0$. Since H is non-degenerate, we can choose a basis $\{X_1, ..., X_p\}$ for $\mathcal{X}(P)$ so $H(X_i, X_j) = \varepsilon_i \delta_{ij}$, where $\varepsilon_i \neq 0$ for $1 \leq i \leq p$ and so that $X = X_1$. Let $i, j \geq 2$. We show that $\mathrm{Rank}(\mathcal{J}(X)) = p - 1$ by computing:

$$g(\mathcal{J}(X)X_i, X_j) = L(X, X)L(X_i, X_j) - L(X, X_i)L(X, X_j) = \varepsilon_i \varepsilon_1 \delta_{ij}.$$

Suppose finally that $H(X, X) = 0$. We can choose a basis $\{X_1, ..., X_p\}$ for $\mathcal{X}(P)$ so $X = X_1$ and so

$$H(X_1, X_2) = 1 \quad \text{and} \quad H(X_1, X_i) = 0 \quad \text{for} \quad i \neq 2.$$

We show that $\mathrm{Rank}(\mathcal{J}(X)) = 1$ by computing similarly that

$$g(\mathcal{J}(X)X_2, X_i) = -\delta_{2i} \quad \text{and} \quad g(\mathcal{J}(X)X_i, X_j) = 0 \text{ for } i \neq 2.$$

The desired result now follows. $\qquad\qquad\qquad\qquad\qquad\qquad\qquad\square$

Proof of Theorem 2.5.1 (1) and (2). Fix $P \in \mathbb{R}^{2p}$. Let H be non-degenerate. Decompose

$$Z_i = X_i + \tilde{X}_i$$

for $X_i := \rho_{\mathcal{X}} Z_i \in \mathcal{X}(P)$ and $\tilde{X}_i := \rho_{\tilde{\mathcal{X}}} Z_i \in \tilde{\mathcal{X}}(P)$. Then:

$$\begin{aligned} &g(\mathcal{J}(Z_1)Z_2, Z_3) \\ &= L(Z_1, Z_1)L(Z_2, Z_3) - L(Z_1, Z_2)L(Z_1, Z_3) \\ &= H(X_1, X_1)H(X_2, X_3) - H(X_1, X_2)H(X_1, X_3). \end{aligned}$$

Suppose $Z_1 \in S^{\pm}(T_P M, g)$. Since $\tilde{\mathcal{X}}$ is totally isotropic, $X_1 \neq 0$. Because $\mathcal{J}(X_1)^2 = 0$, the rank of $\mathcal{J}(X_1)$ determines the Jordan normal form of $\mathcal{J}(Z_1)$.

If H is definite, Lemma 2.5.3 shows that $\mathrm{Rank}(\mathcal{J}(Z_1)) = p - 1$ so $\mathcal{M}$ is timelike and spacelike Jordan Osserman. If $p = 2$, then Lemma 2.5.3 implies that $\mathrm{Rank}(\mathcal{J}(Z_1)) = 1$ and again $\mathcal{M}$ is timelike and spacelike Jordan Osserman. Finally, if H is indefinite and if $p > 2$, then Alternatives (2) and (3) in Lemma 2.5.3 are possible and distinct so $\mathcal{M}$ is neither timelike nor spacelike Jordan Osserman. This proves Assertions (1) and (2) of Theorem 2.5.1. $\qquad\qquad\qquad\qquad\qquad\qquad\qquad\square$

Before establishing Theorem 2.5.1 (3), we must generalize Lemma 2.5.3.

Lemma 2.5.4 *Let $\pi := \mathrm{Span}\{Z_1, ..., Z_r\}$, let $\mathcal{J} := \mathcal{J}(Z_1) + ... + \mathcal{J}(Z_r)$, and let H be positive definite. Then*

$$\mathrm{Rank}(\mathcal{J}) = \begin{cases} 0 & \text{if} \quad \dim(\rho_{\mathcal{X}}\pi) = 0, \\ p - 1 & \text{if} \quad \dim(\rho_{\mathcal{X}}\pi) = 1, \\ p & \text{if} \quad \dim(\rho_{\mathcal{X}}\pi) \geq 2. \end{cases}$$

Proof. Let $X_i := \rho_{\mathcal{X}}(Z_i)$; since $\mathcal{J}(Z_i) = \mathcal{J}(X_i)$, $X_i = 0$ implies that $\mathcal{J}(Z_i) = 0$. It is convenient to consider the associated bilinear forms:

$$\Xi_i(W_1, W_2) = g(\mathcal{J}(Z_i)W_1, W_2) = R(W_1, Z_i, Z_i, W_2)$$
$$= H(X_i, X_i)H(\rho_{\mathcal{X}}W_1, \rho_{\mathcal{X}}W_2) - H(X_i, \rho_{\mathcal{X}}W_1)H(X_i, \rho_{\mathcal{X}}W_2),$$
$$\Xi := \sum_i \Xi_i.$$

Since H is positive definite, $\Xi_i(W, W)$ and $\Xi(W, W)$ are positive semi-definite quadratic forms. If $X_i \neq 0$, $\mathrm{Rank}\{\mathcal{J}(X_i)\} = p - 1$ by Lemma 2.5.3. Furthermore, if $X \in \mathcal{X}$, then

$$\Xi_i(X, X) = 0 \quad \Leftrightarrow \quad X \in \mathrm{Span}\{X_i\}.$$

If $\dim(\rho_{\mathcal{X}}\pi) = 0$, then $X_i = 0$ for all i and $\mathcal{J} = 0$. This is the first possibility of the Lemma. Suppose $\dim(\rho_{\mathcal{X}}\pi) = 1$. We suppose without loss of generality that $X_1 \neq 0$; let $X_i = c_i X_1$ for $i \geq 1$. Then $\mathcal{J} = (\sum_i c_i^2)\mathcal{J}(X_1)$ has rank $p - 1$ which is the second possibility.

Suppose $\dim(\rho_{\mathcal{X}}\pi) \geq 2$. We suppose, without loss of generality, that $\{X_1, X_2\}$ is a linearly independent set. One has that

$$\mathrm{Range}(\mathcal{J}) \subseteq \tilde{\mathcal{X}} \quad \text{so} \quad \mathrm{Rank}(\mathcal{J}) \leq \dim(\tilde{\mathcal{X}}) = p.$$

On the other hand, if $X \in \mathcal{X}$ and if $\Xi(X, X) = 0$, then $\Xi_i(X, X) = 0$ for $1 \leq i \leq r$. In particular

$$X \in \mathrm{Span}\{X_1\} \cap \mathrm{Span}\{X_2\} = \{0\}.$$

This implies Ξ is positive definite on $\mathcal{X}$ and hence $\mathrm{Rank}(\mathcal{J}) = p$. $\qquad\square$

Proof of Theorem 2.5.1 (3). Let $\{Z_1, ..., Z_r\}$ be an orthonormal basis for a spacelike subspace $\pi \in \mathrm{Gr}_{0,r}(\mathcal{M})$. Let $X_i := \rho_{\mathcal{X}}Z_i$. Because π is spacelike, $\pi \cap \tilde{\mathcal{X}} = \{0\}$. Since $\ker \rho_{\mathcal{X}} = \tilde{\mathcal{X}}$, $\dim(\rho_{\mathcal{X}}\pi) = r$. We have

$$\mathcal{J}(\pi) = \mathcal{J}(X_1) + ... + \mathcal{J}(X_r).$$

Since $\mathcal{J}(\pi)^2 = 0$, the Jordan normal form of $\mathcal{J}(\pi)$ is determined by the rank. We apply Lemma 2.5.4 to see that the rank is $p - 1$ if $r = 1$ and that the rank is p if $r > 1$. This shows $\mathcal{M}$ is Jordan Osserman of type $(0, r)$. One shows similarly that $\mathcal{M}$ is Jordan Osserman of type $(r, 0)$. Theorem 1.9.2 then shows that $\mathcal{M}$ is Jordan Osserman of types $(p, p - r)$ and $(p - r, p)$. Thus $\mathcal{M}$ is Jordan Osserman of type (r, s) if (r, s) is on the boundary of the square, or, equivalently, if there exists t with $1 \leq t \leq p - 1$ so that

$$(r, s) = (0, t), \quad (r, s) = (p, p - t), \quad (r, s) = (t, 0), \quad (r, s) = (p - t, p).$$

We complete the proof by showing $\mathcal{M}$ is not Jordan Osserman for the remaining admissible values of (r, s).

Consider the elements:

$$X_i := \partial_{x_i} - \tfrac{1}{2} \sum_j g(\partial_{x_i}, \partial_{x_j}) \partial_{\tilde{x}_j}.$$

We note $\rho_{\mathcal{X}} X_i = \partial_{x_i}$ and $\{X_1, ..., X_p, \partial_{\tilde{x}_1}, ..., \partial_{\tilde{x}_p}\}$ is a *hyperbolic basis*; this means that:

$$\begin{aligned}
(X_i, X_j) &= (\partial_{\tilde{x}_i}, \partial_{\tilde{x}_j}) = 0 \quad \text{and} \\
(X_i, \partial_{\tilde{x}_j}) &= \delta_{ij} \quad \text{for} \quad 1 \leq i, j \leq p.
\end{aligned} \tag{2.5.b}$$

Let $0 < r < p$ and $0 < s < p$. We must show that $\mathcal{M}$ is not Jordan Osserman of type (r, s). By Theorem 1.9.2, we may assume $r + s \leq p$. We assume $0 < r \leq s < p$ as the situation when $0 < s \leq r < p$ is similar. We distinguish two cases:

1) Suppose $s = 1$. Then $r = 1$. We use Eq. (2.5.b) to define the following subspaces of type $(1, 1)$ with the indicated orthonormal bases and Jacobi operators:

$$\begin{aligned}
\pi_1 &:= \mathrm{Span}\{X_1 - \tfrac{1}{2}\partial_{\tilde{x}_1}, X_1 + \tfrac{1}{2}\partial_{\tilde{x}_1}\}, \\
\mathcal{J}(\pi_1) &= -\mathcal{J}(X_1) + \mathcal{J}(X_1) = 0, \\
\pi_2 &:= \mathrm{Span}\{\varepsilon X_1 - \tfrac{1}{2}\varepsilon^{-1}\partial_{\tilde{x}_1}, X_2 + \tfrac{1}{2}\partial_{\tilde{x}_2}\}, \\
\mathcal{J}(\pi_2) &= -\varepsilon^2 \mathcal{J}(X_1) + \mathcal{J}(X_2).
\end{aligned}$$

As $\mathcal{J}(\pi_1) = 0$, $\mathrm{Rank}(\mathcal{J}(\pi_1)) = 0$. Since

$$\mathrm{Rank}(\mathcal{J}(X_2)) = \mathrm{Rank}(\mathcal{J}(\partial_{x_2})) = p - 1,$$

$\mathrm{Rank}(\mathcal{J}(\pi_2)) \geq \mathrm{Rank}(\mathcal{J}(X_2)) \geq p - 1$ if ε is small. Consequently, $\mathcal{M}$ is not Jordan Osserman of type $(1, 1)$.

2) Suppose $1 \leq r \leq s < p$, $s \geq 2$, and $r + s \leq p$; necessarily $p \geq 3$. For $\alpha \neq 0$ and $\beta \neq 0$, we use Eq. (2.5.b) to define timelike subspaces $\pi^-(\alpha)$ and spacelike subspaces $\pi^+(\beta)$ with the indicated orthonormal bases and

Jacobi operators:

$$\pi^-(\alpha) := \mathrm{Span}\{\alpha X_1 - \tfrac{1}{2}\alpha^{-1}\partial_{\tilde{x}_1}, ..., \alpha \tilde{X}_r - \tfrac{1}{2}\alpha^{-1}\partial_{\tilde{x}_r}\},$$
$$\mathcal{J}_\alpha^- := -\alpha^2\{J(\partial_{x_1}) + ... + J(\partial_{x_r})\},$$
$$\pi^+(\beta) := \mathrm{Span}\{\beta \tilde{X}_{r+1} + \tfrac{1}{2}\beta^{-1}\partial_{\tilde{x}_{r+1}}, ..., \beta \tilde{X}_{r+s} + \tfrac{1}{2}\beta^{-1}\partial_{\tilde{x}_{r+s}}\},$$
$$\mathcal{J}_\beta^+ := \beta^2\{J(\partial_{x_{r+1}}) + ... + J(\partial_{x_{r+s}})\}.$$

Let $\pi(\alpha,\beta) := \pi^-(\alpha) \oplus \pi^+(\beta) \in \mathrm{Gr}_{r,s}(\mathcal{M})$. The associated Jacobi operator $J(\pi(\alpha,\beta)) = \mathcal{J}_\alpha^- + \mathcal{J}_\beta^+$. Since $\mathrm{Range}(J) \subset \tilde{\mathcal{X}}$, one sees that

$$\mathrm{Rank}(J(\pi(\alpha,\beta))) \leq p \quad \text{for all} \quad \alpha, \beta.$$

We have $\rho_{\mathcal{X}}(X_i) = \partial_{x_i}$. We apply Lemma 2.5.4; $J(\cdot)$ is supported on $\mathcal{X}$. Since $s \geq 2$, the bilinear form defined by $\mathcal{J}_\beta^+$ is positive semi-definite of rank p. Thus $\mathrm{Rank}\{J(\pi(1,\beta))\} = p$ for β large. Note that the bilinear form defined by $\mathcal{J}_\alpha^-$ is negative semi-definite of rank at least $p - 1$. Thus for α large, $J(\pi(\alpha,1))$ determines a quadratic form of signature (u,v) for $u \geq p - 1$ and $u + v \leq p$. Thus by continuity, there must exist (α,β) with $\alpha \neq 0$ and $\beta \neq 0$ so $J(\pi(\alpha,\beta))$ determines a degenerate quadratic form on $\mathcal{X}$. For such values of (α,β), $\mathrm{Rank}\{J(\pi(\alpha,\beta))\} < p$. This shows that $\mathrm{Rank}\{J(\pi(\alpha) \oplus \pi(\beta))\}$ is not constant and hence $\mathcal{M}$ is not Jordan Osserman of type (r,s). $\qquad\square$

Proof of Theorem 2.5.1 (4). We study the skew-symmetric curvature operator using arguments of Gilkey and Zhang (2002b). Let $\{Z_1, Z_2\}$ be an orthonormal basis for a non-degenerate 2-plane π. Since $\mathcal{R}(\pi)^2 = 0$, $\mathrm{Rank}(\mathcal{R}(\pi))$ determines the Jordan normal form of $\mathcal{R}(\pi)$. We expand

$$Z_\nu = X_\nu + \tilde{X}_\nu,$$
$$\mathcal{R}(\pi) = \mathcal{R}(X_1, X_2), \qquad\qquad (2.5.c)$$
$$g(\mathcal{R}(\pi)Z_3, Z_4) = H(X_1, X_4)H(X_2, X_3) - H(X_1, X_3)H(X_2, X_4).$$

If π is spacelike or timelike, then π contains no null vectors and thus $\{X_1, X_2\}$ are linearly independent vectors. We extend the set $\{X_1, X_2\}$ to a basis $\{X_1, ..., X_p\}$ for $\mathcal{X}(P)$. Since H is non-degenerate, we can choose a basis $\{X_1^*, ..., X_p^*\}$ for $\mathcal{X}(P)$ so $H(X_i, X_j^*) = \delta_{ij}$. By Eq. (2.5.c),

$$g(\mathcal{R}(X_1, X_2)X_i^*, X_j^*) = \begin{cases} 1 & \text{if } i = 2, j = 1, \\ -1 & \text{if } i = 1, j = 2, \\ 0 & \text{otherwise.} \end{cases}$$

It now follows that

$$\mathrm{Rank}(\mathcal{R}(\pi)) = \mathrm{Rank}(\mathcal{R}(X_1, X_2)) = 2\,. \qquad (2.5.\mathrm{d})$$

Thus $\mathcal{M}$ is spacelike Jordan Ivanov–Petrova and timelike Jordan Ivanov–Petrova.

To see that $\mathcal{M}$ is not mixed Ivanov–Petrova, consider the 2-planes:

$$\pi_1 := \mathrm{Span}\{\partial_{x_1}, \partial_{\tilde{x}_1}\}, \quad \text{and}$$
$$\pi_2(\varepsilon) := \mathrm{Span}\{\varepsilon\partial_{x_1} + \varepsilon^{-1}\partial_{\tilde{x}_1}, \varepsilon\partial_{x_2} - \varepsilon^{-1}\partial_{\tilde{x}_2}\},$$

respectively, where ε is a real parameter. The matrices giving the induced inner products on π_1 and $\pi_2(\varepsilon)$ are given by:

$$A_1 := \begin{pmatrix} 0 & 1 \\ 1 & g(\partial_{x_1}, \partial_{x_1}) \end{pmatrix}, \quad \text{and}$$

$$A_2 := \begin{pmatrix} 2 + \varepsilon^2 g(\partial_{x_1}, \partial_{x_1}) & \varepsilon^2 g(\partial_{x_1}, \partial_{x_2}) \\ \varepsilon^2 g(\partial_{x_1}, \partial_{x_2}) & -2 + \varepsilon^2 g(\partial_{x_2}, \partial_{x_2}) \end{pmatrix}$$

respectively. Since $\det(A_1) = -1$ and $\det(A_2) = -4 + O(\varepsilon^2)$, π_1 and $\pi_2(\varepsilon)$ are mixed 2-planes for ε small. Since

$$\mathcal{R}(\pi_1) = 0 \quad \text{and} \quad \mathcal{R}(\pi_2(\varepsilon)) = c(\varepsilon)\mathcal{R}(\partial_{x_1}, \partial_{x_2}) \neq 0,$$

$\mathcal{R}(\pi_1)$ and $\mathcal{R}(\pi_2(\varepsilon))$ are not Jordan equivalent and hence $\mathcal{M}$ is not mixed Jordan Ivanov–Petrova. $\qquad\square$

Proof of Theorem 2.5.1 (6). We recall the definition of Section 1.8.1. If π is a spacelike or timelike k-plane, let $\{e_1, ..., e_k\}$ be an orthonormal basis for π. The *Stanilov operator* is defined by setting:

$$\Theta(\pi) := \sum_{i,j} \mathcal{R}(e_i, e_j)\mathcal{R}(e_i, e_j)\,.$$

Since $\mathcal{R}(\pi_1)\mathcal{R}(\pi_2) = 0$ for any 2-planes π_i, $\Theta(\pi) = 0$ and hence these manifolds are k-Stanilov for any k. $\qquad\square$

2.5.2 Curvature homogeneity

Proof of Theorem 2.5.2. Assume that H is a non-degenerate bilinear form of signature (r, s) at every point P of $\mathbb{R}^{2p}$. Fix $P \in \mathbb{R}^{2p}$. Diagonalize the quadratic form H at P to choose $\bar{X}_i = \sum_j a_{ij}\partial_{x_j}$ so

$$H(\bar{X}_i, \bar{X}_j) = \begin{cases} 0 & \text{if} \quad i \neq j, \\ -1 & \text{if} \quad i \leq r, \\ +1 & \text{if} \quad i > r. \end{cases}$$

Let a^{ij} be the inverse matrix. Set

$$\tilde{X}_i := \sum_j a^{ji}\partial_{\tilde{x}_j} \quad \text{and} \quad X_i := \bar{X}_i - \sum_j \tfrac{1}{2}g(\bar{X}_i, \bar{X}_j)\tilde{X}_j.$$

The non-zero components of g and of R are then given by:

$$g(X_i, \tilde{X}_i) = 1 \quad \text{and} \quad R(X_i, X_j, X_k, X_l) = \varepsilon_i\varepsilon_j\{\delta_{il}\delta_{jk} - \delta_{ik}\delta_{jl}\}$$

where $\varepsilon_i = -1$ for $i \leq r$ and $\varepsilon_i = +1$ for $i > r$. This provides a normalized basis for T_PM and shows that (T_PM, g_P, R_P) is determined by (r, s). This shows that $\mathcal{M}_{r,s}$ is a 0-model for $\mathcal{M}$. Since $R_P = R_L$ where $\ker(L) = \tilde{\mathcal{X}}$ is totally isotropic, (T_PM, g_P, R_P) is indecomposable by Lemma 1.6.3. By Lemma 2.5.1, $\mathcal{M}_{r,s}$ is a symmetric space. This completes the proof of Assertion (1).

We have $R = A_L$ is determined by the quadratic form L. If ϕ is an isomorphism of $\mathfrak{M}_0(\mathcal{M}, P)$, then $\phi^*A_L = A_L$. Since $\text{Rank}(L) \geq 3$, Lemma 1.6.5 shows that $\phi^*L = \pm L$; Assertion (2) now follows. We have

$$\ker(R) = \{\xi \in T_PM : R(\xi, \xi_1, \xi_2, \xi_3) = 0 \quad \forall \quad \xi_i \in T_PM\} = \tilde{\mathcal{X}}.$$

Let π be the natural projection from T_PM to $\bar{V} := T_PM/\ker(R)$. There are structures $\bar{L}$ and $\bar{A}_\nu$ on $\bar{V}$ so that

$$L = \pi^*\bar{L} \quad \text{and} \quad \nabla^\nu R_P = \pi^*\bar{A}_\nu.$$

We have $\bar{A} = A_{\bar{L}}$. We have $\bar{\phi}^*\bar{L} = \pm\bar{L}$. We can use $\bar{L}$ to define an inner product on $\bar{V}$. Assertion (3) follows from the observation that

$$\alpha^2 = \|\bar{A}_1\|_{\bar{L}}^2.$$

To prove Assertion (4), it suffices to construct f so that $\alpha(f, \cdot)$ is constant on no open subset of $\mathbb{R}^p$; the fact that such f are generic will then follow using standard arguments. We suppose H positive definite in the interests of simplicity. Let $f_{;i} = \partial_{x_i}f$, $f_{;ij} := \partial_{x_i}\partial_{x_j}f$, and so forth. We use

Lemma 2.5.2 to see:

$$R(\partial_{x_i}, \partial_{x_j}, \partial_{x_k}, \partial_{x_l}) = f_{;il}f_{;jk} - f_{;ik}f_{;jl},$$
$$\nabla R(\partial_{x_i}, \partial_{x_j}, \partial_{x_k}, \partial_{x_l}; \partial_{x_n}) = \partial_{x_n}\{f_{;il}f_{;jk} - f_{;ik}f_{;jl}\}.$$

Let $\Theta = \Theta(x_1)$ be a smooth function on $\mathbb{R}$ so that $|\Theta_{;11}| < 1$. Set

$$f(x) := \tfrac{1}{2}\{x_1^2 + ... + x_p^2\} + \Theta(x_1).$$

We may then compute, up to the usual $\mathbb{Z}_2$ symmetries, that the non-zero components of R and of ∇R are:

$$\begin{aligned}
R(\partial_{x_1}, \partial_{x_i}, \partial_{x_i}, \partial_{x_1}) &= 1 + \Theta_{;11} &&\text{for}\quad 2 \le i \le p, \\
R(\partial_{x_i}, \partial_{x_j}, \partial_{x_j}, \partial_{x_i}) &= 1 &&\text{for}\quad 2 \le i < j \le p, \\
\nabla R(\partial_{x_1}, \partial_{x_i}, \partial_{x_i}, \partial_{x_1}; \partial_{x_1}) &= \Theta_{;111} &&\text{for}\quad 2 \le i \le p.
\end{aligned}$$

Consequently after taking into account to normalize the basis for the tangent bundle suitably, we have

$$\alpha_f = \frac{4(p-1)\Theta_{;111}^2}{(1 + \Theta_{;11})^3}.$$

It is now clear the metric g is not locally homogeneous for generic Θ. $\square$

2.6   Plane Wave Manifolds with Flat Factors

We follow the discussion of Gilkey, Ivanova, and Zhang (2002) and of Gilkey, Ivanova, and Zhang (2003) to extend the examples discussed in Section 2.5 with a flat factor; these manifolds have interesting spectral properties.

Definition 2.6.1   Let $f \in C^\infty(\mathbb{R}^p)$. Let $(x_1, ..., x_p, \tilde{x}_1, ..., \tilde{x}_p, y_1, ..., y_{a+b})$ be coordinates on $\mathbb{R}^{2p+a+b}$. Let $\mathcal{N} := (\mathbb{R}^{2p+a+b}, g)$ be the manifold of signature $(p + a, p + b)$ where

$$g(\partial_{x_i}, \partial_{x_i}) := \partial_{x_i}f \cdot \partial_{x_j}f, \quad g(\partial_{x_i}, \partial_{\tilde{x}_i}) := 1$$
$$g(\partial_{y_i}, \partial_{y_i}) := \begin{cases} -1 & \text{if} & 1 \le i \le a, \\ +1 & \text{if} & a+1 \le i \le a+b. \end{cases}$$

Then $\mathcal{N} := \mathcal{M} \times \mathbb{R}^{(a,b)}$ is the isometric product of the manifold $\mathcal{M}$ of Definition 2.5.2 with a flat factor $\mathbb{R}^{(a,b)}$ of signature (a, b) where

$$\mathcal{M} := (\mathbb{R}^{2p}, g|_{\mathbb{R}^{2p}}).$$

If $a = b = 0$, then $\mathcal{N} = \mathcal{M}$; these manifolds were treated in Section 2.5. The new phenomena arise when $a > 0$, $b = 0$, when $a = 0$, $b > 0$, and when $a > 0$, $b > 0$. It is clear that these are generalized plane wave manifolds.

Theorem 2.6.1 *Let $\mathcal{N}$ be as in Definition 2.6.1. Let $H_{ij} := \partial_{x_i}\partial_{x_j} f$ for $1 \leq i, j \leq p$. Assume that H is positive definite. Then*

(1) $\mathcal{N}$ is not mixed Jordan Ivanov–Petrova.
(2) Suppose that $a > 0$ and $b = 0$. Then $\mathcal{N}$ is:

    (a) Not timelike Jordan Osserman.
    (b) Not timelike Jordan Ivanov–Petrova.
    (c) Spacelike Jordan Osserman and spacelike Jordan Ivanov–Petrova.

(3) Suppose that $a = 0$ and $b > 0$. Then $\mathcal{N}$ is:

    (a) Timelike Jordan Osserman and timelike Jordan Ivanov–Petrova.
    (b) Not spacelike Jordan Osserman.
    (c) Not spacelike Jordan Ivanov–Petrova.

(4) Suppose that $a > 0$ and $b > 0$. Then $\mathcal{N}$ is:

    (a) Not timelike Jordan Osserman.
    (b) Not timelike Jordan Ivanov–Petrova.
    (c) Not spacelike Jordan Osserman.
    (d) Not spacelike Jordan Ivanov–Petrova.

(5) $\mathcal{N}$ is Jordan Osserman

    (a) of types $(r, 0)$ and $(p + a - r, p + b)$ if $a = 0$ and if $0 < r \leq p$;
    (b) of types $(0, s)$ and $(p + a, p + b - s)$ if $b = 0$ and if $0 < s \leq p$;
    (c) of types $(r, 0)$ and $(p + a - r, p + b)$ if $a > 0$ and if $a + 2 \leq r \leq p + a$;
    (d) of types $(0, s)$ and $(p + a, p + b - s)$ if $b > 0$ and if $b + 2 \leq s \leq p + b$.

(6) $\mathcal{N}$ is not Jordan Osserman of type (r, s) otherwise.

Remark 2.6.1 This shows that timelike Jordan Osserman (respectively timelike Jordan Ivanov–Petrova) and spacelike Jordan Osserman (respectively spacelike Jordan Ivanov–Petrova) are different notions. It also provides additional examples showing the dependence upon (r, s) in the notion of Jordan Osserman of varying types.

Proof of Theorem 2.6.1 (1)-(4). Let $\mathcal{N} := \mathcal{M} \times \mathbb{R}^{(a,b)}$. We decompose

$$T\mathcal{N} := \mathcal{X} \oplus \tilde{\mathcal{X}} \oplus \tilde{\mathcal{Y}} \quad \text{for}$$

$$\mathcal{X} := \mathrm{Span}\{\partial_{x_i}\}, \quad \tilde{\mathcal{X}} := \mathrm{Span}\{\partial_{\tilde{x}_i}\}, \quad \mathcal{Y} := \mathrm{Span}\{\partial_{y_j}\}.$$

Let $\pi_{\mathcal{X}}$, $\pi_{\tilde{\mathcal{X}}}$, and $\pi_{\mathcal{Y}}$ be the associated projections. We set $\pi_{\mathcal{M}} := \pi_{\mathcal{X}} + \pi_{\tilde{\mathcal{X}}}$. Let $\mathcal{R}$ and $\mathcal{R}_{\mathcal{M}}$ be the curvature tensors on $\mathcal{N}$ and $\mathcal{M}$ respectively. Then

$$\mathcal{R} = \pi_{\mathcal{M}}^* \mathcal{R}_{\mathcal{M}}.$$

One may generalize Eq. (2.5.a) to have:

$$\begin{aligned}
\mathcal{J}(\cdot) : \mathcal{X} \to \tilde{\mathcal{X}}, \quad & \mathcal{J}(\cdot) : \tilde{\mathcal{X}} \to 0, \quad & \mathcal{J}(\cdot) : \mathcal{Y} \to 0, \\
\mathcal{R}(\cdot) : \mathcal{X} \to \tilde{\mathcal{X}}, \quad & \mathcal{R}(\cdot) : \tilde{\mathcal{X}} \to 0, \quad & \mathcal{R}(\cdot) : \mathcal{Y} \to 0.
\end{aligned} \tag{2.6.a}$$

It now follows that

$$\mathcal{J}(Z)^2 = 0 \quad \text{and} \quad \mathcal{R}(Z_1, Z_2)^2 = 0.$$

Consequently the Jordan normal form of these operators is determined by their rank. Assertion (1) follows from the proof of Theorem 2.5.1 given previously; the extra factor of $\mathbb{R}^{(a,b)}$ plays no role.

Fix $P \in N$. Suppose that $b = 0$. Let $Z \in S^+(T_P N)$ be spacelike. Let $X := \pi_{\mathcal{X}} Z$, $\tilde{X} := \pi_{\tilde{\mathcal{X}}} Z$, and $Y := \pi_{\mathcal{Y}} Z$. If $X = 0$, $g(Z, Z) = g(Y, Y) \leq 0$ which is false. Thus $X \neq 0$ and by Lemma 2.5.3

$$\text{Rank}\{\mathcal{J}(Z)\} = \text{Rank}\{\mathcal{J}_{\mathcal{M}}(X)\} = p - 1.$$

Thus $\mathcal{N}$ is spacelike Jordan Osserman.

A similar argument shows that if σ is a spacelike 2-plane, then $\pi_{\mathcal{M}} \sigma$ is spacelike 2-plane in $\mathbb{R}^{2p}$. Thus by Eq. (2.5.d),

$$\text{Rank}\{\mathcal{R}(\sigma)\} = \text{Rank}\{\mathcal{R}_{\mathcal{M}}(\pi_{\mathcal{X}} \sigma)\} = 2.$$

It now follows that $\mathcal{N}$ is spacelike Jordan Ivanov–Petrova. This proves Assertion (2c).

Suppose $b > 0$. Let $Z_1 := \partial_{x_1} + \partial_{\tilde{x}_1}$ and $Z_2 := \partial_{y_{a+b}}$. These are spacelike vectors. Furthermore,

$$\text{Rank}\{\mathcal{J}(Z_1)\} = p - 1 \quad \text{and} \quad \text{Rank}\{\mathcal{J}(Z_2)\} = 0.$$

Thus $\mathcal{N}$ is not spacelike Jordan Osserman. Let $Z_3 := \partial_{x_2} + \partial_{\tilde{x}_2}$. Define spacelike 2-planes

$$\sigma_1 := \text{Span}\{Z_1, Z_3\} \quad \text{and} \quad \sigma_2 := \text{Span}\{Z_1, Z_2\}.$$

One then has that

$$\text{Rank}\{\mathcal{R}(\sigma_1)\} = 2 \quad \text{and} \quad \text{Rank}\{\mathcal{R}(\sigma_2)\} = 0.$$

Thus $\mathcal{N}$ is not spacelike Jordan Ivanov–Petrova. This proves Assertions (3b) and (3c). The proof of Assertions (2a), (2b), (3a), (4a), (4b), (4c), and (4d) is similar. $\qquad\square$

We now turn to the higher order Jacobi operator.

Proof of Theorem 2.6.1 (5). If $\{e_1^-, ..., e_r^-, e_1^+, ..., e_s^+\}$ is an orthonormal basis for a non-degenerate subspace σ of signature (r, s), then

$$\mathcal{J}(\sigma) := -\mathcal{J}(e_1^-) - ... - \mathcal{J}(e_r^-) + \mathcal{J}(e_1^+) + ... + \mathcal{J}(e_s^+).$$

By Eq. (2.6.a), $(\mathcal{J}(\sigma))^2 = 0$. Thus the Jordan normal form is determined by the rank; $\mathcal{N}$ will be Jordan Osserman of type (r, s) if and only if $\text{Rank}\{\mathcal{J}(\sigma)\}$ is constant on $\text{Gr}_{r,s}\,\mathcal{N}$.

Suppose $s = 0$ so σ is timelike. Then

$$\pi_{\mathcal{X}}(\sigma) = \text{Span}\{\pi_{\mathcal{X}}(e_1^-), ..., \pi_{\mathcal{X}}(e_r^-)\}, \text{ and}$$
$$\mathcal{J}(\sigma) = -\mathcal{J}\pi_{\mathcal{X}}(e_1^-) - ... - \mathcal{J}\pi_{\mathcal{X}}(e_r^-).$$

By Lemma 2.5.4,

$$\text{Rank}\{\mathcal{J}(\sigma)\} = \begin{cases} 0 & \text{if } \dim\{\pi_{\mathcal{X}}(\sigma)\} = 0, \\ p - 1 & \text{if } \dim\{\pi_{\mathcal{X}}(\sigma)\} = 1, \\ p & \text{if } \dim\{\pi_{\mathcal{X}}(\sigma)\} \geq 2. \end{cases}$$

Suppose that $a = 0$. Then $\ker\{\pi_{\mathcal{X}}\} = \tilde{\mathcal{X}} \oplus \mathcal{Y}$ contains no spacelike vectors. Thus if σ is spacelike, then $\ker\pi_{\mathcal{X}} \cap \sigma = \{0\}$. Consequently, $\dim\{\pi_{\mathcal{X}}(\sigma)\} = r$ is independent of σ so $\mathcal{N}$ is Jordan Osserman of type $(r, 0)$. Dually, by Theorem 1.9.2, $\mathcal{N}$ is Jordan Osserman of type $(p-r, p+b)$. This proves Theorem 2.6.1 (5a); the proof of Assertion (5b) is similar.

Suppose $r \geq a + 2$ and $a > 0$. If τ is a maximal spacelike subspace of $\tilde{\mathcal{X}} \oplus \mathcal{Y}$, then $\dim \tau \leq a$. Thus

$$\dim\{\ker\pi_{\mathcal{X}} \cap \sigma\} \leq a \quad \text{so} \quad \dim\{\pi_{\mathcal{X}}(\sigma)\} \geq r - a \geq 2$$

and hence $\text{Rank}\{\mathcal{J}(\sigma)\} = p$. Thus $\mathcal{N}$ is Jordan Osserman of type $(r, 0)$ and dually, it is Jordan Osserman of type $(p - r, b + p)$. This establishes Assertion (5c); Assertion (5d) follows similarly. This completes the proof of Assertion (5). $\qquad\square$

Proof of Theorem 2.6.1 (6). We consider several cases. Let

$$\sigma := \text{Span}\{Z_1, ..., Z_r\}$$

and let $\mathcal{J} := \mathcal{J}(Z_1) + ... + \mathcal{J}(Z_r)$. By Lemma 2.5.3,

$$\mathrm{Rank}(\mathcal{J}) = \begin{cases} 0 & \text{if} & \dim(\rho_{\mathcal{X}}\sigma) = 0, \\ p-1 & \text{if} & \dim(\rho_{\mathcal{X}}\sigma) = 1, \\ p & \text{if} & \dim(\rho_{\mathcal{X}}\sigma) \geq 2. \end{cases} \qquad (2.6.\text{b})$$

Suppose $a > 0$ and $r \leq a + 1$. We must show that $\mathcal{N}$ is not Jordan Osserman of type $(r, 0)$; we may then use duality to see that $\mathcal{N}$ is not Jordan Osserman of type $(p + a - r, p + b)$. Let

$$\{Z_1^-, ..., Z_p^-, Y_1^-, ..., Y_a^-\}$$

be an orthonormal basis for a maximal timelike subspace of $T_P M$ where $Z_i^- \in \mathcal{X} + \tilde{\mathcal{X}}$ and $Y_j \in \mathcal{Y}$. Let

$$\sigma_{u,v} := \mathrm{Span}\{Z_1^-, ..., Z_u^-, Y_1^-, ..., Y_v^-\} \in \mathrm{Gr}_{(u+v,0)}(T_P M).$$

Suppose first that $0 < r \leq a$. By Eq. (2.6.b):

$$\dim\{\pi_{\mathcal{X}}(\sigma_{1,r-1})\} = 1 \Rightarrow \mathrm{Rank}\{\mathcal{J}(\sigma_{1,r-1})\} = p - 1,$$
$$\dim\{\pi_{\mathcal{X}}(\sigma_{0,r})\} = 0 \quad \Rightarrow \mathrm{Rank}\{\mathcal{J}(\sigma_{0,r})\} = 0.$$

Therefore $\mathcal{N}$ is not Jordan Osserman of type $(r, 0)$. Suppose next $r = a+1$. Since $a > 0$, $r \geq 2$ and

$$\dim\{\pi_{\mathcal{X}}(\sigma_{2,a-1})\} = 2 \Rightarrow \mathrm{Rank}\{\mathcal{J}\sigma_{2,a-1})\} = p,$$
$$\dim\{\pi_{\mathcal{X}}(\sigma_{1,a})\} = 1 \quad \Rightarrow \mathrm{Rank}\{\mathcal{J}\sigma_{1,a})\} = p - 1.$$

Thus $\mathcal{N}$ is not Jordan Osserman of type $\sigma_{0,r}$. The case $b > 0$ and $r < b+2$ is similar and is omitted in the interests of brevity.

Finally, suppose $1 \leq r \leq p + a - 1$ and $1 \leq s \leq p + b - 1$. Let

$$\{Y_1^-, ..., Y_a^-, Y_1^+, ..., Y_b^+\}$$

be an orthonormal basis for $\mathbb{R}^{(a,b)}$. We define maps

$$T_{a,b}\sigma := \sigma \oplus \mathrm{Span}\{Y_1^-, ..., Y_a^-, Y_1^+, ..., Y_b^+)$$

from $\mathrm{Gr}_{\alpha,\beta}\mathcal{M}$ to $\mathrm{Gr}_{\alpha+a,\beta+b}\mathcal{N}$. We then have

$$\mathcal{J}_{\mathcal{N}}(T_{a,b}\sigma) = \mathcal{J}_{\mathcal{N}}(\sigma) \quad \text{for all} \quad \sigma \in \mathrm{Gr}_{\alpha,\beta}\mathcal{M}. \qquad (2.6.\text{c})$$

Suppose $\mathcal{N}$ is Jordan Osserman of type (r, s). Expand

$$r = \alpha + u \quad \text{for} \quad 1 \leq \alpha \leq p - 1 \quad \text{and} \quad 0 \leq u \leq a,$$
$$s = \beta + v \quad \text{for} \quad 1 \leq \beta \leq p - 1 \quad \text{and} \quad 0 \leq v \leq b.$$

If $\mathcal{N}$ is Jordan Osserman of type (r, s), then we may use Eq. (2.6.c) to see that $\mathcal{M}$ is Jordan Osserman of type (α, β). This contradicts Theorem 2.5.1 and thereby completes the proof of Theorem 2.6.1. $\qquad\square$

2.7 Nikčević Manifolds

In previous sections, we have discussed generalized plane wave manifolds of neutral signature and of signature $(2, 4)$. In this section, we present results of Gilkey, Nikčević, and Videv (2004) and of Gilkey and Nikčević (2004a) where the timelike directions dominate. We also fix a few minor technical mistakes in those papers.

Definition 2.7.1 Fix $s \geq 2$. Let indices i, j, k range from 1 through s. Let $\vec{u} := (u_1, ..., u_s)$, $\vec{t} := (t_1, ..., t_s)$, and $\vec{v} := (v_1, ..., v_s)$ give coordinates $(\vec{u}, \vec{t}, \vec{v})$ on $\mathbb{R}^{3s}$. Introduce distributions

$$\mathcal{U} := \mathrm{Span}_i\{\partial_{u_i}\}, \quad \mathcal{T} := \mathrm{Span}_i\{\partial_{t_i}\}, \quad \mathcal{V} := \mathrm{Span}_i\{\partial_{v_i}\}.$$

Let $f_i \in C^\infty(\mathbb{R})$. Set:

$$F(\vec{u}) := f_1(u_1) + ... + f_s(u_s),$$
$$|u|^2 := \sum_{i=1}^{s} u_i^2, \quad \text{and} \quad u \cdot t := \sum_{i=1}^{s} u_i t_i.$$

Let $\mathcal{M} := (\mathbb{R}^{3s}, g)$ be the pseudo-Riemannian manifold of signature $(2s, s)$ where the non-zero components of the metric are given by:

$$g(\partial_{u_i}, \partial_{u_i}) := -2F(\vec{u}) - 2u \cdot t, \quad g(\partial_{u_i}, \partial_{v_i}) := 1, \quad g(\partial_{t_i}, \partial_{t_i}) := -1.$$

Let $\mathrm{Al}(n)$ be the affine linear group on $\mathbb{R}^n$. In Section 2.7.1, we establish the following result:

Lemma 2.7.1 *The manifold of Definition 2.7.1 is a generalized plane wave manifold of signature $(2s, s)$.*

(1) The possibly non-zero entries in R and ∇R are given by:

(a) $R(\partial_{u_i}, \partial_{u_j}, \partial_{u_j}, \partial_{u_i}) = \partial_{u_i}^2(f_i) + \partial_{u_j}^2(f_j) + |u|^2$ *for $i \neq j$.*
(b) $R(\partial_{u_i}, \partial_{u_j}, \partial_{u_j}, \partial_{t_i}) = 1$ *for $i \neq j$.*
(c) $\nabla R(\partial_{u_i}, \partial_{u_j}, \partial_{u_j}, \partial_{u_i}; \partial_{u_i}) = \partial_{u_i}^3(f_i) + 4u_i$ *for $i \neq j$.*
(d) $\nabla R(\partial_{u_i}, \partial_{u_j}, \partial_{u_j}, \partial_{u_k}; \partial_{u_i}) = u_k$ *for $\{i, j, k\}$ distinct.*

(2) We have $\mathcal{R}(z_1, z_2)\mathcal{R}(z_3, z_4)\mathcal{R}(z_5, z_6) = 0$ and $\mathcal{J}(z_1)\mathcal{J}(z_2)\mathcal{J}(z_3) = 0$.

(3) $\ker(R) = \mathcal{V}$ *and* $\ker(R)^\perp = \mathcal{T} + \mathcal{V}$.

(4) Let ϕ be an isometry of $\mathcal{M}$. There exists $A_1 \in \mathrm{Al}(s)$, a smooth map $A_2 : \mathbb{R}^s \to \mathrm{Al}(s)$, and a smooth map $A_3 : \mathbb{R}^{2s} \to \mathrm{Al}(s)$ so that

$$\phi(\vec{u}, \vec{t}, \vec{v}) = (A_1 \vec{u}, A_2(\vec{u})\vec{t}, A_3(\vec{u}, \vec{t})\vec{v}) .$$

In Section 2.7.2, we shall prove the following result:

Theorem 2.7.1   *Let $\mathcal{M}$ be as in Definition 2.7.1. Then $\mathcal{M}$ is 0-curvature homogeneous and indecomposable.*

Let $\mathcal{M}$ be as in Definition 2.7.1. Set

$$\alpha_{\mathcal{M}} := \sum_{i_1, i_2, i_3, i_4, i_5 = 1}^{s} \nabla R(\partial_{u_{i_1}}, \partial_{u_{i_2}}, \partial_{u_{i_3}}, \partial_{u_{i_4}}; \partial_{u_{i_5}})^2 .$$

The following result will be established in Section 2.7.3:

Theorem 2.7.2

(1) Let $\mathcal{M}_i$ be as in Definition 2.7.1. If $\mathfrak{M}_1(\mathcal{M}_1, P_1)$ is isomorphic to $\mathfrak{M}_1(\mathcal{M}_2, P_2)$, then $\alpha_{\mathcal{M}_1}(P_1) = \alpha_{\mathcal{M}_2}(P_2)$.

(2) If $s \geq 3$ and if $\mathcal{M}$ is as in Definition 2.7.1, then $\mathcal{M}$ is not 1-curvature homogeneous.

The case $s = 2$ is exceptional in many ways. We shall derive the following consequences of Theorem 2.7.2:

Corollary 2.7.1   *Let $\mathcal{M}$ be as in Definition 2.7.1 where $s = 2$.*

(1) The following assertions are equivalent:

    (a) $\mathcal{M}$ is 1-curvature homogeneous.
    (b) $f_1^{(4)} = f_2^{(4)} = -4$.
    (c) $\mathcal{M}$ is homogeneous.

(2) Let $f_i(u_i) = -\frac{2}{3} u_i^3$ define $\mathcal{S}$. The following conditions are equivalent:

    (a) $\mathcal{M}$ is isometric to $\mathcal{S}$.
    (b) $\mathcal{M}$ is a symmetric space.
    (c) $f_1^{(3)}(t) = f_2^{(3)}(t) = -4t$.

We conclude our discussion of these manifolds in Section 2.7.4 by establishing the following result concerning the spectral geometry of this family:

Theorem 2.7.3   *Let $\mathcal{M}$ be as in Definition 2.7.1.*

(1) M is Jordan Osserman of type (a, b) if and only if
$(a, b) = (0, k)$ or $(a, b) = (2s, s - k)$ for $1 \le k \le s$ or
$(a, b) = (k, 0)$ or $(a, b) = (2s - k, s)$ for $s + 2 \le k \le 2s$.

(2) M is spacelike Jordan Ivanov–Petrova of rank 4.
M is not timelike Jordan Ivanov–Petrova.

(3) M is spacelike k-Stanilov if and only if $2 \le k \le s$.
M is timelike k-Stanilov if and only if $k = 2s$.

We remark that since M has vanishing Ricci tensor, similar assertions hold if we replace the words "Osserman" by "conformal Osserman", or if we replace the words "Ivanov–Petrova" by "conformal Ivanov–Petrova", or if we replace the words "Stanilov" by "conformal Stanilov". We also note that it is possible to show that M is not Jordan Stanilov of type (a, b) for $1 \le a \le 2s$ and $1 \le b \le s$ where $2 \le a + b \le 3s - 1$. We omit details in the interests of brevity.

2.7.1 *The curvature tensor*

Proof of Lemma 2.7.1. Let $i \ne j$. The non-zero Christoffel symbols are:

$$g(\nabla_{\partial_{u_i}} \partial_{u_i}, \partial_{u_i}) = -\partial_{u_i}(f_i) - t_i,$$

$$g(\nabla_{\partial_{u_i}} \partial_{u_i}, \partial_{u_j}) = \partial_{u_j}(f_j) + t_j,$$

$$g(\nabla_{\partial_{u_i}} \partial_{u_j}, \partial_{u_i}) = g(\nabla_{\partial_{u_j}} \partial_{u_i}, \partial_{u_i}) = -\partial_{u_j}(f_j) - t_j,$$

$$g(\nabla_{\partial_{u_i}} \partial_{u_i}, \partial_{t_i}) = u_i,$$

$$g(\nabla_{\partial_{u_i}} \partial_{t_i}, \partial_{u_i}) = g(\nabla_{\partial_{t_i}} \partial_{u_i}, \partial_{u_i}) = -u_i,$$

$$g(\nabla_{\partial_{u_i}} \partial_{u_i}, \partial_{t_j}) = u_j,$$

$$g(\nabla_{\partial_{u_i}} \partial_{t_j}, \partial_{u_i}) = g(\nabla_{\partial_{t_j}} \partial_{u_i}, \partial_{u_i}) = -u_j.$$

Consequently, the non-zero covariant derivatives are given by:

$$\nabla_{\partial_{u_i}} \partial_{u_i} = -(\partial_{u_i}(f_i) + t_i)\partial_{v_i} + \sum_{k \ne i}(\partial_{u_k}(f_k) + t_k)\partial_{v_k} - \sum_k u_k \partial_{t_k},$$

$$\nabla_{\partial_{u_i}} \partial_{u_j} = -(\partial_{u_j}(f_j) + t_j)\partial_{v_i} - (\partial_{u_i}(f_i) + t_i)\partial_{v_j}, \qquad (2.7.\text{a})$$

$$\nabla_{\partial_{u_i}} \partial_{t_i} = \nabla_{\partial_{t_i}} \partial_{u_i} = -u_i \partial_{v_i}, \quad \text{and} \quad \nabla_{\partial_{u_i}} \partial_{t_j} = \nabla_{\partial_{t_j}} \partial_{u_i} = -u_j \partial_{v_i}.$$

Since $f_i = f_i(u_i)$, the covariant derivative has the required triangular form relative to the ordering of the variables $(u_1, ..., u_s, t_1, ..., t_s, v_1, ..., v_s)$. Thus M is a generalized plane wave manifold.

It is immediate from Eq. (2.7.a) that $\mathcal{R}(\xi_1, \xi_2)\xi_3 = 0$ if any of the ξ_ν belong to $\mathcal{V}$, or if at least two of the variables belong to $\text{Span}\{\partial_{t_i}\}$. Also $\mathcal{R}(\partial_{u_i}, \partial_{u_j})\partial_{u_k} = 0$ if the indices $\{i, j, k\}$ are distinct. Thus the only curvatures that can be non-zero are $R(\partial_{u_i}, \partial_{u_j}, \partial_{u_j}, \partial_{u_i})$ and $R(\partial_{u_i}, \partial_{u_j}, \partial_{u_j}, \partial_{t_i})$.

In contrast to previous examples that have been studied, the Christoffel symbols play a crucial role in the computation of the curvature tensor and of ∇R; the interaction term $-\sum_k u_k \partial_{t_k}$ in $\nabla_{\partial_{u_i}} \partial_{u_i}$ is in many ways the crucial term. Assertions (1a) and (1b) of Lemma 2.7.1 follow from the computation for $i \neq j$ that:

$$\nabla_{\partial_{u_i}} \nabla_{\partial_{u_j}} \partial_{u_j} = \partial_{u_i}^2(f_i)\partial_{v_i} - \partial_{t_i} + |u|^2 \partial_{v_i},$$

$$\nabla_{\partial_{u_j}} \nabla_{\partial_{u_i}} \partial_{u_j} = -\partial_{u_j}^2(f_j)\partial_{v_i},$$

$$\mathcal{R}(\partial_{u_i}, \partial_{u_j})\partial_{u_j} = \{\partial_{u_i}^2 f_i + \partial_{u_j}^2 f_j + |u|^2\}\partial_{v_i} - \partial_{t_i}.$$

We have similarly that $\nabla_{\xi_1}\mathcal{R}(\xi_2, \xi_3)\xi_4 = 0$ if at least one of the ξ_i belongs to $\mathcal{T} + \mathcal{V}$. Since $\partial_{u_i}\partial_{u_j}F = 0$ for $i \neq j$, we establish Assertions (1c) and (1d) of Lemma 2.7.1 by showing that the only non-zero components of ∇R are given for $i \neq j$ by:

$$\nabla R(\partial_{u_i}, \partial_{u_j}, \partial_{u_j}, \partial_{u_i}; \partial_{u_i})$$
$$= \partial_{u_i} R(\partial_{u_i}, \partial_{u_j}, \partial_{u_j}, \partial_{u_i}) - 2R(\nabla_{\partial_{u_i}}\partial_{u_i}, \partial_{u_j}, \partial_{u_j}, \partial_{u_i})$$
$$- 2R(\partial_{u_i}, \nabla_{\partial_{u_i}}\partial_{u_j}, \partial_{u_j}, \partial_{u_i})$$
$$= \partial_{u_i}^3(f_i) + 2u_i + 2R(\sum_k u_k \partial_{t_k}, \partial_{u_j}, \partial_{u_j}, \partial_{u_i}) = \partial_{u_i}^3(f_i) + 4u_i$$

and for $\{i, j, k\}$ distinct by:

$$\nabla R(\partial_{u_i}, \partial_{u_j}, \partial_{u_j}, \partial_{u_k}; \partial_{u_i}) = -R(\nabla_{\partial_{u_i}}\partial_{u_i}, \partial_{u_j}, \partial_{u_j}, \partial_{u_k})$$
$$= R(\sum_\ell u_\ell \partial_{t_\ell}, \partial_{u_j}, \partial_{u_j}, \partial_{u_k}) = u_k.$$

We use Assertion (1a) of Lemma 2.7.1 to see that

$$\begin{aligned}
\mathcal{R}(\cdot)\mathcal{U} &\subset \mathcal{T} + \mathcal{V}, & \mathcal{J}(\cdot)\mathcal{U} &\subset \mathcal{T} + \mathcal{V}, \\
\mathcal{R}(\cdot)\mathcal{T} &\subset \mathcal{V}, & \mathcal{J}(\cdot)\mathcal{T} &\subset \mathcal{V}, \\
\mathcal{R}(\cdot)\mathcal{V} &= \{0\}, & \mathcal{J}(\cdot)\mathcal{V} &= \{0\}.
\end{aligned}$$

Assertion (2) of Lemma 2.7.1 now follows; Assertion (3) is immediate from Assertion (1) as only the term $R(\partial_{u_i}, \partial_{v_j}, \partial_{v_j}, \partial_{t_i}) = 1$ is needed for the argument. Assertion (4) now follows from Lemma 2.2.2 and the discussion given above. $\qquad \square$

2.7.2 *Curvature homogeneity*

We consider the following models:

Definition 2.7.2   Let $\{\bar{U}_1, ..., \bar{U}_s, \bar{T}_1, ..., \bar{T}_s\}$ be a basis for $\mathbb{R}^{2s}$. Set

$$\bar{\mathcal{T}} := \mathrm{Span}\{\bar{T}_i\}\,.$$

Let $\bar{\mathfrak{M}}_0^w := (\mathbb{R}^{2s}, \bar{A})$ be the weak 0-model where the non-zero components of $\bar{A}$ are given by

$$\bar{A}(\bar{U}_i, \bar{U}_j, \bar{U}_j, \bar{T}_i) := 1 \quad \text{for} \quad i \neq j\,.$$

Definition 2.7.3   Let $\{U_1, ..., U_s, T_1, ..., T_s, V_1, ..., V_s\}$ be a basis for $\mathbb{R}^{3s}$. Let $\mathfrak{M}_0 := (\mathbb{R}^{3s}, \langle \cdot, \cdot \rangle, A)$ be the 0-model where the non-zero components of $\langle \cdot, \cdot \rangle$ are given by:

$$\langle U_i, V_i \rangle = \langle V_i, U_i \rangle := 1, \quad \langle T_i, T_i \rangle := -1,$$

and the non-zero components of A are given by:

$$A(U_i, U_j, U_j, T_i) := 1 \quad \text{for} \quad i \neq j\,.$$

Note that if π is the natural projection from $\mathbb{R}^{3s}$ to $\mathbb{R}^{2s}$ which is defined by $\pi(U_i) := \bar{U}_i$, $\pi(T_i) := \bar{T}_i$, and $\pi(V_i) = 0$, then

$$\pi^* \bar{A} = A\,.$$

Lemma 2.7.2

(1) Let $\bar{v}_1 \in \mathbb{R}^{2s}$. Suppose there is $0 \neq \bar{v}_2 \in \mathbb{R}^{2s}$ so $\bar{A}(\bar{v}_1, \bar{v}_2, \bar{w}_1, \bar{w}_2) = 0$ and $\bar{A}(\bar{v}_1, \bar{w}_1, \bar{w}_2, \bar{v}_2) = 0$ for all $\bar{w}_1, \bar{w}_2 \in \mathbb{R}^{2s}$. Then $\bar{v}_1 \in \bar{\mathcal{T}}$.
(2) If ϕ is an isomorphism of $\bar{\mathfrak{M}}_0^w$, then $\phi \bar{\mathcal{T}} \subset \bar{\mathcal{T}}$.
(3) $\bar{\mathfrak{M}}_0^w$ is indecomposable.
(4) $\mathfrak{M}_0$ is indecomposable.
(5) $\mathfrak{M}_0$ is a 0-model for any manifold $\mathcal{M}$ as in Definition 2.7.1.

Proof.   Let $\bar{v}_1 \in \mathbb{R}^{2s}$. Suppose there exists a non-zero vector $\bar{v}_2 \in \mathbb{R}^{2s}$ so that $\bar{A}(\bar{v}_1, \bar{v}_2, \bar{w}_1, \bar{w}_2) = 0$ and $\bar{A}(\bar{v}_1, \bar{w}_1, \bar{w}_2, \bar{v}_2) = 0$ for all $\bar{w}_1, \bar{w}_2 \in \mathbb{R}^{2s}$. Expand

$$\bar{v}_1 = \sum_i (a_i \bar{U}_i + b_i \bar{T}_i) \quad \text{and} \quad \bar{v}_2 = \sum_i (c_i \bar{U}_i + d_i \bar{T}_i)\,.$$

Suppose $\bar{v}_1 \notin \bar{\mathcal{T}}$; we argue for a contradiction. Since $\bar{v}_1 \notin \bar{\mathcal{T}}$, $a_i \neq 0$ for some i; without loss of generality we may suppose $a_1 \neq 0$. Let $i \neq 1$. Then

$$0 = \bar{A}(\bar{v}_1, \bar{U}_i, \bar{T}_1, \bar{v}_2) = -a_1 c_i,$$
$$0 = \bar{A}(\bar{v}_1, \bar{U}_i, \bar{T}_i, \bar{v}_2) = \sum_{i \neq j} a_j c_j \, .$$

Consequently $a_1 c_i = \sum_{i \neq j} a_j c_j = 0$. Since $a_1 \neq 0$, $c_i = 0$ for $i \neq 1$. Thus $0 = \sum_{i \neq j} a_j c_j = a_1 c_1$. Since $a_1 \neq 0$, $c_1 = 0$. This shows

$$\bar{v}_2 = d_1 \bar{T}_1 + \ldots + d_s \bar{T}_s \, .$$

We may therefore compute

$$0 = \bar{A}(\bar{v}_1, \bar{v}_2, \bar{U}_i, \bar{U}_1) = a_1 d_i,$$
$$0 = \bar{A}(\bar{v}_1, \bar{U}_i, \bar{U}_i, \bar{v}_2) = \sum_{j \neq i} a_j d_j \, .$$

Thus $0 = a_1 d_i = \sum_{j \neq i} a_j d_j$. Since $a_1 \neq 0$, $d_i = 0$ for $i \neq 1$. Since then $\sum_{j \neq i} a_j d_j = a_1 d_1$, we also have $d_1 = 0$. This implies $\bar{v}_2 = 0$; this contradiction establishes Assertion (1). Since Assertion (1) gives a basis free definition of $\bar{\mathcal{T}}$, Assertion (2) follows.

Suppose we have a non-trivial decomposition $\mathbb{R}^{2s} = \bar{V}^1 \oplus \bar{V}^2$ which induces a decomposition $\bar{A} = \bar{A}^1 \oplus \bar{A}^2$. Choose $0 \neq \bar{v}^i \in \bar{V}^i$. We then have

$$\bar{A}(\bar{v}^1, \bar{v}^2, \eta_1, \eta_2) = \bar{A}(\bar{v}^1, \eta_1, \eta_2, \bar{v}^2) = 0 \quad \forall \quad \eta_i \in \mathbb{R}^{2s} \, .$$

This implies $\bar{v}^1 \in \bar{\mathcal{T}}$ so $\bar{V}^1 \subset \bar{\mathcal{T}}$. Similarly $\bar{V}^2 \subset \bar{\mathcal{T}}$ and thus $\mathbb{R}^{2s} \subset \bar{\mathcal{T}}$ which is false. Assertion (3) follows. Assertion (4) follows from Lemma 1.6.4, from Assertion (3), and from the observation that $\ker(A) = \operatorname{Span}\{V_i\}$ is totally isotropic.

Fix $P \in \mathbb{R}^{3s}$. Let ε_i and ϱ_i be constants to be specified presently. Set

$$U_i := \partial_{u_i} + \varepsilon_i \partial_{t_i} + \varrho_i \partial_{v_i}, \quad T_i := \partial_{t_i} + \varepsilon_i \partial_{v_i}, \quad \text{and} \quad V_i := \partial_{v_i} \, .$$

Let $i \neq j$. Since $g(U_i, T_i) = \varepsilon_i - \varepsilon_i = 0$, the possibly non-zero entries of g and R are given by

$$g(U_i, U_i) = g(\partial_{u_i}, \partial_{u_i}) - \varepsilon_i^2 + 2\varrho_i,$$
$$g(T_i, T_i) = -1, \quad g(U_i, V_i) = 1,$$
$$R(U_i, U_j, U_j, T_i) = 1, \quad \text{and}$$
$$R(U_i, U_j, U_j, U_i) = \partial_i^2 f_i + \partial_j^2 f_j + |u|^2 + 2\varepsilon_i + 2\varepsilon_j \, .$$

We set

$$\varepsilon_i := -\tfrac{1}{2}\partial_i^2 f_i - \tfrac{1}{4}|u|^2 \quad \text{and} \quad \varrho_i := \tfrac{1}{2}\{\varepsilon_i^2 - g(\partial_{u_i}, \partial_{u_i})\}\,.$$

This ensures that $g(U_i, U_i) = 0$ and $R(U_i, U_j, U_j, U_i) = 0$ and establishes the existence of a basis with the normalizations of Definition 2.7.3. This shows that $\mathfrak{M}_0$ is a 0-model for $\mathcal{M}$. $\qquad\square$

Proof of Theorem 2.7.1. By Lemma 2.7.2, $\mathfrak{M}_0$ is a 0-model for $\mathcal{M}$. By Lemma 2.7.2, $\mathfrak{M}_0$ is indecomposable. $\qquad\square$

2.7.3 *Local isometry invariants*

These manifolds are generalized plane wave manifolds; thus once again, we must find local invariants which are not of Weyl type.

Let $O(s) \subset M_s(\mathbb{R})$ be the usual orthogonal group of $s \times s$ matrices; $\kappa_{ij} \in O(s)$ if and only if $\sum_k \kappa_{ik}\kappa_{jk} = \delta_{ij}$. Let $G(\mathfrak{M}_0)$ be the isomorphism group of the 0-model $\mathfrak{M}_0$. Define a degenerate positive semi-definite form h on $\mathbb{R}^{3s}$ whose only non-zero components are

$$h(U_i, U_j) = \delta_{ij}\,.$$

Lemma 2.7.3 *Let $\mathfrak{M}_0 = (\mathbb{R}^{3s}, \langle\cdot,\cdot\rangle, A)$ be as in Definition 2.7.3.*

(1) $\psi \in G(\mathfrak{M}_0)$ if and only if there exists $\kappa \in O(s)$ so that

$$\psi = \left\{ \begin{pmatrix} \kappa & 0 & 0 \\ \star & \kappa & 0 \\ \star & \star & \kappa \end{pmatrix} \right\}\,.$$

(2) If $\psi \in G(\mathfrak{M}_0)$, then $\psi^* h = h$.

(3) Let $\mathfrak{M}_1 := (\mathbb{R}^{3s}, \langle\cdot,\cdot\rangle, A, A_1)$ where A_1 is a covariant derivative curvature tensor with $A(\xi_1, \xi_2, \xi_3, \xi_4; \xi_5) = 0$ if any entry ξ_ν belongs to $\mathrm{Span}\{T_i, V_i\}$. Set

$$\alpha_{\mathfrak{M}_1} := \sum_{i_1, i_2, i_3, i_4, i_5} \{A_1(U_{i_1}, U_{i_2}, U_{i_3}, U_{i_4}; U_{i_5})\}^2\,.$$

Then $\alpha_{\mathfrak{M}_1}$ is an invariant of $\mathfrak{M}_1$.

(4) Let σ be a spacelike or timelike k-plane in $\mathbb{R}^{3s}$. There exists $\psi \in G(\mathfrak{M}_0)$ and an orthonormal basis $\{X_1, ..., X_k\}$ for $\psi\sigma$ so that $X_i = a_i U_i + T + V$.

Proof. Let $\psi \in G(\mathfrak{M}_0)$. The following subspaces are invariantly defined and are preserved by ψ:

$$\ker(R) := \{\eta \in \mathbb{R}^{3s} : A(\zeta_1, \zeta_2, \zeta_3, \eta) = 0 \text{ for all } \zeta_1, \zeta_2, \zeta_3 \in \mathbb{R}^3\}$$
$$= \operatorname{Span}\{V_1, ..., V_s\},$$
$$\ker(R)^\perp := \{\eta \in \mathbb{R}^{3s} : \langle \eta, \zeta \rangle = 0 \text{ for all } \zeta \in \ker(R)\}$$
$$= \operatorname{Span}\{T_1, ..., T_s, V_1, ..., V_s\}.$$

Since $V_i \in \ker(R)$, $\psi V_i \in \ker(R)$; since $T_i \in \ker(R)^\perp$, $\psi T_i \in \ker(R)^\perp$. Thus

$$\psi U_i = \sum_j \{\kappa_{1,ij} U_j + \kappa_{2,ij} T_j + \kappa_{3,ij} V_j\},$$
$$\psi T_i = \sum_j \{\kappa_{4,ij} T_j + \kappa_{5,ij} V_j\}, \quad \text{and} \quad \psi V_i = \sum_j \kappa_{6,ij} V_j.$$

We verify that $\kappa_4 \in O(s)$ by checking

$$-\delta_{ij} = \langle \psi T_i, \psi T_j \rangle = \sum_{k,l} \kappa_{4,ik} \kappa_{4,jl} \langle T_k, T_l \rangle = -\sum_k \kappa_{4,ik} \kappa_{4,jk}.$$

If $\kappa \in O(s)$, define $\psi_\kappa = \kappa \oplus \kappa \oplus \kappa$ by

$$\psi_\kappa U_i := \sum_j \kappa_{ij} U_j, \quad \psi_\kappa T_i := \sum_j \kappa_{ij} T_j, \quad \psi_\kappa V_i := \sum_j \kappa_{ij} V_j.$$

Clearly $\psi_\kappa \in G(\mathfrak{M}_0)$. By replacing the original isomorphism ψ by $\psi_{\kappa_4^{-1}}\psi$, we may suppose that $\kappa_4 = \mathrm{id}$ in the proof of the Lemma. Consequently we may assume without loss of generality

$$\psi U_i = \sum_j \{\kappa_{1,ij} U_j + \kappa_{2,ij} T_j + \kappa_{3,ij} V_j\},$$
$$\psi T_i = T_i + \sum_j \kappa_{5,ij} V_j, \quad \text{and} \quad \psi V_i = \sum_j \kappa_{6,ij} V_j.$$

We wish to show $\kappa_1 = \mathrm{id}$. Suppose $s = 2$. We compute:

$$1 = R(\psi U_1, \psi U_2, \psi U_2, \psi T_1) = (\kappa_{1,11}\kappa_{1,22} - \kappa_{1,12}\kappa_{1,21})\kappa_{1,22},$$
$$1 = R(\psi U_1, \psi U_2, \psi T_2, \psi U_1) = (\kappa_{1,11}\kappa_{1,22} - \kappa_{1,12}\kappa_{1,21})\kappa_{1,11},$$
$$0 = R(\psi U_1, \psi U_2, \psi U_1, \psi T_1) = (\kappa_{1,11}\kappa_{1,22} - \kappa_{1,12}\kappa_{1,21})\kappa_{1,12},$$
$$0 = R(\psi U_1, \psi U_2, \psi T_2, \psi U_1) = (\kappa_{1,11}\kappa_{1,22} - \kappa_{1,12}\kappa_{1,21})\kappa_{1,21}.$$

From this it follows that $\kappa_{12} = \kappa_{21} = 0$ and $\kappa_{11} = \kappa_{22}$. Consequently $\kappa_{11}^3 = 1$ and $\kappa_1 = \mathrm{id}$ as desired.

Suppose that $s \geq 3$. Fix j. We have

$$\dim \left\{ \mathrm{Span}_{k \neq j} \{U_k\} \right\} = s - 1,$$
$$\dim \left\{ \mathrm{Span}_{k \neq j; 1 \leq \ell \leq s} \{\psi U_k, \psi T_\ell, \psi V_\ell\} \right\} \geq 3s - 1,$$
$$\dim \left\{ \{ \mathrm{Span}_{k \neq j} \{U_k\} \cap \mathrm{Span}_{k \neq j; 1 \leq \ell \leq s} \{\psi U_k, \psi T_\ell, \psi V_\ell\} \right\}$$
$$\geq (s - 1) + (3s - 1) - 3s = s - 2 > 0.$$

Thus we may choose

$$0 \neq u \in \mathrm{Span}_{k \neq j} \{U_k\} \cap \mathrm{Span}_{k \neq j; 1 \leq \ell \leq s} \{\psi U_k, \psi T_\ell, \psi V_\ell\}.$$

Since $u \in \mathrm{Span}_{k \neq j; 1 \leq \ell \leq s} \{\psi U_k, \psi T_\ell, \psi V_\ell\}$,

$$0 = A(\psi U_i, u, u, \psi T_j).$$

Expand $u = \sum_{k \neq j} \varepsilon_k U_k$. As $\psi T_j = T_j + \sum_k \kappa_{5,jk} V_k$, we may compute

$$0 = A(\psi U_i, u, u, \psi T_j) = A \left(\sum_k \kappa_{1,ik} U_k, \sum_{a \neq j} \varepsilon_a U_a, \sum_{b \neq j} \varepsilon_b U_b, T_j \right)$$
$$= \kappa_{1,ij} \sum_{a \neq j} \varepsilon_a^2.$$

This shows $\kappa_{1,ij} = 0$ for $i \neq j$ so κ_1 is diagonal. Since

$$1 = A(\psi U_i, \psi U_j, \psi U_j, \psi T_i) = \kappa_{1,ii} \kappa_{1,jj} \kappa_{1,jj},$$
$$1 = A(\psi U_j, \psi U_i, \psi U_i, \psi T_j) = \kappa_{1,jj} \kappa_{1,ii} \kappa_{1,ii},$$

$\kappa_{1,ii} = 1$ as desired. We use the identity $\langle \psi U_i, \psi V_j \rangle = \delta_{ij}$ to see $\kappa_6 = \mathrm{id}$ as well. This completes the proof of Assertion (1) of Lemma 2.7.3.

Assertion (2) is now immediate from Assertion (1).

Let $\mathbb{R}^s := \mathbb{R}^{3s} / \ker(R)^\perp$ and let $\pi : \mathbb{R}^{3s} \to \mathbb{R}^s$ be the associated projection; $\mathbb{R}^s = \mathrm{Span}\{\pi U_i\}$. Define a positive definite inner product $\bar{h}$ on $\mathbb{R}^s$ by

$$h(\pi U_i, \pi U_j) = \delta_{ij}.$$

Any isomorphism of $\mathfrak{M}_0$ preserves $\ker(R)^\perp$ and induces, by Assertion (2), an isometry of $(\mathbb{R}^s, \bar{h})$. Let A_1 satisfy the hypotheses of Assertion (3). Then there is an algebraic covariant derivative curvature tensor $\bar{A}_1$ on $\mathbb{R}^s$ so $\pi^* \bar{A}_1 = A_1$. Assertion (3) follows since

$$\alpha_{\mathfrak{M}_1} = \|\bar{A}_1\|_{\bar{h}}^2.$$

Let σ be a spacelike (respectively timelike) k-plane. Then $\langle \cdot, \cdot \rangle$ is definite on σ. We can diagonalize h with respect to $\langle \cdot, \cdot \rangle$ to choose an orthonormal basis X_i for σ so that for $\leq i, j \leq k$ one has:

$$h(\pi X_i, \pi X_j) = a_i^2 \delta_{ij} \, .$$

Since $\{a_i^{-1} \pi X_i\}_{a_i \neq 0}$ forms an orthonormal set in $(\mathbb{R}^s, \bar{h})$, we may choose $\kappa \in O(s)$ so that

$$\kappa \pi X_i = a_i \pi U_i \quad \text{for} \quad 1 \leq i \leq k \, .$$

We take $\psi = \psi_\kappa$ to establish Assertion (4) of Lemma 2.7.3. $\square$

Proof of Theorem 2.7.2. Assertion (1) of Theorem 2.7.2 is an immediate consequence of Lemma 2.7.1 and of Lemma 2.7.3.

Suppose that $s \geq 3$. If $\mathcal{M}$ is 1-curvature homogeneous, then $\alpha = c$ must be constant. We may estimate:

$$u_1^2 = \nabla R(\partial_{u_1}, \partial_{u_2}, \partial_{u_2}, \partial_{u_3}; \partial_{u_3})^2 \leq \alpha_{\mathcal{M}} = c \, .$$

Since u_1 is not bounded, such an estimate is impossible. Thus $\mathcal{M}$ is not 1-curvature homogeneous. Assertion (2) of Theorem 2.7.2 follows. $\square$

Remark 2.7.1 We can construct additional invariants of the metric by considering the norms of higher order covariant derivatives of the curvature tensor by setting

$$\alpha_{\mathcal{M},k} := \sum_{i_1, i_2, i_3, i_4, j_1, \ldots, j_k} R(\partial_{i_1}^u, \partial_{i_2}^u, \partial_{i_3}^u, \partial_{i_4}^u; \partial_{j_1}^u, \ldots, \partial_{j_k}^u)^2 \, .$$

Proof of Corollary 2.7.1. Let $\mathcal{M}$ be as in Definition 2.7.1 where $s = 2$. We have

$$\alpha_{\mathcal{M}} = \mu \left\{ (f_1^{(3)}(u_1) + 4u_1)^2 + (f_2^{(3)}(u_2) + 4u_2)^2 \right\}$$

where μ is a suitable positive integer that plays no role in the analysis. Suppose that $\mathcal{M}$ is 1-curvature homogeneous. Then $\alpha_{\mathcal{M}}$ is constant. This implies $f_i^{(3)}(u_i) + 4u_i = a_i$ or equivalently that $f_i^{(4)} = -4$ for $i = 1, 2$. On the other hand, if this condition is satisfied, then $\mathcal{M}$ is 1-curvature homogeneous since the extra terms of Lemma 2.7.1 (1d) do not appear when $s = 2$. Since $\nabla^2 R = 0$, $\mathcal{M}$ is ∞-curvature homogeneous. The metric is polynomial and hence real analytic. Thus Theorem 2.2.2 may be applied to see $\mathcal{M}$ is homogeneous. Clearly if $\mathcal{M}$ is homogeneous, it is 1-curvature homogeneous. Assertion (1) now follows.

Let $\mathcal{S}$ be defined by taking $f_i(u_i) = -\frac{2}{3}u_i^3$. If $\mathcal{M}$ is isometric to $\mathcal{S}$, then $\mathcal{S}$ is a symmetric space by Theorem 2.7.1. If $\mathcal{M}$ is a symmetric space, then necessarily $\nabla R = 0$ and hence $\alpha = 0$. If $\alpha = 0$, then $\sum_i \{\partial_{u_i}^3(f_i) + 4u_i\}^2 = 0$. This implies each term vanishes separately and hence $f_i^{(3)}(u_i) + 4u_i = 0$ for all i. Finally, if this condition is satisfied, then $\nabla R = 0$ since the extra terms of Lemma 2.7.1 (1d) do not appear. By Theorem 2.7.1 the 0-model and the 0-model of S are isomorphic. Since $\nabla R = 0$, this isomorphism induces an isomorphism of the ∞-models. Since f_i is a cubic polynomial, the metric is real analytic. Thus Theorem 2.2.2 yields the desired isomorphism between $\mathcal{M}$ and $\mathcal{S}$. This establishes Assertion (2). $\qquad\square$

2.7.4 *The spectral geometry of the curvature tensor*

Proof of Theorem 2.7.3 (1). By Lemma 2.7.2, it suffices to establish the corresponding assertions for the 0-model $\mathcal{M}_0$. By Theorem 1.9.2, $\mathfrak{M}_0$ is Jordan Osserman of type (a, b) if and only if $\mathfrak{M}_0$ is Jordan Osserman of type $(2s - a, s - b)$. We shall study the cases $(k, 0)$ and $(0, k)$ separately. This analysis also deals with the cases $(2s - k, s)$ and $(2s, s - k)$. We will then study the cases (a, b) for $1 \leq a \leq 2s - 1$ and $1 \leq b \leq s - 1$.

Let σ be a spacelike k-plane where $1 \leq k \leq s$. We apply Lemma 2.7.3. By replacing σ by $\psi\sigma$ where ψ is an isomorphism of the 0-model $\mathfrak{M}_0$, we may assume without loss of generality that σ has an orthonormal basis $\{X_1, ..., X_k\}$ where

$$X_i = a_i U_i + \mathcal{T} + \mathcal{V} \quad \text{for} \quad 1 \leq i \leq k.$$

Since σ is spacelike, $\dim\{\pi\sigma\} = k$. Thus the a_i are all non-zero.

Suppose $k = 1$. We have $\{X_1, U_2, ..., U_s, T_1, ..., T_s, V_1, ..., V_s\}$ is a basis for $\mathbb{R}^{3s}$. Furthermore

$$
\begin{aligned}
\mathcal{J}(X_1)X_1 &= 0, & \mathcal{J}(X_1)U_i &= -a_1^2 T_i && \text{for} \quad i \geq 2, \\
\mathcal{J}(X_1)T_1 &= 0, & \mathcal{J}(X_1)T_i &= a_1^2 V_i && \text{for} \quad i \geq 2, \\
\mathcal{J}(X_1)V_1 &= 0, & \mathcal{J}(X_1)V_i &= 0 && \text{for} \quad i \geq 2.
\end{aligned}
$$

This shows that $\mathcal{J}(X_1)^3 = 0$,

$$\text{Rank}\{\mathcal{J}(X_1)\} = 2s - 2 \quad \text{and} \quad \text{Rank}\{\mathcal{J}(X_1)^2\} = s - 1.$$

This determines the Jordan normal form of $\mathcal{J}(X_1)$ and shows that $\mathfrak{M}_0$ is spacelike Jordan Osserman, or equivalently Jordan Osserman of type $(0, 1)$.

Next, we suppose that $k \geq 2$. We have

$$\mathcal{J}(\pi) = \sum_{i=1}^{k} \mathcal{J}(X_i)\,.$$

We may express

$$\mathcal{J}(\pi)U_j = -\sum_{i \neq j} a_i^2 T_j + \mathcal{V}, \quad \mathcal{J}(\pi)T_j = -\sum_{i \neq j} a_i^2 V_j, \quad \mathcal{J}(\pi)V_j = 0\,.$$

Since $k \geq 2$, one has that $\sum_{i \neq j} a_j^2 \neq 0$. We have $\mathcal{J}(\pi)^3 = 0$ while

$$\mathrm{Rank}\{\mathcal{J}(\pi)\} = 2s \quad \text{and} \quad \mathrm{Rank}\{\mathcal{J}(\pi)^2\} = s\,.$$

Thus $\mathfrak{M}_0$ is spacelike Jordan Osserman of type $(k,0)$ for $2 \leq k \leq s$ as well.

Let σ be a timelike k-plane. Let $\ell(\pi) := \dim\{\pi\sigma\}$ where $\pi : \mathbb{R}^{3s} \to \mathbb{R}^s$ is the projection defined in the proof of Lemma 2.7.3. We use Lemma 2.7.3 to see that after replacing σ by $g\sigma$ where $g \in G(\mathfrak{M}_0)$, we may assume without loss of generality that we have chosen an orthonormal basis $\{X_i\}$ for σ of the form

$$X_i = a_i U_i + \mathcal{T} + \mathcal{V}\,.$$

Furthermore, the number of indices i with $a_i \neq 0$ is equal to ℓ. The calculations above show that $\mathcal{J}(\pi)^3 = 0$ while

$$\mathrm{Rank}\{\mathcal{J}(\pi)\} = \begin{cases} 0 & \text{if} \quad \ell = 0, \\ 2s - 2 & \text{if} \quad \ell = 1, \\ 2s & \text{if} \quad \ell \geq 2, \end{cases}$$

$$\mathrm{Rank}\{\mathcal{J}(\pi)^2\} = \begin{cases} 0 & \text{if} \quad \ell = 0, \\ s - 1 & \text{if} \quad \ell = 1, \\ s & \text{if} \quad \ell \geq 2\,. \end{cases}$$

If $1 \leq k \leq s$, we can choose $\sigma_i \in \mathrm{Gr}_{k,0}(\mathbb{R}^{3s})$ so that $\ell(\sigma_1) = 1$ and $\ell(\sigma_2) = 0$. Thus $\mathfrak{M}_0$ is not Jordan Osserman of type $(k,0)$. If $k = s+1$, we can choose $\sigma_i \in \mathrm{Gr}_{k,0}(\mathbb{R}^{3s})$ so that $\ell(\sigma_1) = 2$ and $\ell(\sigma_2) = 1$. Thus $\mathfrak{M}_0$ is not Jordan Osserman of type $(s+1,0)$. If $k \geq s+2$ and if $\sigma \in \mathrm{Gr}_{k,0}(\mathbb{R}^{3s})$, then $\ell(\sigma) \geq 2$ and hence the Jordan normal form is constant. This proves $\mathfrak{M}_0$ is Jordan Osserman of type $(k,0)$ for $k \geq s+2$.

We complete the proof of Assertion (1) of Theorem 2.7.3 by showing $\mathfrak{M}_0$ is not Jordan Osserman of type (a,b) if $1 \leq a \leq s-1$ and $1 \leq b \leq 2s-1$.

The following orthonormal set contains $s - 2$ spacelike and $2s - 2$ timelike vectors:

$$\{(U_3 + V_3)/\sqrt{2}, ..., (U_s + V_s)/\sqrt{2},$$
$$(U_3 - V_3)/\sqrt{2}, ..., (U_s - V_s)/\sqrt{2}, T_1, ..., T_s\}.$$

Choose a subcollection $\{e_3, ..., e_{a+b}\}$ so that $a - 1$ of these vectors are spacelike and $b - 1$ of these vectors are timelike. Let

$$\sigma_1 := \mathrm{Span}\{e_3, ..., e_{a+b}\},$$
$$e_{1,\varepsilon}^+ := (\varepsilon U_1 + \varepsilon^{-1} V_1)/\sqrt{2}, \quad e_{2,\varrho}^- := (\varrho U_2 - \varrho^{-1} V_2)/\sqrt{2},$$
$$\sigma(\varepsilon, \varrho) := \mathrm{Span}\{e_{1,\varepsilon}^+, e_{2,\varepsilon}^-\} \oplus \sigma_1.$$

We then have $\mathcal{J}(\sigma_1) = \sum_i a_i \mathcal{J}(U_i)$ where $a_i = 0, \pm 1$. We compute:

$$\mathcal{J}(\sigma)U_i = - \begin{cases} (\sum_{j \neq 1} a_j - \varrho^2)T_1 & \text{if} \quad i = 1, \\ (\sum_{j \neq 2} a_j + \varepsilon^2)T_2 & \text{if} \quad i = 2, \\ (\sum_{j \neq i} a_j + \varepsilon^2 - \varrho^2)T_i & \text{if} \quad i \geq 3, \end{cases}$$

$$\mathcal{J}(\sigma)T_i = \begin{cases} (\sum_{j \neq 1} a_j - \varrho^2)V_1 & \text{if} \quad i = 1, \\ (\sum_{j \neq 2} a_j + \varepsilon^2)V_2 & \text{if} \quad i = 2, \\ (\sum_{j \neq i} a_j + \varepsilon^2 - \varrho^2)V_i & \text{if} \quad i \geq 3, \end{cases}$$

$$\mathcal{J}(\sigma)V_i = \quad 0.$$

If $s \geq 3$, then we can take $i = 3$. Since $\sum_{j \neq i} a_i + \varepsilon^2 - \varrho^2$ has non-trivial zeros and at these zeros the rank of $\mathcal{J}$ drops, $\mathfrak{M}_0$ is not Jordan Osserman of type (a, b).

We are left with the cases $s = 2$ and $(a, b) = (1, 1)$ or $(a, b) = (1, 2)$. These cases are dual; thus by Theorem 1.9.2, we may assume without loss of generality that $(a, b) = (1, 1)$. We let

$$\sigma_1 := \mathrm{Span}\{(U_1 + V_1)/\sqrt{2}, (U_1 - V_1)/\sqrt{2}\},$$
$$\sigma_2 := \mathrm{Span}\{(U_1 + V_1)/\sqrt{2}, T_1\}.$$

We then have $\mathcal{J}(\sigma_1) = 0$ and $\mathcal{J}(\sigma_2) \neq 0$. Thus $\mathfrak{M}_0$ is not Jordan Osserman of type $(1, 1)$. $\qquad \square$

Proof of Theorem 2.7.3 (2). We must show $\mathfrak{M}_0$ is spacelike Jordan Ivanov–Petrova of rank 4, but not timelike Jordan Ivanov–Petrova. Let π

be an oriented spacelike 2-plane. We apply Lemma 2.7.3 to renormalize π so there exists an orthonormal basis for π so that $\pi = \mathrm{Span}\{X_1, X_2\}$ where

$$X_1 = U_1 + \mathcal{T} + \mathcal{V} \quad \text{and} \quad X_2 = U_2 + \mathcal{T} + \mathcal{V}.$$

Thus one has that:

$$\begin{aligned}
&\mathcal{R}(\pi) : U_1 \rightarrow T_2 + \mathcal{V}, &\quad &\mathcal{R}(\pi) : T_1 \rightarrow -V_2, \\
&\mathcal{R}(\pi) : U_2 \rightarrow -T_1 + \mathcal{V}, &\quad &\mathcal{R}(\pi) : T_2 \rightarrow V_1, &\quad &(2.7.\mathrm{b}) \\
&\mathcal{R}(\pi) : U_i \rightarrow \mathrm{Span}\{V_1, V_2\}, &\quad &\mathcal{R}(\pi) : T_i \rightarrow 0.
\end{aligned}$$

Consequently,

$$\mathrm{Rank}(\mathcal{A}(\pi)) = 4, \quad \mathrm{Rank}(\mathcal{A}(\pi)^2) = 2, \quad \text{and} \quad \mathrm{Rank}(\mathcal{A}(\pi)^3) = 0.$$

Thus the Jordan normal form of $\mathcal{A}(\pi)$ is independent of the particular spacelike 2-plane π chosen so $\mathfrak{M}_0$ is spacelike Ivanov–Petrova rank 4.

To show that $\mathfrak{M}_0$ is not timelike Ivanov–Petrova, we consider the following timelike 2-planes:

$$\pi_1 := \mathrm{Span}\{T_1, T_2\} \quad \text{and} \quad \pi_2 := \mathrm{Span}\{(U_1 - V_1), (U_2 - V_2)\}.$$

Since $\mathcal{A}(\pi_1) = 0$ and $\mathcal{A}(\pi_2) \neq 0$, $\mathfrak{M}_0$ is not timelike Ivanov–Petrova. $\square$

Proof of Theorem 2.7.3 (3). We recall the definition of the Stanilov operator Θ from Section 1.8.1. Let $\{e_1, ..., e_k\}$ be a basis for a non-degenerate plane π. Let $g_{ij} := \langle e_i, e_j \rangle$. The *Stanilov operator* is defined by setting:

$$\Theta(\pi) := \sum_{i,j,k,l} g^{ij} g^{kl} \mathcal{R}(e_i, e_k) \mathcal{R}(e_j, e_\ell).$$

Thus by Lemma 2.7.1, $\Theta(\pi)^2 = 0$. Since $\Theta(\pi)(\mathcal{T} + \mathcal{V}) = 0$, the Jordan normal form of $\Theta(\pi)$ is determined by

$$\dim\{\Theta(\pi)(\mathcal{U})\}.$$

We first show $\mathfrak{M}_0$ is Jordan Stanilov of type $(0, k)$ for $2 \leq k \leq s$. Let σ be a spacelike k-plane. We apply Lemma 2.7.3 to normalize σ so that we may choose an orthonormal basis $\{X_i\}$ for π so

$$X_i = a_i U_i + \mathcal{T} + \mathcal{V} \quad \text{for} \quad 1 \leq i \leq k.$$

Let $1 \leq i, j \leq k$. We use Eq. (2.7.b) to see that:

$$\mathcal{A}(X_i, X_j)^2 U_\ell = \begin{cases} a_i^2 a_j^2 V_\ell & \text{if } \ell = i, j, \\ 0 & \text{otherwise}. \end{cases}$$

Since $\Theta(\pi) = \sum_{i,j} A(X_i, X_j)^2 = \sum_{i,j} a_i^2 a_j^2 A(U_i, U_j)^2$, we show $\mathfrak{M}_0$ is k-spacelike Jordan Stanilov by computing:

$$\Theta(\pi)U_i = \begin{cases} \sum_{j\neq i, 1 \leq j \leq k} a_i^2 a_j^2 V_i & \text{if } i \leq k, \\ 0 & \text{if } k+1 \leq i \leq s. \end{cases}$$

Suppose that π is a timelike k-plane. Let $\ell(\pi)$ be the rank of the bilinear form h restricted to π as discussed in the proof of Lemma 2.7.3. We apply Lemma 2.7.3 to assume without loss of generality there exists an orthonormal basis for π so

$$X_i = a_i U_i + \mathcal{T} + \mathcal{V}$$

where $a_i \neq 0$ for $1 \leq i \leq \ell$. Since

$$\Theta(\pi) = \sum_{i=1}^{\ell} \sum_{j=1}^{\ell} a_i^2 a_j^2 A(U_i, U_j)^2,$$

one has that

$$\mathrm{Rank}\{\Theta(\pi)\} = \ell.$$

If $\pi \in \mathrm{Gr}_{2s,0}(\mathbb{R}^{3s})$, then $\ell(\pi) = s$ and hence $\mathfrak{M}_0$ is Jordan Stanilov of type $(2s, 0)$. Since ℓ is not constant on $\mathrm{Gr}_{k,0}(\mathbb{R}^{3s})$ for $2 \leq k < 2s$, $\mathfrak{M}_0$ is not Jordan Stanilov of type $(k, 0)$ for $2 \leq k < 2s$. $\qquad\square$

2.8 Dunn Manifolds

We shall omit the proofs of most of the assertions in this section and content ourselves here for the most part with a summary of the relevant results; further details are available from Dunn (2006).

Definition 2.8.1 Let the index i range from 1 through s. Let $\varepsilon_i = \pm 1$ be a choice of signs and let $f_i \in C^\infty(\mathbb{R})$. Let $\{u_0, ..., u_s, v_0, ..., v_s, t_1, ..., t_s\}$ be coordinates on $\mathbb{R}^{3s+2}$. Let $\mathcal{M} := (\mathbb{R}^{3s+2}, g)$ where

$$g(\partial_{u_0}, \partial_{u_i}) := 2f_i(u_i)t_i, \qquad g(\partial_{u_i}, \partial_{u_i}) := -2u_0 t_i,$$
$$g(\partial_{u_i}, \partial_{v_i}) = g(\partial_{u_0}, \partial_{v_0}) := 1, \quad g(\partial_{t_i}, \partial_{t_i}) := \varepsilon_i.$$

Assume that there are a of the ε_i which are $+1$ and b of the ε_i which are -1; one has that $a + b = s$.

Let $\mathrm{Al}(n)$ be the group of invertible affine linear maps of $\mathbb{R}^n$.

Lemma 2.8.1   *Let $\mathcal{M}$ be as in Definition 2.8.1.*

(1) $\mathcal{M}$ is a generalized plane wave manifold of signature $(s+1+a, s+1+b)$.

(2) The possibly non-zero entries in $\nabla^\nu R$ are

(a) $R(\partial_{u_0}, \partial_{u_i}, \partial_{u_i}, \partial_{u_0}) = f_i(u_i)^2 \varepsilon_i$.

(b) $R(\partial_{u_0}, \partial_{u_i}, \partial_{u_i}, \partial_{t_i}) = f_i'(u_i) + 1$.

(c) $\nabla R(\partial_{u_0}, \partial_{u_i}, \partial_{u_i}, \partial_{u_0}; \partial_{u_i}) = 2f_i(u_i)\varepsilon_i(2f_i'(u_i) + 1)$.

(d) $\nabla R(\partial_{u_0}, \partial_{u_i}, \partial_{u_i}, \partial_{t_i}; \partial_{u_i}) = f_i''(u_i)$.

(e) $\nabla^2 R(\partial_{u_0}, \partial_{u_i}, \partial_{u_i}, \partial_{t_i}; \partial_{u_i}, \partial_{u_i}) = f^{(3)}(u_i)$.

(f) $\nabla^2 R(\partial_{u_0}, \partial_{u_i}, \partial_{u_i}, \partial_{u_0}; \partial_{u_i}, \partial_{u_i})$
$= 2\varepsilon_i \left[2(f_i'(u_i))^2 + f_i'(u_i) + 3f_i(u_i)f_i''(u_i) \right]$.

(g) For arbitrary ℓ, we have

$$\nabla^\ell R(\partial_{u_0}, \partial_{u_i}, \partial_{u_i}, \partial_{t_i}; \partial_{u_i}, \dots, \partial_{u_i}) = f^{(\ell+1)}(u_i).$$

(3) If $f_i(u_i) = 0$, or $f_i(u_i) = -\frac{1}{2}u_i + a_i$ where a_i is a constant, then $\mathcal{M}$ is a symmetric space.

(4) Suppose that $f_i' + 1 \neq 0$. Let ϕ be an isometry of $\mathcal{M}$. There exists $A_1 \in \mathrm{Al}(s+1)$, a smooth map $A_2 : \mathbb{R}^{s+1} \to \mathrm{Al}(s+1)$, and a smooth map $A_3 : \mathbb{R}^{2s+2} \to \mathrm{Al}(s)$ so that

$$\phi(\vec{u}, \vec{t}, \vec{v}) = (A_1\vec{u}, A_2(\vec{u})\vec{t}, A_3(\vec{u}, \vec{t})\vec{v}).$$

Proof.   The Christoffel symbols are given by:

$$\nabla_{\partial_{u_0}} \partial_{u_i} = \nabla_{\partial_{u_i}} \partial_{u_0} = -t_i \partial_{v_i} - f_i(u_i)\epsilon_i \partial_{t_i},$$
$$\nabla_{\partial_{u_i}} \partial_{u_i} = (2f_i'(u_i) + 1)t_i \partial_{v_0} + u_0 \epsilon_i \partial_{t_i},$$
$$\nabla_{\partial_{u_0}} \partial_{t_i} = \nabla_{\partial_{t_i}} \partial_{u_0} = f_i(u_i)\partial_{v_i},$$
$$\nabla_{\partial_{u_i}} \partial_{t_i} = \nabla_{\partial_{t_i}} \partial_{u_i} = f_i(u_i)\partial_{v_0} - u_0 \partial_{v_i}.$$

We see that $\mathcal{M}$ is a generalized plane wave manifold relative to the ordering

$$\{\partial_{u_0}, \dots, \partial_{u_s}, \partial_{v_0}, \dots, \partial_{v_s}, \partial_{t_1}, \dots, \partial_{t_s}\}.$$

Assertions (2) and (3) now follow easily.

To prove Assertion (4), we suppose $f_i'(u_i) + 1 \neq 0$. We apply Lemma 2.2.2. Let

$$\mathcal{X} := \mathrm{Span}\{\partial_{u_i}\}, \quad \mathcal{Y} := \mathrm{Span}\{\partial_{t_i}\}, \quad \mathcal{Z} := \mathrm{Span}\{\partial_{v_i}\}.$$

We then have

$$\nabla_{\partial_{u_i}}\partial_{u_j} \in \mathcal{Y} + \mathcal{Z}, \quad \nabla_{\partial_{t_i}}\partial_{t_j} \in \mathcal{Z}, \nabla_{\partial_{v_i}}\partial_{v_j} = 0\,.$$

We also have $\mathcal{Z} = \ker(R)$ and $\mathcal{Y} + \mathcal{Z} = \mathcal{Z}^{\perp}$. Thus both these subspaces are preserved by any isometry. Assertion (4) follows. $\qquad\square$

2.8.1 *Models and the structure groups*

Let $\{U_0, ..., U_s, V_0, ..., V_s, T_1, ..., T_s\}$ be a basis for $\mathbb{R}^{3s+2}$. We consider the 0-model $\mathfrak{M}_0 := (\mathbb{R}^{3s+2}, \langle \cdot, \cdot \rangle, A_0)$ where

$$\langle U_i, V_i \rangle = 1, \quad \langle T_i, T_i \rangle = \varepsilon_i, \quad A_0(U_0, U_i, U_i, T_i) = 1\,.$$

The following subspaces are invariantly defined and hence are necessarily preserved by any isomorphism of $\mathfrak{M}_0$:

$$\mathcal{K} := \{\xi \in V : A_0(\xi_1, \xi_2, \xi_3, \xi) = 0 \ \forall\ \xi_i\} = \mathrm{Span}\{V_0, ..., V_s\}$$
$$\mathcal{K}^{\perp} = \mathrm{Span}\{T_1, ..., T_s, V_0, ..., V_s\}\,.$$

Assume $f_i' + 1 \neq 0$ for $1 \leq i \leq s$. Set

$$\gamma_2 := \sum_{i=1}^{s} \left[\frac{\varepsilon_i}{(f_i' + 1)^2} \left(4(f_i')^2 + 2f_i' + 6f_i f_i'' - \frac{(f_i)^2 f_i'''}{f_i' + 1} \right) \right],$$
$$\beta_2 := \sum_{i=1}^{s} \frac{f_i^{(3)}(1 + f_i')}{\{f_i''\}^2} \quad \text{if} \quad f_i'' \neq 0 \quad \text{for} \quad 1 \leq i \leq s\,.$$

Theorem 2.8.1 *Assume that $f_i'(u_i) + 1 \neq 0$ for $1 \leq i \leq s$. Let $s \geq 2$.*

(1) $\mathfrak{M}_0$ is a 0-model for $\mathcal{M}_F$.
(2) If Θ is an isomorphism of $\mathfrak{M}_0$, then there exists a permutation σ of the indices $\{1, ..., s\}$ and constants a_0, b_i with $|a_0| b_i^2 = 1$ so that

$$\begin{aligned}
\Theta U_0 &= a_0 U_0 + \Xi_0 & \text{for some} \quad &\Xi_0 \in \mathcal{K}, \\
\Theta U_i &= b_i U_{\sigma(i)} + \Xi_i & \text{for some} \quad &\Xi_i \in \mathcal{K}^{\perp}, \\
\Theta T_i &= \mathrm{sign}(a_0) T_{\sigma(i)} + \bar{\Xi}_i & \text{for some} \quad &\bar{\Xi}_i \in \mathcal{K}\,.
\end{aligned}$$

(3) γ_2 is an invariant of $\mathfrak{M}_2(\mathcal{M}_F, P)$.
(4) If $f_i'' \neq 0$ for $1 \leq i \leq s$, then β_2 is an invariant of $\mathfrak{M}_2(\mathcal{M}_F, P)$.
(5) If $f_i'' \neq 0$ for $1 \leq i \leq s$ then $\mathcal{M}_F$ is not 2-curvature homogeneous.

Proof.   Assume $f_i(u_i) + 1 \neq 0$ for $1 \leq i \leq k$. To prove Assertion (1), set

$$
\begin{aligned}
&U_0 := \partial_{u_0} + \sum_j a_j \partial_{t_j}, \quad U_i := b_i \partial_{u_i} + \beta_i \partial_{v_0} + \tilde{\beta}_i \partial_{v_i}, \\
&T_i := \kappa_i \partial_{t_i} + \gamma_i \partial_{v_i}, \qquad\quad V_0 := \partial_{v_0}, \\
&V_i = b_i^{-1} \partial_{v_i} .
\end{aligned}
\tag{2.8.a}
$$

The potentially non-zero curvatures are then:

$$
\begin{aligned}
R(U_0, U_i, U_i, U_0) &= b_i^2 \{ f_i(u_i)^2 \varepsilon_i + 2a_i(f_i'(u_i) + 1) \}, \\
R(U_0, U_i, U_i, T_i) &= b_i^2 (f_i'(u_i) + 1) \varepsilon_i \kappa_i .
\end{aligned}
$$

To ensure that $R(U_0, U_i, U_i, U_0) = 0$ and $R(U_0, U_i, U_i, T_i) = +1$, we set

$$
\begin{aligned}
a_i &:= -\frac{f_i(u_i)^2 \varepsilon_i}{2(f_i'(u_i)+1)}, \\
\kappa_i &:= \varepsilon_i \operatorname{sign}(f_i'(u_i) + 1), \\
b_i &:= |f_i'(u_i) + 1|^{-1/2} .
\end{aligned}
\tag{2.8.b}
$$

The potentially non-zero inner products are

$$
\begin{aligned}
(U_0, V_0) &= 1, & (U_0, T_i) &= \kappa_i a_i + \gamma_i, \\
(U_0, U_i) &= b_i g_F(\partial_{u_0}, \partial_{u_i}) + \beta_i, & (T_i, T_i) &= 1, \\
(U_i, U_i) &= b_i^2 g_F(\partial_{u_i}, \partial_{u_i}) + 2b_i \tilde{\beta}_i, & (U_i, V_i) &= 1 .
\end{aligned}
$$

We complete the proof of the Assertion (1) by setting:

$$
\begin{aligned}
\gamma_i &:= -\kappa_i a_i, \quad \beta_i := -b_i g_F(\partial_{u_0}, \partial_{u_i}), \\
\tilde{\beta}_i &:= -\tfrac{1}{2} b_i g_F(\partial_{u_i}, \partial_{u_i}) .
\end{aligned}
$$

It will be convenient to compute ∇R and $\nabla^2 R$ on the basis constructed above. We use Eqs. (2.8.a) and (2.8.b) to see

$$
\begin{aligned}
&\nabla R(U_0, U_i, U_i, U_0; U_i) \\
&= \nabla R\left(\partial_{u_0} + \sum_j a_j \partial_{t_j}, b_i \partial_{u_i}, b_i \partial_{u_i}, \partial_{u_0} + \sum_j a_j \partial_{t_j}; b_i \partial_{u_i} \right) \\
&= b_i^3 \left[\nabla R(\partial_{u_0}, \partial_{u_i}, \partial_{u_i}, \partial_{u_0}; \partial_{u_i}) + 2a_i \nabla R(\partial_{u_0}, \partial_{u_i}, \partial_{u_i}, \partial_{t_i}; \partial_{u_i}) \right] \\
&= |f_i' + 1|^{-3/2} \left[2 f_i \varepsilon_i (2 f_i' + 1) - \frac{f_i^2 \varepsilon_i}{f_i' + 1} f_i'' \right] \\
&= \frac{f_i \varepsilon_i}{(f_i' + 1)^{5/2}} \left[2(2 f_i' + 1)(f_i' + 1) - f_i f_i'' \right]
\end{aligned}
$$

and

$$\nabla R(U_0, U_i, U_i, T_i; U_i)$$
$$= \nabla R(\partial_{u_0} + \sum_j a_j \partial_{t_j}, b_i \partial_{u_i}, b_i \partial_{u_i}, \kappa \partial_{t_i}; b_i \partial_{u_i}) \qquad (2.8.c)$$
$$= b_i^3 \kappa_i f_i'' = \frac{f_i^{(2)} \kappa_i}{|f_i' + 1|^{3/2}} \, .$$

We also have that

$$\nabla^2 R(U_0, U_i, U_i, S_i; U_i, U_i)$$
$$= \nabla^2 R(\partial_{u_0} + \sum_j a_j \partial_{t_j}, b_i \partial_{u_i}, b_i \partial_{u_i}, \kappa_i \partial_{t_i}; b_i \partial_{u_i}, b_i \partial_{u_i}) \qquad (2.8.d)$$
$$= b_i^4 \kappa_i \nabla^2 R(\partial_{u_0}, \partial_{u_i}, \partial_{u_i}, \partial_{t_i}; \partial_{u_i}, \partial_{u_i}) = \kappa_i f_i^{(3)}(f_i' + 1)^{-2}$$

and that

$$\nabla^2 R(U_0, U_i, U_i, U_0; U_i, U_i)$$
$$= \nabla^2 R(\partial_{u_0} + \sum_j a_j \partial_{t_j}, b_i \partial_{u_i}, b_i \partial_{u_i}, \partial_{u_0} + \sum_j a_j \partial_{t_j}; b_i \partial_{u_i}, b_i \partial_{u_i})$$
$$= b_i^4 \left[\nabla^2 R(\partial_{u_0}, \partial_{u_i}, \partial_{u_i}, \partial_{u_0}; \partial_{u_i}, \partial_{u_i}) \right.$$
$$\left. + 2a_i \nabla^2 R(\partial_{u_0}, \partial_{u_i}, \partial_{u_i}, \partial_{t_i}; \partial_{u_i}, \partial_{u_i}) \right] \qquad (2.8.e)$$
$$= b_i^4 \left[2\varepsilon_i \left(2(f_i')^2 + f_i' + 3f_i f_i'' \right) - f_i^2 f_i^{(3)} \varepsilon_i (f_i' + 1)^{-1} \right]$$
$$= \varepsilon_i (f_i' + 1)^{-2} \left\{ 4(f_i')^2 + 2f_i' + 6f_i f_i'' - (f_i)^2 f_i'''(f_i' + 1)^{-1} \right\} .$$

To prove the Assertion (2), we note $\Theta T_i \in \mathcal{K}^\perp$. We expand:

$$\Theta U_0 = a_0 U_0 + \sum_j (b_{0j} U_j + d_{0j} T_j) + \mathcal{K},$$
$$\Theta T_i = \qquad \sum_j f_{ij} T_j + \mathcal{K}, \qquad (2.8.f)$$
$$\Theta U_i = a_i U_0 + \sum_j b_{ij} U_j + \mathcal{K}^\perp .$$

For any $\xi_1, \xi_2 \in V$, we have that:

$$0 = A_0(\xi_1, U_0, U_0, \xi_2) = A_0(\Theta\xi_1, \Theta U_0, \Theta U_0, \Theta\xi_2) . \qquad (2.8.g)$$

Choose ξ_i so $\Theta\xi_1 = U_0$ and $\Theta\xi_2 = T_j$. We then have

$$0 = A_0(U_0, \Theta U_0, \Theta U_0, T_j) = b_{0j}^2 .$$

Consequently $b_{0j} = 0$. We have $\Theta\mathcal{K} = \mathcal{K}$. As $1 = \langle U_0, V_0 \rangle = \langle \Theta U_0, \Theta V_0 \rangle$, as $\Theta U_0 = a_0 U_0 + \mathcal{K}^\perp$, and as $\Theta V_0 \in \mathcal{K}$, $a_0 \neq 0$. Choosing $\Theta\xi_1 = \Theta\xi_2 = U_i$ in Eq. (2.8.g) we have:

$$0 = A_0(U_i, \Theta U_0, \Theta U_0, U_i) = 2a_0 d_{0j} .$$

Since $a_0 \neq 0$, $d_{0j} = 0$ so Eq. (2.8.f) becomes

$$\Theta U_0 = a_0 U_0 + \mathcal{K}, \quad \Theta T_i = \sum_j f_{ij} T_j + \mathcal{K},$$

$$\Theta U_i = a_i U_0 + \sum_j b_{ij} U_j + \mathcal{K}^\perp.$$

Since $\Theta V_i \in \mathcal{K}$, the matrix b_{ij} is invertible. Suppose $b_{ij} \neq 0$. Choose ξ_1 so $\Theta \xi_1 = T_j$. Since $s \geq 2$, we may choose $1 \leq k \leq s$ with $k \neq i$. Thus

$$0 = A_0(U_0, U_i, U_k, \xi_1) = A_0(\Theta U_0, \Theta U_i, \Theta U_k, \Theta \xi_1) = a_0 b_{ij} b_{kj}.$$

Thus if $b_{ij} \neq 0$, $b_{kj} = 0$ for $i \neq k$. So in the matrix b_{ij}, each column has at most one non-zero entry. Since the matrix b_{ij} is invertible, each column has exactly one non-zero entry. So one has:

$$\Theta U_0 = a_0 U_0 + \mathcal{K}, \quad \Theta T_i = \sum_j f_{ij} T_j + \mathcal{K},$$

$$\Theta U_i = a_i U_0 + b_i U_{\sigma(i)} + \mathcal{K}^\perp.$$

The relation $\delta_{ij} = A_0(\Theta U_0, \Theta U_i, \Theta U_i, \Theta T_j)$ shows $f_{ij} = 0$ for $j \neq \sigma(i)$. Since ΘT_j is a unit vector, this coefficient is ± 1. Thus

$$\Theta U_0 = a_0 U_0 + \mathcal{K}, \quad \Theta T_i = \pm T_{\sigma(i)} + \mathcal{K}, \quad \Theta U_i = a_i U_0 + b_i U_{\sigma(i)} + \mathcal{K}^\perp.$$

Since $1 = A_0(\Theta U_0, \Theta U_i, \Theta U_i, \Theta T_i)$, we have

$$\pm b_i^2 a_0 = 1.$$

Finally given $1 \leq j \leq s$, we may choose $1 \leq i \leq s$ so $i \neq j$. As

$$0 = A_0(\Theta U_i, \Theta U_j, \Theta U_j, \Theta T_i)$$

we have $a_i b_j = 0$ and hence $a_i = 0$. Assertion (2) follows as we have

$$|a_0| b_i^2 = 1 \quad \text{and} \quad \Theta T_i = \text{sign}(a_0) T_{\sigma(i)}.$$

To establish Assertion (3), we use Eqs. (2.8.d) and (2.8.e) to see

$$\gamma_2 = \sum_{i=1}^{s} \nabla^2 R(U_0, U_i, U_i, U_0; U_i, U_i).$$

If Θ is an isomorphism of $\mathfrak{M}_0$, then Assertion (3) now yields

$$\nabla^2 R(\Theta U_0, \Theta U_i, \Theta U_i, \Theta U_0; \Theta U_i, \Theta U_i)$$
$$= a_0^2 b_i^4 \nabla^2 R(U_0, U_{\sigma(i)}, U_{\sigma(i)}, U_0; U_{\sigma(i)}, U_{\sigma(i)}).$$

Since $a_0^2 b_i^4 = 1$, summing over i shows γ_2 is an invariant of the 2-model.

To prove Assertion (4), we use Eq. (2.8.c) to see that

$$\beta_2 = \sum_{j=1}^{s} \frac{\nabla^2 R(U_0, U_j, U_j, T_j; U_j, U_j)}{(\nabla R(U_0, U_j, U_j, T_j; U_j))^2} \, .$$

The proof that this is independent of the normalized basis chosen and hence is an invariant of the 2-model now follows along similar lines as that used to prove Assertion (4).

Assertion (5) now follows from Assertions (3) and (4) with a bit of work; there are no functions f_i with $f_i' + 1 \neq 0$, $f_i'' \neq 0$, γ_2 constant, and β_2 constant. We refer to Dunn (2006) for further details. $\qquad\square$

2.8.2 *Invariants which are not of Weyl type*

Theorem 2.8.1 gives information about the structure group of the 0-model $\mathfrak{M}_0$ that can be used to define other invariants which are not of Weyl type. Let $\mathbb{RP}^{s-1}$ denote the projective space of real lines in $\mathbb{R}^s$. The symmetric group acts on $\mathbb{RP}^{s-1}$ by permuting the coordinates of $\mathbb{R}^s$. Let $\mathbb{SRP}^{s-1}$ be the quotient of $\mathbb{RP}^{s-1}$ under this group action and let $\pi : \mathbb{R}^s - \{0\} \to \mathbb{SRP}^{s-1}$ be the natural projection. Extend

$$\pi : \mathbb{R}^s \to \mathbb{SRP}^{s-1} \cup \{\star\}$$

by setting $\pi(0) = \star$. Let

$$\begin{aligned}
\Gamma &:= \pi\big(\nabla R(U_0, U_1, U_1, T_1; U_1), \ldots, \nabla R(U_0, U_s, U_s, T_s; U_s)\big) \\
&= \pi\big(f_1''(f_1' + 1)^{-3/2}, \ldots, f_s''(f_s' + 1)^{-3/2}\big) \in \mathbb{SRP}^{s-1} \cup \{\star\}, \\
\Gamma_0 &:= \pi\big(\nabla R(U_0, U_1, U_1, U_0; U_1), \ldots, \nabla R(U_0, U_s, U_s, U_0; U_s)\big) \\
&= \pi\big(f_1 \varepsilon_1 (f_1' + 1)^{-5/2}(2(2f_1' + 1)(f_1' + 1) - f_1 f_1''), \ldots \\
&\qquad \tfrac{f_s \varepsilon_s}{(f_s' + 1)^{5/2}}(2(2f_s' + 1)(f_s' + 1) - f_s f_s'')\big) \in \mathbb{SRP}^{s-1} \cup \{\star\} \, .
\end{aligned}$$

There is a unique line $\mathbb{I} := (1, \ldots, 1) \cdot \mathbb{R} \in \mathbb{RP}^{s-1}$ which is fixed by the action of the symmetric group. If $\Gamma \neq 0$ (respectively if $\Gamma_0 \neq 0$), then we can take the cosine of the angle between $\Gamma \cdot \mathbb{R}$ (respectively $\Gamma_0 \cdot \mathbb{R}$) and $\mathbb{I}$ to define

$$\Xi = \frac{\sum_j \frac{f_j''}{(f_j'+1)^{3/2}}}{\sqrt{s}\sqrt{\sum_j \frac{(f_j'')^2}{(f_j'+1)^3}}} \quad \text{and}$$

$$\Xi_0 = \frac{\sum_j \frac{f_i}{(f_i'+1)^{5/2}}[2(2f_i'+1)(f_i'+1)-f_i f_i'']}{\sqrt{s}\sqrt{\sum_j \left[\frac{f_i^2}{(f_i'+1)^5}[2(2f_i'+1)(f_i'+1)-f_i f_i'']^2\right]}}.$$

We set $\Xi = \star$ if $\Gamma = \{0\}$ and we set $\Xi_0 = \star$ if $\Gamma_0 = \{0\}$. The following result now follows from Theorem 2.8.1 and from the discussion given above; again, we refer to Dunn (2006) for further details:

Theorem 2.8.2 *Suppose $f_i' + 1 \neq 0$. Then Γ, Γ_0, Ξ, and Ξ_0 are invariants of $\mathfrak{M}_1(\mathcal{M}_F)$.*

2.9 k-Curvature Homogeneous Manifolds I

We shall follow the discussion in Gilkey and Nikčević (2004d) throughout this section. We recall that by Theorem 1.4.2, there exists an integer $k_{p,q}$, which is called the *Singer number*, so that if $\mathcal{M}$ is a complete simply connected pseudo-Riemannian manifold of signature (p, q) which is $k_{p,q}$-curvature homogeneous, then $\mathcal{M}$ is homogeneous. The following result provides a lower bound if $\min(p, q) \geq 3$; the case where $\min(p, q) = 2$ was dealt with in Section 2.3, see Theorem 2.3.7 for details.

Theorem 2.9.1 *Let $r := \min(p, q) \geq 3$. Then $k_{p,q} \geq r$.*

This section contains a number of other results which are of independent interest and which play a role in the proof of Theorem 2.9.1. Let $\ell \geq 0$. We shall construct a family of generalized plane wave Fiedler manifolds of signature $(\ell + 3, \ell + 3)$ so that certain manifolds in this family are $(\ell + 2)$-curvature homogeneous but are not $(\ell + 3)$-affine curvature homogeneous. These manifolds are 0-modeled on a decomposable symmetric space $\mathcal{N}_0$ and are i-modeled on a homogeneous space $\mathcal{N}_i$ for $1 \leq i \leq \ell + 2$. The homogeneous spaces $\mathcal{N}_i$ are j-modeled on $\mathcal{N}_j$ for $j \leq i$ but not j-modeled on $\mathcal{N}_j$ for $j > i$. This filtration illustrates the fact that the j-model does not capture the full geometry and thus these examples are of interest in their own right.

Certain elements in this family are not homogeneous but contain a proper open subset which is homogeneous; this does not occur in the Riemannian setting as if $\mathcal{M}$ is a Riemannian space which is homogeneous, then $\mathcal{M}$ is complete.

Definition 2.9.1 Let $\ell \geq 0$. Let indices i and j range from 0 through ℓ. Let $(x, y, z_0, ..., z_\ell, \tilde{x}, \tilde{y}, \tilde{z}_0, ..., \tilde{z}_\ell)$ be coordinates on $\mathbb{R}^{6+2\ell}$. Let $F(y, \vec{z})$ be an *affine warping function* of $\vec{z}$; this means that:

$$F(y, \vec{z}) := f(y) + f_0(y)z_0 + ... f_\ell(y)z_\ell\,.$$

Let $\mathcal{M} = \mathcal{M}_F = (\mathbb{R}^{6+2\ell}, g)$ where $g = g_F$ has non-zero components:

$$g(\partial_x, \partial_x) := -2F(y, \vec{z}) \quad \text{and} \quad g(\partial_x, \partial_{\tilde{x}}) = g(\partial_y, \partial_{\tilde{y}}) = g(\partial_{z_i}, \partial_{\tilde{z}_i}) := 1\,.$$

We remark that if one took $\ell = -1$, so that the $\vec{z}$ variables were not present, one would obtain the manifolds of signature $(2, 2)$ discussed in Section 2.3 where $g(\partial_x, \partial_x) = -2f(y)$, and $g(\partial_x, \partial_{\tilde{x}}) = g(\partial_y, \partial_{\tilde{y}}) = 1$. Thus the manifolds presented here are a higher signature generalization of those results. We begin our investigation with:

Lemma 2.9.1 *Let $\mathcal{M}$ be as in Definition 2.9.1. Then $\mathcal{M}$ is a generalized plane wave manifold of signature $(3 + \ell, 3 + \ell)$. Let $\{\xi_1, ..., \xi_{\nu+2}\}$ be coordinate vector fields. The possibly non-zero components of $\nabla^\nu R$ are:*

$$\nabla^\nu R(\partial_x, \xi_1, \xi_2, \partial_x; \xi_3, ..., \xi_{2+\nu}) = \xi_1 ... \xi_{\nu+2} F\,.$$

Proof. The manifolds of Definition 2.9.1 are obtained by an appropriate specialization of the warping functions ψ of Definition 2.5.1. Lemma 2.9.1 now follows from Lemma 2.5.1. $\qquad\square$

We note that since F is an affine function of the $\vec{z}$ variables, that $\nabla^\nu R$ vanishes unless either all the ξ_μ are ∂_y or unless all but one of the ξ_μ are ∂_y and the remaining variable is ∂_{z_i} for some i. This will play an important role in our analysis.

Definition 2.9.2 Specialize the warping function F of Definition 2.9.1 to create the following examples:

(1) Let $\mathcal{N}_i$ be specified by $F := yz_0 + ... + y^{i+1}z_i$ for $0 \leq i \leq \ell$.
(2) Let $\mathcal{N}_{\ell+1}$ be specified by $F := yz_0 + ... + y^{\ell+1}z_\ell + y^{\ell+3}$.
(3) Let $\mathcal{N}_{\ell+2}$ be specified by $F := yz_0 + y^2z_1 + ... + y^{\ell+1}z_\ell + e^y$.

The manifolds $\mathcal{N}_\mu$ form a sequence of homogeneous spaces partially sharing the same models:

Theorem 2.9.2   *Let $\mathcal{N}_\mu$ be as in Definition 2.9.2 for $0 \le k \le \ell + 2$.*

(1) $\mathcal{N}_0$ is a symmetric space.
(2) If $1 \le k \le \ell + 2$, then $\mathcal{N}_k$ is not symmetric but is homogeneous.
(3) If $k < n$, then $\mathcal{N}_n$ is k-modeled on $\mathcal{N}_k$.
(4) If $n < k$, then $\mathcal{N}_n$ not k-modeled on $\mathcal{N}_k$.

We remark that if $\mu < \ell$, then $\mathcal{N}_\mu$ is a product manifold and hence is decomposable. In Section 2.10, we shall discuss a similar family which is comprised of indecomposable manifolds.

We now turn our attention to affine geometry. Recall that the *affine k-model* is given by:

$$\mathfrak{F}_k(\mathcal{M}, P) := (T_P M, \mathcal{R}, ..., \nabla^k \mathcal{R}).$$

We specialize Definition 2.9.1 to define the following family:

Definition 2.9.3   Let $f \in C^\infty(\mathbb{R})$ where $f^{(\ell+3)} > 0$ and $f^{(\ell+4)} > 0$. Let

$$\alpha_\mu := f^{(\ell+\mu+2)}(f^{(\ell+3)})^{\mu-2}(f^{(\ell+4)})^{1-\mu} \quad \text{for} \quad \mu \ge 3.$$

Let $\mathcal{M} = \mathcal{M}_f$ be as in Definition 2.9.1 where

$$F(y, \vec{z}) := f(y) + y z_0 + ... + y^{\ell+1} z_\ell.$$

Theorem 2.9.1 will follow in the special case $p = q$ from:

Theorem 2.9.3   *Let $\mathcal{M}$ be as in Definition 2.9.3*

(1) $\mathcal{M}$ is $(\ell + 2)$-modeled on $\mathcal{N}_{\ell+2}$.
(2) α_μ is an invariant of the $(\mu + \ell)$-affine model $\mathfrak{F}_{\mu+\ell}(\mathcal{M}, P)$.
(3) If f_1 and f_2 are real analytic and if $\alpha_\mu(f_1)(P_1) = \alpha_\mu(f_2)(P_2)$ for all $\mu \ge 3$, then there is an isometry $\Phi : \mathcal{M}_{f_1} \to \mathcal{M}_{f_2}$ with $\Phi(P_1) = P_2$.
(4) The following assertions are equivalent:

> *(a) $\mathcal{M}$ is $(\ell + 3)$-curvature homogeneous.*
> *(b) $\mathcal{M}$ is $(\ell + 3)$-affine curvature homogeneous.*
> *(c) α_3 is constant.*
> *(d) $f^{(\ell+3)}(y) = a e^{by}$ for some $a > 0$ and $b > 0$.*
> *(e) $\mathcal{M}$ is homogeneous.*

We can also study some local examples:

Theorem 2.9.4 *Let $n \in \mathbb{N}$ with $n \geq \ell + 5$. Let $\mathcal{L}_n$ be as in Definition 2.9.3 where $f(y) = y^n$. Let $O = \{P : y > 0\}$. Let $\mathcal{O}_n = (O, g_f|_O)$. Then:*

(1) $\mathcal{L}_n$ is ℓ-modeled on $\mathcal{N}_\ell$.
(2) $\mathcal{L}_n$ is not $(\ell + 1)$-curvature homogeneous.
(3) $\mathcal{O}_n$ is homogeneous and $(\ell + 2)$-modeled on $\mathcal{N}_{\ell+2}$.
(4) $\mathcal{O}_{n_1}$ is not locally isometric to $\mathcal{O}_{n_2}$ for $n_1 \neq n_2$.

The following result classifies the local isometry type of $(\ell + 3)$-affine curvature homogeneous manifolds in this family. Instead of taking $\mathbb{R}^{6+2\ell}$ as the underlying manifold, we can apply the construction of Definition 2.9.3 to any open subset of $\mathbb{R}^{6+2\ell}$. Let $\mathcal{O}_a$ be defined by the warping function $f(y) := y^a$ on the set $y > 0$ for any $a \in \mathbb{R}$.

Theorem 2.9.5 *Let $f = f(y)$ be an analytic function on a connected open subset $U \subset \mathbb{R}^{2\ell+6}$ with $f^{(\ell+3)} > 0$ and $f^{(\ell+4)} > 0$. Let $\mathcal{U}$ be as in Definition 2.9.3. If $\mathcal{U}$ is $(\ell + 3)$-affine curvature homogeneous, then $\mathcal{U}$ is locally isometric to either $\mathcal{N}_{\ell+2}$ or to $\mathcal{O}_a$ for some a and $\mathcal{U}$ is locally homogeneous.*

Here is a brief guide to the remainder of this Section. In Section 2.9.1, we give k-models for this family. In Section 2.9.2, we discuss invariants for this family which are not of Weyl type. These two sections lead to the proof of Theorems 2.9.2–2.9.5. In Section 2.9.3, we change the signature by taking a product with a flat factor to complete the proof of Theorem 2.9.1. The argument is quite analogous to that performed in Section 2.3.3 to pass from signature $(2, 2)$ to $(2, s)$ and $(s, 2)$ where $s > 2$.

2.9.1 *Models*

We introduce the following models:

Definition 2.9.4 Let $\{X, Y, Z_0, ..., Z_\ell, \bar{X}, \bar{Y}, \bar{Z}_0, ..., \bar{Z}_\ell\}$ be a basis for $\mathbb{R}^{2\ell+6}$. Give $\mathbb{R}^{2\ell+6}$ the hyperbolic inner product:

$$\langle X, \bar{X} \rangle = \langle Y, \bar{Y} \rangle = \langle Z_i, \bar{Z}_i \rangle := 1 \,. \tag{2.9.a}$$

For $0 \leq i \leq \ell$, define algebraic curvature tensors and algebraic covariant derivative curvature tensors by setting:

$$A_i(X, Y, Z_i, X; Y, ..., Y) = A_i(X, Y, Y, X; Z_i, Y, ..., Y)$$
$$= ... = A_i(X, Y, Y, X; Y, ..., Y, Z_i) := 1.$$

As exceptional cases define

$$A_i(X, Y, Y, X; Y, ..., Y) := 1 \quad \text{for} \quad i = \ell + 1, \ell + 2.$$

We use the inner product to raise indices and define the associated curvature operators $\mathcal{A}_i$ for $0 \le i \le \ell + 2$. For $0 \le k \le \ell + 2$, define models

$$\mathfrak{M}_k := (\mathbb{R}^{2\ell+6}, \langle \cdot, \cdot \rangle, A_0, ..., A_k) \quad \text{and} \quad \mathfrak{F}_k := (\mathbb{R}^{2\ell+6}, \mathcal{A}_0, ..., \mathcal{A}_k).$$

We begin our study by showing:

Lemma 2.9.2 *Let $L \le \ell$. Let $\mathcal{M}$ be as in Definition 2.9.1 where we take $F = yz_0 + ... + y^{L+1}z_L + f(y, z_{L+1}, ..., z_\ell)$; if $L = \ell$, take $f = f(y)$.*

(1) $\mathfrak{M}_L$ is a L-model for $\mathcal{M}$.
(2) If $L = \ell$ and if $f^{(\ell+3)} > 0$, then $\mathfrak{M}_{\ell+1}$ is an $(\ell + 1)$-model for $\mathcal{M}$.
(3) If $L = \ell$, if $f^{(\ell+3)} > 0$, and if $f^{(\ell+4)} > 0$, then $\mathfrak{M}_{\ell+2}$ is an $(\ell+2)$-model for $\mathcal{M}$.

Proof. Let $0 \le k \le L \le \ell$. By Lemma 2.9.1,

$$\nabla^k R(\partial_x, \partial_y, \partial_{z_i}, \partial_x; \partial_y, ..., \partial_y) = \partial_y^{k+1}(y^{i+1})$$
$$= \begin{cases} 0 & \text{if} \quad i < k \le L, \\ (k+1)! & \text{if} \quad i = k \le L. \end{cases} \tag{2.9.b}$$

We shall exploit the upper triangular form of Eq. (2.9.b) to prove Assertion (1). Let $a^i(y, \vec{z})$ and $b_i^j(y, \vec{z})$ be smooth functions that will be chosen presently. Set

$$X := \partial_x - \tfrac{1}{2}g(\partial_x, \partial_x)\partial_{\tilde{x}},$$

$$Z_i := \begin{cases} \partial_{z_i} + \displaystyle\sum_{j=0}^{L} b_i^j \partial_{z_j} & \text{if} \quad L < i \le \ell, \\[2mm] b_i^i \partial_{z_i} + \displaystyle\sum_{j=0}^{i-1} b_i^j \partial_{z_j} & \text{if} \quad 0 \le i \le L, \end{cases}$$

$$Y := \partial_y + a^0 \partial_{z_0} + ... + a^L \partial_{z_L}.$$

We shall take $b_i^i \neq 0$ for $0 \leq i \leq L$. The space $\mathrm{Span}\{X, Z_0, ..., Z_L, Y\}$ will then be a maximal totally isotropic subspace. Consequently, we may take

$$\tilde{X} = \partial_{\tilde{x}}, \quad \tilde{Y} = \partial_{\tilde{y}}, \quad \tilde{Z}_i \in \mathrm{span}\{\partial_{\tilde{y}}, \partial_{\tilde{z}_i}\}$$

so that the metric is hyperbolic; this means that $\langle \cdot, \cdot \rangle$ has the normalizations of Eq. (2.9.a).

To ensure that $\nabla^k R(X, Y, Y, X; Y, ..., Y) = 0$ for $0 \leq k \leq L$, we impose the conditions

$$0 = \nabla^k R(\partial_x, \partial_y, \partial_y, \partial_x; \partial_y, ...)$$
$$+ (k+1) \sum_{i=0}^{L} a^k \nabla^k R(\partial_x, \partial_y, \partial_{z_i}, \partial_x; \partial_y, ..., \partial_y)$$
$$= \partial_y^{k+2} F + (k+1) \sum_{i=k}^{L} a^k \partial_y^{k+1}\{y^{i+1}\}.$$

We set $k = L$ to determine a^L. Once a^L is determined, we set $k = L - 1$ to determine a^{L-1}. We continue in this fashion to determine all the coefficients a^k and thereby define Y.

Similarly, we must impose the relations:

$$\nabla^k R(X, Y, Z_j, X; Y, ...) = \delta_{jk} \quad \text{for} \quad 0 \leq j \leq \ell, 0 \leq k \leq L. \qquad (2.9.c)$$

Let $L < i \leq \ell$. We use Eqs. (2.9.b) and (2.9.c) to see:

$$0 = \nabla^k R(\partial_x, \partial_y, \partial_y, \partial_{z_i}, \partial_x; \partial_y, ..., \partial_y)$$
$$+ \sum_{j=k}^{L} b_i^j \nabla^k R(\partial_x, \partial_y, \partial_y, \partial_{z_j}, \partial_x; \partial_y, ..., \partial_y).$$

We take $k = L$ to determine b_i^L. Then we take $k = L - 1$ to determine b_i^{L-1}. Thus all the coefficients b_j^k may be determined recursively by this identity.

Next let $0 \leq i \leq L$. We determine b_i^i from the identity:

$$1 = b_i^i \nabla^i R(\partial_x, \partial_y, \partial_y, \partial_{z_i}, \partial_x; \partial_y, ..., \partial_y).$$

Fix $0 \leq k < i$. We have

$$0 = \sum_{j=k}^{i} b_i^j \nabla^k R(\partial_x, \partial_y, \partial_y, \partial_{z_j}, \partial_x; \partial_y, ..., \partial_y).$$

We take $k = i - 1$ to determine b^i_{i-1}; the remaining coefficients are then determined recursively. This creates a basis with the normalizations of Definition 2.9.4 which completes the proof of Assertion (1) of Lemma 2.9.2.

We suppose $L = \ell$ and that $f = f(y)$. The non-zero components of $\nabla^\nu R$ for $\nu > \ell$ are given by

$$\nabla^\nu R(\partial_x, \partial_y, \partial_y, \partial_x; \partial_y...\partial_y) = f^{(\nu+2)} \, .$$

There is still a bit of freedom left in the choice of basis we can use to normalize these coefficients. Let ε_0 and ε_1 be non-zero functions. We set

$$X^1 = \varepsilon_0 X, \quad Y^1 = \varepsilon_1 Y, \quad Z^1_i = \varepsilon_0^{-2}\varepsilon_1^{-i-1} Z_i \text{ for } \quad 0 \le i \le \ell$$
$$\bar{X}^1 = \varepsilon_0^{-1}\bar{X}, \quad \bar{Y}^1 = \varepsilon_1^{-1}\bar{Y}, \quad \bar{Z}^1_i = \varepsilon_0^2\varepsilon_1^{i+1}\bar{Z}_i \quad \text{ for } \quad 0 \le i \le \ell \, .$$

The normalizations of Definition 2.9.4 are preserved for $\{g, R, ..., \nabla^\ell R\}$. Furthermore, one has that:

$$\nabla^{\ell+1} R(X^1, Y^1, Y^1, X^1; Y^1...Y^1) = \varepsilon_0^2\varepsilon_1^{\ell+3} f^{(\ell+3)},$$
$$\nabla^{\ell+2} R(X^1, Y^1, Y^1, X^1; Y^1...Y^1) = \varepsilon_0^2\varepsilon_1^{\ell+4} f^{(\ell+4)} \, .$$

If $f^{(\ell+3)} > 0$, we may set $\varepsilon_1 = 1$ and $\varepsilon_0 = (f^{(\ell+3)})^{-1/2}$ to create a normalized basis and establish $\mathfrak{M}_{\ell+1}$ is a $(\ell + 1)$-model for $\mathcal{M}$. This proves Assertion (2). If $f^{(\ell+3)} > 0$ and $f^{(\ell+4)} > 0$, we may set

$$\varepsilon_1 := \frac{f^{(\ell+3)}}{f^{(\ell+4)}} \quad \text{and} \quad \varepsilon_0 := \frac{1}{\{\varepsilon_1^{\ell+3} f^{(\ell+3)}\}^{\frac{1}{2}}} \, .$$

This shows that $\mathfrak{M}_{\ell+2}$ is a $(\ell + 2)$-model for $\mathcal{M}$ and completes the proof of the Lemma. $\qquad\square$

2.9.2 *Affine invariants*

Let $\mathfrak{F}_\nu(\mathcal{M}, P) := (T_P M, \mathcal{R}_P, ..., \nabla^\nu \mathcal{R}_P)$ be the ν-affine model. We can now define affine invariants:

Lemma 2.9.3 *Let $f_i = f_i(y)$. Assume that $f_i^{(\ell+4)} > 0$. Let $\mathcal{M}_i$ be as in Definition 2.9.1 where $F = yz_0 + ... + y^{\ell+1}z_\ell + f_i(y)$. For $\mu \ge 3$, set*

$$\alpha_\mu(f) := f^{(\ell+\mu+2)}(f^{(\ell+3)})^{\mu-2}(f^{(\ell+4)})^{1-\mu} \, .$$

(1) If $\mathfrak{F}_{\mu+\ell}(\mathcal{M}_{f_1}, P_1) \approx \mathfrak{F}_{\mu+\ell}(\mathcal{M}_{f_2}, P_2)$, then $\alpha_\mu(f_1)(P_1) = \alpha_\mu(f_2)(P_2)$.
(2) If f_1 and f_2 are real analytic and if $\alpha_\mu(f_1)(P_1) = \alpha_\mu(f_2)(P_2)$ for all $\mu \ge 3$, then there is an isometry $\Phi : \mathcal{M}_{f_1} \to \mathcal{M}_{f_2}$ with $\Phi(P_1) = P_2$.

Proof. We generalize the argument given in Section 2.3. Choose X and Y in $T_P M$ so that $\nabla^{\ell+2}\mathcal{R}(X,Y;Y,...,Y)Y \neq 0$. Expand

$$X = a_0\partial_x + a_1\partial_y + \mathrm{Span}\{\partial_{z_i},\partial_{\tilde{z}_i},\partial_{\tilde{x}},\partial_{\tilde{y}}\},$$
$$Y = b_0\partial_x + b_1\partial_y + \mathrm{Span}\{\partial_{z_i},\partial_{\tilde{z}_i},\partial_{\tilde{x}},\partial_{\tilde{y}}\}\,.$$

If $\mu > \ell$, then

$$\nabla^\mu\mathcal{R}(X,Y;Y,...,Y)Y = f^{(\mu+2)}(a_0 b_1 - a_1 b_0)b_1^\mu(b_1\partial_{\tilde{x}} - b_0\partial_{\tilde{y}})\,.$$

In particular, taking $\mu = \ell + 2$ shows that $b_1(a_0 b_1 - a_1 b_0) \neq 0$. Choose a linear function θ so

$$\theta\{b_1\partial_{\tilde{x}} - b_0\partial_{\tilde{y}}\} \neq 0\,.$$

Set

$$\begin{aligned}
\Theta_\mu :&= \theta\{\nabla^{\ell+\mu}\mathcal{R}(X,Y;Y,...,Y)Y\} \\
&= f^{(\ell+\mu+2)}(a_0 b_1 - a_1 b_0)b_1^{\ell+\mu}\theta\{b_1\partial_{\tilde{x}} - b_0\partial_{\tilde{y}}\}\,.
\end{aligned}$$

One may then compute:

$$\begin{aligned}
\Theta_\mu\Theta_1^{\mu-2}\Theta_2^{1-\mu} &= f^{(\ell+\mu+2)}(f^{(\ell+3)})^{\mu-2}(f^{(\ell+4)})^{1-\mu} \\
&\quad \times \{(a_0 b_1 - a_1 b_0)\theta(b_1\partial_{\tilde{x}} - b_0\partial_{\tilde{y}})\}^{1+(\mu-2)+(1-\mu)} \\
&\quad \times b_1^{\ell+\mu+(\ell+1)(\mu-2)+(\ell+2)(1-\mu)} \\
&= f^{(\ell+\mu+2)}(f^{(\ell+3)})^{\mu-2}(f^{(\ell+4)})^{1-\mu} = \alpha_\mu\,.
\end{aligned}$$

This shows that α_μ is determined by the affine $(\ell+\mu)$-model and establishes Assertion (1). Conversely, if we normalize a basis as in Definition 2.9.4, then one has for $\mu \geq 3$ that

$$\nabla^{\ell+\mu}R(X,Y,Y,X;Y,...,Y) = \alpha_\mu\,.$$

Thus if $\alpha_\mu(f_1,P_1) = \alpha_\mu(f_2,P_2)$ for $\mu \geq 3$, then $\mathfrak{M}_\infty(\mathcal{M}_{f_1},P_1)$ and $\mathfrak{M}_\infty(\mathcal{M}_{f_2,P_2})$ are isomorphic. The requisite isomorphism of Assertion (2) now follows from Theorem 2.2.2. $\square$

Proof of Theorem 2.9.2. The manifold $\mathcal{N}_0$ is defined by $F = yz_0$. Since all the third derivatives of F vanish and since $\mathcal{N}_0$ is a generalized plane wave manifold, $\mathcal{N}_0$ is a symmetric space. By Lemma 2.9.2, $\mathfrak{M}_\mu$ is a μ-curvature model for $\mathcal{N}_\mu$; a-fortori, $\mathfrak{M}_\nu$ is a ν-curvature model for $\mathcal{N}_\mu$ if $\nu < \mu$. Since $\nabla R \neq 0$, the spaces $\mathcal{N}_\mu$ are not symmetric. Since $\nabla^\mu R_\nu = 0$ and $\nabla^\mu R_\mu \neq 0$ for $0 \leq \mu < \nu \leq \ell + 2$, $\mathfrak{M}_\mu$ is not a μ model for $\mathcal{R}_\nu$ if $\nu < \mu$. If $\mu < \ell + 2$, then $\nabla^k R = 0$ for $k > \mu$ and hence $\mathfrak{M}_\mu$ is k-curvature homogeneous for all k. Theorem 2.2.2 now implies $\mathcal{M}_\mu$ is homogeneous. The manifold $\mathcal{M}_{\ell+2}$ is defined by taking $f(y)$ to be an exponential. It now follows that $\alpha_i = 1$ for $i \geq 3$ and hence by Lemma 2.9.3, $\mathcal{M}_{\ell+2}$ is homogeneous. $\qquad\square$

Proof of Theorem 2.9.3. Assertion (1) follows from Lemma 2.9.2 and Assertions (2) and (3) follow from Lemma 2.9.3.

Clearly (4a) $\Rightarrow$ (4b) $\Rightarrow$ (4c). If α_3 is constant, then

$$f^{(\ell+5)} f^{(\ell+3)} = c(f^{(\ell+4)})^2$$

for some constant c. Lemma 1.5.5 then implies either that $f^{(\ell+3)} = ae^{by}$ or that $f^{(\ell+3)} = a(y + b)^c$. This latter case is not possible since we require $f^{(\ell+3)} > 0$ for all y. Consequently $f^{(\ell+3)} = ae^{by}$. This shows α_μ is constant for all $\mu \geq 3$ and hence $\mathcal{M}$ is homogeneous. Thus (4c) $\Rightarrow$ (4d) $\Rightarrow$ (4e). Clearly (4e) $\Rightarrow$ (4a). $\qquad\square$

Proof of Theorem 2.9.4. Assertion (1) follows from Lemma 2.9.2 (1). Since $\nabla^{\ell+1} R = 0$ if and only if $y = 0$, $\mathcal{N}_n$ is not $(\ell + 1)$-curvature homogeneous; this establishes Assertion (2).

The proof of Lemma 2.9.2 involved local computations. Since $f^{(\ell+3)} > 0$ and $f^{(\ell+4)} > 0$ on O, Assertion (3) follows. If $f = y^n$ for $n \geq \ell + 5$, then

$$\alpha_3 = \frac{n(n-1)...(n-\ell-4)y^{n-\ell-5} \cdot n(n-1)...(n-\ell-2)y^{n-\ell-3}}{n(n-1)...(n-\ell-3)y^{n-\ell-4} \cdot n(n-1)...(n-\ell-3)y^{n-\ell-4}}$$

$$= \frac{n-\ell-4}{n-\ell-3}.$$

This determines n and hence $\mathcal{O}_{n_1}$ is not isometric to $\mathcal{O}_{n_2}$ for $n_1 \neq n_2$. $\qquad\square$

Proof of Theorem 2.9.5. We apply Lemma 1.5.5 to see that if α_3 is constant, then $f^{(p+3)} = a(y + b)^k$ or $f^{(p+3)} = ae^{by}$. The first possibility yields $\alpha_3 = \frac{k-1}{k}$ and the second possibility yields $\alpha_3 = 1$. Thus α_i for $i \geq 4$ is determined by α_3 and the required local isometry now follows from a suitable local version of Theorem 1.4.2. $\qquad\square$

2.9.3   *Changing the signature*

We take product with a flat factor to complete the proof of Theorem 2.9.1. We suppose without loss of generality that $(p, q) = (p, p + a)$ for $a > 0$ as the case $(p, q) = (q + a, q)$ is similar.

Lemma 2.9.4 *Let $\mathcal{M}$ be defined by setting $f(y) = e^y + e^{2y}$ in Definition 2.9.2. Let $\mathcal{M}_a := \mathcal{M} \times \mathbb{R}^{(0,a)}$. Then $\mathcal{M}_a$ is a $(\ell+2)$-curvature homogeneous generalized plane wave manifold of signature $(\ell+3, \ell+3+a)$ which is not $(\ell+3)$-affine curvature homogeneous.*

Proof. The arguments given above extend immediately to show that $\mathcal{M}_a$ is $(\ell+2)$-curvature homogeneous. Furthermore, exactly the same arguments show α_3 remains an affine invariant; see, for example, the discussion in Section 2.3.5. Furthermore, α_3 is constant if and only if the warping function satisfies $f^{(\ell+2)} = ae^{by}$. Since this is not the case, $\mathcal{M}_a$ is not $(\ell+3)$-affine curvature homogeneous. $\qquad\square$

2.9.4 *Indecomposability*

We complete this section by studying the decomposability of the symmetric space $\mathcal{S}$ or, equivalently, of the model $\mathfrak{M}_0$. It is clear that $\mathfrak{M}_0$ is decomposable if $\ell > 0$. On the other hand, we have that:

Lemma 2.9.5 *If $\ell = 0$, then $\mathfrak{M}_0$ is indecomposable.*

Proof. We set $\ell = 0$ so $\mathbb{R}^6 = \mathrm{Span}\{X, Y, Z_0, \tilde{X}, \tilde{Y}, \tilde{Z}_0\}$. Since the only non-zero curvature entry is given by $A_0(X, Y, Z_0, X) = 1$,

$$\ker(A_0) = \mathrm{Span}\{\tilde{X}, \tilde{Y}, \tilde{Z}_0\}$$

is totally isotropic. By Lemma 1.6.4 it suffices to show that the associated weak model $\bar{\mathfrak{M}}_0^w := (\mathbb{R}^3, \bar{A}_0)$ is indecomposable where $\bar{A}_0(\bar{X}, \bar{Y}, \bar{Z}_0, \bar{X}) = 1$. Suppose to the contrary that there is a non-trivial decomposition of the form $\mathbb{R}^3 = V^1 \oplus V^2$ which induces a corresponding decomposition $\bar{A}_0 = \bar{A}_0^1 \oplus \bar{A}_0^2$. Assume the notation chosen so

$$\dim(V^1) = 2 \quad \text{and} \quad \dim(V^2) = 1\,.$$

Let $0 \neq \xi \in V_2$. Since $\dim(V_2) = 1$, $\bar{A}_0^2 = 0$ so $\bar{A}_0(\eta_1, \eta_2, \eta_3, \xi) = 0$ for all $\eta_i \in \mathbb{R}^3$. We expand $\xi = a\bar{X} + b\bar{Y} + c\bar{Z}$. We then have

$$a = \bar{A}_0(\xi, \bar{Y}, \bar{Z}_0, \bar{X}) = 0,$$
$$b = \bar{A}_0(\bar{X}, \xi, \bar{Z}_0, \bar{X}) = 0,$$
$$c = \bar{A}_0(\bar{X}, \bar{Y}, \xi, \bar{X}) = 0\,.$$

Thus $\xi = 0$ which is false. $\qquad\square$

2.10   *k*-Curvature Homogeneous Manifolds II

In Section 2.9, we constructed $(\ell + 2)$-curvature homogeneous manifolds of neutral signature $(\ell + 3, \ell + 3)$ which were not $(\ell + 3)$-affine curvature homogeneous. If $\ell > 0$, these manifolds were modeled on a decomposable symmetric space. In this section, we present results of Gilkey and Nikčević (2005b) and of Gilkey and Nikčević (2005c) where the 0-model space is indecomposable. We also study the isometry groups of these spaces. Ensuring that the 0-model space is an indecomposable symmetric spaces requires raising the dimension of the underlying space; our examples will have neutral signature $(2\ell + 3, 2\ell + 3)$, they will be $(\ell + 2)$-curvature homogeneous, and they will not be $(\ell + 3)$-affine curvature homogeneous. These examples are all generalized plane wave Fiedler manifolds.

Definition 2.10.1   For $\ell \geq 1$, let

$$(x, y, z_1, ..., z_\ell, \tilde{y}, \tilde{z}_1, ..., \tilde{z}_\ell, x^*, y^*, z_1^*, ..., z_\ell^*, \tilde{y}^*, \tilde{z}_1^*, ..., \tilde{z}_\ell^*)$$

be coordinates on $\mathbb{R}^{4\ell+6}$. Let indices i, j range from 1 through ℓ. Let

$$F(y, \vec{z}) := f(y) + f_1(y)z_1 + ... + f_\ell(y)z_\ell$$

be an affine function of $\vec{z}$ which depends smoothly on y. We construct a generalized plane wave manifold $\mathcal{M} := (\mathbb{R}^{4\ell+6}, g)$ of signature $(3+2\ell, 3+2\ell)$ by defining:

$$g(\partial_x, \partial_x) := -2\{F(y, \vec{z}) + y\tilde{y} + z_1\tilde{z}_1 + ... + z_\ell\tilde{z}_\ell\}, \quad \text{and}$$
$$g(\partial_x, \partial_{x^*}) = g(\partial_y, \partial_{y^*}) = g(\partial_{\tilde{y}}, \partial_{\tilde{y}^*}) = g(\partial_{z_i}, \partial_{z_i^*}) = g(\partial_{\tilde{z}_i}, \partial_{\tilde{z}_i^*}) := 1.$$

A word on notation. The *dual variables* $\{x^*, y^*, z_i^*, \tilde{y}^*, \tilde{z}_i^*\}$ enter only rather trivially; the span of these variables is a parallel totally isotropic distribution of maximal dimension. The dependence of the metric on the variables $\{\tilde{y}, \tilde{z}_1, ..., \tilde{z}_\ell\}$ is fixed and ensures that the 0-model space is an indecomposable symmetric space; the variable y^* plays the role that z_0 played in Section 2.9. The crucial variables are $\{x, y, z_1, ..., z_\ell\}$. The following Lemma is an immediate consequence of Lemma 2.5.1

Lemma 2.10.1   *Let $\mathcal{M}$ be as in Definition 2.10.1. Then $\mathcal{M}$ is a generalized plane wave manifold. The possibly non-zero components of R are:*

$$R(\partial_x, \partial_y, \partial_{\tilde{y}}, \partial_x) = R(\partial_x, \partial_{z_i}, \partial_{\tilde{z}_i}, \partial_x) = 1,$$
$$R(\partial_x, \partial_y, \partial_{z_j}, \partial_x) = \partial_y\partial_{z_j}F,$$
$$R(\partial_x, \partial_y, \partial_y, \partial_x) = \partial_y^2 F.$$

If $\xi_1, \xi_2, ..., \xi_{\nu+2}$ are coordinate vector fields and if $\nu > 0$, then the possibly non-zero entries of $\nabla^\nu R$ are $\nabla^\nu R(\partial_x, \xi_1, \xi_2, \partial_x; \xi_3, ..., \xi_{\nu+2}) = \xi_1 ... \xi_{\nu+2} F$. This can be non-zero if at most one of the ξ_i is not equal to ∂_y and instead equals ∂_{z_i} for some i.

We remark that Lemma 2.2.1 and Corollary 2.2.1 can be used to get information about the isometries and Killing vector fields of $\mathcal{M}$ where we take $\mathcal{Y} := \ker(R)$ and where:

$$\mathcal{X} = \operatorname{Span}\{\partial_x, \partial_y, \partial_{\tilde{y}}, \partial_{z_i}, \partial_{\tilde{z}_i}\},$$
$$\mathcal{Y} = \operatorname{Span}\{\partial_{x^*}, \partial_{y^*}, \partial_{\tilde{y}^*}, \partial_{z_i^*}, \partial_{\tilde{z}_i^*}\}.$$

In analogy to Definition 2.9.2, we specialize the warping function F of Definition 2.10.1 to define certain homogeneous manifolds:

Definition 2.10.2 In Definition 2.10.1, let

(1) $\mathcal{N}_0$ be defined by $F := 0$,
(2) $\mathcal{N}_\nu$ be defined by $F := z_1 y^2 + ... + z_\nu y^{\nu+1}$ if $1 \le \nu \le \ell$,
(3) $\mathcal{N}_{\ell+1}$ be defined by $F := z_1 y^2 + ... + z_\ell y^{\ell+1} + y^{\ell+3}$,
(4) $\mathcal{N}_{\ell+2}$ be defined by $F := z_1 y^2 + ... + z_\ell y^{\ell+1} + e^y$.

In analogy with the results of Section 2.9, we have:

Theorem 2.10.1 *The manifolds $\mathcal{N}_\mu$ of Definition 2.10.2 are generalized plane wave manifolds of signature $(2\ell + 3, 2\ell + 3)$.*

(1) $\mathcal{N}_0$ is an indecomposable symmetric space.
(2) $\mathcal{N}_\mu$ is a homogeneous space which is not symmetric for $1 \le \mu \le \ell + 2$.
(3) If $\mu < \nu$, then $\mathcal{N}_\nu$ is μ-modeled on $\mathcal{N}_\mu$.
(4) If $\nu < \mu$, then $\mathcal{N}_\nu$ is not μ-modeled on $\mathcal{N}_\mu$.

We also define the following family:

Definition 2.10.3 Let $f \in C^\infty(\mathbb{R})$ where $f^{(\ell+3)} > 0$ and $f^{(\ell+4)} > 0$. Let $\mathcal{M} = \mathcal{M}_f$ be as in Definition 2.10.1 for $F(y, \vec{z}) := f(y) + y^2 z_1 + ... + y^{\ell+1} z_\ell$. Set $\alpha_\mu := f^{(\ell+\mu+2)} (f^{(\ell+3)})^{\mu-2} (f^{(\ell+4)})^{1-\mu}$ for $\mu \ge 3$.

Theorem 2.10.2 *Let $\mathcal{M}$ be as in Definition 2.10.3. Then:*

(1) $\mathcal{M}$ is $(\ell + 2)$-modeled on $\mathcal{N}_{\ell+2}$.
(2) α_k is an invariant of the affine $(k + \ell)$-model $\mathfrak{F}_{k+\ell}(\mathcal{M}, P)$.
(3) If f_1 and f_2 are real analytic and if $\alpha_k(f_1)(P_1) = \alpha_k(f_2)(P_2)$ for $k \ge 3$, then there is an isometry $\Phi : \mathcal{M}_{f_1} \to \mathcal{M}_{f_2}$ with $\Phi(P_1) = P_2$.
(4) The following assertions are equivalent:

(a) $\mathcal{M}$ is $(\ell+3)$-curvature homogeneous.

(b) $\mathcal{M}$ is $(\ell+3)$-affine curvature homogeneous.

(c) α_3 is constant.

(d) $f^{(\ell+3)}(y) = ae^{by}$ for some $a > 0$ and $b > 0$.

(e) $\mathcal{M}$ is homogeneous.

We shall also study the isometry groups $G(\cdot)$ of these manifolds and of their models. A byproduct of our study is the following result that shows, not surprisingly, that the symmetric space $\mathcal{N}_0$ has the largest isometry group.

Theorem 2.10.3   *Let $\mathcal{M}$ be as in Definition 2.10.3 where α_3 is non-constant, or, equivalently, $f^{(\ell+3)}(y) \neq ae^{by}$. Adopt the notation of Definition 2.10.2.*

(1) $\dim\{G(\mathcal{N}_0)\} = 4\ell + 6 + (\ell+1)(3+2\ell) + 2\ell + 3 + (\ell+1)(2\ell+1)$.

(2) If $1 \leq \mu \leq \ell$, then $\dim\{G(\mathcal{N}_\mu)\} = 4\ell + 6 + (\ell+1)(3+2\ell)$
$$+ 2\ell + 3 + 2\ell + 2 + \tfrac{1}{2}\mu(\mu-1) + (\ell-\mu)(2\ell-2\mu-1).$$

(3) $\dim\{G(\mathcal{N}_{\ell+1})\} = 4\ell + 6 + (\ell+1)(3+2\ell) + 2\ell + 3 + 2\ell + 1 + \tfrac{1}{2}\ell(\ell-1)$.

(4) $\dim\{G(\mathcal{N}_{\ell+2})\} = 4\ell + 6 + (\ell+1)(3+2\ell) + 2\ell + 2 + 2\ell + 1 + \tfrac{1}{2}\ell(\ell-1)$.

(5) $\dim\{G(\mathcal{M})\} = 4\ell + 5 + (\ell+1)(3+2\ell) + 2\ell + 2 + 2\ell + 1 + \tfrac{1}{2}\ell(\ell-1)$.

The remainder of this section is devoted to the proof of these results. In Section 2.10.1, we prove Theorems 2.10.1 and 2.10.2. The main new feature is showing that the 0-model is indecomposable, see Lemma 2.10.3 for details.

In Section 2.10.2, we establish Theorem 2.10.3. In Lemma 2.10.4, we relate the isometry group of the full manifold to the isotropy subgroups and reduce the proof of Theorem 2.10.3 from a geometric question to an algebraic question by relating the dimension of the isometry groups of $\mathcal{N}_\mu$ and $\mathcal{M}$ to the dimensions of the isometry groups of certain models. The dimension of the isometry group of the appropriate models is then studied in Lemmas 2.10.5 and 2.10.6 to complete the proof.

2.10.1   *Models*

We introduce the associated models we shall be studying in this section:

Definition 2.10.4   Let

$$\{X, Y, Z_1, ..., Z_\ell, \tilde{Y}, \tilde{Z}_1, ..., \tilde{Z}_\ell, X^*, Y^*, Z_1^*, ..., Z_\ell^*, \tilde{Y}^*, \tilde{Z}_1^*, ..., \tilde{Z}_\ell^*\}$$

be a basis for $\mathbb{R}^{4\ell+6}$. Define a hyperbolic inner product on $\mathbb{R}^{4\ell+6}$ by pairing ordinary variables with the corresponding dual variables:

$$\langle X, X^* \rangle = \langle Y, Y^* \rangle = \langle \tilde{Y}, \tilde{Y}^* \rangle = \langle Z_i, Z_i^* \rangle = \langle \tilde{Z}_i, \tilde{Z}_i^* \rangle := 1\,.$$

Let $\mathfrak{M}_\mu := (V, \langle \cdot, \cdot \rangle, A_0, A_1, ..., A_\mu)$ for $0 \le \mu \le \ell + 2$ where the algebraic curvature tensor A_0 has non-zero components

$$A_0(X, Y, \tilde{Y}, X) = A_0(X, Z_i, \tilde{Z}_i, X) := 1,$$

where the non-zero components of the algebraic covariant derivative curvature tensors for $1 \le i \le \ell$ are

$$A_i(X, Y, Z_i, X; Y, ..., Y) = ... = A_i(X, Y, Y, X; Y, ..., Y, Z_i) := 1,$$

and where as exceptional cases we have

$$A_k(X, Y, Y, X; Y, ..., Y) := 1 \quad \text{for} \quad k = \ell + 1, \ell + 2\,.$$

Lemma 2.9.2 generalizes to this setting to become:

Lemma 2.10.2 *Let $L \le \ell$. Let $\mathcal{M}$ be as in Definition 2.10.1 where we take $F = z_1 y^2 + ... + z_L y^{L+1} + f(y, z_{L+1}, ..., z_\ell)$; if $L = \ell$, take $f = f(y)$. Then:*

(1) $\mathfrak{M}_L$ is a L-model for $\mathcal{M}$.
(2) If $L = \ell$ and if $f^{(\ell+3)} > 0$, then $\mathfrak{M}_{\ell+1}$ is an $\ell + 1$-model for $\mathcal{M}$.
(3) If $L = \ell$, if $f^{(\ell+3)} > 0$, and if $f^{(\ell+4)} > 0$, then $\mathfrak{M}_{\ell+2}$ is an $\ell + 2$-model for $\mathcal{M}$.

Proof. By Lemma 2.10.1,

$$R(\partial_x, \partial_y, \partial_{\tilde{y}}, \partial_x) = R(\partial_x, \partial_{z_i}, \partial_{\tilde{z}_i}, \partial_x) = 1,$$
$$R(\partial_x, \partial_y, \partial_{z_i}, \partial_x) = \partial_y \partial_{z_i} F,$$
$$R(\partial_x, \partial_y, \partial_y, \partial_x) = \partial_y^2 F\,.$$

We set

$$
\begin{aligned}
X &:= \partial_x + F \partial_{x^*}, & X^* &:= \partial_{x^*}, \\
Y &:= \partial_y - \tfrac{1}{2}(\partial_y^2 F)\partial_{\tilde{y}}, & Y^* &:= \partial_{y^*}, \\
\tilde{Y} &:= \partial_{\tilde{y}}, & \tilde{Y}^* &:= \partial_{\tilde{y}^*} + \sum_i (\partial_y \partial_{z_i} F)\partial_{z_i^*} + \tfrac{1}{2}(\partial_y^2 F)\partial_{\tilde{y}}, \\
Z_i &:= \partial_{z_i} - (\partial_y \partial_{z_i} F)\partial_{\tilde{y}}, & Z_i^* &:= \partial_{z_i^*}, \\
\tilde{Z}_i &:= \partial_{\tilde{z}_i}, & \tilde{Z}_i^* &:= \partial_{\tilde{z}_i^*}\,.
\end{aligned}
$$

The only non-zero components of R and of g are then

$$g(X, X^*) = g(Y, Y^*) = g(\tilde{Y}, \tilde{Y}^*) = g(Z_i, Z_i^*) = g(\tilde{Z}_i, \tilde{Z}_i^*),$$
$$R(X, Y, \tilde{Y}, X) = R(X, Z_i, \tilde{Z}_i, X) = 1.$$

This shows that $\mathfrak{M}_0$ is a 0-model for $\mathcal{M}$. Normalizing the higher covariant derivatives then follows exactly the same lines as those used in Section 2.9 so we omit the proof in the interests of brevity. $\qquad\square$

One new feature for these examples is the indecomposability of $\mathfrak{M}_0$:

Lemma 2.10.3 $\mathfrak{M}_0$ *is indecomposable.*

Proof.   Suppose we have a non-trivial orthogonal direct sum decomposition $\mathbb{R}^{4\ell+6} = V_1 \oplus V_2$ which decomposes the curvature tensor A_0. Let π_i be the associated projections on V_i. Since

$$1 = \langle X, X^* \rangle = \langle \pi_1 X, X^* \rangle + \langle \pi_2 X, X^* \rangle$$

we may assume without loss of generality $\langle \pi_1 X, X^* \rangle \neq 0$. Set $\alpha := \pi_1(X)$. Let $\beta \in (X^*)^\perp \cap V_2$. Then $A_0(\alpha, \cdot, \beta, \alpha) = 0$ as $\alpha \in V_1$ and $\beta \in V_2$. Since β does not involve X,

$$0 = A_0(\alpha, Z_i, \beta, \alpha) = \langle \alpha, X^* \rangle^2 \langle \beta, \tilde{Z}_i^* \rangle, \quad \text{and}$$
$$0 = A_0(\alpha, \tilde{Z}_i, \beta, \alpha) = \langle \alpha, X^* \rangle^2 \langle \beta, Z_i^* \rangle.$$

Consequently $\langle \beta, X^* \rangle = 0$, $\langle \beta, Z_i^* \rangle = 0$, and $\langle \beta, \tilde{Z}_i^* \rangle = 0$. Thus

$$\beta \in \mathrm{Span}\{X^*, Z_0^*, ..., Z_\ell^*, \tilde{Z}_0^*, ..., \tilde{Z}_\ell^*\}$$

so $(X^*)^\perp \cap V_2$ is totally isotropic. Since the restriction of $\langle \cdot, \cdot \rangle$ to V_2 is non-degenerate and since

$$\dim\{(X^*)^\perp \cap V_2\} \geq \dim\{V_2\} - 1,$$

we conclude that $\dim\{V_2\} = 2$. Furthermore there must exist an element of V_2 not in $(X^*)^\perp$. We can therefore interchange the roles of V_1 and V_2 to see that $\dim\{V_1\} = 2$. This shows that

$$4\ell + 6 = \dim\{V_1\} + \dim\{V_2\} = 4$$

which provides the desired contradiction. $\qquad\square$

The arguments given in Section 2.9 now extend without change to establish Theorem 2.10.1 and Theorem 2.10.2; we omit details in the interests of brevity.

2.10.2 *Isometry groups*

The proof of Theorem 2.10.3 will be based on several Lemmas. We first reduce the geometric problem to an algebraic one:

Lemma 2.10.4 *Let $P \in \mathbb{R}^{4\ell+6}$. Let $0 \leq \mu \leq \ell + 2$. Let $\mathcal{M}$ be as in Definition 2.10.3 where f is real analytic. Assume that $f^{(\ell+3)} > 0$, that $f^{(\ell+4)} > 0$, and that $\alpha_3(f)$ is non-constant. Let $\mathcal{N}_\mu$ be as in Definition 2.10.2.*

(1) $\dim\{G(\mathcal{N}_\mu)\} = 4\ell + 6 + \dim\{G_P(\mathcal{N}_\mu)\}$.
(2) $\dim\{G(\mathcal{M})\} = 4\ell + 6 - 1 + \dim\{G_P(\mathcal{M})\}$.
(3) $G_P(\mathcal{N}_\mu) = G(\mathfrak{M}_\infty(\mathcal{N}_\mu, P))$.
(4) $G_P(\mathcal{M}) = G(\mathfrak{M}_\infty(\mathcal{M}, P))$.
(5) $G(\mathfrak{M}_\infty(\mathcal{N}_\mu, P)) = G(\mathfrak{M}_\mu)$.
(6) $G(\mathfrak{M}_\infty(\mathcal{M}, P)) = G(\mathfrak{M}_{\ell+2})$.
(7) $\dim\{G(\mathcal{N}_\mu)\} = 4\ell + 6 + \dim\{G(\mathfrak{M}_\mu)\}$.
(8) $\dim\{G(\mathcal{M})\} = 5 + 4\ell + \dim\{G(\mathfrak{M}_{\ell+2})\}$.

Proof. We apply Lemma 1.6.1 to the canonical action of the group of isometries on the underlying manifold. Assertion (1) follows as $\mathcal{N}_\mu$ is a homogeneous space. We consider the invariants α_i of Theorem 2.10.2. These functions are constant on the hyperplanes $y = c$; thus the group of isometries acts transitively on such a hyperplane. Consequently

$$\dim\{G(\mathcal{M})\} \geq \dim\{G_P(\mathcal{M})\} + 4\ell + 6 - 1\,.$$

Since $\mathcal{M}$ is not a homogeneous space, equality holds.

Assertions (3) and (4) follow from Theorem 2.2.2 since we are in the real analytic context.

Restriction induces injective maps

$$r : G(\mathfrak{M}_\infty(\mathcal{N}_\mu, P)) \to G(\mathfrak{M}_\mu(\mathcal{N}_\mu, P)), \quad \text{and}$$
$$r : G(\mathfrak{M}_\infty(\mathcal{M}, P)) \to G(\mathfrak{M}_\mu(\mathcal{M}, P))\,.$$

Since $\nabla^j R = 0$ for $j > \mu$ for $\mathcal{N}_\mu$ if $\mu \leq \ell + 1$, any isomorphism of the μ-model is an isomorphism of the ∞-model which establishes Assertion (5) for $0 \leq \mu \leq \ell + 1$.

We complete the proof of Assertions (5) and (6) by dealing with the manifolds $\mathcal{N}_{\ell+2}$ and $\mathcal{M}$. Choose a basis $\mathcal{B}$ for $T_P M$ giving the normalizations of Definition 2.10.4. If $g \in G(\mathfrak{M}_{\ell+2}(\mathcal{N}_{\ell+2}, P))$, then $g\mathcal{B}$ also satisfies the normalizations described above. Expand

$$gX = a_0 X + a_1 Y + \mathrm{Span}\{Z_1, ...\},$$
$$gY = b_0 X + b_1 Y + \mathrm{Span}\{Z_1, ...\}.$$

For $i \geq \ell + 1$, we have

$$\nabla^i R(gX, gY, gY, gX; gX, ..., gX) = (a_0 b_1 - a_1 b_0)^2 a_1^i,$$
$$\nabla^i R(gX, gY, gY, gX; gY, ..., gY) = (a_0 b_1 - a_1 b_0)^2 b_1^i.$$

Since

$$0 = \nabla^{\ell+1} R(gX, gY, gY, gX; gX, ..., gX),$$
$$1 = \nabla^{\ell+1} R(gX, gY, gY, gX; gY, ..., gY),$$
$$1 = \nabla^{\ell+2} R(gX, gY, gY, gX; gY, ..., gY)$$

we have $a_1 = 0$, $b_1 = 1$, and $a_0 = \pm 1$. Consequently g preserves $\nabla^i R$ for all $i \geq \ell + 1$ which establishes Assertions (5) and (6). Assertions (7) and (8) now follow from Assertions (1)-(6). $\qquad\qquad\square$

Let $\mathbb{R}^{3+2\ell} := \mathrm{Span}\{X, Y, Z_i, \tilde{Y}, \tilde{Z}_i\}$; we may restrict A_i to $\mathbb{R}^{3+2\ell}$ without losing information on the curvature. We introduce the affine models by restricting the domain and suppressing the metric:

$$\mathfrak{F}_\mu := (\mathbb{R}^{3+2\ell}, A_0, A_1, ..., A_\mu).$$

There is a natural action of $G(\mathfrak{F}_\mu)$ on $\mathbb{R}^{3+2\ell}$; let $G_X(\mathfrak{F}_\mu)$ be the isotropy subgroup of this action. We also consider the double isotropy group,

$$G_{X,Y}(\mathfrak{F}_\mu) = \{g \in G(\mathfrak{F}_\mu) : gX = X \text{ and } gY = Y\}.$$

Lemma 2.10.5

(1) If $0 \leq \mu \leq \ell + 2$, then $\dim\{G(\mathfrak{M}_\mu)\} = \dim\{G(\mathfrak{F}_\mu)\} + (\ell + 1)(3 + 2\ell)$.

(2) If $\mu \leq \ell + 1$, then $\dim\{G(\mathfrak{F}_\mu)\} = \dim\{G_X(\mathfrak{F}_\mu)\} + 2\ell + 3$.

(3) We have that $\dim\{G(\mathfrak{F}_{\ell+2})\} = \dim\{G_X(\mathfrak{F}_{\ell+2})\} + 2\ell + 2$.

(4) $\dim\{G_X(\mathfrak{F}_0)\} = (\ell + 1)(2\ell + 1)$.

(5) If $1 \leq \mu \leq \ell$, then $\dim\{G_X(\mathfrak{F}_\mu)\} = \dim\{G_{X,Y}(\mathfrak{F}_\mu)\} + 2\ell + 2$.

(6) If $\mu = \ell + 1, \ell + 2$, then $\dim\{G_X(\mathfrak{F}_\mu)\} = \dim\{G_{X,Y}(\mathfrak{F}_\mu)\} + 2\ell + 1$.

(7) $G_{X,Y}(\mathfrak{F}_\ell) = G_{X,Y}(\mathfrak{F}_{\ell+1}) = G_{X,Y}(\mathfrak{F}_{\ell+2})$.

Proof. Let $\mathfrak{o}(s)$ be Lie algebra of skew-symmetric $s \times s$ real matrices. Set

$$\mathcal{S} := (S_1, ..., S_{3+2\ell}) = (X, Y, Z_1..., Z_\ell, \tilde{Y}, \tilde{Z}_1..., \tilde{Z}_\ell),$$
$$\mathcal{S}^* := (S_1^*, ..., S_{3+2\ell}^*) = (X^*, Y^*, Z_1^*, ..., Z_\ell^*, \tilde{Y}^*, \tilde{Z}_1^*, ..., \tilde{Z}_\ell^*),$$
$$\ker(R) := \{\xi \in \mathbb{R}^{4\ell+6} : A_0(\xi, \eta_1, \eta_2, \eta_3) = 0 \ \forall \ \eta_i \in \mathbb{R}^{4\ell+6}\}$$
$$= \mathrm{Span}_{S \in \mathcal{S}^*}\{S^*\}.$$

Let $g \in G(\mathfrak{M}_\mu)$. The space $\ker(R)$ is preserved by g. Thus

$$gS_i = \sum_{k=1}^{3+2\ell}\left\{g_{0,ik}S_k + g_{1,ik}S_k^*\right\} \quad \text{and} \quad gS_i^* = \sum_{k=1}^{3+2\ell} g_{2,ik}S_k^*.$$

We have $\langle gS_i, gS_j \rangle = 0$ and $\langle gS_i, gS_j^* \rangle = \delta_{ij}$. Thus

$$\sum_{k=1}^{3+2\ell}\left\{g_{0,ik}g_{1,jk} + g_{1,ik}g_{0,jk}\right\} = 0 \quad \text{and} \quad \sum_{k=1}^{3+2\ell} g_{0,ik}g_{2,jk} = \delta_{ij}$$

for all i, j. Set $\gamma := g_0 g_1^t$. One then has

$$g_0 \in G(\mathfrak{F}_\mu), \quad \gamma + \gamma^t = 0, \quad \text{and} \quad g_0 g_2^t = \mathrm{id}.$$

On the other hand, if this identity is satisfied then $g \in G(\mathfrak{M}_\mu)$. The map $g \to (g_0, \gamma)$ yields an identification of

$$G(\mathfrak{M}_\mu) = G(\mathfrak{F}_\mu) \times \mathfrak{o}(3 + 2\ell)$$

as a twisted product. Assertion (2) follows as

$$\dim\{\mathfrak{o}(3 + 2\ell)\} = \tfrac{1}{2}(3 + 2\ell)(2 + 2\ell).$$

Assertions (2) and (3) will follow from Lemma 1.6.1 and from the following relations:

$$G(\mathfrak{F}_\mu)X = \{\xi \in \mathbb{R}^{3+2\ell} : \langle \xi, X^* \rangle \neq 0\} \text{ if } k \leq \ell + 1,$$
$$G(\mathfrak{F}_{\ell+2})X = \{\xi \in \mathbb{R}^{3+2\ell} : \langle \xi, X^* \rangle = \pm 1\}. \tag{2.10.a}$$

We first show $\supset$ holds in Eq. (2.10.a). Let $\xi \in V$ with $a := \langle \xi, X^* \rangle \neq 0$. Define a linear transformation of $\mathbb{R}^{3+2\ell}$ by setting $gX = \xi$ and by setting

$$\varepsilon_0 := (a^2)^{-1/(\ell+3)}, \quad gY = \varepsilon_0 Y, \quad g\tilde{Y} = a^{-2}\varepsilon_0^{-1}\tilde{Y},$$
$$\varepsilon_i := \{a^2\varepsilon_0^{i+1}\}^{-1}, \quad gZ_i := \varepsilon_i Z_i, \quad gZ_i^* = \varepsilon_i^{-1}a^{-2}\tilde{Z}_i.$$

The non-zero components of A_i for $0 \leq i \leq \ell + 2$ are then given by

$$A_0(gX, gY, g\tilde{Y}, gX) = a^2\varepsilon_0 a^{-2}\varepsilon_0^{-1} = 1,$$
$$A_0(gX, gZ_i, g\tilde{Z}_i, gX) = a^2\varepsilon_i\varepsilon_i^{-1}a^{-2} = 1,$$
$$A_1(gX, gY, gZ_1, gX; gY) = a^2\varepsilon_0^2\varepsilon_1 = 1,$$
$$A_1(gX, gY, gY, gX; gZ_1) = a^2\varepsilon_0^2\varepsilon_1 = 1,...$$
$$A_\ell(gX, gY, gZ_\ell, gX; gY, ..., gY)$$
$$= A_\ell(gX, gY, gY, gX; gZ_\ell, gY, ...) = ...$$
$$= A_\ell(gX, gY, gY, gX; gY, ..., gZ_\ell) = a^2\varepsilon_0^{\ell+1}\varepsilon_\ell = 1,$$
$$A_{\ell+1}(gX, gY, gY, gX; gY, ..., gY) = a^2\varepsilon_0^{\ell+3} = 1,$$
$$A_{\ell+2}(gX, gY, gY, gX; gY, ..., gY) = a^2\varepsilon_0^{\ell+4} = \varepsilon_0.$$

Thus $g \in G(\mathfrak{F}_{\ell+1})$. Furthermore, $g \in G(\mathfrak{F}_{\ell+2})$ if $a^2 = 1$. Consequently:

$$\{\xi \in \mathbb{R}^{3+2\ell} : \langle \xi, X^* \rangle \neq 0\} \subset G(\mathfrak{F}_\mu) \cdot X,$$
$$\{\xi \in \mathbb{R}^{3+2\ell} : \langle \xi, X^* \rangle = \pm 1\} \subset G(\mathfrak{F}_{\ell+2}) \cdot X. \tag{2.10.b}$$

We must establish the reverse inclusions to complete the proof. Let $\mathcal{J}_\xi(\eta_1, \eta_2) := A_0(\xi, \eta_1, \eta_2, \xi)$ be the *Jacobi form* defined by $\xi \in V$. Let $a = \langle \xi, X^* \rangle$. Expand

$$\xi = aX + b^0 Y + \tilde{b}^0\tilde{Y} + \sum_{i=1}^{\ell}\{b^i Z_i + \tilde{b}^i\tilde{Z}_i\}.$$

We have the following cases

(1) If $a = 0$, then $\mathcal{J}_\xi = 0$ on $\mathrm{Span}\{Y, \tilde{Y}, Z_i, \tilde{Z}_i\}$ so $\mathrm{Rank}(\mathcal{J}_\xi) \leq 1$.
(2) If $a \neq 0$, then $\mathcal{J}_\xi(Y, \tilde{Y}) \neq 0$ so $\mathrm{Rank}(\mathcal{J}_\xi) \geq 2$.

If $g \in G(\mathfrak{F}_\mu)$, then $\mathrm{Rank}\{\mathcal{J}_\xi\} = \mathrm{Rank}\{\mathcal{J}_{g\xi}\}$. Consequently

$$\langle \xi, X^* \rangle = 0 \Leftrightarrow \mathrm{Rank}(\mathcal{J}_\xi) \leq 1 \Leftrightarrow \mathrm{Rank}(\mathcal{J}_{g\xi}) \leq 1 \Leftrightarrow \langle g\xi, X^* \rangle = 0.$$

This relation implies we have

$$G(\mathfrak{F}_\mu) \cdot X \subset \{\xi \in \mathbb{R}^{3+2\ell} : \langle \xi, X^* \rangle \neq 0\},$$
$$G(\mathfrak{F}_\mu) \cdot \mathrm{Span}\{Y, Z_i, \tilde{Z}_i\} = \mathrm{Span}\{Y, Z_i, \tilde{Z}_i\}. \tag{2.10.c}$$

Suppose $k = \ell + 2$. Since $\mathrm{Rank}(\mathcal{J}_Y) = 0$, $\mathrm{Rank}(\mathcal{J}_{gY}) = 0$ so $\langle gY, X^* \rangle = 0$. Expand

$$gX = aX + a_0 Y + \tilde{a}_0\tilde{Y} + a^i Z_i + \tilde{a}^i\tilde{Z}_i, \quad \text{and}$$
$$gY = \quad\quad b_0 Y + \tilde{b}_0\tilde{Y} + b^i Z_i + \tilde{b}^i\tilde{Z}_i.$$

Then

$$1 = A_{\ell+1}(gX, gY, gY, gX; gY, ..., gY) = a^2 b_0^{\ell+3}, \quad \text{and}$$
$$1 = A_{\ell+2}(gX, gY, gY, gX; gY, ..., gY) = a^2 b_0^{\ell+4}.$$

This shows that $a^2 = 1$ and $b_0 = 1$ so

$$G(\mathfrak{F}_{\ell+2})X \subset \{\xi \in \mathbb{R}^{3+2\ell} : \langle \xi, X^* \rangle = \pm 1\},$$
$$G(\mathfrak{F}_{\ell+2})Y \subset \{\xi \in \mathbb{R}^{3+2\ell} : \langle \xi, X^* \rangle = 0, \text{ and } \langle \xi, Y^* \rangle = 1\}. \tag{2.10.d}$$

We use Eqs. (2.10.b), (2.10.c), and (2.10.d) to derive Eq. (2.10.a). Assertions (2) and (3) of Lemma 2.10.5 then follow.

As noted above, the Jacobi form $\mathcal{J}_X(\cdot, \cdot) = A_0(X, \cdot, \cdot, X)$ defines a nonsingular bilinear form of signature $(\ell + 1, \ell + 1)$ on

$$W : = \operatorname{Span}\{Y, Z_1, ..., Z_\ell, \tilde{Y}, \tilde{Z}_1, ..., \tilde{Z}_\ell\}$$
$$= \{\xi \in \mathbb{R}^{3+2\ell} : \operatorname{Rank}(\mathcal{J}_\xi) \leq 1\}.$$

Let $O(W, \mathcal{J}_X)$ be the associated orthogonal group. If $g \in G_X(\mathfrak{F}_\mu)$, then we have $gW = W$ by Eq. (2.10.c). Since $gX = X$, we may safely identify g with $g|_W$. Furthermore,

$$\mathcal{J}_X(\xi, \eta) = \mathcal{J}_{gX}(g\xi, g\eta) = \mathcal{J}_X(g\xi, g\eta) \quad \text{so} \quad G_X(\mathfrak{F}_\mu) \subset O(\mathcal{J}_X).$$

Conversely, if g is a linear map of W which preserves $\mathcal{J}_X$, we may extend g to $\mathbb{R}^{3+2\ell}$ by defining $gX = X$ and thereby obtain an element of $G_X(\mathfrak{F}_0)$. Thus $G_X(\mathfrak{F}_0) = O(W, \mathcal{J}_X)$. Assertion (4) now follows since

$$\dim\{O(W, \mathcal{J}_X)\} = \tfrac{1}{2} \dim W(\dim W - 1) = \tfrac{1}{2}(2 + 2\ell)(1 + 2\ell).$$

Assertions (5) and (6) will follow from Lemma 1.6.1 if we can establish the following relations:

$$G_X(\mathfrak{F}_\mu) \cdot Y = \{\xi \in W : \langle \xi, Y^* \rangle \neq 0\} \quad \text{for} \quad 1 \leq \mu \leq \ell,$$
$$G_X(\mathfrak{F}_{\ell+1}) \cdot Y = \{\xi \in W : \langle \xi, Y^* \rangle^{\ell+3} = 1\}, \tag{2.10.e}$$
$$G_X(\mathfrak{F}_{\ell+2}) \cdot Y = \{\xi \in W : \langle \xi, Y^* \rangle = 1\}.$$

We argue as follows to establish the relations of Eq. (2.10.e). If $\xi \in W$, let $S_\xi(\eta) := A_1(X, \xi, \xi, X; \eta)$. Expand

$$\xi = b_0 Y + \tilde{b}_0 \tilde{Y} + b^i Z_i + \tilde{b}^i \tilde{Z}_i. \tag{2.10.f}$$

We then have that

$$S_\xi(X) = 0, \quad S_\xi(\tilde{Z}_i) = 0,$$
$$S_\xi(Y) = 2b_0 b_1, \quad S_\xi(Z_1) = b_0^2,$$
$$S_\xi(Z_i) = 0 \quad \text{for} \quad i \geq 2.$$

Thus $S_\xi = 0$ if and only if $b_0 = \langle \xi, Y^* \rangle = 0$. Thus for $\mu \geq 1$ one has:

$$G_X(\mathfrak{F}_\mu)Y \subset \{\xi \in W : \langle \xi, Y^* \rangle \neq 0\},$$
$$G_X(\mathfrak{F}_\mu)\operatorname{Span}\{Z_i, \tilde{Y}, \tilde{Z}_i\} \subset \operatorname{Span}\{Z_i, \tilde{Y}, \tilde{Z}_i\}. \tag{2.10.g}$$

Since $a = 1$, the analysis used to prove Lemma 2.10.5 (2,3) shows $b_0^{\ell+3} = 1$ if $k = \ell + 1$ and $b_0 = 1$ if $k = \ell + 2$. This establishes the inclusion $\subset$ in Eq. (2.10.e).

We complete the verification of Eq. (2.10.e) by establishing the reverse inclusion in Eq. (2.10.e). Expand ξ in the form given in Eq. (2.10.f). Assume $b_0 \neq 0$. Let $\varepsilon_i := b_0^{-i-1}$ and define a linear transformation g of $\mathbb{R}^{3+2\ell}$ fixing X by setting:

$$gX := X, \qquad gY := \xi, \qquad g\tilde{Y} := b_0^{-1}\tilde{Y},$$
$$gZ_i := \varepsilon_i\{Z_i - b_0^{-1}\tilde{b}_i\tilde{Y}\}, \quad \text{and}$$
$$g\tilde{Z}_i := \varepsilon_i^{-1}\{\tilde{Z}_i - b_0^{-1}b_i\tilde{Y}\}.$$

The possibly non-zero components of A_0 are then given by

$$A_0(gX, gY, g\tilde{Y}, gX) = 1,$$
$$A_0(gX, gY, gZ_i, gX) = \varepsilon_i\{\tilde{b}_i - b_0 b_0^{-1}\tilde{b}_i\} = 0,$$
$$A_0(gX, gZ, g\tilde{Z}_i, gX) = \varepsilon_i^{-1}\{b_i - b_0 b_0^{-1}b_i\} = 0,$$
$$A_0(gX, gZ_i, g\tilde{Z}_i, gX) = \varepsilon_i^{-1}\varepsilon_i = 1.$$

The non-zero components of A_i for $1 \leq i \leq \ell$ are given by

$$A_i(gX, gY, gZ_i, gX; gY, ..., gY) = ...$$
$$= A_i(gX, gY, gY, gX; gY, ..., gZ_i) = b_0^{i+1}\varepsilon_i = 1.$$

We also have that

$$A_{\ell+1}(gX, gY, gY, gX; gY, ..., gY) = b_0^{\ell+3},$$
$$A_{\ell+2}(gX, gY, gY, gX; gY, ..., gY) = b_0^{\ell+4}.$$

If $b_0^{\ell+3} = 1$, then $g \in G(\mathfrak{F}_{\ell+1})$; if $b_0 = 1$, then $g \in G(\mathfrak{F}_{\ell+2})$. This establishes the reverse inclusions in Eq. (2.10.e) and completes the proof of Assertions (5) and (6); Assertion (7) of Lemma 2.10.5 is immediate. $\qquad\square$

We complete the proof of Theorem 2.10.3 by showing

Lemma 2.10.6 *Let* $1 \leq \mu \leq \ell$.

(1) If $g \in G_{X,Y}(\mathfrak{F}_\mu)$, *then* $g\tilde{Y} = \tilde{Y}$.
(2) $G_{X,Y}(\mathfrak{F}_\mu) = \Lambda^2(\mathbb{R}^\mu) \oplus O(\ell - \mu, \ell - \mu)$.
(3) $\dim G_{X,Y}(\mathfrak{F}_\mu) = \frac{1}{2}\mu(\mu - 1) + (\ell - \mu)(2\ell - 2\mu - 1)$.

Proof. Let $g \in G_{X,Y}(\mathfrak{F}_\mu)$. Let $\xi \in \mathrm{Span}\{Z_i, \tilde{Y}, \tilde{Z}_i\}$. We may use Eq. (2.10.g) and the relation $A_0(X, Y, g\xi, X) = A_0(X, Y, \xi, X)$, to see

$$g\tilde{Y} = \tilde{Y} + \sum_{j=1}^{\ell}\left\{a^j Z_j + \tilde{a}^j \tilde{Z}_j\right\},$$

$$gZ_i = \sum_{j=1}^{\ell}\left\{b_i^j Z_j + \tilde{b}_i^j \tilde{Z}_j\right\},$$

$$g\tilde{Z}_i = \sum_{j=1}^{\ell}\left\{c_i^j Z_j + \tilde{c}_i^j \tilde{Z}_j\right\}.$$

Consequently $\mathrm{Span}\{gZ_i, g\tilde{Z}_{\tilde{i}}\} = \mathrm{Span}\{Z_i, Z_{\tilde{i}}\}$ and the relation

$$A_0(X, gZ_i, g\tilde{Y}, X) = A_0(Xg\tilde{Z}_{\tilde{i}}, g\tilde{Y}, X) = 0$$

implies $a^i = a^{\tilde{i}} = 0$; Assertion (1) follows.

To simplify the notation in the proof of Assertion (2), we clear the previous notation and set:

$$\begin{aligned}
S &= (S_1, ..., S_\mu) = (Z_1, ..., Z_\mu), \\
T &= (T_1, ..., T_{\ell-\mu}) = (Z_{\mu+1}, ..., Z_\ell), \\
\tilde{S} &= (\tilde{S}_1, ..., \tilde{S}_\mu) = (\tilde{Z}_1, ..., \tilde{Z}_\mu), \\
\tilde{T} &= (\tilde{T}_1, ..., \tilde{T}_{\ell-\mu}) = (\tilde{Z}_{\mu+1}, ..., \tilde{Z}_\ell).
\end{aligned}$$

We let indices a, b (respectively $\tilde{a}$, $\tilde{b}$) range from 1 through μ and index elements of S (respectively $\tilde{S}$); we let indices α, β (respectively $\tilde{\alpha}$, $\tilde{\beta}$) range from 1 through $\ell - \mu$ and index elements of T (respectively $\tilde{T}$). Suppose that $g \in G_{X,Y}(\mathfrak{F}_\mu)$. Since

$$A_a(X, Y, gS_b, X; Y, ..., Y) = \delta_b^a,$$
$$A_a(X, Y, gS_{\tilde{b}}, X; Y, ..., Y) = 0,$$
$$A_a(X, Y, gT_\alpha; Y, ..., Y) = 0,$$
$$A_a(X, Y, gT_{\tilde{\alpha}}; Y, ..., Y) = 0,$$

one has that

$$gS_a = S_a + g_a^{\tilde{b}}\tilde{S}_{\tilde{b}} + g_a^\beta T_\beta + g_a^{\tilde{\beta}}\tilde{T}_{\tilde{\beta}},$$
$$gS_{\tilde{a}} = g_{\tilde{a}}^{\tilde{b}}\tilde{S}_{\tilde{b}} + g_{\tilde{a}}^\beta T_\beta + g_{\tilde{a}}^{\tilde{\beta}}\tilde{T}_{\tilde{\beta}},$$
$$gT_\alpha = g_\alpha^{\tilde{b}}\tilde{S}_{\tilde{b}} + g_\alpha^\beta T_\beta + g_\alpha^{\tilde{\beta}}\tilde{T}_{\tilde{\beta}},$$
$$gT_{\tilde{\alpha}} = g_{\tilde{\alpha}}^{\tilde{b}}\tilde{S}_{\tilde{b}} + g_{\tilde{\alpha}}^\beta T_\beta + g_{\tilde{\alpha}}^{\tilde{\beta}}\tilde{T}_{\tilde{\beta}}.$$

The non-zero components of A_0 are

$$A_0(X, gS_a, gS_{\tilde{a}}, X) = 1 \quad \text{and} \quad A_0(X, gT_\alpha, gT_{\tilde{\alpha}}, X) = 1.$$

Consequently we have that

$$g_{\tilde{a}}^{\tilde{b}} = \delta_{\tilde{a}}^{\tilde{b}}, \qquad g_a^{\tilde{b}} + g_b^{\tilde{a}} = 0, \qquad g_\alpha^{\tilde{b}} = 0, \qquad g_{\tilde{\alpha}}^{\tilde{\beta}} = 0.$$

Thus

$$gS_a = S_a + g_a^{\tilde{b}}\tilde{S}_{\tilde{b}} + g_a^\beta T_\beta + g_a^{\tilde{\beta}}\tilde{T}_{\tilde{\beta}},$$
$$gS_{\tilde{a}} = \tilde{S}_{\tilde{a}} + g_{\tilde{a}}^\beta T_\beta + g_{\tilde{a}}^{\tilde{\beta}}\tilde{T}_{\tilde{\beta}},$$
$$gT_\alpha = g_\alpha^\beta T_\beta + g_\alpha^{\tilde{\beta}}\tilde{T}_{\tilde{\beta}},$$
$$gT_{\tilde{\alpha}} = g_{\tilde{\alpha}}^\beta T_\beta + g_{\tilde{\alpha}}^{\tilde{\beta}}\tilde{T}_{\tilde{\beta}}.$$

Since g is non-singular, we must have

$$\text{Span}\{gT_\alpha, gT_{\tilde{\alpha}}\} = \text{Span}\{T_\alpha, T_{\tilde{\alpha}}\}.$$

Since $A_0(X, gS_{\tilde{a}}, gT_\alpha, X) = 0$ and $A_0(X, gS_{\tilde{a}}, gT_{\tilde{\alpha}}, X) = 0$,

$$g_a^\beta = 0, \quad g_a^{\tilde{\beta}} = 0, \quad g_{\tilde{a}}^\beta = 0, \quad \text{and} \quad g_{\tilde{a}}^{\tilde{\beta}} = 0$$

so the variables decouple and we have, where $g_a^{\tilde{b}} + g_b^{\tilde{a}} = 0$,

$$
\begin{aligned}
gS_a &= S_a + g_a^{\tilde{b}}\tilde{S}_{\tilde{b}}, \\
gS_{\tilde{a}} &= \tilde{S}_{\tilde{a}}, \\
gT_\alpha &= g_\alpha^\beta T_\beta + g_\alpha^{\tilde{\beta}}\tilde{T}_{\tilde{\beta}}, \\
gT_{\tilde{\alpha}} &= g_{\tilde{\alpha}}^\beta T_\beta + g_{\tilde{\alpha}}^{\tilde{\beta}}\tilde{T}_{\tilde{\beta}}.
\end{aligned}
$$

Furthermore, the matrix

$$
h = \begin{pmatrix} g_\alpha^\beta & g_\alpha^{\tilde{\beta}} \\ g_{\tilde{\alpha}}^\beta & g_{\tilde{\alpha}}^{\tilde{\beta}} \end{pmatrix}
$$

must belong to $O(\ell,\ell)$. Conversely, it is immediate that such a matrix g belongs to $G_{X,Y}(\mathfrak{F}_\mu)$. $\qquad\square$

We can now complete the proof of Theorem 2.10.3.

(1) By Lemma 2.10.6,
    (a) If $1 \le \mu \le \ell$, then $\dim\{G_{X,Y}(\mathfrak{F}_\mu)\} = \frac{1}{2}\mu(\mu-1)+(\ell-\mu)(2\ell-2\mu-1)$.
    (b) If $\mu = \ell+1, \ell+2$, then $\dim\{G_{X,Y}(\mathfrak{F}_\mu)\} = \frac{1}{2}\ell(\ell-1)$.

(2) By Lemma 2.10.5 (4-7):
    (a) $\dim\{G_X(\mathfrak{F}_0)\} = (\ell+1)(2\ell+1)$.
    (b) If $1 \le \mu \le \ell$, then $\dim\{G_X(\mathfrak{F}_\mu)\} = 2\ell + 2 + \frac{1}{2}\mu(\mu-1)$
$$+ (\ell-\mu)(2\ell-2\mu-1).$$
    (c) If $\mu = \ell+1, \ell+2$, then $\dim\{G_X(\mathfrak{F}_\mu)\} = 2\ell + 1 + \frac{1}{2}\ell(\ell-1)$.

(3) By Lemma 2.10.5 (2,3):
    (a) $\dim\{G(\mathfrak{F}_0)\} = 2\ell + 3 + (\ell+1)(2\ell+1)$.
    (b) If $1 \le \mu \le \ell$, then $\dim\{G(\mathfrak{F}_\mu)\} = 2\ell + 3 + 2\ell + 2 + \frac{1}{2}\mu(\mu-1)$
$$+ (\ell-\mu)(2\ell-2\mu-1).$$
    (c) $\dim\{G_X(\mathfrak{F}_{\ell+1})\} = 2\ell + 3 + 2\ell + 1 + \frac{1}{2}\ell(\ell-1)$.
    (d) $\dim\{G_X(\mathfrak{F}_{\ell+2})\} = 2\ell + 2 + 2\ell + 1 + \frac{1}{2}\ell(\ell-1)$.

(4) By Lemma 2.10.5 (1), one has:
    (a) $\dim\{G(\mathfrak{M}_0)\} = (\ell+1)(3+2\ell) + 2\ell + 3 + (\ell+1)(2\ell+1)$.
    (b) If $1 \le \mu \le \ell$, then $\dim\{G(\mathfrak{M}_\mu)\} = (\ell+1)(3+2\ell) + 2\ell + 3$
$$+ 2\ell + 2 + \frac{1}{2}\mu(\mu-1) + (\ell-\mu)(2\ell-2\mu-1).$$
    (c) $\dim\{G(\mathfrak{M}_{\ell+1})\} = (\ell+1)(3+2\ell) + 2\ell + 3 + 2\ell + 1 + \frac{1}{2}\ell(\ell-1)$.
    (d) $\dim\{G(\mathfrak{M}_{\ell+2})\} = (\ell+1)(3+2\ell) + 2\ell + 2 + 2\ell + 1 + \frac{1}{2}\ell(\ell-1)$.

(5) By Lemma 2.10.4, one has

(a) $\dim\{G(\mathcal{N}_0)\} = 4\ell + 6 + (\ell + 1)(3 + 2\ell) + 2\ell + 3 + (\ell + 1)(2\ell + 1).$

(b) If $1 \leq \mu \leq \ell$, then $\dim\{G(\mathcal{N}_\mu)\} = 4\ell + 6 + (\ell + 1)(3 + 2\ell)$
$$+ 2\ell + 3 + 2\ell + 2 + \tfrac{1}{2}\mu(\mu - 1) + (\ell - \mu)(2\ell - 2\mu - 1).$$

(c) $\dim\{G(\mathcal{N}_{\ell+1})\} = 4\ell + 6 + (\ell + 1)(3 + 2\ell) + 2\ell + 3 + 2\ell + 1 + \tfrac{1}{2}\ell(\ell - 1).$

(d) $\dim\{G(\mathcal{N}_{\ell+2})\} = 4\ell + 6 + (\ell + 1)(3 + 2\ell) + 2\ell + 2 + 2\ell + 1 + \tfrac{1}{2}\ell(\ell - 1).$

(e) $\dim\{G(\mathcal{M})\} = 4\ell + 5 + (\ell + 1)(3 + 2\ell) + 2\ell + 2 + 2\ell + 1 + \tfrac{1}{2}\ell(\ell - 1).$

Chapter 3

Other Pseudo-Riemannian Manifolds

3.1 Introduction

In Chapter 3, we continue our study of specific examples; the manifolds of Chapter 3 for the most part are not generalized plane wave manifolds.

A pseudo-Riemannian manifold $\mathcal{M} = (M, g)$ is said to be a *generalized Walker manifold*, see Walker (1950), if it admits a parallel totally isotropic distribution of maximal dimension. The manifolds discussed in Sections 2.3 and 2.5 are generalized Walker manifolds. In Section 3.2, we present results of Gilkey and Nikčević (2005d) relating to a family of 3-dimensional generalized Walker Lorentz manifolds. Let $\{x, y, \tilde{x}\}$ be coordinates on $\mathbb{R}^3$, let $f \in C^\infty(\mathbb{R})$, and let $\mathcal{M} := (\mathbb{R}^3, g)$ where

$$g(\partial_x, \partial_x) = -2f(y) \quad \text{and} \quad g(\partial_x, \partial_{\tilde{x}}) = g(\partial_y, \partial_y) = 1 \,.$$

These manifolds are VSI Walker manifolds but not generalized plane wave manifolds. The geodesic structure of these manifolds is examined and their curvature tensors are determined. Some of these manifolds are geodesically complete; others exhibit Ricci blowup. There are examples in this family which are 1-curvature homogeneous but not locally homogeneous; 2-affine curvature homogeneity implies local homogeneity for this family. There are examples which are 1-affine curvature homogeneous but not 1-curvature homogeneous; thus 1-affine curvature homogeneity and 1-curvature homogeneity are not equivalent notions. All the members in this family are modeled on indecomposable symmetric spaces.

The geometry of neutral signature 4-dimensional manifolds is a particularly rich one and we divide our discussion of this topic into two sections. In Section 3.3 we summarize some work by various authors dealing with this topic. Section 3.3.1 presents results of Blažić, Bokan, and Rakić (2001a)

181

and of García-Río, Kupeli, and Vázquez-Lorenzo (2002) which give a normal form for Osserman algebraic curvature tensors in signature $(2,2)$. A family of Walker metrics of signature $(2,2)$ on $\mathbb{R}^4$ is introduced in Definition 3.3.1. Various subfamilies of this class are then used to illustrate certain geometric phenomena. Work of García-Río, Rakić, and Vázquez-Abal (2005) dealing with Kähler Osserman manifolds in this family is discussed in 3.3.2. Results of Díaz-Ramos, García-Río, and Vázquez-Lorenzo (2006) which show that there are Jordan Osserman manifolds which are not nilpotent and whose Jacobi operator is not diagonalizable are given in Section 3.3.3. Conformal Osserman manifolds of signature $(2,2)$ are studied following results of Brozos-Vázquez, García-Río, and Vázquez-Lorenzo (2005) in Section 3.3.4; it is somewhat surprising that there are conformal Osserman manifolds whose conformal Jacobi operator has complex eigenvalues. Our discussion of neutral signature $(2,2)$ walker manifolds is concluded in Section 3.4 when questions of completeness and Ricci blowup are examined for many of the examples which were presented in Section 3.3.

In Section 3.5, we study Fiedler manifolds. Although there are members of this family which are generalized plane wave manifolds, the family is more general. All members of this family are Jacobi–Tsankov; this means that $J(x)J(y) = J(y)J(x)$ for any two tangent vectors. We will study this property subsequently. There are members in this family which are Ivanov–Petrova and which are nilpotent of order 3. There are manifolds in this family which are Jacobi nilpotent of arbitrarily high order. There are also members of this family which are Szabó nilpotent of arbitrarily high order. We also create an example where it is possible to normalize the components of the metric and of ∇R but not possible to normalize the components of R.

3.2 Lorentz Manifolds

Consider the following family of 3-dimensional Lorentz manifolds:

Definition 3.2.1   Let $\{x, y, \tilde{x}\}$ be coordinates on $\mathbb{R}^3$, let $f \in C^\infty(\mathbb{R})$, and let $\mathcal{M} := (\mathbb{R}^3, g)$ where g is the Lorentz metric on $\mathbb{R}^3$ given by:

$$g(\partial_x, \partial_x) := -2f(y) \quad \text{and} \quad g(\partial_x, \partial_{\tilde{x}}) = g(\partial_y, \partial_y) := 1\,.$$

Lemma 3.2.1   *Let $\mathcal{M}$ be as in Definition 3.2.1.*

(1) The possibly non-zero Christoffel symbols are given by:

$$g(\nabla_{\partial_x}\partial_x, \partial_y) = f', \qquad g(\nabla_{\partial_x}\partial_y, \partial_x) = g(\nabla_{\partial_y}\partial_x, \partial_x) = -f',$$
$$\nabla_{\partial_x}\partial_x = f'\partial_y, \qquad \nabla_{\partial_x}\partial_y = \nabla_{\partial_y}\partial_x = -f'\partial_{\tilde{x}}.$$

(2) $\nabla^\nu R(\cdot) = 0$ if any entry is $\partial_{\tilde{x}}$.
(3) All the scalar invariants of $\mathcal{M}$ vanish.

Proof.   Assertion (1) follows directly from Eq. (1.2.f); Assertion (2) follows from Assertion (1). Since $f'(y)$ depends on y in general, the Levi-Civita connection does not have the triangular form for a generalized plane wave manifold and, as will be noted presently in Remark 3.2.1, $\mathcal{M}$ need not in fact be a generalized plane wave manifold. Thus it is not possible to apply Theorem 2.2.1 to establish Assertion (3). Instead, we proceed directly. Let

$$X := \partial_x + f\partial_{\tilde{x}}, \qquad Y := \partial_y, \qquad \tilde{X} := \partial_{\tilde{x}},$$
$$e_1^+ := (X + \tilde{X})/\sqrt{2}, \quad e_1^- = (X - \tilde{X})/\sqrt{2}, \quad e_2^+ := Y.$$

Then $\{e_1^-, e_1^+, e_2^+\}$ is an orthonormal frame for $T(\mathbb{R}^3)$. Furthermore,

$$\nabla^k R(..., e_1^+, ...) = \nabla^k R(..., e_1^-, ...) = \nabla^k R(..., \partial_x, ...)/\sqrt{2}.$$

Since e_1^- is timelike and since e_1^+ is spacelike, terms where $e_i = e_1^+$ and terms where $e_i = e_1^-$ appear with opposite signs in any Weyl summation and cancel. Thus all the terms must be e_2^+ which yields zero. $\square$

Various properties of certain manifolds in this family have been studied by Bueken and Djoric (2000), by Bueken and Vanhecke (1997), by Cahen, Leroy, Parker, Tricerri, and Vanhecke (1991), by Chaichi, García-Río, and Vázquez-Abal (2005), by Koutras and McIntosh (1996), and by Pravda, Coley, and Milson (2002), to give but a few of the many possible references. For example the existence of 1-curvature homogeneous 3-dimensional Lorentzian manifolds which are not locally homogeneous follows from the discussion in Bueken and Djoric (2000) and the existence of 3-dimensional VSI Lorentzian manifolds follows from work of Pravda, Coley, and Milson (2002).

We say that a pseudo-Riemannian manifold $\mathcal{M}$ exhibits *Ricci blowup* if there exists a geodesic $\gamma : [0, T) \rightarrow M$ where $T < \infty$ and where $\limsup_{t\rightarrow T} |\rho(\dot{\gamma}, \dot{\gamma})| = \infty$. Such a manifold is necessarily geodesically incomplete and does not embed as an open subset of a geodesically complete manifold.

Definition 3.2.2   We specialize the manifolds of Definition 3.2.1 setting $\mathcal{N}_i^{\pm} := \mathcal{M}_{f_i^{\pm}}$ where

$$
\begin{aligned}
f_0^-(y) &:= -\tfrac{1}{2}y^2, & f_0^+(y) &:= \tfrac{1}{2}y^2, \\
f_1^-(y) &:= -e^{-y}, & f_1^+(y) &:= e^y, \\
f_2^-(y) &:= -e^{-y} + y, & f_2^+(y) &:= e^y + y, \\
f_3^-(y) &:= -e^{-y} - e^{-2y}, & f_3^+(y) &:= e^y + e^{2y}.
\end{aligned}
$$

The following is the main result of this section:

Theorem 3.2.1

(1) $\mathcal{N}_0^-$ is a geodesically complete indecomposable symmetric space.

(2) $\mathcal{N}_1^-$ is 0-curvature modeled on $\mathcal{N}_0^-$, is locally homogeneous, and has Ricci blowup.

(3) $\mathcal{N}_2^-$ is 0-curvature modeled on $\mathcal{N}_0^-$, is 1-curvature modeled on $\mathcal{N}_1^-$, is not 2-affine curvature homogeneous, and has Ricci blowup.

(4) $\mathcal{N}_3^-$ is 0-curvature modeled on $\mathcal{N}_0^-$, is not 1-curvature homogeneous, is 1-affine curvature modeled on $\mathcal{N}_1$, is not 2-affine curvature homogeneous, and has Ricci blowup.

(5) $\mathcal{N}_0^+$ is a geodesically complete indecomposable symmetric space.

(6) $\mathcal{N}_1^+$ is 0-curvature modeled on $\mathcal{N}_0^+$, is geodesically complete, and is homogeneous.

(7) $\mathcal{N}_2^+$ is 0-curvature modeled on $\mathcal{N}_0^+$, is 1-curvature modeled on $\mathcal{N}_1^+$, is not 2-affine curvature homogeneous, and is geodesically complete.

(8) $\mathcal{N}_3^+$ is 0-curvature modeled on $\mathcal{N}_0^+$, is not 1-curvature homogeneous, is 1-affine curvature modeled on $\mathcal{N}_1^+$, is not 2-affine curvature homogeneous, and is geodesically complete.

Remark 3.2.1   By Theorem 3.2.1, certain manifolds of Definition 3.2.1 are geodesically incomplete and thus are not generalized plane wave manifolds. Thus Lemma 3.2.1 does not follow from Theorem 2.2.1. All the manifolds of Definition 3.2.1 are generalized Walker manifolds. The distribution $\mathrm{Span}\{\partial_{\tilde{x}}\}$ is a parallel totally isotropic distribution of maximal dimension.

Here is a brief outline to the remainder of this section; we shall follow the treatment in Gilkey and Nikčević (2005d). In Section 3.2.1, we study the geodesics and curvature tensor of a manifold in this family. In Section 3.2.2, we study questions of geodesic completeness and of Ricci blowup. In

Section 3.2.3, we turn our attention to curvature homogeneity and complete the proof of Theorem 3.2.1.

3.2.1 *Geodesics and curvature*

Lemma 3.2.2 *Let $\mathcal{M}$ be as in Definition 3.2.1.*

(1) The possibly non-zero components of R, of ∇R, of $\nabla^2 R$, and of ρ are:

$$\mathcal{R}(\partial_x, \partial_y)\partial_y = f'' \partial_{\tilde{x}}, \qquad \mathcal{R}(\partial_x, \partial_y)\partial_x = -f'' \partial_y,$$
$$R(\partial_x, \partial_y, \partial_y, \partial_x) = f'', \qquad \rho(\partial_x, \partial_x) = f'',$$
$$\nabla_{\partial_y} \mathcal{R}(\partial_x, \partial_y)\partial_y = f''' \partial_{\tilde{x}}, \qquad \nabla_{\partial_y} \mathcal{R}(\partial_x, \partial_y)\partial_x = -f''' \partial_y,$$
$$\nabla R(\partial_x, \partial_y, \partial_y, \partial_x; \partial_y) = f''', \qquad \nabla_{\partial_y}\nabla_{\partial_y}\mathcal{R}(\partial_x, \partial_y)\partial_y = f'''' \partial_{\tilde{x}},$$
$$\nabla_{\partial_y}\nabla_{\partial_y}\mathcal{R}(\partial_x, \partial_y)\partial_x = -f''''\partial_y, \qquad \nabla_{\partial_x}\nabla_{\partial_x}\mathcal{R}(\partial_x, \partial_y)\partial_y = -f' f''' \partial_{\tilde{x}},$$
$$\nabla_{\partial_x}\nabla_{\partial_x}\mathcal{R}(\partial_x, \partial_y)\partial_x = f' f''' \partial_y, \qquad \nabla^2 R(\partial_x, \partial_y, \partial_y, \partial_x; \partial_y, \partial_y) = f'''',$$
$$\nabla^2 R(\partial_x, \partial_y, \partial_y, \partial_x; \partial_x, \partial_x) = -f' f''' \,.$$

(2) $\gamma(t) = (x(t), y(t), \tilde{x}(t))$ is a geodesic if and only if

(a) $x(t) = x_0 + x_1 t$.
(b) $y''(t) = -x_1^2 f'(y(t))$.
(c) $\tilde{x}(t) = \tilde{x}_0 + t\tilde{x}_0' + 2x_1 \int_0^t \int_0^s f'(y(u))y'(u)\,du\,ds$.

Proof. The curvature tensor can be computed from Lemma 3.2.1. We omit detailed calculation to establish Assertion (1) in the interests of brevity.

By Lemma 3.2.1 (1), the geodesic equation becomes

$$x''(t) = 0, \quad y''(t) = -f'(y(t))x'(t)x'(t), \quad \tilde{x}''(t) = 2f'(y(t))y'(t)x'(t)\,.$$

We solve the first equation to obtain the conclusion of Assertion (2a). The remaining equations then become

$$y''(t) = -x_1^2 f'(y(t)) \quad \text{and} \quad \tilde{x}''(t) = 2x_1 f'(y(t))y'(t)\,.$$

Assertion (2b) now follows; once y is determined, we use the second relation to determine $\tilde{x}$ and establish Assertion (2c). $\qquad\qquad\square$

Assertions (1) and (5) of Theorem 3.2.1 will follow from:

Lemma 3.2.3 *Let $\mathcal{N}_0^{\pm}$ be defined by $f_{\pm}(y) := \pm\frac{1}{2}y^2$. Then:*

(1) $\mathcal{N}_0^{\pm}$ is a geodesically complete indecomposable local symmetric space.
(2) For any $P \in \mathbb{R}^3$, $\exp_P^{\mathcal{N}_0^+}$ is neither surjective nor injective.

(3) For every $P \in \mathbb{R}^3$, $\exp_P^{\mathcal{N}_0^-}$ is a global diffeomorphism from $T_P\mathbb{R}^3$ to $\mathbb{R}^3$.

Proof.   Since $f(y) = \pm\frac{1}{2}y^2$ is quadratic, $\nabla R = 0$. Thus $\mathcal{N}_0^+$ and $\mathcal{N}_0^-$ are local symmetric spaces. Let $\{X, Y, \tilde{X}\}$ be a basis for $\mathbb{R}^3$. Consider the 0-models $\mathfrak{M}_\pm := (\mathbb{R}^3, \langle\cdot,\cdot\rangle, A_\pm)$ where

$$\langle X, \tilde{X}\rangle = \langle Y, Y\rangle = 1, \quad \text{and} \quad A_\pm(X, Y, Y, X) = \pm 1\,.$$

Setting $X = \partial_x + f\partial_{\tilde{x}}$, $Y = \partial_y$, and $\tilde{X} = \partial_{\tilde{x}}$ then shows that $\mathfrak{M}_\pm$ is a 0-model for $\mathcal{N}_0^\pm$. Thus to show $\mathcal{N}_0^\pm$ is indecomposable, it suffices to establish the corresponding assertion for $\mathfrak{M}_\pm$.

Suppose to the contrary that $\mathbb{R}^3 = V_1 \oplus V_2$ is an orthogonal direct sum decomposition which decomposes $A_\pm$. Choose the notation so $\dim(V_1) = 2$ and $\dim(V_2) = 1$. Since $\dim(V_2) = 1$, $V_2 \in \ker(A_\pm) = \text{Span}\{\partial_{\tilde{x}}\}$. This implies that the metric is totally isotropic on V_2. This is not possible as $\mathbb{R}^3 = V_1 \oplus V_2$ is an orthogonal direct sum decomposition. This contradiction shows that $\mathcal{N}_0^+$ and $\mathcal{N}_0^-$ are indecomposable symmetric spaces; they are not irreducible as $\text{Span}\{\partial_{\tilde{x}}\}$ is a parallel distribution and hence invariant under holonomy.

Suppose first $\mathcal{M} = \mathcal{N}_0^+$ so $f(y) = \frac{1}{2}y^2$. We apply Lemma 3.2.2 (2). The geodesic equation $y''(t) = -x_1^2 f'(y(t))$ then becomes $y'' = -x_1^2 y(t)$. We solve this equation with initial data $y(0) = y_0$ and $\dot{y}(0) = y_1$:

$$y(t) = \begin{cases} y_0 + y_0't & \text{if } x_1 = 0, \\ y_0\cos(x_1 t) + \frac{1}{x_1}y_0'\sin(x_1 t) & \text{if } x_1 \neq 0\,. \end{cases}$$

This shows that $\mathcal{N}_0^+$ is geodesically complete. Furthermore, a geodesic with $x(0) = x_0$ and $x(1) = x_0 + 2\pi$ has the form:

$$\gamma(t) = (x_0 + 2\pi t, y_0\cos(2\pi t) + \tfrac{1}{2\pi}y_0'\sin(2\pi t), \tilde{x}(t))\,.$$

Thus $y(1) = y(0)$ and the exponential map is not surjective; neither is it injective as y_0' plays no role. This establishes the Assertions of the Lemma concerning $\mathcal{N}_0^+$.

Next, we study $f(y) = -\frac{1}{2}y^2$ so $y''(t) = x_1^2 y(t)$. This yields

$$y(t) = \begin{cases} y_0 + y_0't & \text{if } x_1 = 0, \\ \frac{1}{2}y_0\{e^{x_1 t} + e^{-x_1 t}\} + \frac{1}{2x_1}y_0'\{e^{x_1 t} - e^{-x_1 t}\} & \text{if } x_1 \neq 0\,. \end{cases}$$

This shows that $\mathcal{S}_-$ is geodesically complete. We take $P := (x_0, y_0, \tilde{x}_0)$ as the initial point. Suppose $Q := (x_1, y_1, \tilde{x}_1)$ is given. The exponential map is given by setting $t = 1$. Thus $x(t) = x_0 + t(x_1 - x_0)$. If $x_1 - x_0 = 0$, then set $y(t) = y_0 + t(y_1 - y_0)$. If $x_1 - x_0 \neq 0$, we determine y_0' uniquely by solving the equation:

$$y_1 = \tfrac{1}{2}y_0\{e^{x_1-x_0} + e^{x_0-x_1}\} + \frac{1}{2(x_1-x_0)}y_0'\{e^{x_1-x_0} - e^{x_0-x_1}\}.$$

Once x and y have been determined, we use Lemma 3.2.2 (2c) to solve for $\tilde{x}_0'$. This shows that $\mathcal{N}_0^-$ is geodesically complete and that the exponential map is a diffeomorphism. $\qquad\square$

3.2.2   *Ricci blowup*

Recall that a pseudo-Riemannian manifold $\mathcal{M}$ is said to exhibit *Ricci blowup* if there exists a geodesic γ in $\mathcal{M}$ with domain $[0, T)$ for $T < \infty$ where $\limsup_{t \to T} |\rho(\dot\gamma, \dot\gamma)| = \infty$. Such a manifold is necessarily incomplete.

The Assertions of Theorem 3.2.1 concerning geodesic completeness and Ricci blowup for the manifolds $\mathcal{N}_i^{\pm}$ with $i = 1, 2, 3$ will follow from:

Lemma 3.2.4   *Assume that $f'(y) > 0$ for all $y \in \mathbb{R}$.*

(1) If $\exists\, C > 0$ so $f'(y) \leq C|y|$ for $y \leq -1$, $\mathcal{M}$ is geodesically complete.
(2) If $\exists\, \epsilon, \delta > 0$ so $f'(y) \geq \epsilon|y|^{1+\delta}$ for $y \leq -1$, $\mathcal{M}$ has Ricci blowup.

Proof.   We suppose that $f' > 0$ and that $f'(y) \leq C|y|$ for $y \leq -1$. Let γ be a geodesic with $\dot{x}(0) = x_1$. If $x_1 = 0$, then γ extends for all time so we assume $x_1 \neq 0$. Set $h = -x_1^2 f'$. The geodesic equation given in Lemma 3.2.2 (2) then becomes $y''(t) = h(y(t))$. Choose a maximal domain $[0, T)$ for the solution to this ordinary differential equation with initial condition $y(0) = y_0$ and $y'(0) = y_0'$. If $T < \infty$, Lemma 1.5.6 shows that

$$\lim_{y \to T} y(t) = -\infty \quad \text{and} \quad \limsup_{y \to T} \left|\frac{h(y(t))}{y(t)}\right| = \infty$$

which contracts the assumption that $0 < f' \leq C|y|$. Thus $T = \infty$ and $\mathcal{M}$ is geodesically complete. Assertion (1) now follows.

Suppose $f'(y) > 0$ for all y and that $f'(y) \geq \varepsilon|y|^{1+\delta}$ for $y \leq -1$. Choose a geodesic with $x(0) = 0$, $x'(0) = 1$, $y(0) = -1$, and $y'(0) = -1$. The geodesic equation then becomes

$$y'' = -f'(y).$$

Thus by Lemma 1.5.7 for some finite T, $\lim_{t \to T} y(t) = -\infty$. This shows that $\mathcal{M}$ is geodesically incomplete.

If $|f''(y)| \leq K$ on $(-\infty, 0]$, then $f'(y) \leq K|y| + f'(y(0))$ on $(-\infty, 0]$ which is false. Thus $|f''(y)|$ is not bounded on $(-\infty, 0]$. Since $y(t) \to -\infty$ as $t \to T$, $f''(y(t))$ is not bounded on $[0, T)$. We have $\rho(\dot{\gamma}, \dot{\gamma}) = f''(y(t))$. Consequently, $\mathcal{M}$ has Ricci blowup. $\qquad\square$

3.2.3 *Curvature homogeneity*

We shall need a technical lemma related to the structure of $\nabla^k R$ when the defining function f is a pure exponential. Let

$$\nabla^k R(\vec{\xi}) := \nabla^k R(\xi_1, \xi_2, \xi_3, \xi_4; \xi_5, ..., \xi_{4+k}) \quad \text{for} \quad \vec{\xi} = (\xi_1, ..., \xi_{4+k}) \,.$$

We suppose $\xi_i = \partial_x$ or $\xi_i = \partial_y$; there is no need to take $\xi_i = \partial_{\tilde{x}}$ since $\nabla^k R(\vec{\xi})$ vanishes if any $\xi_i = \partial_{\tilde{x}}$.

Lemma 3.2.5   *Let $\Theta(\vec{\xi})$ denote the number of times that $\xi_i = \partial_x$. Let $\mathcal{M}$ be as in Definition 3.2.1 where $f = ae^{by}$. Then there exists a coefficient $\gamma(a, b; \vec{\xi})$ so*

$$\nabla^k R(\vec{\xi}) = \gamma(a, b; \vec{\xi}) e^{\frac{1}{2}\Theta(\vec{\xi})by} \,.$$

Proof . We proceed by induction on k. We define

$$\gamma(a, b; \partial_x, \partial_y, \partial_y, \partial_x) = ab^2,$$
$$\gamma(a, b; \partial_x, \partial_y, \partial_y, \partial_x; \partial_y) = ab^3,$$
$$\gamma(a, b; \partial_x, \partial_y, \partial_y, \partial_x; \partial_y, \partial_y) = ab^4,$$
$$\gamma(a, b; \partial_x, \partial_y, \partial_y, \partial_x; \partial_x, \partial_x) = -a^2b^4 \,.$$

Lemma 3.2.5 for $k \leq 2$ then follows from Lemma 3.2.2 if we introduce the appropriate $\pm$ signs when considering the $\mathbb{Z}_2$ symmetries in the first 4-indices and if we set the other entries to zero.

We prove the Lemma inductively, the discussion given above shows it holds when $k = 0, 1, 2$. By Eq. (1.2.d),

$$\nabla^k R(\vec{\xi}) = \xi_{k+4} \nabla^{k-1} R(\xi_1, ..., \xi_{k+3}) \qquad (3.2.a)$$

$$- \sum_{i=1}^{k+3} \nabla^{k-1} R(\xi_1, ..., \xi_{i-1}, \nabla_{\xi_{k+4}}\xi_i, \xi_{i+1}, ..., \xi_{k+3}) \,. \qquad (3.2.b)$$

We distinguish two cases:

(1) Suppose $\xi_{k+4} = \partial_y$. Since $\nabla_{\partial_y}\xi_i$ is a multiple of $\partial_{\tilde{x}}$, the terms in Eq. (3.2.b) vanish and only the term in Eq. (3.2.a) must be considered.

Let $\vec{\eta} := (\xi_1, ..., \xi_{k+3})$. Then

$$\nabla^k R(\vec{\xi}) = \partial_y \nabla^{k-1} R(\vec{\eta}) = \partial_y \gamma(a, b; \vec{\eta}) e^{\frac{1}{2}\Theta(\vec{\eta})by}$$
$$= \frac{b\Theta(\vec{\eta})}{2} \gamma(a, b; \vec{\eta}) e^{\frac{1}{2}\Theta(\vec{\eta})by} .$$

This has the desired form since $\Theta(\vec{\eta}) = \Theta(\vec{\xi})$.

(2) Suppose that $\xi_{k+4} = \partial_x$. Since $\xi_{k+4} \nabla^{k-1} R(\xi_1, ..., \xi_{k+3}) = 0$, the term in (3.2.a) vanishes. Replace ξ_i by ∂_y in position i to define

$$\vec{\eta}_i := (\xi_1, ..., \xi_{i-1}, \partial_y, \xi_{i+1}, ..., \xi_{k+3}) .$$

Since $\nabla_{\partial_x} \partial_y$ is a multiple of $\partial_{\tilde{x}}$, we can ignore terms where $\xi_i = \partial_y$. Furthermore, as $\nabla_{\partial_x} \partial_x = f' \partial_y = abe^{by} \partial_y$, the terms in (3.2.b) yield

$$\nabla^k R(\vec{\xi}) = -abe^{by} \sum_{i:\xi_i=\partial_x} \nabla^{k-1} R(\vec{\eta}_i)$$
$$= -abe^{by} \sum_{i:\xi_i=\partial_x} \gamma_{\vec{\eta}_i}(a, b) e^{\frac{1}{2}\Theta(\vec{\eta}_i)by} .$$

Since $\Theta(\vec{\eta}_i) = \Theta(\vec{\xi}) - 2$, $e^{by} e^{\frac{1}{2}\Theta(\vec{\eta}_i)} = e^{\frac{1}{2}\Theta(\vec{\eta})}$. Consequently this has the desired form. $\qquad\square$

We make the following

Definition 3.2.3 Assume that $f'' \neq 0$.

(1) Set $\alpha_1 := \frac{f' f'''}{f'' f''}$.

(2) We say that a basis $\mathcal{B} = \{X, Y, \tilde{X}\}$ for $\mathbb{R}^3$ is *normalized* if:

    (a) The non-zero components of g are $g(X, \tilde{X}) = g(Y, Y) = 1$.
    (b) The non-zero components of R are $R(X, Y, Y, X) = \text{sign}(f'')$.
    (c) We have $\nabla R(\xi_1, \xi_2, \xi_3, \xi_4; X) = 0$ for all $\xi_1, \xi_2, \xi_3, \xi_4$.

(3) We say that a basis $\mathcal{B} = \{X, Y, \tilde{X}\}$ for $\mathbb{R}^3$ is *affine normalized* if the non-zero components of $\mathcal{R}$ and $\nabla\mathcal{R}$ are:

    (a) $\mathcal{R}(X, Y)Y = \text{sign}(f'')\tilde{X}$.
    (b) $\mathcal{R}(X, Y)X = -\text{sign}(f'')Y$.
    (c) $\nabla_Y \mathcal{R}(X, Y)Y = \tilde{X}$.
    (d) $\nabla_Y \mathcal{R}(X, Y)X = -Y$.

The invariant α_1 of Definition 3.2.3 (1) is similar to the invariant α_2 described Lemma 2.3.1 and plays a similar role in our development. The remaining assertions of Theorem 3.2.1 concern curvature homogeneity; they will follow from the following result.

Lemma 3.2.6   *Let $\mathcal{M}$ be as in Definition 3.2.1 where $f'' \neq 0$.*

(1) Let $\varepsilon := \text{sign}\{f''\}$. Then $\mathcal{M}$ is 0-curvature modeled on $\mathcal{N}_0^\varepsilon$.
(2) Assume additionally that $f'''(y) \neq 0$. Then:

> *(a) $\mathcal{M}$ is 1-affine curvature homogeneous.*
> *(b) $\mathcal{M}$ is 1-curvature homogeneous if and only if $f'' = ae^{by}$.*
> *(c) α_1 is an invariant of $\mathfrak{F}_2(\mathcal{M}, P)$.*
> *(d) The following assertions are equivalent:*
>
>> *i. $\mathcal{M}$ is locally homogeneous.*
>> *ii. $\mathcal{M}$ is k-curvature homogeneous for all k.*
>> *iii. $\mathcal{M}$ is 2-curvature homogeneous.*
>> *iv. $\mathcal{M}$ is 2-affine curvature homogeneous.*
>> *v. $f' = ae^{by}$.*

Proof.   We may define a basis normalized as in Definition 3.2.3 (2) and establish Assertion (1) of Lemma 3.2.6 by setting

$$X := |f''|^{-1/2}\{\partial_x + f\partial_{\tilde{x}}\}, \quad Y := \partial_y, \quad \tilde{X} := |f''|^{1/2}\partial_{\tilde{x}}. \tag{3.2.c}$$

We construct a basis which is affine normalized as in Definition 3.2.3 (3) by rescaling the coordinate frame. Let a_1, a_2, and a_3 be constants to be determined. By Lemma 3.2.2,

$$\mathcal{R}(a_1\partial_x, a_2\partial_y)a_2\partial_y = a_1 a_2^2 a_3^{-1} f'' a_3 \partial_{\tilde{x}},$$
$$\mathcal{R}(a_1\partial_x, a_2\partial_y)a_1\partial_x = -a_1^2 f'' a_2 \partial_y,$$
$$\nabla_{a_2\partial_y}\mathcal{R}(a_1\partial_x, a_2\partial_y)a_2\partial_y = a_1 a_2^3 a_3^{-1} f''' a_3 \partial_{\tilde{x}},$$
$$\nabla_{a_2\partial_y}\mathcal{R}(a_1\partial_x, a_2\partial_y)a_1\partial_x = -a_1^2 a_2 f''' a_2 \partial_y.$$

Assume that $f''(y)$ and $f'''(y)$ never vanish. We define an affine normalized basis and prove Assertion (2a) by setting

$$X := a_1\partial_x, \qquad Y := a_2\partial_y, \qquad \tilde{X} := a_3\partial_{\tilde{x}} \qquad \text{where}$$
$$a_1 := \{|f''|\}^{-1/2}, \, a_2 := |f''|\{f'''\}^{-1}, \, a_3 := a_1 a_2^2 |f''|. \tag{3.2.d}$$

We note for future reference that

$$\nabla_X\nabla_X\mathcal{R}(X, Y)Y = a_1^3 a_2^2 \nabla_{\partial_x}\nabla_{\partial_x}\mathcal{R}(\partial_x, \partial_y)\partial_y = -a_1^3 a_2^2 f' f''' \partial_{\tilde{x}}$$
$$= -f' f'''\{f''\}^{-2}\tilde{X}. \tag{3.2.e}$$

We study the relevant symmetry group to construct additional invariants of the 1-model. Let $\mathcal{B} = \{X, Y, \tilde{X}\}$ be the normalized basis defined in

Eq. (3.2.c). Suppose that $\mathcal{B}_1 = \{X_1, Y_1, \tilde{X}_1\}$ is another normalized basis. Expand:

$$X_1 = a_{11}X + a_{12}Y + a_{13}\tilde{X},$$
$$Y_1 = a_{21}X + a_{22}Y + a_{23}\tilde{X},$$
$$\tilde{X}_1 = a_{31}X + a_{32}Y + a_{33}\tilde{X}.$$

Since $R(\xi_1, \xi_2, \xi_3, \tilde{X}_1) = 0$ for any ξ_1, ξ_2, ξ_3, we have $a_{31} = a_{32} = 0$. Since $\nabla R(\xi_1, \xi_2, \xi_3, \xi_4; X_1) = 0$ for any $\xi_1, \xi_2, \xi_3, \xi_4$, $a_{12} = 0$. Thus

$$X_1 = a_{11}X + a_{13}\tilde{X}, \quad Y_1 = a_{21}X + a_{22}Y + a_{23}\tilde{X}, \quad \tilde{X}_1 = a_{33}\tilde{X}.$$

Because $g(X_1, \tilde{X}_1) = 1$, $a_{33}a_{11} = 1$. As $g(Y_1, \tilde{X}_1) = 0$, $a_{21} = 0$. Since $g(X_1, X_1) = 0$, $a_{13} = 0$. Consequently,

$$X_1 = a_{11}X, \quad Y_1 = a_{22}Y + a_{23}\tilde{X}, \quad \tilde{X}_1 = a_{11}^{-1}\tilde{X}.$$

As $g(X_1, Y_1) = 0$, $a_{23} = 0$. Since $g(Y_1, Y_1) = 1$, $a_{22}^2 = 1$. Because $R(X_1, Y_1, Y_1, X_1) = \operatorname{sign}(f'')$, $a_{11}^2 a_{22}^2 = 1$. Therefore

$$X_1 = a_{11}X, \quad Y_1 = a_{22}Y, \quad \tilde{X}_1 = a_{11}^{-1}\tilde{X} \quad \text{where} \quad a_{11}^2 = a_{22}^2 = 1.$$

In particular we may use Eq. (3.2.c) and Lemma 3.2.2 to see:

$$|\nabla R(X_1, Y_1, Y_1, X_1; Y_1)| = |\nabla R(X, Y, Y, X; Y)| = |f'''\{f''\}^{-1}|$$

is an invariant of the 1-model. This is constant if and only if $f''' = cf''$; this means that $f'' = ae^{by}$. Assertion (2b) now follows.

We now establish Assertion (2d); Assertion (2c) will follow from our investigations. The following implications are clear:

$$(2\text{d-i}) \Rightarrow (2\text{d-ii}) \Rightarrow (2\text{d-iii}) \Rightarrow (2\text{d-iv}).$$

We investigate the implication (2d-iv)$\Rightarrow$(2d-v) and we establish Assertion (2c). Let $\mathcal{B} := \{X, Y, \tilde{X}\}$ be the affine normalized basis defined in Eq. (3.2.d). Let

$$\mathcal{I}_0 := \operatorname{Span}_{\xi_1, \xi_2, \xi_3}\{\mathcal{R}(\xi_1, \xi_2)\xi_3\} = \operatorname{Span}\{Y, \tilde{X}\},$$
$$\mathcal{K}_0 := \{\eta : \mathcal{R}(\eta, \xi_1)\xi_2 = 0 \text{ for all } \xi_1, \xi_2\} = \operatorname{Span}\{\tilde{X}\},$$
$$\mathcal{K}_1 := \{\eta : \nabla_\eta \mathcal{R}(\xi_1, \xi_2)\xi_3 = 0 \text{ for all } \xi_1, \xi_2, \xi_3\} = \operatorname{Span}\{X, \tilde{X}\}.$$

Suppose that $\mathcal{B}_1 := \{X_1, Y_1, \tilde{X}_1\}$ is another affine normalized basis. Since the spaces given above are invariantly defined, we may expand

$$X_1 = a_{11}X + a_{13}\tilde{X}, \quad Y_1 = a_{22}Y + a_{23}\tilde{X}, \quad \tilde{X}_1 = a_{33}\tilde{X}.$$

We have $\mathcal{R}(X_1, Y_1)X_1 = a_{11}^2 \mathcal{R}(X, Y_1)X$ is a multiple of Y. Since the basis is normalized, it is also a multiple of Y_1. Thus $a_{23} = 0$. We now compute

$$\mathcal{R}(X_1, Y_1)Y_1 = a_{11}a_{22}^2 a_{33}^{-1} \operatorname{sign}(f'')\tilde{X}_1, \ \mathcal{R}(X_1, Y_1)X_1 = -a_{11}^2 \operatorname{sign}(f'')Y_1,$$
$$\nabla_{Y_1}\mathcal{R}(X_1, Y_1)Y_1 = a_{11}a_{22}^3 a_{33}^{-1}\tilde{X}_1, \qquad \nabla_{Y_1}\mathcal{R}(X_1, Y_1)X_1 = -a_{11}^2 a_{22}Y_1.$$

Because the basis is normalized, $a_{11}^2 = 1$, $a_{22} = 1$, $a_{33} = a_{11}$. Consequently Eq. (3.2.e) yields

$$\nabla_{X_1}\nabla_{X_1}\mathcal{R}(X_1, Y_1)Y_1 = a_{11}^3 a_{22}^2 \nabla_X \nabla_X \mathcal{R}(X, Y)Y$$
$$= -a_{11}^3 a_{22}^2 f' f''' \{f''\}^{-2}\tilde{X} = -a_{11}^3 a_{22}^2 a_{33}^{-1} f' f''' \{f''\}^{-2}\tilde{X}_1$$
$$= -f' f''' \{f''\}^{-2}\tilde{X}_1.$$

This shows that $f' f''' \{f''\}^{-2}$ is an invariant of the affine 2-model and proves Assertion (2c). Consequently if $\mathcal{M}$ is 2-affine curvature homogeneous, then

$$f' f''' \{f''\}^{-2} = c.$$

We apply Lemma 1.5.5. Either $f' = ae^{by}$ or $f' = a(y+b)^c$ for some $\{a, b, c\}$. This final choice is ruled out as f''' and f'' are assumed to be globally defined and non-zero. Thus we may conclude $f' = ae^{by}$; this establishes the implication:

$$(\text{2d-iv}) \Rightarrow (\text{2d-v}).$$

Finally, we suppose that $f' = ae^{by}$; we take $f = \frac{a}{b}e^{by}$. Consider the normalized basis $\{X, Y, \tilde{X}\}$ defined in Eq. (3.2.c):

$$X = |ab|^{-1/2}e^{-by/2}\{\partial_x + \tfrac{a}{b}e^{by}\partial_{\tilde{x}}\}, \quad Y = \partial_y, \quad \tilde{X} = |ab|^{1/2}e^{by/2}\partial_{\tilde{x}}.$$

Let $\nabla^k R(\vec{\eta}) = \nabla^k R(\eta_1, \eta_2, \eta_3, \eta_4; \eta_5, ..., \eta_{4+k})$ for $\vec{\eta} = (\eta_1, ..., \eta_{4+k})$ where $\eta_i = X$ or $\eta_i = Y$ for $1 \leq i \leq 4+k$. Let $\vec{\xi}$ be the corresponding string where X and Y are replaced by ∂_x and ∂_y. We set $\Theta(\vec{\eta}) = \Theta(\vec{\xi})$ to be the number of times that X or equivalently ∂_x appear. We apply Lemma 3.2.5

to see:

$$\nabla^k R(\vec{\eta}) = |ab|^{-\Theta(\eta)/2} e^{-\Theta(\vec{\eta})by/2} \nabla^k R(\vec{\xi})$$
$$= |ab|^{-\Theta(\eta)/2} e^{-\Theta(\vec{\eta})by/2} \gamma(a, b; \vec{\xi}) e^{\Theta(\vec{\xi})by/2}$$
$$= |ab|^{-\Theta(\eta)/2} \gamma(a, b; \vec{\xi}) \,.$$

This shows that $\mathcal{M}$ is k-curvature homogeneous for all k; a local version of Theorem 1.4.2 now shows that $\mathcal{M}$ is locally homogeneous as desired. Consequently, (2d-v)$\Rightarrow$(2d-i). $\qquad\qquad\square$

3.3 Signature $(2, 2)$ Walker Manifolds

In this section, we present results which deal with 4-dimensional pseudo-Riemannian manifolds of neutral signature $(2, 2)$. In Section 3.3.1 we review the basic structure theorem for Osserman algebraic curvature tensors of signature $(2, 2)$ which is due to Blažić, Bokan, and Rakić (2001a) and to García-Río, Kupeli, and Vázquez-Lorenzo (2002). We shall study a family of *Walker metrics* of signature $(2, 2)$ on $\mathbb{R}^4$ generalizing the examples of Definition 2.3.2:

Definition 3.3.1 Let $\psi_{ij}(x_1, x_2, x_3, x_4) = \psi_{ji}(x_1, x_2, x_3, x_4)$, where $i, j = 3, 4$, define a smooth map from $\mathbb{R}^4$ to $S^2(\mathbb{R}^2)$. Let $g = g_\psi$ be defined by:

$$g(\partial_{x_1}, \partial_{x_3}) = g(\partial_{x_2}, \partial_{x_4}) = 1,$$
$$g(\partial_{x_i}, \partial_{x_j}) = \psi_{ij} \quad \text{for} \quad i, j = 3, 4 \,.$$

Define a Hermitian almost complex structure J on $\mathcal{M} := (\mathbb{R}^4, g)$ by setting:

$$J\partial_{x_1} = \partial_{x_2}, \qquad\qquad J\partial_{x_2} = -\partial_{x_1},$$
$$J\partial_{x_3} = -\psi_{34}\partial_{x_1} + \tfrac{1}{2}(\psi_{44} - \psi_{33})\partial_{x_2} + \partial_{x_4},$$
$$J\partial_{x_4} = \tfrac{1}{2}(\psi_{44} - \psi_{33})\partial_{x_1} + \psi_{34}\partial_{x_2} - \partial_{x_3} \,.$$

Here is a brief outline to Section 3.3. In Section 3.3.2, we present work of García-Río, Rakić, and Vázquez-Abal (2005) dealing with Kähler Osserman manifolds in this family. In Section 3.3.3, we present work of Díaz-Ramos, García-Río, and Vázquez-Lorenzo (2006) giving Jordan Osserman manifolds which are not nilpotent and whose Jacobi operator is not diagonalizable. In Section 3.3.4, we present work of Brozos-Vázquez, García-Río, and Vázquez-Lorenzo (2005) dealing with the conformal Jacobi operator;

this yields conformal Osserman manifolds whose conformal Jacobi operator has complex eigenvalues.

3.3.1 *Osserman curvature tensors of signature* $(2,2)$

Zero is always an eigenvalue of the Jacobi operator since $J(x)x = 0$. If x is spacelike or timelike, let $\tilde{J}(x)$ be the restriction of $J(x)$ to $x^\perp$; $\mathcal{M}$ is spacelike Osserman if and only if the *reduced Jacobi operator $\tilde{J}$* has constant eigenvalues on $S^+(\mathcal{M})$. As discussed previously in Theorem 1.9.1, this means equivalently that $\mathcal{M}$ is timelike Osserman and that $\tilde{J}$ has constant eigenvalues on $S^-(\mathcal{M})$.

There is a complete algebraic classification of Osserman curvature tensors of signature $(2,2)$ which is given by Blažić, Bokan, and Rakić (2001a) and by García-Río, Kupeli, and Vázquez-Lorenzo (2002). We may summarize those results as follows:

Theorem 3.3.1 *Let $\mathfrak{M}$ be a 0-model of signature $(2,2)$. Then $\mathfrak{M}$ is Osserman if and only if one of the following four possibilities holds:*

(1) Type I-a: The reduced Jacobi operator is diagonalizable with eigenvalues $\{\alpha, \beta, \gamma\}$; $\tilde{J}(x)$ is conjugate to the following matrix for any $x \in S^+(V, \langle \cdot, \cdot \rangle)$:

$$\begin{pmatrix} \alpha & 0 & 0 \\ 0 & \beta & 0 \\ 0 & 0 & \gamma \end{pmatrix}.$$

This occurs if and only if there is an orthonormal basis $\{e_1, e_2, e_3, e_4\}$ for $\mathbb{R}^4$ so the non-zero components of A are given by:

$$\begin{aligned}
A_{1221} &= A_{4334} = \alpha, & A_{1331} &= A_{4224} = -\beta, \\
A_{1441} &= A_{3223} = -\gamma, & A_{1234} &= \tfrac{1}{3}(-2\alpha + \beta + \gamma), \\
A_{1423} &= \tfrac{1}{3}(\alpha + \beta - 2\gamma), & A_{1342} &= \tfrac{1}{3}(\alpha - 2\beta + \gamma).
\end{aligned}$$

(2) Type I-b: The reduced Jacobi operator has 2 complex conjugate eigenvalues $\{\alpha \pm \sqrt{-1}\beta\}$ and a real eigenvalue γ; $\tilde{J}(x)$ is conjugate to the following matrix for any $x \in S^+(V, \langle \cdot, \cdot \rangle)$:

$$\begin{pmatrix} \alpha & \beta & 0 \\ -\beta & \alpha & 0 \\ 0 & 0 & \gamma \end{pmatrix}.$$

This occurs if and only if there is an orthonormal basis $\{e_1, e_2, e_3, e_4\}$ for $\mathbb{R}^4$ so the non-zero components of A are given by:

$$
\begin{aligned}
A_{1221} &= A_{4334} = \alpha, & A_{1331} &= A_{4224} = -\alpha, \\
A_{1441} &= A_{3223} = -\gamma, & A_{2113} &= A_{2443} = -\beta, \\
A_{1224} &= A_{1334} = \beta, & A_{1234} &= \tfrac{1}{3}(-\alpha + \gamma), \\
A_{1423} &= \tfrac{2}{3}(\alpha - \gamma), & A_{1342} &= \tfrac{1}{3}(-\alpha + \gamma).
\end{aligned}
$$

(3) Type II: The minimal polynomial for the reduced Jacobi operator has α as a double root; $\tilde{J}(x)$ is conjugate to the following matrix for any $x \in S^+(V, \langle \cdot, \cdot \rangle)$:

$$
\begin{pmatrix}
\pm(\alpha - \tfrac{1}{2}) & \pm\tfrac{1}{2} & 0 \\
\mp\tfrac{1}{2} & \pm(\alpha + \tfrac{1}{2}) & 0 \\
0 & 0 & \beta
\end{pmatrix}.
$$

This occurs if and only if there is an orthonormal basis $\{e_1, e_2, e_3, e_4\}$ for $\mathbb{R}^4$ so the non-zero components of A are given by:

$$
\begin{aligned}
A_{1221} &= A_{4334} = \pm(\alpha - \tfrac{1}{2}), & A_{1331} &= A_{4224} = \mp(\alpha + \tfrac{1}{2}), \\
A_{1441} &= A_{3223} = -\beta, & A_{2113} &= A_{2443} = \mp\tfrac{1}{2}, \\
A_{1224} &= A_{1334} = \pm\tfrac{1}{2}, & A_{1234} &= \tfrac{1}{3}(\pm(-\alpha + \tfrac{3}{2}) + \beta), \\
A_{1423} &= \tfrac{2}{3}(\pm\alpha - \beta), & A_{1342} &= \tfrac{1}{3}(\pm(-\alpha - \tfrac{3}{2}) + \beta).
\end{aligned}
$$

(4) Type III: The minimal polynomial for the reduced Jacobi operator has α as a triple root; $\tilde{J}(x)$ is conjugate to the following matrix for any $x \in S^+(V, \langle \cdot, \cdot \rangle)$:

$$
\begin{pmatrix}
\alpha & 0 & \frac{1}{\sqrt{2}} \\
0 & \alpha & \frac{1}{\sqrt{2}} \\
-\frac{1}{\sqrt{2}} & \frac{1}{\sqrt{2}} & \alpha
\end{pmatrix}.
$$

This occurs if and only if there is an orthonormal basis $\{e_1, e_2, e_3, e_4\}$ for $\mathbb{R}^4$ so the non-zero components of A are given by:

$$
\begin{aligned}
A_{1221} &= A_{4334} = \alpha, & A_{1331} &= A_{4224} = -\alpha, \\
A_{1441} &= A_{3223} = -\alpha, & A_{2114} &= A_{2334} = -\tfrac{1}{\sqrt{2}}, \\
A_{3114} &= -A_{3224} = \tfrac{1}{\sqrt{2}}, & A_{1223} &= A_{1443} = A_{1332} = \tfrac{1}{\sqrt{2}}, \\
A_{1442} &= -\tfrac{1}{\sqrt{2}}.
\end{aligned}
$$

Let W be the associated Weyl algebraic curvature tensor discussed previously in Section 1.3.3. Let $\star$ be the Hodge operator. We say A is *self-dual* (respectively *anti self-dual*) if $\star W = W$ (respectively if $\star W = -W$).

The following consequence can then be drawn from the classification given in Theorem 3.3.1; we refer to Brozos-Vázquez, García-Río, and Vázquez-Lorenzo (2005) for details:

Corollary 3.3.1 *Let $\mathfrak{M}$ be a 0-model of neutral signature $(2,2)$. The following conditions are equivalent:*

(1) $\mathfrak{M}$ is Osserman.

(2) $\mathfrak{M}$ is spacelike Jordan Osserman.

(3) $\mathfrak{M}$ is timelike Jordan Osserman.

(4) A is either self-dual Einstein or anti self-dual Einstein.

We remark that the examples given in Chapter 2 show that (1), (2), and (3) are different concepts in the higher signature setting.

3.3.2 *Indefinite Kähler Osserman manifolds*

We begin by recalling the following result of Blažić, Bokan, and Rakić (2001a):

Theorem 3.3.2 *Let $\mathcal{M}$ be a pseudo-Riemannian manifold of signature $(2,2)$. The following conditions are equivalent:*

(1) $\mathcal{M}$ is Osserman.

(2) $\mathcal{M}$ is locally isomorphic to one of the following manifolds:

> *(a) A manifold of constant sectional curvature.*
>
> *(b) A Kähler manifold of constant holomorphic sectional curvature.*
>
> *(c) A para-Kähler manifold of constant para-holomorphic sectional curvature.*
>
> *(d) A manifold whose Jacobi operator is nondiagonalizable, and its characteristic polynomial has a triple zero α.*

The authors note that such a manifold admits a foliation by 2-dimensional totally geodesic isotropic submanifolds. Thus the study of Kähler geometry is very relevant and one has García-Río, Rakić, and Vázquez-Abal (2005):

Theorem 3.3.3 *Adopt the notation of Definition 3.3.1 to give the pseudo-Riemannian manifold $\mathcal{M} := (\mathbb{R}^4, g_\psi)$ a Hermitian almost complex structure J. Then $\mathcal{M}$ is Kähler Osserman if and only if there exist smooth functions $\Xi = \Xi(x_3, x_4)$ and $\Psi_{ij} = \Psi_{ij}(x_3, x_4)$ so that:*

(1) $\psi_{33} = \quad x_1 \partial_{x_4} \Xi(x_3, x_4) + x_2 \partial_{x_3} \Xi(x_3, x_4) + \Psi_{33}(x_3, x_4).$

(2) $\psi_{44} = -x_1\partial_{x_4}\Xi(x_3,x_4) - x_2\partial_{x_3}\Xi(x_3,x_4) + \Psi_{44}(x_3,x_4)$.

(3) $\psi_{34} = -x_1\partial_{x_3}\Xi(x_3,x_4) + x_2\partial_{x_4}\Xi(x_3,x_4) + \Psi_{34}(x_3,x_4)$.

If $\partial_{x_4}^2\Xi + \partial_{x_3}^2\Xi \neq 0$, then the Jacobi operator is 3-step nilpotent.

3.3.3 *Jordan Osserman manifolds which are not nilpotent*

We follow the discussion in Díaz-Ramos, García-Río, and Vázquez-Lorenzo (2006) to present manifolds of signature $(2,2)$ which are Osserman but not nilpotent and whose Jacobi operators are not diagonalizable. We specialize the examples of Definition 3.3.1:

Definition 3.3.2 Let $\{x_1, x_2, x_3, x_4\}$ be coordinates on $M = \mathbb{R}^4$ and let $f = f(x_4)$ be non-constant. Let $\mathcal{M}_f := (\mathbb{R}^4, g_f)$ have neutral signature $(2,2)$ where

$$g_f(\partial_{x_1}, \partial_{x_3}) := 1, \qquad\qquad g_f(\partial_{x_2}, \partial_{x_4}) := 1,$$
$$g_f(\partial_{x_3}, \partial_{x_3}) := 4kx_1^2 - \tfrac{1}{4k}f(x_4)^2, \qquad g_f(\partial_{x_4}, \partial_{x_4}) := 4kx_2^2,$$
$$g_f(\partial_{x_3}, \partial_{x_4}) := 4kx_1x_2 + x_2 f(x_4) - \tfrac{1}{4k}f'(x_4).$$

One then has that the curvatures are

$$R_{1313} = R_{2424} = -4k, \qquad R_{1324} = R_{1423} = -2k,$$
$$R_{1334} = kx_2(4kx_1 + f(x_4)), \quad R_{1434} = 4k^2x_2^2,$$
$$R_{2334} = \tfrac{1}{4}f(x_4)^2 - 4k^2x_1^2, \qquad R_{2434} = \tfrac{1}{2}f'(x_4) - kx_2(4kx_1 + f(x_4)),$$
$$R_{3434} = \tfrac{1}{4k}f'(x_4)^2 + 2kx_1x_2 f'(x_4) - 2kx_2^2 f(x_4)^2$$
$$- x_1 f''(x_4) - f(x_4)\left(8k^2x_1x_2^2 - \tfrac{5}{2}x_2 f'(x_4) - \tfrac{1}{4k}f''(x_4)\right).$$

Set $\mathcal{P}_f := 24kf(x_4)f'(x_4)x_2 - 12kf''(x_4)x_1 + 3f(x_4)f''(x_4) + 4f'(x_4)^2$. We refer to the paper cited above for the proof of the following result:

Theorem 3.3.4   [Díaz-Ramos, García-Río, Vázquez-Lorenzo].

(1) The manifold $\mathcal{M}_f$ is Osserman with eigenvalues $\{0, 4k, k, k\}$.

(2) The Jacobi operators are diagonalizable at P if and only if $\mathcal{P}_f(P) = 0$.

(3) Let O be an open subset of $\mathbb{R}^4$ where $\mathcal{P}_f \neq 0$. Let $\mathcal{O} := (\mathbb{R}^3 \times O, g_f)$.

(a) $\mathcal{O}$ is spacelike and timelike Jordan Osserman.

(b) k is a double root of the minimal polynomial of the Jacobi operator.

(4) The manifold $\mathcal{M}_f$ is not 2-Osserman.

(5) The manifold $\mathcal{M}_f$ is not locally symmetric.

(6) The manifold $\mathcal{M}_f$ is nilpotent Szabó. It is neither spacelike nor timelike pointwise Jordan Szabó.

(7) The manifold $\mathcal{M}_f$ does not have vanishing scalar invariants.

It has been pointed out to us that this example seems to contradict Theorem 3.3.2 since the reduced Jacobi operator of $\mathcal{O}$ has two different eigenvalues and it is not diagonalizable if $k \neq 0$. Note that $\mathcal{O}$ is not locally isometric to (a)-(d) in Theorem 3.3.2.

3.3.4 *Conformally Osserman manifolds*

Let W and $\mathcal{W}$ be the Weyl conformal curvature tensor and curvature operator, respectively, for a pseudo-Riemannian manifold $\mathcal{M}$ as discussed in Section 1.3.3. Then W is an algebraic curvature tensor in its own right, so we can form the associated conformal Jacobi operator $\mathcal{J}_W$ as discussed in Section 1.7.3. One says that $\mathcal{M}$ is *spacelike conformally Jordan Osserman* if the eigenvalues of $\mathcal{J}_W$ are constant on $S_P^+(\mathcal{M})$ for each point $P \in M$; the eigenvalues being permitted to vary from point to point. We showed in Theorem 1.9.9 that this is a conformally invariant notion. Examples of such manifolds were given previously in Corollary 2.5.1. We now summarize work of Brozos-Vázquez, García-Río, and Vázquez-Lorenzo (2005) giving additional 4-dimensional neutral signature manifolds.

The following is an immediate consequence of Corollary 3.3.1:

Theorem 3.3.5 *Let $\mathcal{M}$ be a pseudo-Riemannian manifold of neutral signature $(2,2)$. Then $\mathcal{M}$ is conformally Osserman if and only if there is a choice of orientation for $\mathcal{M}$ such that it is self-dual or anti self-dual.*

One can consider the following family of *Walker manifolds*

Definition 3.3.3 Let $\mathcal{M} := (\mathbb{R}^4, g)$ where g is defined by:

$$g(\partial_{x_1}, \partial_{x_3}) = g(\partial_{x_2}, \partial_{x_4}) := 1,$$
$$g(\partial_{x_3}, \partial_{x_4}) = \psi_{34}(x_1, x_2, x_3, x_4).$$

Let $\psi_{ij/k} = \partial_{x_k}\psi_{ij}$ and $\psi_{ij/kl} := \partial_{x_k}\partial_{x_l}\psi_{ij}$. One then has that:

$$
\begin{aligned}
R_{2334} &= \tfrac{1}{4}\{\psi_{34/2}^2 - 2\psi_{34/23}\}, & R_{1314} &= -\tfrac{1}{2}\psi_{34/11}, \\
R_{1334} &= \tfrac{1}{4}\{\psi_{34/1}\psi_{34/2} - 2\psi_{34/13}\}, & R_{2324} &= -\tfrac{1}{2}\psi_{34/22}, \\
R_{2434} &= \tfrac{1}{4}(-\psi_{34/1}\psi_{34/2} + 2\psi_{34/24}\}, & R_{1324} &= -\tfrac{1}{2}\psi_{34/12}, \\
R_{3434} &= \tfrac{1}{2}(-\psi_{34}\psi_{34/1}\psi_{34/2} + 2\psi_{34/34}\}, & R_{1423} &= -\tfrac{1}{2}\psi_{34/12}, \\
R_{1434} &= \tfrac{1}{4}\{-\psi_{34/1}\psi_{34/1} + 2\psi_{34/14}\}.
\end{aligned}
$$

As the scalar curvature is given by $\tau = 2\psi_{34/12}$, these need not be generalized plane wave manifolds. We shall see presently they are not in general complete. We refer to Brozos-Vázquez, García-Río, and Vázquez-Lorenzo (2005) for the proof of the following result:

Theorem 3.3.6 *Let M be as in Definition 3.3.3. Then M is Osserman if and only if ψ_{34} has one of the following forms:*

(1) $\psi_{34} = \psi_{34}(x_3, x_4)$.
(2) $\psi_{34} = -\frac{2}{x_4+\alpha}x_1 + c(x_3, x_4)$.
(3) $\psi_{34} = -\frac{2}{x_3+\alpha}x_2 + c(x_3, x_4)$.
(4) $\psi_{34} = -\frac{2}{x_4+a(x_3)}x_1 - \frac{2}{x_3+b(x_4)}x_2 + c(x_3, x_4)$ where the functions a and b satisfy the relation $a'(x_3 + b)^2 + b'(x_4 + a)^2 = 2(x_3 + b)(x_4 + a)$ and where the function c is arbitrary.

We note that if $\psi_{34} = \psi_{34}(x_3, x_4)$ is as in Assertion (1), then M belongs to the family considered in Section 2.2. One also has that the following result (see Remark 4.4 of Brozos-Vázquez, García-Río, and Vázquez-Lorenzo (2005)) showing all possible algebraic types are geometrically realizable:

Theorem 3.3.7 *Let M be as in Definition 3.3.3.*

(1) Let $\psi_{34} = x_1^2 - x_2^2$. Then M is conformally Osserman of type I-a with eigenvalues $\{0, 0, \frac{1}{2}, -\frac{1}{2}\}$.
(2) Let $\psi_{34} = x_1^2 + x_2^2$. Then M is conformally Osserman of type I-b with eigenvalues $\{0, 0, \frac{\sqrt{-1}}{2}, -\frac{\sqrt{-1}}{2}\}$.
(3) Let $\psi_{34} = x_1x_4 + x_3x_4$. Then M is conformally nilpotent of order 2 and is Osserman of type II.
(4) Let $\psi_{34} = x_1^2$. Then M is conformally nilpotent of order 3 and is Osserman of type III.
(5) M is not Osserman in (1)-(4) above.

The Jordan normal form of the conformal Jacobi operator can change depending on the point considered. Let $\mathcal{P}_P$ be the minimal polynomial of the conformal Jacobi operator. The following examples of nilpotent conformally Osserman manifolds illustrate this behaviour. We refer to Remark 4.5 of Brozos-Vázquez, García-Río, and Vázquez-Lorenzo (2005):

Theorem 3.3.8 *Let M be as in Definition 3.3.3 be defined by:*

(1) Let $\psi_{34} = x_2 x_4^2 + x_3^2 x_4$. Then $\mathcal{M}$ is conformally Osserman and

$$\mathcal{P}_P = \begin{cases} \lambda^3 & \text{if } x_4(P) \neq 0, \\ \lambda^2 & \text{if } x_4(P) = 0, \ x_3(P) \neq 0, \\ \lambda & \text{if } x_3(P) = x_4(P) = 0. \end{cases}$$

(2) Let $\psi_{34} = x_2 x_4^2 + x_3 x_4$. Then $\mathcal{M}$ is conformally Osserman and

$$\mathcal{P}_P = \begin{cases} \lambda^3 & \text{if } x_4(P) \neq 0, \\ \lambda^2 & \text{if } x_4(P) = 0. \end{cases}$$

(3) Let $\psi_{34} = x_1 x_3^2$. Then $\mathcal{M}$ is conformally Osserman and

$$\mathcal{P}_P = \begin{cases} \lambda^3 & \text{if } x_3(P) \neq 0, \\ \lambda & \text{if } x_3(P) = 0. \end{cases}$$

(4) Let $\psi_{34} = x_1 x_3 + x_2 x_4$. Then $\mathcal{M}$ is conformally Osserman and

$$\mathcal{P}_P = \begin{cases} \lambda^2 & \text{if } \{x_1 x_3 + x_2 x_4\}(P) \neq 0, \\ \lambda & \text{if } \{x_1 x_3 + x_2 x_4\}(P) = 0. \end{cases}$$

Let $\mathrm{Spec}_P(\mathcal{M})$ be the spectrum of the conformal Jacobi operator on the unit spacelike vectors for a conformally Osserman manifold. The eigenvalues can change. They can even change from real to complex. We refer to Remark 4.5 of Brozos-Vázquez, García-Río, and Vázquez-Lorenzo (2005):

Theorem 3.3.9   *Let $\mathcal{M}$ be as in Definition 3.3.3. Then:*

(1) If $\psi_{34} = x_1^4 + x_1^2 - x_2^4 - x_2^2$, then $\mathcal{M}$ is conformally Osserman and

$$\mathrm{Spec}_P = \left\{ 0, 0, \pm\tfrac{1}{2}\sqrt{(6x_1^2 + 1)(6x_2^2 + 1)} \right\}.$$

(2) If $\psi_{34} = x_1^4 + x_1^2 + x_2^4 + x_2^2$, then $\mathcal{M}$ is conformally Osserman and

$$\mathrm{Spec}_P = \left\{ 0, 0, \pm\tfrac{1}{2}\sqrt{-(6x_1^2 + 1)(6x_2^2 + 1)} \right\}.$$

(3) If $\psi_{34} = x_1^3 - x_2^3$, then $\mathcal{M}$ is conformally Osserman and

$$\mathrm{Spec}_P = \left\{ 0, 0, \pm\tfrac{3}{2}\sqrt{x_1 x_2} \right\}.$$

3.4 Geodesic Completeness and Ricci Blowup

In this section, we study questions of geodesic completeness related to the manifolds introduced in Section 3.3. Recall that $\mathcal{M}$ is said to be *geodesically complete* if all geodesics extend for infinite time and that $\mathcal{M}$ is said to exhibit *Ricci blowup* if there exists a geodesic defined for $t \in [0, T)$ such that $\lim_{t \to T} |\rho(\dot{\gamma}, \dot{\gamma})| = \infty$. In Section 3.4.1, we study the geodesic equation for the manifolds of Definition 3.3.3. In Section 3.4.2, we examine these questions for the manifolds of Section 3.3.4 which are conformally Osserman.

3.4.1 *The geodesic equation*

We now study questions of geodesic completeness and Ricci blowup. Let $\psi_{ij}(x_1, x_2, x_3, x_4) = \psi_{ji}(x_1, x_2, x_3, x_4)$ for $i, j = 3, 4$ define a smooth map from $\mathbb{R}^4$ to $S^2(\mathbb{R}^2)$. Following Definition 3.3.1, let $\mathcal{M} = (\mathbb{R}^4, g)$ where

$$g_\psi(\partial_{x_1}, \partial_{x_3}) = g(\partial_{x_2}, \partial_{x_4}) = 1,$$
$$g_\psi(\partial_{x_i}, \partial_{x_j}) = \psi_{ij} \text{ for } i, j = 3, 4 \,.$$

Let $\psi_{ij/k} := \partial_{x_k}\psi_{ij}$. We begin by showing:

Lemma 3.4.1 *Adopt the notation established above. The geodesic equation for $\mathcal{M}$ is:*

(1) $0 = \ddot{x}_1 + \dot{x}_1\dot{x}_3\psi_{33/1} + \dot{x}_1\dot{x}_4\psi_{34/1} + \dot{x}_2\dot{x}_3\psi_{33/2} + \dot{x}_2\dot{x}_4\psi_{34/2}$
    $+ \frac{1}{2}\dot{x}_3\dot{x}_3(\psi_{33/3} + \psi_{34}\psi_{33/2} + \psi_{33}\psi_{33/1})$
    $+ \frac{1}{2}\dot{x}_4\dot{x}_4(2\psi_{34/4} - \psi_{44/3} + \psi_{34}\psi_{44/2} + \psi_{33}\psi_{44/1})$
    $+ \dot{x}_3\dot{x}_4(\psi_{33/4} + \psi_{34}\psi_{34/2} + \psi_{33}\psi_{34/1}) \,.$

(2) $0 = \ddot{x}_2 + \dot{x}_1\dot{x}_3\psi_{34/1} + \dot{x}_1\dot{x}_4\psi_{44/1} + \dot{x}_2\dot{x}_3\psi_{34/2} + \dot{x}_2\dot{x}_4\psi_{44/2}$
    $+ \frac{1}{2}\dot{x}_3\dot{x}_3(2\psi_{34/3} - \psi_{33/4} + \psi_{44}\psi_{33/2} + \psi_{34}\psi_{33/1})$
    $+ \frac{1}{2}\dot{x}_4\dot{x}_4(\psi_{44/4} + \psi_{44}\psi_{44/2} + \psi_{34}\psi_{44/1})$
    $+ \dot{x}_3\dot{x}_4(\psi_{44/3} + \psi_{44}\psi_{34/2} + \psi_{34}\psi_{34/1}) \,.$

(3) $0 = \ddot{x}_3 - \frac{1}{2}\dot{x}_3\dot{x}_3\psi_{33/1} - \dot{x}_3\dot{x}_4\psi_{34/1} - \frac{1}{2}\dot{x}_4\dot{x}_4\psi_{44/1} \,.$

(4) $0 = \ddot{x}_4 - \frac{1}{2}\dot{x}_3\dot{x}_3\psi_{33/2} - \dot{x}_3\dot{x}_4\psi_{34/2} - \frac{1}{2}\dot{x}_4\dot{x}_4\psi_{44/2} \,.$

Proof. Following the discussion in García-Río, Rakić, and Vázquez-Abal (2005) one has that:

$$\nabla_{\partial_{x_1}}\partial_{x_3} = \nabla_{\partial_{x_3}}\partial_{x_1} = \tfrac{1}{2}\psi_{33/1}\partial_{x_1} + \tfrac{1}{2}\psi_{34/1}\partial_{x_2},$$
$$\nabla_{\partial_{x_1}}\partial_{x_4} = \nabla_{\partial_{x_4}}\partial_{x_1} = \tfrac{1}{2}\psi_{34/1}\partial_{x_1} + \tfrac{1}{2}\psi_{44/1}\partial_{x_2},$$

$$\nabla_{\partial_{x_2}}\partial_{x_3} = \nabla_{\partial_{x_3}}\partial_{x_2} = \tfrac{1}{2}\psi_{33/2}\partial_{x_1} + \tfrac{1}{2}\psi_{34/2}\partial_{x_2},$$

$$\nabla_{\partial_{x_2}}\partial_{x_4} = \nabla_{\partial_{x_4}}\partial_{x_2} = \tfrac{1}{2}\psi_{34/2}\partial_{x_1} + \tfrac{1}{2}\psi_{44/2}\partial_{x_2},$$

$$\nabla_{\partial_{x_3}}\partial_{x_3} = \tfrac{1}{2}(\psi_{33/3} + \psi_{34}\psi_{33/2} + \psi_{33}\psi_{33/1})\partial_{x_1}$$
$$+\tfrac{1}{2}(2\psi_{34/3} - \psi_{33/4} + \psi_{44}\psi_{33/2} + \psi_{34}\psi_{33/1})\partial_{x_2}$$
$$-\tfrac{1}{2}\psi_{33/1}\partial_{x_3} - \tfrac{1}{2}\psi_{33/2}\partial_{x_4},$$

$$\nabla_{\partial_{x_3}}\partial_{x_4} = \nabla_{\partial_{x_4}}\partial_{x_3} = \tfrac{1}{2}(\psi_{33/4} + \psi_{34}\psi_{34/2} + \psi_{33}\psi_{34/1})\partial_{x_1}$$
$$+\tfrac{1}{2}(\psi_{44/3}+\psi_{44}\psi_{34/2}+\psi_{34}\psi_{34/1})\partial_{x_2} -\tfrac{1}{2}\psi_{34/1}\partial_{x_3}-\tfrac{1}{2}\psi_{34/2}\partial_{x_4},$$

$$\nabla_{\partial_{x_4}}\partial_{x_4} = \tfrac{1}{2}(2\psi_{34/4} - \psi_{44/3} + \psi_{34}\psi_{44/2} + \psi_{33}\psi_{44/1})\partial_{x_1}$$
$$+\tfrac{1}{2}(\psi_{44/4}+\psi_{44}\psi_{44/2}+\psi_{34}\psi_{44/1})\partial_{x_2} -\tfrac{1}{2}\psi_{44/1}\partial_{x_3}-\tfrac{1}{2}\psi_{44/2}\partial_{x_4}.$$

The Lemma now follows. $\square$

We shall use this in Section 3.4.2 to study the Conformally Osserman manifolds discussed in Section 3.3.4 and the Jordan Osserman manifolds discussed in Section 3.3.3.

3.4.2 *Conformally Osserman manifolds*

In this section, we will examine which manifolds of Section 3.3.4 are geodesically complete and which are geodesically incomplete, and which exhibit Ricci blowup. Assertion (1) of the following Lemma follows by specializing Lemma 3.4.1 to the manifolds given in Definition 3.3.3; Assertion (2) follows from the work of Díaz-Ramos, García-Río, and Vázquez-Lorenzo (2006); we adopt the notation of Definition 3.3.3.

Lemma 3.4.2 *Let $\mathcal{M} := (\mathbb{R}^4, g)$ where $g(\partial_{x_1}, \partial_{x_3}) = g(\partial_{x_2}, \partial_{x_4}) = 1$ and where $g(\partial_{x_3}, \partial_{x_4}) = \psi_{34}(x_1, x_2, x_3, x_4)$. Then:*

(1) The geodesic equation is:
$$0 = \ddot{x}_1 + \dot{x}_1\dot{x}_4\psi_{34/1} + \dot{x}_2\dot{x}_4\psi_{34/2} + \dot{x}_3\dot{x}_4\psi_{34}\psi_{34/2} + \dot{x}_4\dot{x}_4\psi_{34/4},$$
$$0 = \ddot{x}_2 + \dot{x}_1\dot{x}_3\psi_{34/1} + \dot{x}_2\dot{x}_3\psi_{34/2} + \dot{x}_3\dot{x}_3\psi_{34/3} + \dot{x}_3\dot{x}_4\psi_{34}\psi_{34/1},$$
$$0 = \ddot{x}_3 - \dot{x}_3\dot{x}_4\psi_{34/1}, \quad and \quad 0 = \ddot{x}_4 - \dot{x}_3\dot{x}_4\psi_{34/2}.$$

(2) The Ricci tensor is:
$$\rho_{13} = \tfrac{1}{2}\psi_{34/12}, \qquad \rho_{33} = \tfrac{1}{2}\{-\psi_{34/2}^2 + 2\psi_{34/23}\},$$
$$\rho_{14} = \tfrac{1}{2}\psi_{34/11}, \qquad \rho_{24} = \tfrac{1}{2}\psi_{34/12},$$
$$\rho_{23} = \tfrac{1}{2}\psi_{34/22}, \qquad \rho_{44} = \tfrac{1}{2}\{-\psi_{34/1}^2 + 2\psi_{34/14}\},$$
$$\rho_{34} = \tfrac{1}{2}\{\psi_{34/1}\psi_{34/2} + 2\psi_{34}\psi_{34/12} - \psi_{34/13} - \psi_{34/24}\}.$$

We study the manifolds of Theorem 3.3.7, of Theorem 3.3.8, and of Theorem 3.3.9. Only one of these manifolds is complete.

Case 1: Suppose that $\psi_{34/1} = p(x_1)$ and that $\psi_{34/4} = 0$ where $p(x_1) \geq x_1$ for $x_1 \geq 1$. This is the case for the warping functions of Theorem 3.3.7 (1, 2, 4) and of Theorem 3.3.9. We set $x_2(t) = 0$, $x_3(t) = 0$ and $x_4(t) = -t$. The geodesic equations then become:

$$\ddot{x}_1 - \dot{x}_1 \psi_{34/1} = 0, \quad \ddot{x}_2 = 0, \quad \ddot{x}_3 = 0, \quad \ddot{x}_4 = 0\,.$$

Thus the geodesic equations reduce to the ordinary differential equation.

$$\ddot{x}_1 = \dot{x}_1 p(x_1)\,.$$

We suppose the associated manifold is complete and argue for a contradiction. Let $\gamma(t)$ satisfy initial condition $x_1(0) = 1$ and $\dot{x}_1(0) = 1$. We suppose $\gamma(t)$ persists for all time. We then have $\ddot{x}_1(0) \geq 1$. It now follows that $\ddot{x}_1(t) \geq 1$ for all t and that $\dot{x}_1(t)$ and $x_1(t)$ are increasing functions of t. Let $t_1 = 0$ and let $t_n = t_{n-1} + \frac{3}{n^2}$. We consider the Proposition:

$$P_n : \quad x_1(t_n) \geq n \quad \text{and} \quad \dot{x}_1(t_n) \geq n^2\,.$$

Clearly P_1 is true since $x_1(t_1) = 1$ and $\dot{x}_1(t_1) = 1$. We may estimate

$$x_1(t_{n+1}) \geq x_1(t_n) + \dot{x}_1(t_n)\tfrac{3}{n^2} \geq n + n^2\tfrac{3}{n^2} \geq n + 1,$$
$$\dot{x}_1(t_{n+1}) \geq \dot{x}_1(t_n) + \ddot{x}_1(t_n)\tfrac{3}{n^2} \geq n^2 + x_1(t_n)\dot{x}_1(t_n)\tfrac{3}{n^2}$$
$$\geq n^2 + n^3\tfrac{3}{n^2} \geq n^2 + 3n \geq (n+1)^2\,.$$

Consequently, Proposition P_n holds for all n. Since $\sum \frac{3}{n^2} < \infty$,

$$T = \lim_{n \to \infty} t_n$$

exists and is finite. This must be false as

$$x_1(T) = \lim_{n \to \infty} x_1(t) = \infty\,.$$

Consequently, γ persists only for finite time. Since

$$\lim_{t \to T} x_2(t) = 0, \quad \lim_{t \to T} x_3(t) = 0, \quad \lim_{t \to T} x_4(t) = -T,$$

we must have $\lim_{t \to T} x_1(T) = \infty$. We use Lemma 3.4.2 to compute the Ricci tensor; only ρ_{44} is non-zero and we have:

$$\rho(\dot{\gamma}, \dot{\gamma}) = -\tfrac{1}{2}p(x_1)^2 \leq -\tfrac{1}{2}x_1^2\,.$$

Consequently

$$\lim_{t \to T} \rho(\dot{\gamma}, \dot{\gamma}) = -\infty\,.$$

This shows that these manifolds exhibit Ricci blowup. They do not embed as an open subset of a geodesically complete manifold; they are intrinsically incomplete.

Case 2. Let $\psi_{34} = x_1 x_3 + x_2 x_4$ be the warping function of Theorem 3.3.8 (4). The geodesic equations then become:

(1) $0 = \ddot{x}_1 + \dot{x}_1 \dot{x}_4 x_3 + \dot{x}_2 \dot{x}_4 x_4 + \dot{x}_3 \dot{x}_4 (x_1 x_3 + x_2 x_4) x_4 + x_2 \dot{x}_4 \dot{x}_4$.
(2) $0 = \ddot{x}_2 + \dot{x}_1 \dot{x}_3 x_3 + \dot{x}_2 \dot{x}_3 x_4 + \dot{x}_3 \dot{x}_3 x_1 + \dot{x}_3 \dot{x}_4 (x_1 x_3 + x_2 x_4) x_3$.
(3) $0 = \ddot{x}_3 - \dot{x}_3 \dot{x}_4 x_3$.
(4) $0 = \ddot{x}_4 - \dot{x}_3 \dot{x}_4 x_4$.

We start with initial conditions $x_3(0) = \dot{x}_3(1) = x_4(1) = \dot{x}_4(1) = 1$. Symmetry implies that $x_3(t) = x_4(t) = h(t)$ where h solves the equation

$$h''(t) = h'(t) h'(t) h(t) \, .$$

We apply the same argument as that used in Case 1 to show that $h \to \infty$ at finite time. We have $h(t)$ and h' are monotonically increasing as $h'' > 0$. Let $t_1 = 0$ and $t_n = t_{n-1} + \frac{3}{n^2}$. Consider the Proposition:

$$P_n : h(t_n) \geq n \quad \text{and} \quad \dot{h}(t_n) \geq n^2 \, .$$

Clearly P_1 is true since $h(t_1) = 1$ and $\dot{h}(t_1) = 1$. We estimate

$$h(t_{n+1}) \geq h(t_n) + \dot{h}(t_n)\tfrac{3}{n^2} \geq n + n^2 \tfrac{3}{n^2} \geq n + 1,$$
$$\dot{h}(t_{n+1}) \geq \dot{h}(t_n) + \ddot{h}(t_n)\tfrac{3}{n^2} \geq n^2 + h(t_n)\dot{h}(t_n)\dot{h}(t_n)\tfrac{3}{n^2}$$
$$\geq n^2 + n^5 \tfrac{3}{n^2} \geq n^2 + 3n \geq (n+1)^2 \, .$$

This shows that Proposition P_n is true for all n. The remainder of the argument is the same as in Case 1 to show that $x_3(t) = x_4(t) \to \infty$ as $t \to T$ for some finite time T. By Lemma 3.4.2, $\rho_{ij} = 0$ unless $3 \leq i, j \leq 4$. Thus

$$\rho(\dot{\gamma}, \dot{\gamma}) = \rho_{33}\dot{x}_3 \dot{x}_3 + \rho_{44}\dot{x}_4 \dot{x}_4 + 2\rho_{34}\dot{x}_3 \dot{x}_4$$
$$= -\tfrac{1}{2}x_4^2 \dot{x}_3^2 - \tfrac{1}{2}x_3^2 \dot{x}_4^2 + 2\dot{x}_3 \dot{x}_4 (x_3 x_4 - 1)$$
$$= -\tfrac{1}{2}(\dot{x}_3 x_4 - x_3 \dot{x}_4)^2 - 2\dot{x}_3 \dot{x}_4 \, .$$

Since $\dot{x}_3 = \dot{x}_4 \to \infty$, $\rho(\dot{\gamma}, \dot{\gamma}) \to -\infty$ and $\mathcal{M}$ exhibits Ricci blowup.

Case 3. $\psi_{34} = x_1 x_3^2$. This is the warping function of Theorem 3.3.8 (3). The final two geodesic equations become:

$$\ddot{x}_3 = \dot{x}_3 \dot{x}_4 x_3^2 \quad \text{and} \quad \ddot{x}_4 = 0 \, .$$

Setting $x_4 = t$ then yields the equation $\ddot{x}_3 = \dot{x}_3 x_3^2$. The same argument as that given in Case 1 then shows there is some finite t so that $x_3(t) \to \infty$ as $t \to T$. We compute

$$\rho(\dot{\gamma}, \dot{\gamma}) = 2\dot{x}_3\dot{x}_4\rho_{34} + \dot{x}_4\dot{x}_4\rho_{44} = -2\dot{x}_3\dot{x}_4 x_3 - \tfrac{1}{2}\dot{x}_4\dot{x}_4 x_3^4\,.$$

Thus $\lim_{t \to T} \rho(\dot{\gamma}, \dot{\gamma}) = -\infty$ and $\mathcal{M}$ exhibits Ricci blowup.

Case 4. $\psi_{34} = x_2 x_4^2 + x_3^2 x_4$ or $\psi_{34} = x_2 x_4^2 + x_3 x_4$. These are the warping functions of Theorem 3.3.8(1, 2). The final two geodesic equations become:

$$0 = \ddot{x}_3 \quad \text{and} \quad \ddot{x}_4 = \dot{x}_3\dot{x}_4 x_4^2\,.$$

The same analysis as that performed in Case 3 shows $\mathcal{M}$ is geodesically incomplete. We take $x_3 = t$ so the equation becomes $\ddot{x}_4 = \dot{x}_4 x_4^2$. Thus $\lim_{t \to T} x_4 = \infty$ at some finite time. We have

$$\rho(\dot{\gamma}, \dot{\gamma}) = \dot{x}_3\dot{x}_3\rho_{33} + 2\dot{x}_3\dot{x}_4\rho_{34} - \tfrac{1}{2}\dot{x}_3\dot{x}_3\psi_{34/2}^2 - \dot{x}_3\dot{x}_4\psi_{34/24}\,.$$

This tends to $-\infty$ as $t \to T$ and thus $\mathcal{M}$ exhibits Ricci blowup.

Case 5. $\psi_{34} = x_1 x_4 + x_3 x_4$. This is the warping function of Theorem 3.3.7 (3). The geodesic equations in the last two variables are:

$$\ddot{x}_3 - \dot{x}_3\dot{x}_4 x_4 = 0 \quad \text{and} \quad \ddot{x}_4 = 0\,.$$

Consequently $x_4 = a + bt$ is linear. Set $X_3 := \dot{x}_3$; one then has $x_3 = \int X_3$. Since $\dot{X}_3 - b(a + bt)X_3 = 0$, $X_3 = ce^{b(at + b\frac{1}{2}t^2)}$. This determines x_3. The equation for x_1 takes the form:

$$\ddot{x}_1 + \dot{x}_1\dot{x}_4 x_4 + \dot{x}_4\dot{x}_4 x_1 = 0\,.$$

By changing the parameter t, there are really only two cases to be considered. These are $x_4 = a$ and $x_4 = t$. If $x_4 = a$, we get the equation $\ddot{x}_1 = 0$ which has linear solutions. If $x_4 = t$, we get the equation

$$\ddot{x}_1 + t\dot{x}_1 + x_1 = 0\,.$$

We set $x_1 := fe^{-\frac{1}{2}t^2}$ where $f = \int ce^{\frac{1}{2}t^2}$ solves $f'' - tf' = 0$. Then

$$\dot{x}_1 = (f' - tf)e^{-\frac{1}{2}t^2},$$
$$\ddot{x}_1 = (f'' - 2tf' + t^2f - f),$$
$$\ddot{x}_1 + t\dot{x}_1 + x_1 = (f'' - 2tf' + t^2f - f + tf' - t^2f + f)e^{-\frac{1}{2}t^2} = 0\,.$$

The final equation takes the form $\ddot{x}_2 + F(x_1, x_3, x_4, \dot{x}_1, \dot{x}_2, \dot{x}_4) = 0$ which can be solved. Consequently, this manifold is geodesically complete.

3.4.3 *Jordan Osserman Walker manifolds*

We adopt the notation of Section 3.3.3 and set

$$g(\partial_{x_1}, \partial_{x_3}) := 1, \qquad\qquad g(\partial_{x_2}, \partial_{x_4}) := 1,$$
$$g(\partial_{x_3}, \partial_{x_3}) = \psi_{33} := 4kx_1^2 - \tfrac{1}{4k} f(x_4)^2, \quad g(\partial_{x_4}, \partial_{x_4}) = \psi_{44} := 4kx_2^2,$$
$$g(\partial_{x_3}, \partial_{x_4}) = \psi_{34} := 4kx_1 x_2 + x_2 f(x_4) - \tfrac{1}{4k} f'(x_4) \,.$$

We suppose that $k \neq 0$ and that $f(x_4)$ does not vanish identically. Choose ξ_4 so $f(\xi_4) \neq 0$. Choose ξ_1 so $16k^2 \xi_1^2 = f(\xi_4)^2$; normalize the choice of sign so $k\xi_1 > 0$. As an ansatz, we set $x_1 = \xi_1$ and $x_4 = \xi_4$ are constant. The geodesic equations in $\ddot{x}_1$ and $\ddot{x}_4$ then become $\ddot{x}_1 = \ddot{x}_4 = 0$ which are satisfied. The remaining geodesic equations then take the form:

$$0 = \ddot{x}_2 + F(x_1, x_2, x_3, x_4, \dot{x}_2, \dot{x}_3, \dot{x}_4) = 0,$$
$$0 = \ddot{x}_3 - 4kx_1 \dot{x}_3 \dot{x}_3 \,.$$

The arguments given above show this leads to blowup at finite time for x_3; thus this manifold is geodesically incomplete. Note that since we are dealing with an Einstein manifold, we do not have Ricci blowup.

3.5 Fiedler Manifolds

In this section, we follow the discussion Fiedler and Gilkey (2003) and, using a single warping function, define the class of *Fiedler manifolds*. As noted earlier, the manifolds discussed in Sections 2.3, 2.4, 2.9, 2.10, and 3.2 are Fiedler manifolds so the results of this section apply to those examples.

Definition 3.5.1 Let $(x, u_1, ..., u_\nu, y)$ be coordinates on $\mathbb{R}^{\nu+2}$. Let $f \in C^\infty(\mathbb{R}^\nu)$ and let $\Xi = \Xi_{ab}$ be an invertible symmetric $\nu \times \nu$ matrix of signature (r, s). We say that $\mathcal{M} := (\mathbb{R}^{\nu+2}, g)$ is a *Fiedler manifold* if g is the pseudo-Riemannian metric of signature $(r+1, s+1)$ given by:

$$g(\partial_x, \partial_x) := -2f(\vec{u}), \quad g(\partial_x, \partial_y) := 1, \quad g(\partial_{u_a}, \partial_{u_b}) := \Xi_{ab} \,.$$

Since the Lorentzian manifolds of Definition 3.2.1 are Fiedler manifolds and since these manifolds can exhibit Ricci blowup, Fiedler manifolds need not be generalized plane wave manifolds.

One says that $\mathcal{M}$ is *skew Tsankov* at P if

$$\mathcal{R}(x, y)\mathcal{R}(z, w) = \mathcal{R}(z, w)\mathcal{R}(x, y) \quad \forall x, y, z, w \in T_P M \,.$$

One says that $\mathcal{M}$ is *skew Tsankov* if this condition holds for all $P \in M$. In Theorem 3.5.2, we shall show that Fiedler manifolds are skew Tsankov; we will discuss other examples of skew Tsankov manifolds subsequently in Chapter 6 when we shall conduct a systematic study of Stanilov–Tsankov theory.

We say that a 0-model $\mathfrak{M}_0 = (V, \langle \cdot, \cdot \rangle, A)$ is *Jacobi nilpotent of order n* if $\mathcal{J}(\xi)^n = 0$ for all $\xi \in V$ and if there exists $\eta \in V$ so $\mathcal{J}(\eta)^{n-1} \neq 0$; since 0 is the only eigenvalue of the Jacobi operator, $\mathfrak{M}_0$ is necessarily Osserman. We say that a pseudo-Riemannian manifold $\mathcal{M} = (M, g)$ is *Jacobi nilpotent of order n* if $\mathcal{J}(x)^n = 0$ for all tangent vectors x and if there exists a point P of M and a tangent vector $y \in T_P M$ so that $\mathcal{J}(y)^{n-1} \neq 0$. One uses the skew-symmetric curvature operator or the Szabó operator to define the related concepts of *Ivanov–Petrova nilpotent* and *Szabó nilpotent*, respectively.

In Section 3.5.1, we summarize the basic geometric properties of Fiedler manifolds. In the remaining sections, we construct Fiedler manifolds which are Jacobi or Szabó nilpotent of arbitrarily high orders. We also study other geometric properties of these manifolds. The main result of this section is the following:

Theorem 3.5.1   *Let $\nu \geq 1$ be given.*

(1) There exists a Fiedler manifold $\mathcal{M}_{\nu+2}$ of dimension $\nu + 2$ which is Jacobi nilpotent of order ν.

(2) There exists a Fiedler manifold $\mathcal{N}_{\nu+2}$ of dimension $\nu+2$ which is Szabó nilpotent of order ν.

We note that if $\mathcal{M}$ is Jacobi nilpotent of order 1, then by Lemma 1.7.1, the Jacobi operator vanishes identically and hence $\mathcal{M}$ is flat. Similarly, if $\mathcal{M}$ is Szabó nilpotent of order 1, then the Szabó operator vanishes identically and one can show $\nabla R = 0$. Thus Theorem 3.5.1 is trivial in these examples so we shall always assume $n \geq 2$; the case $n = 2$ is somewhat exceptional and is discussed separately in Section 3.5.2.

3.5.1   *Geometric properties of Fiedler manifolds*

We summarize the relevant geometric properties of Fiedler manifolds in the following result:

Theorem 3.5.2   *Let $\mathcal{M}$ be a Fiedler manifold.*

(1) $\mathcal{M}$ has signature $(r + 1, s + 1)$.

(2) The possibly non-zero components of R, of ∇R, and of $\nabla^2 R$ are:

   (a) $R(\partial_x, \partial_{u_a}, \partial_{u_b}, \partial_x) = \partial_{u_a}\partial_{u_b}(f)$.

   (b) $\nabla R(\partial_x, \partial_{u_a}, \partial_{u_b}, \partial_x; \partial_{u_c}) = \partial_{u_a}\partial_{u_b}\partial_{u_c}(f)$.

   (c) $\nabla^2 R(\partial_x, \partial_{u_a}, \partial_{u_b}, \partial_x; \partial_{u_c}, \partial_{u_d}) = \partial_{u_d}\partial_{u_a}\partial_{u_b}\partial_{u_c}(f)$.

   (d) $\nabla^2 R(\partial_x, \partial_{u_a}, \partial_{u_b}, \partial_x; \partial_x, \partial_x) = -\sum_d \Xi^{cd}\partial_{u_d}f \cdot \partial_{u_a}\partial_{u_b}\partial_{u_c}(f)$.

(3) $\mathcal{M}$ is skew Tsankov.

(4) $\mathcal{M}$ is Ivanov–Petrova nilpotent of order $\nu \leq 3$.

Proof.   Assertion (1) is immediate. We argue as follows to prove Assertion (2). Since $d\Xi = 0$, the potentially non-zero Christoffel symbols are:

$$g(\nabla_{\partial_x}\partial_x, \partial_{u_a}) = \partial_{u_a}(f),$$

$$g(\nabla_{\partial_{u_a}}\partial_x, \partial_x) = g(\nabla_{\partial_x}\partial_{u_a}, \partial_x) = -\partial_{u_a}(f).$$

Let Ξ^{ab} be the inverse matrix. Then

$$\nabla_{\partial_x}\partial_x = \sum_{ab} \Xi^{ab}\partial_{u_a}(f)\partial_{u_b}, \tag{3.5.a}$$

$$\nabla_{\partial_x}\partial_{u_a} = \nabla_{\partial_{u_a}}\partial_x = -\partial_{u_a}(f)\partial_y. \tag{3.5.b}$$

The quadratic terms in the Christoffel symbols play no role in the calculation of R. Assertion (2a) now follows. Similarly, the Christoffel symbols play no role in the computation of ∇R and Assertion (2b) follows from Assertion (2a). When we compute $\nabla^2 R$, we use Eq. (3.5.a) in the expansion of $\nabla^2 R(\partial_x, \partial_{u_a}, \partial_{u_b}, \partial_x; \partial_x, \partial_x)$ to derive Assertion (2d).

The non-zero components of the curvature operator are:

$$\mathcal{R}(\partial_x, \partial_{u_a})\partial_{u_b} = \{\partial_{u_a}\partial_{u_b}f\}\partial_y,$$

$$\mathcal{R}(\partial_x, \partial_{u_a})\partial_x = -\sum_{b,c} \Xi^{bc}\{\partial_{u_a}\partial_{u_c}f\}\partial_{u_b}.$$

Thus the only potentially non-zero quadratic terms in the curvature are

$$\mathcal{R}(\partial_x, \partial_{u_d})\mathcal{R}(\partial_x, \partial_{u_a})\partial_x = -\sum_{b,c} \Xi^{bc}\{\partial_{u_a}\partial_{u_c}f\}\{\partial_{u_d}\partial_{u_b}f\}\partial_y.$$

Since this expression is symmetric in the roles of a and d, Assertion (3) follows. Assertion (4) follows as $\mathcal{R}(\pi_1)\mathcal{R}(\pi_2)\mathcal{R}(\pi_3) = 0$ for any 2-planes π_1, π_2, and π_3. $\square$

3.5.2 *Fiedler manifolds of signature* $(2, 2)$

We begin our study of nilpotent Jacobi and nilpotent Szabó manifolds with the following example; we change notation slightly to make it compatible with the notation of Theorem 2.3.6.

Lemma 3.5.1 *Let $\mathcal{N}_1^+$ be the Fiedler manifold of signature $(2, 2)$ defined by the metric $g(\partial_x, \partial_x) := -2e^y$ and $g(\partial_x, \partial_{\tilde{x}}) = g(\partial_y, \partial_{\tilde{y}}) = 1$. Then $\mathcal{N}_1^+$ is a homogeneous generalized plane wave manifold of signature $(2, 2)$ which is Jacobi nilpotent of order 2 and which is Szabó nilpotent of order 2.*

Proof.   We apply Theorem 2.3.6 to see $\mathcal{N}_1^+$ is a homogeneous generalized plane wave manifold. Furthermore, by Theorem 3.5.2, the only non-zero entries of R and of ∇R are given by:

$$R(\partial_x, \partial_y, \partial_y, \partial_x) = e^y \quad \text{and} \quad R(\partial_x, \partial_y, \partial_y, \partial_x; \partial_y) = e^y \,.$$

We normalize the basis to set

$$X := e^{-y/2}\{\partial_x + e^y \partial_{\tilde{x}}\}, \quad Y := \partial_y, \quad \tilde{X} := e^{y/2}\partial_{\tilde{x}}, \quad \tilde{Y} := \partial_{\tilde{y}} \,.$$

The non-zero entries in the metric tensor and in the curvature tensor are then given by:

$$g(X, \tilde{X}) = g(Y, \tilde{Y}) = R(X, Y, Y, X) = \nabla R(X, Y, Y, X; Y) = 1 \,.$$

This implies that

$$\mathcal{R}(X, Y)Y = \tilde{X}, \qquad \mathcal{R}(X, Y)X = -\tilde{Y},$$
$$\nabla_Y \mathcal{R}(X, Y)Y = \tilde{X}, \ \nabla_Y \mathcal{R}(X, Y)X = -\tilde{Y} \,.$$

Consequently, if ξ is any tangent vector field,

$$\mathcal{J}(\xi)\operatorname{Span}\{X, Y\} \subset \operatorname{Span}\{\tilde{X}, \tilde{Y}\}, \quad \mathcal{J}(\xi)\operatorname{Span}\{\tilde{X}, \tilde{Y}\} = \{0\},$$
$$\mathcal{S}(\xi)\operatorname{Span}\{X, Y\} \subset \operatorname{Span}\{\tilde{X}, \tilde{Y}\}, \quad \mathcal{S}(\xi)\operatorname{Span}\{\tilde{X}, \tilde{Y}\} = \{0\} \,.$$

This shows $\mathcal{J}(\xi)^2 = \mathcal{S}(\xi)^2 = 0$. The Lemma now follows since $\mathcal{J}(Y)$ and $\mathcal{S}(Y)$ are non-zero. $\qquad\square$

3.5.3 *Nilpotent Jacobi manifolds of order $2r$*

There are pseudo-Riemannian manifolds of arbitrary high nilpotency n. We first suppose $n = 2r$ is even; the odd case will be dealt with in Section 3.5.4. The case $n = 2$ was dealt with in Lemma 3.5.1 so we shall assume $n \geq 4$ and thus $r \geq 2$.

Theorem 3.5.3   *Let $r \geq 2$. Let $(x, u_1, ..., u_r, v_1, ..., v_r, y)$ be coordinates on $\mathbb{R}^{2r+2}$. Let $\psi \in C^\infty(\mathbb{R})$ with $\psi'' > 0$ and $\psi''' > 0$. Let $\mathcal{M} := (\mathbb{R}^{2r+2}, g)$ be the Fiedler manifold defined by $g(\partial_x, \partial_y) = 1$, $g(\partial_{u_i}, \partial_{v_j}) = \delta_{ij}$, and $g(\partial_x, \partial_x) = -2u_1 v_2 - ... - 2u_{r-1}v_r - 2\psi(u_r)$.*

(1) The possibly non-zero components of R, ∇R, and $\nabla^2 R$ are:

> *(a) $R(\partial_x, \partial_{u_r}, \partial_{u_r}, \partial_x) = \psi''(u_r)$.*
> *(b) $R(\partial_x, \partial_{u_i}, \partial_{v_{i+1}}, \partial_x) = 1$ for $1 \leq i < r$.*
> *(c) $\nabla R(\partial_x, \partial_{u_r}, \partial_{u_r}, \partial_x; \partial_{u_r}) = \psi'''(u_r)$.*
> *(d) $\nabla^2 R(\partial_x, \partial_{u_r}, \partial_{u_r}, \partial_x; \partial_x, \partial_x) = -u_{r-1}\psi'''(u_r)$.*
> *(e) $\nabla^2 R(\partial_x, \partial_{u_r}, \partial_{u_r}, \partial_x; \partial_{u_r}, \partial_{u_r}) = \psi''''(u_r)$.*

(2) $\mathcal{M}$ is a generalized plane wave manifold of signature $(r+1, r+1)$.

(3) Let $\vec{X} := (x_1, u_1)$, $\vec{Y} = (u_2, ..., u_r, v_2, ..., v_r)$, and $\vec{Z} = (v_1, y)$. If $\psi'' \neq 0$ and if ϕ is an isometry of $\mathcal{M}$, then there exists $A_1 \in \mathrm{Al}(2)$, a smooth map $A_2 : \mathbb{R}^2 \to \mathrm{Al}(2r-2)$, and a smooth map $A_3 : \mathbb{R}^{2r} \to \mathrm{Al}(2)$ so that $\phi(\vec{X}, \vec{Y}, \vec{Z}) = (A_1 \vec{X}, A_2(\vec{X})\vec{Y}, A_3(\vec{X}, \vec{Y})\vec{Z})$.

(4) $\mathcal{M}$ is 1-curvature homogeneous but not 2-curvature homogeneous.

(5) $\mathcal{M}$ is $2r$-Jacobi nilpotent.

(6) $\mathcal{M}$ is indecomposable.

Proof.   Assertion (1) follows from Theorem 3.5.2. We use Eqs. (3.5.a) and (3.5.b) to see:

$$\nabla_{\partial_x}\partial_x = u_1\partial_{u_2} + ... + u_{r-1}\partial_{u_r} + v_2\partial_{v_1} + ... + v_r\partial_{v_{r-1}} + \psi'(u_r)\partial_{v_r},$$

$$\nabla_{\partial_x}\partial_{u_i} = \nabla_{\partial_{u_i}}\partial_x = \begin{cases} -v_{i+1}\partial_y & \text{if } i < r, \\ -\psi'\partial_y & \text{if } i = r, \end{cases}$$

$$\nabla_{\partial_x}\partial_{u_i} = \nabla_{\partial_{u_i}}\partial_x = \begin{cases} 0 & \text{if } i = 1, \\ -u_{i-1}\partial_y & \text{if } i > 1. \end{cases}$$

We take the coordinate ordering $(x, u_1, ..., u_r, v_r, ..., v_1, y\}$ to see that $\mathcal{M}$ is a generalized plane wave manifold.

Adopt the notation of Assertion (3) and denote the corresponding distributions by

$$\mathcal{X} := \mathrm{Span}\{\partial_x, \partial_{u_1}\},$$
$$\mathcal{Y} := \mathrm{Span}\{\partial_{u_2}, ..., \partial_{u_r}, \partial_{v_2}, ..., \partial_{v_r}\},$$
$$\mathcal{Z} := \mathrm{Span}\{\partial_y, \partial_{v_1}\}.$$

We verify that the assumptions of Lemma 2.2.2 are met as follows. Let ξ_1, ξ_2 be coordinate vector fields in $\mathcal{X}$, let η_1, η_2 be coordinate vector fields in $\mathcal{Y}$, and let ζ_1, ζ_2 be coordinate vector fields in $\mathcal{Z}$. Then:

$$\nabla_{\xi_1}\xi_2 \in \mathcal{Y} + \mathcal{Z}, \quad \nabla_{\xi_1}\eta_1 \in \mathcal{Z}, \quad \nabla_{\eta_1}\eta_2 \in \mathcal{Z},$$
$$\nabla_{\xi_1}\zeta_2 = \nabla_{\eta_1}\zeta_2 = \nabla_{\zeta_1}\zeta_2 = 0.$$

Thus covariant differentiation has the proper triangular form. Furthermore, since $\psi'' > 0$, Assertions (1a) and (1b) imply that $\mathcal{Z} = \ker\{\mathcal{R}\}$. Thus $\mathcal{Y} + \mathcal{Z} = \ker\{\mathcal{R}\}^{\perp}$. Consequently $\mathcal{Z}$ and $\mathcal{Y} + \mathcal{Z}$ are preserved by any isometry of $\mathcal{M}$. We can now derive Assertion (3) from Lemma 2.2.2.

Let $\mathbb{R}^{2r+2} = \mathrm{Span}\{X, U_1, ..., U_r, V_r, ..., V_1, Y\}$. To prove Assertion (4), we introduce the 1-model $\mathfrak{M}_1 := (V, \langle \cdot, \cdot \rangle, A, A_1)$ where:

$$\langle X, Y \rangle := 1, \quad \langle U_i, V_i \rangle := 1, \quad A(X, U_r, U_r, X) := 1,$$
$$A(X, U_i, V_{i+1}, X) := 1 \quad \text{for} \ \ 1 \le i < r,$$
$$A_1(X, U_r, U_r, X; U_r) := 1.$$

We show that $\mathfrak{M}_1$ is a 1-model for $\mathcal{M}$ as follows. Let $\varepsilon_i \in C^{\infty}(M)$. Define a frame $\{X, Y, U_1, ..., U_r, V_1, ..., V_r\}$ for $T(\mathbb{R}^{2r+2})$ by setting:

$$X := \varepsilon_0\{\partial_x - \tfrac{1}{2}g(\partial_x, \partial_x)\partial_y\}, \qquad Y := \varepsilon_0^{-1}\partial_y,$$
$$U_i = \varepsilon_i\partial_{u_i}, \qquad\qquad\qquad\quad V_i := \varepsilon_i^{-1}\partial_{v_i}.$$

The non-zero entries in g are given by $g(X, Y) = 1$ and $g(U_i, V_i) = 1$. By Assertion (1), the non-zero entries in R and ∇R are

$$R(X, U_r, U_r, X) = \varepsilon_0^2\varepsilon_r^2\psi'',$$
$$R(X, U_i, V_{i+1}, X) = \varepsilon_0^2\varepsilon_i\varepsilon_{i+1}^{-1} \quad \text{for} \ \ 1 \le i < r,$$
$$\nabla R(X, U_r, U_r, X; U_r) = \varepsilon_0^2\varepsilon_r^3\psi'''.$$

We show that $\mathfrak{M}_1$ is a 1-model for $\mathcal{M}$ by setting:

$$\varepsilon_r := \psi''(\psi''')^{-1}, \quad \varepsilon_0 := (\varepsilon_r^2\psi'')^{-1/2},$$
$$\varepsilon_i := \varepsilon_0^{-2}\varepsilon_{i+1} \quad \text{for} \ \ 1 \le i < r.$$

We complete the proof of Assertion (4) by assuming to the contrary that $\mathcal{M}$ is 2-curvature homogeneous and then arguing for a contradiction. We introduce a topological space valued invariant of the 2-model of $\mathcal{M}$ by setting:

$$K_Q := \{\xi \in \mathbb{R}^{2r+2} : \exists \xi_i \in T_Q\mathbb{R}^{2r+2} : \nabla^2 R(\xi_1, \xi_2, \xi_3, \xi_4; \xi_5, \xi) \ne 0\}.$$

We expand $\xi = \xi_0 \partial_x + \xi_1 \partial_{u_1} + ... + \xi_r \partial_{u_r} + \xi_{r+1} \partial_{v_r} + ... + \xi_{2r} \partial_{v_1} + \xi_{2r+1} \partial_y$. By Assertion (1),

$$
K_Q = \begin{cases}
\{\xi \in \mathbb{R}^{2r+2} : \xi_0^2 + \xi_r^2 \neq 0\} & \text{if } \psi'''' \neq 0 \text{ and } u_{r-1} \neq 0, \\
\{\xi \in \mathbb{R}^{2r+2} : \xi_r \neq 0\} & \text{if } \psi'''' \neq 0 \text{ and } u_{r-1} = 0, \\
\{\xi \in \mathbb{R}^{2r+2} : \xi_0 \neq 0\} & \text{if } \psi'''' = 0 \text{ and } u_{r-1} \neq 0, \\
\{0\} & \text{if } \psi'''' = 0 \text{ and } u_{r-1} = 0.
\end{cases}
$$

Since K_Q is an invariant of the 2-model and $\mathcal{M}$ is assumed 2-curvature homogeneous, K_{Q_1} is diffeomorphic to K_{Q_2} for any two points Q_1 and Q_2 of $\mathbb{R}^{2r+2}$.

Choose u_r so $\psi'''(u_r) \neq 0$. Let

$$Q_1 := (0, ..., 1, u_r, 0, ..., 0) \quad \text{and} \quad Q_2 := (0, ..., 0, u_r, 0, ..., 0).$$

Suppose first that $\psi''''(u_r) \neq 0$. Then K_{Q_1} is connected and K_{Q_2} is not connected; this is a contradiction. Suppose $\psi''''(u_r) = 0$. Then K_{Q_1} is disconnected and $K_{Q_2} = \{0\}$. This final contradiction shows that $\mathcal{M}$ is not 2-curvature homogeneous and completes the proof of Assertion (4).

To prove Assertion (5), it suffices to show that the associated 0-model $\mathcal{M}_0$ is $2r$-Jacobi nilpotent. Let $1 \leq i < r$. The non-zero entries of the curvature operator are:

$$
\begin{aligned}
\mathcal{A}(X, U_r)U_r &= Y, & \mathcal{A}(X, U_r)X &= -V_r, \\
\mathcal{A}(X, U_i)V_{i+1} &= Y, & \mathcal{A}(X, U_i)X &= -U_{i+1}, \\
\mathcal{A}(X, V_{i+1})U_i &= Y, & \mathcal{A}(X, V_{i+1})X &= -V_i.
\end{aligned}
$$

If $\xi \in \mathbb{R}^{2r+2}$, then:

$$
\begin{aligned}
&\mathcal{J}(\xi)X \in \mathrm{Span}\{U_2, ..., U_r, V_1, ..., V_r, Y\}, \\
&\mathcal{J}(\xi)U_r \in \mathrm{Span}\{V_r, Y\}, \qquad \mathcal{J}(\xi)Y = \mathcal{J}(\xi)V_1 = 0, \\
&\mathcal{J}(\xi)V_{i+1} \in \mathrm{Span}\{V_i, Y\}, \qquad \mathcal{J}(\xi)U_i \in \mathrm{Span}\{U_{i+1}, Y\} \text{ for } 1 \leq i < r.
\end{aligned}
$$

Consequently

$$
\begin{aligned}
&\mathrm{Range}\,\mathcal{J}(\xi) \subset \mathrm{Span}\{U_2, ..., U_r, V_1, ..., V_r, Y\}, \\
&\mathrm{Range}\,\mathcal{J}(\xi)^2 \subset \mathrm{Span}\{U_3, ..., U_r, V_1, ..., V_r, Y\}, ... \\
&\mathrm{Range}\,\mathcal{J}(\xi)^{r-1} \subset \mathrm{Span}\{U_r, V_1, ..., V_r, Y\}, \\
&\mathrm{Range}\,\mathcal{J}(\xi)^r \subset \mathrm{Span}\{V_1, ..., V_r, Y\}, \\
&\mathrm{Range}\,\mathcal{J}(\xi)^{r+1} \subset \mathrm{Span}\{V_1, ..., V_{r-1}, Y\}, ... \\
&\mathrm{Range}\,\mathcal{J}(\xi)^{2r-1} \subset \mathrm{Span}\{V_1, Y\}, \\
&\mathrm{Range}\,\mathcal{J}(\xi)^{2r} \subset \mathrm{Span}\{0\}.
\end{aligned}
$$

This shows that $\mathcal{J}$ is nilpotent of order at most $2r$. To show that $\mathcal{J}$ is nilpotent of order at least $2r$, we exhibit a chain of length $2r$

$$\mathcal{J}(X) : U_1 \to U_2 \to \dots \to U_r \to V_r \to V_{r-1} \to \dots \to V_1 \to 0 \,.$$

Assertion (5) now follows.

To prove Assertion (6), we suppose to the contrary that $\mathfrak{M}_0$ is decomposable and argue for a contradiction. Suppose there is a non-trivial orthogonal decomposition

$$\mathbb{R}^{2r+2} = W^1 \oplus W^2 \qquad \text{and} \quad A = A^1 \oplus A^2 \,.$$

Clearly neither W^1 nor W^2 can be totally isotropic. Let $X = X^1 + X^2$ for $X^i \in W^i$. Then either $\mathcal{J}(X^1)$ or $\mathcal{J}(X^2)$ is nilpotent of order $2r$; we may assume without loss of generality that the notation is chosen so that $\mathcal{J}(X^1)$ is nilpotent of order $2r$. This shows

$$\dim(W^1) \geq 2r \quad \text{so} \quad \dim(W^2) = 1, 2 \,.$$

Suppose $\dim(W^2) = 1$. Let $W^2 = \operatorname{Span}\{\xi_1^2\}$. Then $A^2 = 0$ so $A(\xi_1^2, \cdot, \cdot, \cdot) = 0$. Consequently $\xi_1^2 \in \operatorname{Span}\{V_1, Y\}$ is a null vector. This implies W^2 is totally isotropic which is false. Thus $\dim(W^2) = 2$. Let $\{\xi_1^2, \xi_2^2\}$ be an orthonormal basis for W^2. If $A^2 = 0$, then $W^2 = \operatorname{Span}\{V_1, Y\}$ is totally isotropic which is false. Thus $A(\xi_1^2, \xi_2^2, \xi_2^2, \xi_1^2) \neq 0$ so $\mathcal{J}(\xi_1^2)\xi_2^2 = a\xi_2^2$ for $a \neq 0$. This contradicts the fact that $\mathcal{J}$ is nilpotent; this contradiction completes the proof. $\qquad\qquad\square$

Remark 3.5.1 If we drop the assumption that $\psi'' > 0$, we may let $\mathcal{S}$ be defined by taking $\psi(u_r) = u_r^2$. Then $\mathcal{S}$ is a symmetric space and $\mathcal{M}$ is 0-modeled on $\mathcal{S}$.

3.5.4 *Nilpotent Jacobi manifolds of order $2r + 1$*

We construct an example motivated by the discussion in Fiedler and Gilkey (2003). We shall only define the symmetric space in the family and shall omit the more general construction in the interests of brevity as our purpose is to exhibit pseudo-Riemannian manifolds which are nilpotent Jacobi of odd order $2r+1$ of signature $(r+1, r+2)$ for $r \geq 1$. Other manifolds can be defined by replacing $-2tu_r$ by $-2t\psi(u_r)$ where $\psi' \neq 0$ in what follows; the analysis would then follow a similar pattern to that described in Section 3.5.3.

Theorem 3.5.4   *Let $(x, u_1, ..., u_r, t, v_1, ..., v_r, y)$ be coordinates on $\mathbb{R}^{2r+3}$ where $r \geq 1$. Let $\mathcal{M} := (\mathbb{R}^{2r+2}, g)$ be the Fiedler manifold defined by $g(\partial_x, \partial_y) = 1$, $g(\partial_{u_i}, \partial_{v_j}) = \delta_{ij}$, $g(\partial_x, \partial_x) = -2tu_r - 2u_1v_2 - ... - 2u_{r-1}v_r$, and $g(\partial_t, \partial_t) = 1$. Then $\mathcal{M}$ is a generalized plane wave indecomposable symmetric space of signature $(r + 1, r + 2)$ which is Jacobi nilpotent of degree $2r + 1$.*

Proof.   We give the proof in the case $r \geq 2$; the case $r = 1$ follows similarly where we set $g(\partial_x, \partial_x) = -2tu_1$. By Eqs. (3.5.a) and (3.5.b),

$$\nabla_{\partial_x}\partial_x = u_r\partial_t + t\partial_{v_r} + u_1\partial_{u_2} + v_2\partial_{v_1} + ... + u_{r-1}\partial_{u_r} + v_r\partial_{v_{r-1}},$$

$$\nabla_{\partial_x}\partial_{u_i} = \nabla_{\partial_{u_i}}\partial_x = -\begin{cases} t\partial_y & \text{if } i = r, \\ v_{i+1}\partial_y & \text{if } i < r, \end{cases}$$

$$\nabla_{\partial_x}\partial_{v_i} = \nabla_{\partial_{v_i}}\partial_x = -\begin{cases} 0 & \text{if } i = 1, \\ u_{i-1}\partial_y & \text{if } 1 < i, \end{cases}$$

$$\nabla_{\partial_x}\partial_t = \nabla_{\partial_t}\partial_x = -u_r\partial_y.$$

This has the proper form for a generalized plane wave manifold relative to the coordinate ordering $(x, u_1, ..., u_r, t, v_r, ..., v_1, y)$.

By Theorem 3.5.2, $\nabla R = 0$ so $\mathcal{M}$ is a local symmetric space. Furthermore, the non-zero components of R are given by

$$R(\partial_x, \partial_{u_r}, \partial_t, \partial_x) = 1 \quad \text{and} \quad R(\partial_x, \partial_{u_i}, \partial_{v_{i+1}}, \partial_x) = 1.$$

Let $\mathbb{R}^{2r+3} := \mathrm{Span}\{X, U_1, ..., U_r, T, V_r, ..., V_1, Y\}$ and let $\mathfrak{M}_0 := (V, \langle \cdot, \cdot \rangle, A)$ where

$$\langle X, Y \rangle = 1, \quad \langle U_i, V_i \rangle = 1, \quad \langle T, T \rangle = 1,$$
$$A(X, U_r, T, X) = 1,$$
$$A(X, U_i, V_{i+1}, X) = 1 \quad \text{for} \quad 1 \leq i < r.$$

Setting $X = \partial_x - \frac{1}{2}g(\partial_x, \partial_x)\partial_y$, $Y = \partial_y$, $U_i = \partial_{u_i}$, $V_i = \partial_{v_i}$, $T = \partial_t$ then shows $\mathfrak{M}_0$ is a 0-model for $\mathcal{M}$. As we have adjusted the metric tensor, the expression for curvature operator simplifies to become:

$$\begin{aligned} \mathcal{A}(X, U_r)T &= Y, & \mathcal{A}(X, U_r)X &= -T, \\ \mathcal{A}(X, T)U_r &= Y, & \mathcal{A}(X, T)X &= -V_r, \\ \mathcal{A}(X, U_i)V_{i+1} &= Y, & \mathcal{A}(X, U_i)X &= -U_{i+1} & \text{for} \quad 1 \leq i < r, \\ \mathcal{A}(X, V_{i+1})U_i &= Y, & \mathcal{A}(X, V_{i+1})X &= -V_i & \text{for} \quad 1 \leq i < r. \end{aligned}$$

The Jacobi operator can now be studied:

$$\mathcal{J}(\xi)X \in \mathrm{Span}\{T, U_2, ..., U_r, V_1, ..., V_r, Y\},$$

$$\mathcal{J}(\xi)T \in \mathrm{Span}\{V_r, Y\},$$

$$\mathcal{J}(\xi)U_r \in \mathrm{Span}\{T, Y\},$$

$$\mathcal{J}(\xi)U_i \in \mathrm{Span}\{U_{i+1}, Y\} \quad \text{if} \quad i < r,$$

$$\mathcal{J}(\xi)V_{i+1} \in \mathrm{Span}\{Y, V_i\} \quad \text{if} \quad i < r.$$

We also have $\mathcal{J}(\xi)V_1 = \mathcal{J}(\xi)Y = 0$. Consequently,

$$\mathrm{Range}\{\mathcal{J}(\xi)\} \subset \mathrm{Span}\{U_2, ..., U_r, T, V_r, ..., V_1, Y\},$$

$$\mathrm{Range}\{\mathcal{J}(\xi)\}^2 \subset \mathrm{Span}\{U_3, ..., U_r, T, V_r, ..., V_1, Y\}, ...$$

$$\mathrm{Range}\{\mathcal{J}(\xi)\}^{r-1} \subset \mathrm{Span}\{U_r, T, V_r, ..., V_1, Y\},$$

$$\mathrm{Range}\{\mathcal{J}(\xi)\}^r \subset \mathrm{Span}\{T, V_r, ..., V_1, Y\},$$

$$\mathrm{Range}\{\mathcal{J}(\xi)\}^{r+1} \subset \mathrm{Span}\{V_r, ..., V_1, Y\},$$

$$\mathrm{Range}\{\mathcal{J}(\xi)\}^{r+2} \subset \mathrm{Span}\{V_{r-1}, ..., V_1, Y\}, ...$$

$$\mathrm{Range}\{\mathcal{J}(\xi)\}^{2r} \subset \mathrm{Span}\{V_1, Y\},$$

$$\mathrm{Range}\{\mathcal{J}(\xi)\}^{2r+1} \subset \mathrm{Span}\{0\}.$$

This shows $\mathfrak{M}_0$ is nilpotent Jacobi of order at most $2r+1$. To prove equality holds, we take $\xi = X$ and create a chain of length $2r + 1$:

$$\mathcal{J}(X) : U_1 \to U_2 \to ... \to U_r \to T \to V_r \to V_{r-1} \to ... \to V_1 \to 0.$$

We complete the proof by showing $\mathfrak{M}_0$ is indecomposable; the proof follows the same lines as those given to establish a similar fact in the previous section. Suppose we have a non-trivial orthogonal decomposition

$$\mathbb{R}^{3+2r} = W^1 \oplus W^2 \quad \text{and} \quad A = A^1 \oplus A^2.$$

Let $X = X^1 + X^2$. Then either $J(X^1)$ or $J(X^2)$ is nilpotent of order $2r+1$; assume the notation chosen so $J(X^1)$ nilpotent of order $2r + 1$ and thus $\dim(W^1) \geq 2r + 1$ so $\dim(W^2) \leq 2$. If $A^2 = 0$, then

$$W^2 \subset \ker(\mathcal{A}) = \mathrm{Span}\{V_1, Y\}$$

is totally isotropic. As this is false, $A^2 \neq 0$ and thus $\dim(W^2) = 2$. It now follows that A^2 has constant sectional curvature on W^2 and in particular the Jacobi operator is not nilpotent on W^2, which is false. This contradiction shows $\mathfrak{M}_0$ is indecomposable. $\qquad\square$

3.5.5 *Szabó nilpotent manifolds of arbitrarily high order*

In this section, we complete the proof of Theorem 3.5.1 by studying the Szabó operator. The case $n = 2$ is dealt with in Lemma 3.5.1. Thus we shall assume $n \geq 3$. Lemma 3.5.2 deals with n-even, and Lemma 3.5.3 deals with n-odd. Our warping functions will be cubic polynomials. Throughout the proof, we shall let $\xi \in \mathbb{R}^m$ and we shall let $\star = \star(\xi)$ denote non-trivial polynomials in the components of ξ; these polynomials are then non-zero for generic ξ.

Lemma 3.5.2 *Let $\{x, u_1, ..., u_r, v_r, ..., v_1, y\}$ be coordinates on $\mathbb{R}^{2r+2}$ for $r \geq 2$. Let $\mathcal{M}$ be the Fiedler manifold defined by taking $g(\partial_x, \partial_y) = 1$, $g(\partial_{u_i}, \partial_{v_i}) = 1$, and $g(\partial_x, \partial_x) = -\frac{1}{3}u_1^3 - \sum_{1 \leq a \leq r-1} u_{a+1}^2 v_a$. Then $\mathcal{M}$ is a generalized plane wave manifold of signature $(r+1, r+1)$ which is Szabó nilpotent of order $2r$.*

Proof. We use Eqs. (3.5.a) and (3.5.b) to see that

$$\nabla_{\partial_x}\partial_x = \tfrac{1}{2}u_1^2\partial_{v_1} + \sum_{a=1}^{r-1}\left\{\tfrac{1}{2}u_{a+1}^2\partial_{u_a} + u_{a+1}v_a\partial_{v_{a+1}}\right\}.$$

The remaining covariant derivatives are all multiples of y with coefficients which do not depend on y. Thus $(x, u_r, ..., u_1, v_1, ..., v_r, y)$ gives $\mathcal{M}$ a generalized plane wave structure. Let $1 \leq a \leq r - 1$. By Theorem 3.5.2, the non-zero components of ∇R are:

$$\nabla R(\partial_x, \partial_{u_1}, \partial_{u_1}, \partial_x; \partial_{u_1}) = 1,$$
$$\nabla R(\partial_x, \partial_{u_{a+1}}, \partial_{u_{a+1}}, \partial_x; \partial_{v_a}) = \nabla R(\partial_x, \partial_{u_{a+1}}, \partial_{v_a}, \partial_x; \partial_{u_{a+1}}) = 1.$$

Again, we set $X := \partial_x - \frac{1}{2}g(\partial_x, \partial_x)\partial_y$, $U_i := \partial_{u_i}$, $V_i := \partial_{v_i}$, and $Y = \partial_y$.

$$\begin{aligned}
\nabla_{U_1}\mathcal{R}(X, U_1)U_1 &= Y, & \nabla_{U_1}\mathcal{R}(X, U_1)X &= -V_1, \\
\nabla_{V_a}\mathcal{R}(X, U_{a+1})U_{a+1} &= Y, & \nabla_{V_a}\mathcal{R}(X, U_{a+1})X &= -V_{a+1}, \\
\nabla_{U_{a+1}}\mathcal{R}(X, U_{a+1})V_a &= -Y, & \nabla_{U_{a+1}}\mathcal{R}(X, U_{a+1})X &= U_a, \\
\nabla_{U_{a+1}}\mathcal{R}(X, V_a)U_{a+1} &= -Y, & \nabla_{U_{a+1}}\mathcal{R}(X, V_a)X &= V_{a+1}.
\end{aligned}$$

We compute

$$\begin{aligned}
\mathcal{S}(\xi)X &\in \mathrm{Span}\{U_1, ..., U_{r-1}, V_1, ..., V_r, Y\}, \\
\mathcal{S}(\xi)U_a &= \star(\xi)U_{a-1} + \mathrm{Span}\{V_a, Y\} \quad \text{if} \;\; 2 \leq a \leq r, \\
\mathcal{S}(\xi)U_1 &= \star(\xi)V_1 + \mathrm{Span}\{Y\}, \\
\mathcal{S}(\xi)V_1 &= \star(\xi)V_2 + \mathrm{Span}\{Y\},
\end{aligned}$$

$$\mathcal{S}(\xi)V_{a-1} = \star(\xi)V_a + \mathrm{Span}\{Y\} \quad \text{if} \quad 2 \le a \le r,$$
$$\mathcal{S}(\xi)V_r = \mathcal{S}(\xi)Y = 0.$$

This pattern shows that $\mathcal{S}^{2r}(\xi) = 0$ for all ξ and that $\mathcal{S}^{2r-1}(\xi)$ is non-zero for generic ξ. $\qquad\square$

We complete our discussion of the Szabó operator by considering the case $n = 2r$.

Lemma 3.5.3 *Let $\{x, u_r, ..., u_1, t, v_1, ..., v_r, y\}$ be coordinates on $\mathbb{R}^{2r+3}$ for $r \ge 1$. Let $\mathcal{M} := (\mathbb{R}^{2r+5}, g)$ be the Fiedler manifold defined by taking $g(\partial_x, \partial_x) = -u_1^2 t - \sum_{1 \le a \le r-1} u_{a+1}^2 v_a$, $g(\partial_x, \partial_y) = 1$, $g(\partial_{u_i}, \partial_{v_i}) = 1$, and $g(\partial_t, \partial_t) = 1$. Then $\mathcal{M}$ is a generalized plane wave manifold of signature $(r+2, r+3)$ which is Szabó nilpotent of order $2r+1$.*

Proof. We adopt the convention that the empty summation is zero and omit the term $u_{a+1}^2 v_a$ if $r = 1$. We use Eqs. (3.5.a) and (3.5.b) to see that

$$\nabla_{\partial_x} \partial_x = u_1 t \partial_{v_1} + \tfrac{1}{2} u_1^2 \partial_t + \sum_{a=1}^{r-1} \left\{ \tfrac{1}{2} u_{a+1}^2 \partial_{u_a} + u_{a+1} v_a \partial_{v_{a+1}} \right\}.$$

The remaining covariant derivatives are all multiples of y with coefficients which do not depend on y. Thus $(x, u_r, ..., u_1, \partial_t, v_1, ..., v_r, y)$ gives $\mathcal{M}$ a generalized plane wave structure.

Let $1 \le a \le r - 1$. By Theorem 3.5.2, the non-zero components of ∇R are:

$$\nabla R(\partial_x, \partial_{u_1}, \partial_{u_1}, \partial_x; \partial_t) = \nabla R(\partial_x, \partial_t, \partial_{u_1}, \partial_x; \partial_{u_1}) = 1,$$
$$\nabla R(\partial_x, \partial_{u_{a+1}}, \partial_{u_{a+1}}, \partial_x; \partial_{v_a}) = \nabla R(\partial_x, \partial_{u_{a+1}}, \partial_{v_a}, \partial_x; \partial_{u_{a+1}}) = 1.$$

Again, we introduce a suitable normalized basis and compute:

$$\nabla_T \mathcal{R}(X, U_1)U_1 = Y, \qquad \nabla_T \mathcal{R}(X, U_1)X = -V_1,$$
$$\nabla_{U_1} \mathcal{R}(X, T)U_1 = Y, \qquad \nabla_{U_1} \mathcal{R}(X, T)X = -V_1,$$
$$\nabla_{U_1} \mathcal{R}(X, U_1)T = Y, \qquad \nabla_{U_1} \mathcal{R}(X, U_1)X = -T,$$
$$\nabla_{V_a} \mathcal{R}(X, U_{a+1})U_{a+1} = Y, \qquad \nabla_{V_a} \mathcal{R}(X, U_{a+1})X = -V_{a+1},$$
$$\nabla_{U_{a+1}} \mathcal{R}(X, U_{a+1})V_a = Y, \qquad \nabla_{U_{a+1}} \mathcal{R}(X, U_{a+1})X = -U_a,$$
$$\nabla_{U_{a+1}} \mathcal{R}(X, V_a)U_{a+1} = Y, \qquad \nabla_{U_{a+1}} \mathcal{R}(X, V_a)X = -V_{a+1}.$$

If $\xi \in \mathbb{R}^{2r+5}$, then:

$$\mathcal{S}(\xi)X \in \mathrm{Span}\{U_{r-1}, ..., U_1, T, V_1, ..., V_r, Y\},$$
$$\mathcal{S}(\xi)U_{a+1} = \star(\xi)U_a + \mathrm{Span}\{T, V_1, ..., V_r, Y\},$$

$$\mathcal{S}(\xi)U_1 = \star(\xi)T + \mathrm{Span}\{V_1, ..., V_r, Y\},$$
$$\mathcal{S}(\xi)T = \star(\xi)V_1 + \mathrm{Span}\{Y\},$$
$$\mathcal{S}(\xi)V_a = \star(\xi)V_{a+1} + \mathrm{Span}\{Y\},$$
$$\mathcal{S}(\xi)V_r = \mathcal{S}(\xi)Y = 0.$$

From these computations it follows that $\mathcal{S}(\xi)^{2r+1} = 0$ and that $\mathcal{S}(\xi)^{2r}U_{r+1}$ is generically non-zero. The Lemma follows. $\qquad\square$

Chapter 4

The Curvature Tensor

4.1 Introduction

Chapter 4 treats other topics related to the Riemann curvature tensor. In Section 4.2, we shall summarize the results from algebraic topology that will be needed subsequently. Some basic results from bundle theory are introduced in Sections 4.2.1 and 4.2.2. Clifford algebras in arbitrary signatures are defined in Section 4.2.3; Section 4.2.4 treats Bott periodicity in the context of Riemannian Clifford modules. In Section 4.2.5 we recall work of Adams (1962) dealing with vector fields on spheres; this is used in Section 4.2.6 to discuss the possible signatures of pseudo-Riemannian metrics on spheres. In Section 4.2.7 we exhibit results of Szabó (1991) concerning equivariant vector fields on spheres. In Section 4.2.8, we state results of Gilkey (2002) generalizing earlier work by Glover, Homer, and Stong (1982) concerning geometrically symmetric vector bundles.

Let $\Psi \in S^2(V^*)$ and let $\Psi_1 \in S^3(V^*)$ be totally symmetric 2-tensors and 3-tensors, respectively. Following Eqs. (1.3.b) and (1.6.a), define an algebraic curvature tensor $A_\Psi \in \mathcal{A}\mathrm{lg}_0(V)$ and an algebraic covariant derivative curvature tensor $A_{1,\Psi,\Psi_1} \in \mathcal{A}\mathrm{lg}_1(V)$ by:

$$A_\Psi(x, y, z, w) := \Psi(x, w)\Psi(y, z) - \Psi(x, z)\Psi(y, w),$$

$$\begin{aligned}
A_{1,\Psi,\Psi_1}(x, y, z, w; v) := {}& \Psi_1(x, w, v)\Psi(y, z) + \Psi(x, w)\Psi_1(y, z, v) \\
& - \Psi_1(x, z, v)\Psi(y, w) - \Psi(x, z)\Psi_1(y, w, v).
\end{aligned}$$

In Section 4.3, we follow the discussion of Díaz-Ramos, Fiedler, García-Río, and Gilkey (2004a) to show that the tensors A_Ψ and A_{1,Ψ,Ψ_1} generate the spaces $\mathcal{A}\mathrm{lg}_0(V)$ and $\mathcal{A}\mathrm{lg}_1(V)$. We also obtain an upper bound on the number of such tensors of this form which are needed to express any fixed $A \in \mathcal{A}\mathrm{lg}_0(V)$ or $A_1 \in \mathcal{A}\mathrm{lg}_1(V)$. The primary technical tool which is used here is the embedding Theorem of Nash (1956).

In Section 4.4, we discuss some results of Gilkey and Ivanova (2001b) and of Gilkey and Ivanova (2002b) for Jordan Osserman algebraic curvature tensors. Theorem 4.4.1 shows that the Jordan normal form of a neutral signature spacelike and timelike Jordan Osserman algebraic curvature tensor can be arbitrarily complicated. By contrast, Theorem 4.4.2 shows that if $p < q$ and if A is spacelike Jordan Osserman, then $\mathcal{J}_A(x)$ is diagonalizable if $x \in S^+(V, \langle \cdot, \cdot \rangle)$; thus the Jordan normal form is trivial if the spacelike directions dominate.

Let $\mathfrak{M}_1 := (V, \langle \cdot, \cdot \rangle, A, A_1)$ be a 1-model. The *Szabó operator* is an analogue of the Jacobi operator which is defined by setting:

$$\mathcal{S}(x) : y \to \mathcal{A}_1(y, x; x)x \,.$$

We say that $\mathfrak{M}_1$ is *Szabó* if the eigenvalues of $\mathcal{S}$ are constant on $S^+(V, \langle \cdot, \cdot \rangle)$ if $q > 0$ or, equivalently by Theorem 1.9.1, on $S^-(V, \langle \cdot, \cdot \rangle)$ if $p > 0$.

Let $\mathfrak{M}_1$ be Szabó. In Theorem 4.5.1 we present work of Szabó (1991) showing that if $\mathfrak{M}_1$ is Riemannian, then $A_1 = 0$. We then present results of Gilkey, Ivanova, and Stavrov (2003) and of Gilkey and Stavrov (2002) in the pseudo-Riemannian setting showing in Theorem 4.5.2 that $\mathcal{S}$ is nilpotent on the null cone $\mathcal{N}$ of V and, furthermore, that if $\mathfrak{M}_1$ is Lorentzian, then $A_1 = 0$. There are non-trivial Szabó 1-models in the higher signature setting. If $\mathcal{M}$ is as in Definition 2.5.1, we show in Theorem 4.5.3 that $\mathcal{M}$ is Szabó. However such a manifold $\mathcal{M}$ is Jordan Szabó if and only if $\mathcal{M}$ is symmetric.

In Section 4.6, we study topics in conformal geometry. Suppose given a 0-model $\mathfrak{M} = (V, \langle \cdot, \cdot \rangle, A)$. If $\{e_i\}$ is an orthonormal basis for V, let $\varepsilon_i := \langle e_i, e_i \rangle$. The Ricci tensor ρ, the scalar curvature τ, and the Weyl tensor W are given by:

$$\rho(x, y) := \sum_{i=1}^{m} \varepsilon_i A(x, e_i, e_i, y), \quad \tau := \sum_{i=1}^{m} \varepsilon_i \rho(e_i, e_i),$$

$$\begin{aligned}
W(x, y, z, w) := {}& A(x, y, z, w) - \tfrac{1}{m-2}\{\rho(x, w)\langle y, z \rangle \\
& + \langle x, w \rangle \rho(y, z) - \rho(x, z)\langle y, w \rangle - \langle x, z \rangle \rho(y, w)\} \\
& + \tfrac{1}{(m-1)(m-2)}\tau\{\langle x, w \rangle \langle y, z \rangle - \langle x, z \rangle \langle y, w \rangle\} \,.
\end{aligned}$$

Since W is an algebraic curvature tensor in its own right, we can form the Weyl 0-model $\mathfrak{M}_W := (V, \langle \cdot, \cdot \rangle, W)$ and let $\mathcal{J}_W$ and $\mathcal{R}_W$ be the associated conformal Jacobi operator and conformal skew-symmetric curvature operator, respectively. In Sections 4.6.2 and 4.6.3, we study the spectral geometry of the conformal Jacobi operator and in Section 4.6.4, we study the spectral geometry of the conformal skew-symmetric curvature operator.

Let $\mathfrak{M} = (V, \langle \cdot, \cdot \rangle, A)$ be a 0-model. Let π be a spacelike or a timelike k-plane. Let $\{e_1, ..., e_k\}$ be an orthonormal basis for π. In Section 1.8.3 we defined the *Stanilov operator*:

$$\Theta(\pi) := \sum_{i=1}^{m} \sum_{j=1}^{m} \mathcal{A}(e_i, e_j) \mathcal{A}(e_i, e_j) \,.$$

In Section 4.7, we follow the treatment in Gilkey, Nikčević, and Videv (2004) to discuss the spectral geometry of this operator.

We adopt the notation of Section 1.6.4. Let J be a Hermitian almost complex structure on $(V, \langle \cdot, \cdot \rangle)$. Let $\mathfrak{M}_J := \{V, \langle \cdot, \cdot \rangle, J, A\}$ be a complex 0-model which is Riemannian. Assume J and A are compatible; this means by Lemma 1.6.6 that the operator $J\mathcal{A}(x, Jx)$ is self-adjoint and complex. In Section 4.8, we study the spectral geometry of this operator; the geometry of the complex Jacobi operator will be discussed in greater detail in Chapter 5.

4.2 Topological Results

4.2.1 *Real vector bundles*

We recall a few notions concerning vector bundles which we will use in what follows. Let $\rho : E \to M$ be a real vector bundle over a smooth manifold M. The *fibers* $E_P := \rho^{-1}(P)$ are real vector spaces which vary smoothly with the point $P \in M$. A *non-degenerate fiber metric* on E is a collection of non-degenerate inner products on each fiber which vary smoothly on M. A *bundle morphism* ψ of E is a collection of smooth linear maps ψ_P of the fibers E_P which vary smoothly with P. We say ψ is invertible if each ψ_P is invertible. We say ψ is *self-adjoint* if each ψ_P is self-adjoint.

Let V be a vector space with a non-degenerate inner product. We can decompose V as a direct sum $V^+ \oplus V^-$ of complementary orthogonal subspaces, where V^+ is a maximal spacelike subspace and V^- is a maximal timelike subspace. There is a similar decomposition possible in the vector bundle setting:

Lemma 4.2.1 *Let E be a vector bundle over a smooth manifold M which is equipped with a non-degenerate fiber metric. Then we can decompose E as*

*a direct sum $E^+ \oplus E^-$ of complementary orthogonal subbundles, where E^+
is a maximal spacelike subbundle and E^- is a maximal timelike subbundle.*

Proof.   As noted above, we can decompose each individual fiber as an
orthogonal direct sum of a maximal spacelike and a maximal timelike subspace. The main technical difficulty is to make the decompositions vary
smoothly with $P \in M$. Let $\langle \cdot, \cdot \rangle$ denote the given indefinite fiber metric
on E. We can use a partition of unity to put an auxiliary positive definite
inner product $(\cdot, \cdot)_e$ on E. Define a bundle morphism ψ of E by means
of the identity $\langle v, w \rangle = (\psi v, w)_e$. Since each linear map ψ_P is self-adjoint
with respect to the positive definite inner product $(\cdot, \cdot)_e$ on each fiber E_P,
ψ_P is diagonalizable and has only real eigenvalues. As the original inner
product $\langle \cdot, \cdot \rangle$ is non-degenerate, each ψ_P is invertible. Let $E_\lambda(\psi_P) \subset E_P$
be the eigenspaces of ψ_P on E_P. We define:

$$E_P^- := \oplus_{\lambda < 0} E_\lambda(\psi_P) \text{ and } E_P^+ := \oplus_{\lambda > 0} E_\lambda(\psi_P).$$

The subspaces E_P^+ and E_P^- are orthogonal, they vary smoothly, and they
have constant rank. Thus they patch together to define smooth orthogonal
complementary subbundles $E^\pm$ of E. The subbundle E^- is a maximal
timelike subbundle and the subbundle E^+ is a maximal spacelike subbundle.
The Lemma now follows. $\square$

4.2.2   *Bundles over projective spaces*

Let $\mathbb{F}$ denote either the field of real numbers $\mathbb{R}$ or the field of complex
numbers $\mathbb{C}$. Let $\mathbb{F}\mathbb{P}^{n-1}$ be the projective space of $\mathbb{F}$-lines through the origin
in $\mathbb{F}^n$:

$$\mathbb{F}\mathbb{P}^{n-1} = \left\{ \mathbb{F}^n - \{0\} \right\} / \left\{ \mathbb{F} - \{0\} \right\}.$$

Let $\mathbb{1}^k := \mathbb{F}\mathbb{P}^{n-1} \times \mathbb{F}^k$ be the trivial $\mathbb{F}$-bundle of dimension k over $\mathbb{F}\mathbb{P}^{n-1}$.
We take $k = n$ and define the *tautological line bundle* $\mathbb{L} \subset \mathbb{1}^n$ and the
orthogonal complement, the hyperplane bundle, $\mathbb{L}^\perp \subset \mathbb{1}^n$ over $\mathbb{F}\mathbb{P}^{n-1}$ by
setting:

$$\mathbb{L} := \{ (\pi, \xi) \in \mathbb{F}\mathbb{P}^{n-1} \times \mathbb{F}^n : \xi \in \pi \},$$
$$\mathbb{L}^\perp := \{ (\pi, \xi) \in \mathbb{F}\mathbb{P}^{n-1} \times \mathbb{F}^n : \xi \perp \pi \}.$$

These bundles play an important role in the classification of real and complex line bundles.

In the following Lemma, we show that no positive rank subbundle of $\mathbb{L}^{\perp}$ embeds as a subbundle of $\mathbb{1}^{k}$ for $k < n$:

Lemma 4.2.2 *Let* $\mathbb{F} = \mathbb{R}$ *or* $\mathbb{F} = \mathbb{C}$. *Let* E_1 *be a positive rank subbundle of* $\mathbb{L}^{\perp}$ *over* $\mathbb{F}\mathbb{P}^{n-1}$. *Then* E_1 *is not isomorphic to a subbundle of* $\mathbb{1}^{k}$ *if* $k < n$.

Proof. We suppose $\mathbb{F} = \mathbb{R}$; the argument is the same if $\mathbb{F} = \mathbb{C}$ except that the Chern classes are used instead of the Stiefel-Whitney classes and other minor modifications are made in the argument.

The Stiefel-Whitney classes are cohomological invariants of real vector bundles; we refer to the discussion in Milnor and Stasheff (1974) for further details. Let $w_1 := w_1(\mathbb{L})$ be the first Stiefel-Whitney class of $\mathbb{L}$. The cohomology ring of the projective space $\mathbb{R}\mathbb{P}^{n-1}$ is the truncated polynomial ring, see again Milnor and Stasheff (1974),

$$H^*(\mathbb{R}\mathbb{P}^{n-1}; \mathbb{Z}_2) = \mathbb{Z}_2[w_1]/(w_1^n = 0).$$

To prove Lemma 4.2.2, we suppose the contrary. We assume that there exist subbundles E_1 of $\mathbb{L}^{\perp}$ and E_2 of $\mathbb{1}^{k}$ of rank $r > 0$ so that E_1 is isomorphic to E_2, and argue for a contradiction. Since we may decompose $\mathbb{L}^{\perp} = E_1 \oplus E_1^{\perp}$,

$$w(E_1)w(E_1^{\perp}) = w(\mathbb{L}^{\perp}) = 1 + w_1 + ... + w_1^{n-1}.$$

In particular, $w_r(E_1) = w_1^r$. Furthermore, as E_1 is isomorphic to E_2, we may conclude $w_r(E_2) = w_1^r$. As $\mathbb{1}^{k} = E_2 \oplus E_2^{\perp}$, we have the following factorization

$$1 = w(\mathbb{1}^{k}) = w(E_2)w(E_2^{\perp}) = (1 + ... + w_1^r)w(E_2^{\perp}).$$

Since $k < n$, the truncation $(w_1^n = 0)$ in $H^*(\mathbb{R}\mathbb{P}^{n-1}; \mathbb{Z}_2)$ plays no role, so we have the factorization

$$1 = (1 + ... + w_1^r)w(E_2^{\perp}) \quad \text{in} \quad \mathbb{Z}_2[w_1],$$

which is impossible. $\qquad\qquad\square$

4.2.3 *Clifford algebras in arbitrary signatures*

Let V be a vector space which is equipped with a non-degenerate inner product $\langle \cdot, \cdot \rangle$ of signature (p, q). The *Clifford algebra* $\mathrm{Clif}(V, \langle \cdot, \cdot \rangle)$ is the universal unital algebra generated by V subject to the Clifford commutation relations:

$$v * w + w * v = -2\langle v, w \rangle 1 .$$

One can also give a somewhat more formal definition. Let

$$\mathcal{T} := \mathbb{R} \oplus V \oplus \{V \otimes V\} \oplus ... \oplus \{V \otimes ... \otimes V\} \oplus ...$$

be the complete tensor algebra on V. Let $\mathcal{I}$ be the 2-sided ideal of $\mathcal{T}$ which is generated by all elements of the form $v \otimes w + w \otimes v + 2\langle v, w \rangle 1$. Then

$$\mathrm{Clif}(V, \langle \cdot, \cdot \rangle) = \mathcal{T}/\mathcal{I}$$

and the Clifford product "$*$" is induced from the tensor product $\otimes$.

The natural inclusion of $V \subset \mathcal{T}$ induces a natural inclusion of V in $\mathrm{Clif}(V, \langle \cdot, \cdot \rangle)$. Let $\{e_1, ..., e_r\}$ be an orthonormal basis for V. Let

$$I = \{1 \leq i_1 < ... < i_p \leq m\}$$

be a multi-index. Set $e_I := e_{i_1} * ... * e_{i_p}$. The Clifford algebra $\mathrm{Clif}(V, \langle \cdot, \cdot \rangle)$ inherits a natural inner product and the resulting collection $\{e_I\}$ is an orthonormal basis for $\mathrm{Clif}(V, \langle \cdot, \cdot \rangle)$. In particular,

$$\dim\{\mathrm{Clif}(V, \langle \cdot, \cdot \rangle)\} = 2^{\dim(V)} .$$

4.2.4 *Riemannian Clifford algebras*

Let $M_k(\mathbb{A})$ be the set of $k \times k$ matrices over a unital algebra $\mathbb{A}$ where $\mathbb{A} = \mathbb{R}$ or $\mathbb{A} = \mathbb{C}$ or $\mathbb{A} = \mathbb{H}$ denotes the real numbers, the complex numbers, or the quaternions, respectively. Give $\mathbb{R}^n$ the usual positive definite Euclidean inner product $\langle \cdot, \cdot \rangle_e$. Let $\mathrm{Clif}(n) := \mathrm{Clif}(\mathbb{R}^n, \langle \cdot, \cdot \rangle_e)$. One has the following structure theorem due to Atiyah, Bott, and Shapiro (1964) which is closely related to *Bott Periodicity*:

Theorem 4.2.1

$$\begin{aligned}
\mathrm{Clif}(0) &= \mathbb{R}, & \mathrm{Clif}(1) &= \mathbb{C}, \\
\mathrm{Clif}(2) &= \mathbb{H}, & \mathrm{Clif}(3) &= \mathbb{H} \oplus \mathbb{H}, \\
\mathrm{Clif}(4) &= M_2(\mathbb{H}), & \mathrm{Clif}(5) &= M_4(\mathbb{C}), \\
\mathrm{Clif}(6) &= M_8(\mathbb{R}), & \mathrm{Clif}(7) &= M_8(\mathbb{R}) \oplus M_8(\mathbb{R}), \\
\mathrm{Clif}(8) &= M_{16}(\mathbb{R}), & \mathrm{Clif}(k+8) &= M_{16}(\mathrm{Clif}(k)) .
\end{aligned}$$

Definition 4.2.1   The *Adams number $\nu(m)$* is defined for powers of 2 by:

$$\nu(1) = 0, \quad \nu(2) = 1, \quad \nu(4) = 3, \quad \nu(8) = 7$$

and then inductively by setting $\nu(16m) = \nu(m) + 8$. If a is odd, one sets

$$\nu(a2^s) = \nu(2^s).$$

We say that a collection $\mathcal{F} = \{J_1, ..., J_\ell\}$ of skew-symmetric real $m \times m$ matrices is a *Clifford family* of rank ℓ on $\mathbb{R}^m$ if they satisfy the *Clifford commutation relations*

$$J_i J_j + J_j J_i = -2\delta_{ij}.$$

Such a family defines a representation of $\mathrm{Clif}(\ell)$ on $\mathbb{R}^m$. One can then draw the following consequence from Theorem 4.2.1; we refer to Karoubi (1978) for further details:

Corollary 4.2.1   *There exists a Clifford family $\mathcal{F}$ of rank ℓ on $\mathbb{R}^m$ if and only if $\ell \le \nu(m)$.*

Such a family played a central role in the proof of Lemma 1.7.2. These families will also be central to the discussion of Chapter 5; we use Corollary 4.2.1 to establish the following estimate:

Lemma 4.2.3   *Let $\mathcal{F}$ be a Clifford family of rank ℓ on a vector space of dimension m. If $\ell \ge 16$, then $m \ge \ell(\ell - 1)$.*

Proof.   If a vector space V of dimension m admits a Clifford family of rank ℓ, work of Adams gives a lower bound for m in terms of the Adams number. Let $a(\ell)$ be this lower bound as described in Corollary 4.2.1. Let $b(\ell) = \ell(\ell - 1)$. The following table gives the values of ℓ, of $a(\ell)$, and of $b(\ell)$ for $6 \le \ell \le 25$:

ℓ	$a(\ell)$	$b(\ell)$	ℓ	$a(\ell)$	$b(\ell)$	ℓ	$a(\ell)$	$b(\ell)$	ℓ	$a(\ell)$	$b(\ell)$
6	8	30	7	8	42	8	16	56	9	32	72
10	64	90	11	64	110	12	128	132	13	128	156
14	128	182	15	128	210	16	256	240	17	512	272
18	1024	306	19	1024	342	20	2048	380	21	2048	420
22	2048	462	23	2048	506	24	4096	552	25	8192	600

It is clear $a(\ell) \ge b(\ell)$ for $16 \le \ell \le 25$. The Adams number $a(\ell)$ is growing exponentially; $b(\ell)$ is growing quadratically. It now follows that $a(\ell) \ge b(\ell)$ for all $\ell \ge 16$. $\qquad\square$

4.2.5   *Vector fields on spheres*

Let S^{m-1} be the unit sphere in $(\mathbb{R}^m, (\cdot, \cdot)_e)$. We refer to work of Adams (1962) for the proof of the following result:

Theorem 4.2.2   **[Adams]** *Suppose given a non-trivial decomposition of the tangent bundle $T(S^{m-1}) = E_0 \oplus ... \oplus E_\ell$ as an orthogonal direct sum of vector bundles of dimension $\mu_i := \dim(E_i)$, where $\mu_0 \geq ... \geq \mu_\ell$. Then $\mu_1 + ... + \mu_\ell \leq \nu(m)$.*

Clifford algebras can be used to show this estimate is sharp. Suppose that m is even. Let $\nu = \nu(m)$. Then Theorem 4.2.1 implies $\mathbb{R}^m$ admits a $\mathrm{Cliff}(\nu)$ structure. Consequently, there exist ν skew-adjoint $m \times m$ matrices $\{e_1, ..., e_\nu\}$ satisfying the Clifford commutation relations. Let

$$E_i(x) := \mathrm{Span}_{\mathbb{R}} \{e_i x\}.$$

The Clifford commutation relations imply that $\{x, e_1 x, ..., e_\nu x\}$ forms an orthonormal set. Consequently one has that:

$$E_i(x) \perp E_j(x) \quad \text{for} \quad i \neq j \quad \text{and} \quad E_i(x) \subset T_x S^{m-1} = \{\xi : \xi \perp x\}.$$

We set $E_0 := TS^{m-1} \cap \{E_1 \oplus ... \oplus E_\nu\}^\perp$ to obtain a maximal decomposition of the tangent space of the sphere

$$TS^{m-1} = E_0 \oplus E_1 \oplus ... \oplus E_\nu$$

and show Theorem 4.2.2 is sharp.

4.2.6   *Metrics of higher signatures on spheres*

We can apply Theorem 4.2.2 to study the question of constructing pseudo-Riemannian metrics of higher signature on spheres:

Theorem 4.2.3   *Let $p + q = m - 1$ where $p \leq q$. There exists a pseudo-Riemannian metric of signature (p, q) on S^{m-1} if and only if $p \leq \nu(m)$.*

Proof.   Suppose $p \leq \nu(m)$. If $p = 0$, there is nothing to show so we suppose $1 \leq p$. Let $\{e_1, ..., e_p\}$ be a family of skew-symmetric matrices on $\mathbb{R}^m$ satisfying the Clifford commutation relations. We let

$$E_-(x) := \mathrm{Span}\{e_1 x, ..., e_p x\} \quad \text{and} \quad E_+(x) := E_-(x)^\perp$$

relative to the standard Euclidean inner product $\langle \cdot, \cdot \rangle_e$ on $\mathbb{R}^m$. Decompose tangent vectors $v_i \in T_x S^{m-1}$ in the form $v_i = v_{i,-} + v_{i,+}$ for $v_{i,\pm} \in E_\pm(x)$.

Define a metric of signature (p, q) on S^{m-1} by:

$$g(v_1, v_2) = -\langle v_{1,-}, v_{2,-}\rangle_e + \langle v_{1,+}, v_{2,+}\rangle_e .$$

Conversely, let g be a metric of signature (p, q) on S^{m-1} where $p \leq q$. By Lemma 4.2.1, we may decompose

$$T(S^{m-1}) = V_- \oplus V_+$$

where g is negative definite on V_- and positive definite on V_+. The desired estimate $p \leq \nu(m)$ now follows from Theorem 4.2.2. $\qquad\square$

4.2.7   *Equivariant vector fields on spheres*

In this section, we present a technical result of Szabó (1991) that we shall need in Section 4.5.

Theorem 4.2.4

(1) Let $\vec{s}$ be a continuous tangent vector field on the sphere S^{m-1}. Then there exists $x \in S^{m-1}$ so that $\vec{s}(x) = -\vec{s}(-x)$.

(2) Let $A(x)$ be a continuous map from S^{m-1} to the space of self-adjoint linear maps of $\mathbb{R}^m$. Assume that $A(-x) = -A(x)$, that $A(x)x = 0$, and that $\dim \ker(A)$ is constant on S^{m-1}. Then $A \equiv 0$.

Proof. Suppose that Assertion (1) is false. Thus there exists a continuous tangent vector field $\vec{s}$ on the sphere S^{m-1} so that $\vec{s}(x) + \vec{s}(-x) \neq 0$ for all $x \in S^{m-1}$. We argue for a contradiction. Set

$$f(x) := \frac{\vec{s}(x) + \vec{s}(-x)}{|\vec{s}(x) + \vec{s}(-x)|} : S^{m-1} \to S^{m-1} .$$

By assumption $f(x) \perp x$. We show f is a degree 1 map by constructing the following homotopy f_ϵ connecting f to the identity map:

$$f_\epsilon(x) = \cos(\epsilon)f(x) + \sin(\epsilon)x \quad \text{for} \quad \epsilon \in [0, \tfrac{1}{2}\pi] .$$

On the other hand, as $f(x) = f(-x)$, f descends to induce a map $[f]$ from real projective $\mathbb{RP}^{m-1}$ to S^{m-1}. This shows that the degree of f is even. This contradiction establishes Assertion (1).

Suppose that A satisfies the hypotheses of Assertion (2). As $A(x)x = 0$, $A(x)$ preserves $x^\perp$. We let $\tilde{A}(x)$ denote the restriction of $A(x)$ to $x^\perp$. Since $\tilde{A}(x)$ is self-adjoint with respect to a positive definite inner product, $\tilde{A}(x)$ is diagonalizable. We let $E_-(x)$, $E_0(x)$, and $E_+(x)$ denote the span of the eigenvectors with negative, zero, and positive eigenvalues. Since

$\dim \ker A(x)$ is constant, E_-, E_0, and E_+ are vector bundles over S^{m-1}. This gives a decomposition of the tangent bundle of the sphere:

$$T(S^{m-1}) = E_- \oplus E_0 \oplus E_+.$$

Suppose that $\dim E_+ > 0$. Fix a point $x_0 \in S^{m-1}$. Since $S^{m-1} - \{x_0\}$ is contractible, the vector bundle E_+ is trivial over $S^{m-1} - \{x_0\}$ and we can choose a continuous unit section s_+ to E_+ on $S^{m-1} - \{x_0\}$. Let ψ_+ be a continuous function on S^{m-1} which vanishes only at x_0. Then $s_1 := \psi_+ s_+$ is a continuous section to E_+ which vanishes only at x_0. The equivariance property $A(-x) = -A(x)$ shows that the section $s_2(x) := s_1(-x)$ is a continuous section to E_- which vanishes only at $-x_0$. As $A(x)$ is self-adjoint, $E_+ \perp E_-$. Let

$$s(x) := s_1(x) + s_2(x).$$

Since s_1 vanishes only at x_0 and s_2 vanishes only at $-x_0$, s is a nowhere vanishing vector field on S^{m-1}. Since $s(-x) = s(x)$, this contradicts Assertion (1). Thus, we conclude $\dim E_+ = 0$. Since $E_-(x) = E_+(-x)$, we also have $\dim E_- = 0$. Consequently, $A \equiv 0$.   □

4.2.8 *Geometrically symmetric vector bundles*

We shall need the following result in Section 4.8 when we discuss complex Ivanov-Petrova manifolds. We shall also need it in Section 5.2 when we discuss complex Osserman manifolds. We refer to Gilkey (2001a) for the proof and omit details as the methods are entirely algebraic topological in nature; a slightly different treatment may be found in Gilkey (2002). It generalizes earlier work by Glover, Homer, and Stong (1982).

We recall the notation of Section 4.2.2. Complex projective space $\mathbb{CP}^{n-1}$ is the set of complex lines in $\mathbb{C}^n$. Let $\mathbb{1}^n$ be the trivial complex n-plane bundle. The complex *tautological line bundle* $\mathbb{L}$ over $\mathbb{CP}^{n-1}$ is the sub-bundle of $\mathbb{1}^n$:

$$\mathbb{1}^n = \mathbb{CP}^{n-1} \times \mathbb{C}^n \quad \text{and} \quad \mathbb{L} = \{(\pi, \xi) \in \mathbb{1}^n : \xi \in \pi\}.$$

We say that a sub-bundle E of $\mathbb{1}^n$ is a *geometrically symmetric vector bundle* if $\tau \subset E(\sigma)$ implies that $\sigma \subset E(\tau)$ for any $\sigma, \tau \in \mathbb{CP}^{n-1}$. The tautological line bundle $\mathbb{L}$ is clearly geometrically symmetric since $\tau = \sigma$ in this instance. The orthogonal complement $\mathbb{L}^\perp$ is also geometrically symmetric because $\sigma \perp \tau \Rightarrow \tau \perp \sigma$.

Theorem 4.2.5 *Suppose given a decomposition $\mathbb{1}^n = \oplus E_i$ where the E_i are complex vector bundles over $\mathbb{CP}^{n-1}$ of dimension k_i. We choose the ordering so that $k_0 \geq ... \geq k_\ell \geq 1$. Suppose at least one of the vector bundles E_i is geometrically symmetric and that $\ell \geq 1$. If n is odd, then $\ell = 1$ and $k_1 = 1$. If n is even, then $\ell = 1$ and $k_1 = 1,2$ or $\ell = 2$ and $k_1 = k_2 = 1$.*

Theorem 4.2.5 is sharp. The decomposition $\mathbb{1}^n = \mathbb{L}^\perp \oplus \mathbb{L}$ where $\mathbb{L}$ is the tautological line bundle provides an example with $\ell = 1$ and $k_1 = 1$ in Theorem 4.2.5; the possibility $\ell = 0$ can be realized by taking $E_0 := \mathbb{1}^n$. If $n = 2\bar{n}$, we can identify $\mathbb{C}^n = \mathbb{H}^{\bar{n}}$ where the complex structure is given by $J = J_1$ and where $\{J_1, J_2, J_3 := J_1 J_2\}$ give the usual quaternion structure on $\mathbb{H}$. We set

$$E_1(x) := \mathrm{Span}_{\mathbb{C}}\{x\} = \mathbb{L} \subset \mathbb{1}^n,$$
$$E_2(x) := \mathrm{Span}_{\mathbb{C}}\{J_2 x\} = J_2\{\mathbb{L}\} \subset \mathbb{1}^n,$$
$$E_0(x) := \{E_1(x) \oplus E_2(x)\}^\perp \subset \mathbb{1}^n.$$

This yields an example of a decomposition where $\ell = 2$ and $k_1 = k_2 = 1$; a decomposition where $\ell = 1$ and $k_1 = 2$ can then be formed by combining E_1 and E_2. Each of these vector bundles is geometrically symmetric.

4.3 Generators for the Spaces $\mathcal{A}\mathrm{lg}_0$ and $\mathcal{A}\mathrm{lg}_1$

In this section, we discuss work of Díaz-Ramos, Fiedler, García-Río, and Gilkey (2004a) which was motivated by work of Díaz-Ramos and García-Río (2004) and which gives generators for the spaces $\mathcal{A}\mathrm{lg}_0(V)$ and $\mathcal{A}\mathrm{lg}_1(V)$ of algebraic curvature tensors and algebraic covariant derivative tensors defined in Section 1.3.1. Additionally, we shall complete the proof of Theorem 1.6.2 and give a different proof of Theorem 1.6.1 (1).

We recall the notational conventions established previously. Let V be a finite dimensional real vector space. Let $S^p(V) \subset \otimes^p V^*$ be the space of totally symmetric p forms on V. If $\Psi \in S^2(V)$ and if $\Psi_1 \in S^3(V)$, we adopt the notation of Eqs. (1.3.b) and (1.6.a) to define $A_\Psi \in \mathcal{A}\mathrm{lg}_0(V)$ and $A_{1,\Psi,\Psi_1} \in \mathcal{A}\mathrm{lg}_1(V)$ by setting:

$$A_\Psi(x,y,z,w) := \Psi(x,w)\Psi(y,z) - \Psi(x,z)\Psi(y,w),$$
$$A_{1,\Psi,\Psi_1}(x,y,z,w;v) := \Psi_1(x,w,v)\Psi(y,z) + \Psi(x,w)\Psi_1(y,z,v)$$
$$- \Psi_1(x,z,v)\Psi(y,w) - \Psi(x,z)\Psi_1(y,w,v).$$

In a formal sense, one can think of Ψ_1 as the covariant derivative of Ψ and one can think of A_{1,Ψ,Ψ_1} as the covariant derivative of A_Ψ. This analogy is, of course, purely formal. It is, however, useful in motivating certain formulas and will play an important role in the subsequent discussion.

Let $A \in \mathcal{Alg}_0(V)$. Suppose there exists a finite collection $\{\Psi_i\}_{1 \leq i \leq \nu}$ of symmetric 2-tensors such that

$$A = \sum_{i=1}^{\nu} \lambda_i A_{\Psi_i}$$

for some suitably chosen constants λ_i; by rescaling we may always assume $\lambda_i = \pm 1$. We then set $\nu(A)$ to be the minimal ν so such a decomposition exists; set $\nu(A) = \infty$ if no such decomposition exists. Similarly, suppose given $A_1 \in \mathcal{Alg}_1(V)$. If there exist finite collections $\tilde{\Psi}_j \in S^2(V)$ and $\tilde{\Psi}_{1,j} \in S^3(V)$ so that

$$A_1 = \sum_{j=1}^{\nu_1} \lambda_{1,j} A_{1,\tilde{\Psi}_j,\tilde{\Psi}_{1,j}}$$

for suitably chosen constants $\lambda_{1,j}$, then we set $\nu_1(A_1)$ to be the minimal number of terms possible; by rescaling the $\tilde{\Psi}_{1,j}$, we may always assume $\lambda_{1,j} = 1$. We shall show presently that such decompositions always exist; this means that $\nu(A) < \infty$ and $\nu(A_1) < \infty$. Set

$$\nu(m) := \sup_{A \in \mathcal{Alg}_0(V)} \nu(A) \quad \text{and} \quad \nu_1(m) := \sup_{A_1 \in \mathcal{Alg}_1(V)} \nu_1(A_1).$$

Let $[\cdot]$ denote the greatest integer function. Theorem 1.6.1 (1) and Theorem 1.6.2 follow from the following result:

Theorem 4.3.1   *Let $m \geq 2$.*

(1) $[\frac{1}{2}m] \leq \nu(m)$ *and* $[\frac{1}{2}m] \leq \nu_1(m)$.
(2) $\nu(m) \leq \frac{1}{2}m(m+1)$ *and* $\nu_1(m) \leq \frac{1}{2}m(m+1)$.

Remark 4.3.1   The bounds of Theorem 4.3.1 are, of course, not sharp. For example, it is known, Díaz-Ramos and García-Río (2004), that $\nu(2) = 1$ and $\nu(3) = 2$.

We shall establish the lower bounds of Assertion (1) in Section 4.3.1. The upper bound for $\nu(m)$ which is given in Assertion (2) is due to Díaz-Ramos and García-Río (2004). In Section 4.3.2, we shall generalize their approach to establish the following simultaneous "diagonalization" result from which Theorem 4.3.1 (2) will follow as a Corollary:

Theorem 4.3.2 *Let V be an m-dimensional vector space. Suppose given $A \in \mathcal{A}lg_0(V)$ and $A_1 \in \mathcal{A}lg_1(V)$. Then there exist tensors $\Psi_i \in S^2(V)$ and $\Psi_{1,i} \in S^3(V)$ satisfying*

$$A = \sum_{i=1}^{\frac{1}{2}m(m+1)} A_{\Psi_i} \quad and \quad A_1 = \sum_{i=1}^{\frac{1}{2}m(m+1)} A_{1,\Psi_i,\Psi_{1,i}} \, .$$

The study of the tensors A_Ψ arose in the original instance from the Osserman conjecture and related matters; we refer to García-Río, Kupeli, and Vázquez-Lorenzo (2002) and to Gilkey (2002) for a more extensive discussion than is possible here.

We fix the following notational convention for the remainder of this section. Let $\langle \cdot, \cdot \rangle$ be an auxiliary positive definite inner product on V. Given an algebraic curvature tensor $A \in \mathcal{A}lg_0(V)$ and an algebraic covariant derivative curvature tensor $A_1 \in \mathcal{A}lg_1(V)$, we let $\mathfrak{M}_1 := (V, \langle \cdot, \cdot \rangle, A, A_1)$ be the associated 1-model. We use $\langle \cdot, \cdot \rangle$ to raise indices and to define the associated curvature operators $\mathcal{A}$ and $\mathcal{A}_1$ by means of the identities:

$$\langle \mathcal{A}(\xi_1, \xi_2)z, w \rangle = A(\xi_1, \xi_2, z, w), \quad and$$
$$\langle \mathcal{A}_1(\xi_1, \xi_2, \xi_3)z, w \rangle = A_1(\xi_1, \xi_2, z, w; \xi_3) \, .$$

4.3.1 *A lower bound for $\nu(m)$ and for $\nu_1(m)$*

Theorem 4.3.1 (1) will follow from the following Lemma:

Lemma 4.3.1 *Let $m = 2\bar{m}$ or let $m = 2\bar{m} + 1$.*

(1) If $\Psi \in S^2(V)$ and if $\Psi_1 \in S^3(V)$, then for any $\xi_1, \xi_2, \xi_3 \in V$ one has:

$$\mathrm{Rank}\{\mathcal{A}_\Psi(\xi_1, \xi_2)\} \leq 2 \quad and \quad \mathrm{Rank}\{\mathcal{A}_{1,\Psi,\Psi_1}(\xi_1, \xi_2, \xi_3)\} \leq 2 \, .$$

(2) If $A \in \mathcal{A}lg_0(V)$ and $A_1 \in \mathcal{A}lg_1(V)$, then for any $\xi_1, \xi_2, \xi_3 \in V$ one has:

$$\mathrm{Rank}\{\mathcal{A}(\xi_1, \xi_2)\} \leq 2\nu(A) \quad and \quad \mathrm{Rank}\{\mathcal{A}_1(\xi_1, \xi_2, \xi_3)\} \leq 2\nu_1(A_1) \, .$$

(3) There exist $A \in \mathcal{A}lg_0(V)$, $A_1 \in \mathcal{A}lg_1(V)$, and $\xi_1, \xi_2, \xi_3 \in V$ so:

$$\mathrm{Rank}\{\mathcal{A}(\xi_1, \xi_2)\} = 2\bar{m} \quad and \quad \mathrm{Rank}\{\mathcal{A}_1(\xi_1, \xi_2, \xi_1)\} = 2\bar{m} \, .$$

Proof. If $\Psi \in S^2(V)$ and $\Psi_1 \in S^3(V)$, let ψ and $\psi_1(\cdot)$ be the associated self-adjoint endomorphisms characterized by the identities

$$\langle \psi x, y \rangle = \Psi(x, y) \quad and \quad \langle \psi_1(z)x, y \rangle = \Psi_1(x, y, z) \, .$$

Assertion (1) follows from the expression:

$$\mathcal{A}_\Psi(\xi_1,\xi_2)y = \{\Psi(\xi_2,y)\psi\}\xi_1 - \{\Psi(\xi_1,y)\psi\}\xi_2, \quad \text{and}$$

$$\mathcal{A}_{1,\Psi,\Psi_1}(\xi_1,\xi_2,\xi_3)y = \{\Psi(\xi_2,y)\psi_1(\xi_3) + \Psi_1(\xi_2,y,\xi_3)\psi\}\xi_1$$
$$- \{\Psi(\xi_1,y)\psi_1(\xi_3) + \Psi_1(\xi_1,y,\xi_3)\psi\}\xi_2\,.$$

Let $B_i := A_{\Psi_i}$ and let $B_{1,j} := A_{1,\tilde\Psi_j,\tilde\Psi_{1,j}}$. Set

$$A = \sum_{i=1}^{\nu(A)} \lambda_i B_i \quad \text{and} \quad A_1 = \sum_{j=1}^{\nu_1(A_1)} \lambda_{1,j} B_{1,j}\,.$$

Assertion (2) follows from Assertion (1) as

$$\mathrm{Rank}\{\mathcal{A}(\cdot)\} = \mathrm{Rank}\left\{\sum_{i=1}^{\nu(A)} \lambda_i \mathcal{B}_i(\cdot)\right\} \le \sum_{i=1}^{\nu(A)} \mathrm{Rank}\{\mathcal{B}_i(\cdot)\} \le 2\nu(A),$$

$$\mathrm{Rank}\{\mathcal{A}_1(\cdot)\} = \mathrm{Rank}\left\{\sum_{i=1}^{\nu_1(A)} \lambda_{1,j} \mathcal{B}_{1,j}(\cdot)\right\} \le \sum_{i=1}^{\nu_1(A)} \mathrm{Rank}\{\mathcal{B}_{1,j}(\cdot)\} \le 2\nu_1(A_1)\,.$$

If $\dim(V) = 2\bar{m}$, let $\{e_1,...,e_{\bar m}, f_1,...,f_{\bar m}\}$ be an orthonormal basis for V; if $\dim(V)$ is odd, the argument is similar and we simply extend A and A_1 to be trivial on the additional basis vector. Define the non-zero components of $\Psi_i \in S^2(V)$ and $\Psi_{1,i} \in S^3(V)$ by:

$$\Psi_i(e_j,e_k) = \Psi_i(f_j,f_k) = \delta_{ij}\delta_{ik},$$
$$\Psi_{1,i}(e_j,e_k,e_l) = \Psi_{1,i}(f_j,f_k,f_l) = \delta_{ij}\delta_{ik}\delta_{il};$$

$\Psi_i(\cdot,\cdot)$ and $\Psi_{1,i}(\cdot,\cdot,\cdot)$ vanish if both an "e" and an "f" appear. Let

$$B_i := A_{\Psi_i}, \quad B_{1,i} := A_{1,\Psi_i,\Psi_{1,i}},$$
$$A := \sum_{i=1}^{\bar m} B_i, \quad A_1 := \sum_{i=1}^{\bar m} B_{1,i},$$
$$\xi_1 := e_1 + ... + e_{\bar m}, \quad \xi_2 := f_1 + ... + f_{\bar m}, \quad \xi_3 := \xi_1 + \xi_2\,.$$

We may then complete the proof of Assertion (3) by computing:

$$\mathcal{A}(\xi_1,\xi_2)e_i = \mathcal{A}_i(e_i,f_i)e_i = -f_i,$$
$$\mathcal{A}(\xi_1,\xi_2)f_i = \mathcal{A}_i(e_i,f_i)f_i = e_i,$$
$$\mathcal{A}_1(\xi_1,\xi_2,\xi_3)e_i = \mathcal{A}_{1,i}(e_i,f_i,e_i+f_i)e_i = -2f_i,$$
$$\mathcal{A}_1(\xi_1,\xi_2,\xi_3)f_i = \mathcal{A}_{1,i}(e_i,f_i,e_i+f_i)f_i = 2e_i\,. \qquad \square$$

4.3.2 *Geometric realizability*

Let $\langle \cdot, \cdot \rangle$ be the standard Euclidean inner product on $\mathbb{R}^m$. Assume given a 1-model $\mathfrak{M}_1 = (\mathbb{R}^m, \langle \cdot, \cdot \rangle, A, A_1)$. We apply Lemma 1.6.2 to find the germ of a Riemannian metric g on $\mathbb{R}^m$ such that:

$$(T_0\mathbb{R}^m, g(0), R(0), \nabla R(0)) = \mathfrak{M}_1 \,.$$

We apply the Embedding Theorem of Nash (1956) to find the germ of a map $f : \mathbb{R}^m \to \mathbb{R}^{m+\kappa}$ realizing the metric g. By writing the submanifold as a graph over its tangent plane, we can choose coordinates (x, y) on $\mathbb{R}^{m+\kappa}$ where $x = (x_1, ..., x_m)$ and $y = (y_1, ..., y_\kappa)$ so that

$$f(x) = (x, f_1(x), ..., f_\kappa(x)) \quad \text{where} \quad df_\nu(0) = 0 \quad \text{for} \quad 1 \leq \nu \leq \kappa \,.$$

Since $f_*(\partial_i^x) = (0, ..., 1, ..., 0, \partial_i^x f_1, ..., \partial_i^x f_\kappa)$, we have

$$g_{ij}(x) = \delta_{ij} + \sum_{\sigma=1}^{\kappa} \partial_i^x f_\sigma \cdot \partial_j^x f_\sigma \,.$$

As $dg_{ij}(0) = 0$, Lemma 1.6.2 can be applied. Let $\Psi_{ij}^\sigma := \partial_i^x \partial_j^x f_\sigma(0)$ for $1 \leq \sigma \leq \kappa$ represent the second fundamental forms. One has that:

$$
\begin{aligned}
R_{ijkl}(0) = {} & \tfrac{1}{2} \sum_{\sigma=1}^{\kappa} \Big\{ (\Psi_{ij}^\sigma \Psi_{kl}^\sigma + \Psi_{il}^\sigma \Psi_{kj}^\sigma) + (\Psi_{ji}^\sigma \Psi_{lk}^\sigma + \Psi_{jk}^\sigma \Psi_{li}^\sigma) \\
& - (\Psi_{ij}^\sigma \Psi_{lk}^\sigma + \Psi_{ik}^\sigma \Psi_{lj}^\sigma) - (\Psi_{ji}^\sigma \Psi_{kl}^\sigma + \Psi_{jl}^\sigma \Psi_{ki}^\sigma) \Big\} \\
= {} & \sum_{\sigma=1}^{\kappa} \Big\{ \Psi_{il}^\sigma \Psi_{jk}^\sigma - \Psi_{ik}^\sigma \Psi_{jl}^\sigma \Big\} \,.
\end{aligned}
$$

This shows that

$$A = \sum_{\sigma=1}^{\kappa} A_{\Psi^\sigma} \quad \text{so} \quad \nu(A) \leq \kappa \,.$$

Let $\Psi_{ijk}^\sigma := \partial_i^x \partial_j^x \partial_k^x f_\sigma(0)$. One may compute similarly that:

$$
\begin{aligned}
R_{ijkl;n}(0) = {} & \tfrac{1}{2} \sum_{\sigma=1}^{\kappa} \Big\{ (\Psi_{jin}^\sigma \Psi_{lk}^\sigma + \Psi_{jkn}^\sigma \Psi_{li}^\sigma + \Psi_{ji}^\sigma \Psi_{lkn}^\sigma + \Psi_{jk}^\sigma \Psi_{lin}^\sigma \\
& + \Psi_{jik}^\sigma \Psi_{ln}^\sigma + \Psi_{jn}^\sigma \Psi_{lik}^\sigma) + (\Psi_{ijn}^\sigma \Psi_{kl}^\sigma + \Psi_{iln}^\sigma \Psi_{kj}^\sigma + \Psi_{ij}^\sigma \Psi_{kln}^\sigma \\
& + \Psi_{il}^\sigma \Psi_{kjn}^\sigma + \Psi_{ijl}^\sigma \Psi_{kn}^\sigma + \Psi_{in}^\sigma \Psi_{kjl}^\sigma) - (\Psi_{jin}^\sigma \Psi_{kl}^\sigma + \Psi_{jln}^\sigma \Psi_{ki}^\sigma \\
& + \Psi_{ji}^\sigma \Psi_{kln}^\sigma + \Psi_{jl}^\sigma \Psi_{kin}^\sigma + \Psi_{jil}^\sigma \Psi_{kn}^\sigma + \Psi_{jn}^\sigma \Psi_{kil}^\sigma) - (\Psi_{ijn}^\sigma \Psi_{lk}^\sigma
\end{aligned}
$$

$$+\Psi^\sigma_{ikn}\Psi^\sigma_{lj} + \Psi^\sigma_{ij}\Psi^\sigma_{lkn} + \Psi^\sigma_{ik}\Psi^\sigma_{ljn} + \Psi^\sigma_{ijk}\Psi^\sigma_{ln} + \Psi^\sigma_{in}\Psi^\sigma_{ljk}\big)\bigg\}$$

$$= \sum_{\sigma=1}^{\kappa}\bigg\{\Psi^\sigma_{iln}\Psi^\sigma_{jk} + \Psi^\sigma_{jkn}\Psi^\sigma_{il} - \Psi^\sigma_{ikn}\Psi^\sigma_{jl} - \Psi^\sigma_{ik}\Psi^\sigma_{jln}\bigg\}.$$

Thus similarly

$$A_1 = \sum_{\sigma=1}^{\kappa} A_{1,\Psi^\sigma,\Psi^\sigma_1} \quad \text{so} \quad \nu_1(A_1) \leq \kappa.$$

In the analytic category we may take $\kappa \leq \frac{1}{2}m(m+1)$ in the Nash Embedding Theorem. The desired upper bound of Theorem 4.3.2 now follows. $\square$

4.4   Jordan Osserman Algebraic Curvature Tensors

In this section, we present two results which deal with Jordan Osserman algebraic curvature tensors. The first result is due to Gilkey and Ivanova (2001b); it shows that in the neutral signature contex there are Jordan Osserman algebraic curvature tensors with arbitrarily complicated Jordan normal forms.

We adopt the following notational conventions. Let $\mathbb{R}^{(p,q)}$ denote $\mathbb{R}^{(p+q)}$ with the canonical inner product of signature (p,q) given by:

$$\langle x, y \rangle := -x_1 y_1 - \dots - x_p y_p + x_{p+1} y_{p+1} + \dots + x_{p+q} y_{p+q}\,.$$

If $n \leq p$, we can choose a timelike embedding of $\mathbb{R}^n$ in $\mathbb{R}^{(p,q)}$. Let $\mathrm{End}(\mathbb{R}^k)$ denote the vector space of linear endomorphisms of $\mathbb{R}^k$; these are the $k \times k$ matrices. If $J \in \mathrm{End}(\mathbb{R}^n)$, extend J to be 0 on $\{\mathbb{R}^n\}^\perp$ to define a linear map $J \oplus 0 \in \mathrm{End}(\mathbb{R}^{(p,q)})$; $J \oplus 0$ is said to be the *stabilization* of J – it is uniquely defined up to conjugacy. Relative to suitable bases, it takes the following form in matrix notation:

$$J \oplus 0 = \begin{pmatrix} J & 0 \\ 0 & 0 \end{pmatrix}.$$

Theorem 4.4.1   *Let $J \in \mathrm{End}(\mathbb{R}^n)$. There exists $\ell = \ell(n)$ and a spacelike and timelike Jordan Osserman algebraic curvature tensor A on $\mathbb{R}^{(\ell,\ell)}$ so that $\mathcal{J}_A(x)$ is conjugate to $\pm J \oplus 0$ for any $x \in S^\pm(\mathbb{R}^{(\ell,\ell)})$.*

However, the situation is very different if $p < q$; in this setting, the spacelike directions in a certain sense dominate the timelike directions as the following result of Gilkey and Ivanova (2002b) shows:

Theorem 4.4.2 *Let A be a spacelike Jordan Osserman algebraic curvature tensor on $\mathbb{R}^{(p,q)}$, where $p < q$. Then $\mathcal{J}_A(x)$ is diagonalizable for any $x \in S^+(\mathbb{R}^{(p,q)})$.*

The remainder of this section is devoted to the proof of these results. In Section 4.4.1, we establish Theorem 4.4.1 and in Section 4.4.2, we establish Theorem 4.4.2.

4.4.1 *Neutral signature Jordan Osserman tensors*

Let $\mathfrak{so}(\mathbb{R}^{(r,s)})$ be the Lie-algebra of the associated orthogonal group on $\mathbb{R}^{(r,s)}$:

$$\mathfrak{so}(\mathbb{R}^{(r,s)}) = \{T \in \operatorname{End}(\mathbb{R}^{(r,s)}) : T + T^* = 0\}.$$

We say that $\mathbb{R}^{(r,s)}$ admits a $\operatorname{Clif}(\mathbb{R}^{(p,q)})$ *module structure* if there exists a linear map $\phi : \mathbb{R}^{(p,q)} \to \mathfrak{so}(\mathbb{R}^{(r,s)})$ so that

$$\phi(v)\phi(w) + \phi(w)\phi(v) = -2\langle v, w \rangle \operatorname{Id} \quad \text{for all} \quad v, w \in \mathbb{R}^{(p,q)}.$$

This means, of course, that ϕ defines a representation of $\operatorname{Clif}(\mathbb{R}^{(p,q)})$ into the algebra of endomorphisms of $\mathbb{R}^{(p,q)}$.

Lemma 4.4.1 *Let (p, q) be given.*

(1) There exists ℓ so that $\mathbb{R}^{(\ell,\ell)}$ admits a $\operatorname{Clif}(\mathbb{R}^{(p,q)})$ module structure.
(2) Let ϕ be a $\operatorname{Clif}(\mathbb{R}^{(p,q)})$ module structure on $\mathbb{R}^{(\ell,\ell)}$. Let $x^\pm \in S^\pm(\mathbb{R}^{(\ell,\ell)})$.

  (a) $v \to \phi(v)x^+$ is an isometric embedding of $\mathbb{R}^{(p,q)}$ in $\mathbb{R}^{(\ell,\ell)}$.
  (b) $v \to \phi(v)x^-$ is a para-isometric embedding of $\mathbb{R}^{(p,q)}$ in $\mathbb{R}^{(\ell,\ell)}$.

Proof. Assertion (1) follows from standard results in the theory of Clifford algebras, see, for example, Karoubi (1978). Since we are not interested in optimal estimates, it is possible to give an elementary proof. We give the exterior algebra $\Lambda^*(\mathbb{R}^{p+q})$, with the canonical positive definite inner product, a $\operatorname{Clif}(\mathbb{R}^{p+q})$ module structure by defining $c(v) = \operatorname{ext}(v) - \operatorname{int}(v)$ where $\operatorname{ext}(v) : \theta \to v \wedge \theta$ is given by left exterior multiplication and where $\operatorname{int}(v)$ is the dual, interior multiplication; $c(v)^2 = -|v|^2 \operatorname{Id}$. Let $\{e_i\}$ be the standard orthonormal basis for $\mathbb{R}^{p+q}$ and let $\phi_i := c(e_i)$. We complexify and let

$$V := \Lambda^*(\mathbb{R}^{p+q}) \otimes_\mathbb{R} \mathbb{C}.$$

We extend the Euclidean inner product to be complex bilinear and we extend ϕ_i to be complex linear. Let $\Re$ denote the real part of a complex number. We set

$$\langle v, w \rangle := \Re(v, w) \quad \text{and} \quad \Phi_i := \begin{cases} \sqrt{-1}\phi_i & \text{for } 1 \leq i \leq p, \\ \phi_i & \text{for } p < i \leq p+q. \end{cases}$$

Then $\langle v, w \rangle$ is a neutral signature metric on the underlying real vector space and $\{\Phi_i\}$ is a collection of real endomorphisms satisfying

$$\langle \Phi_i v, w \rangle = -\langle v, \Phi_i w \rangle,$$
$$\Phi_i \Phi_j + \Phi_j \Phi_i = 0 \quad \text{for} \quad i \neq j,$$
$$\Phi_i^2 = \begin{cases} \text{Id} & \text{for } 1 \leq i \leq p, \\ -\text{Id} & \text{for } p < i \leq p+q. \end{cases}$$

This establishes Assertion (1); this trick of complexification plays a crucial role in other arguments we have given.

Let ϕ give $\mathbb{R}^{(\ell,\ell)}$ a $\text{Clif}(\mathbb{R}^{(p,q)})$ module structure. Let $\{e_i\}$ be an orthonormal basis for $\mathbb{R}^{(p,q)}$ and let $\phi_i := \phi(e_i)$. Suppose $i \neq j$. We use the Clifford commutation relations to compute:

$$\langle \phi_i x, \phi_j x \rangle = -\langle \phi_j \phi_i x, x \rangle = \langle \phi_i \phi_j x, x \rangle = -\langle \phi_j x, \phi_i x \rangle$$

and consequently $\phi_i x \perp \phi_j x$ if $i \neq j$. One also has that

$$\langle \phi_i x, \phi_i x \rangle = -\langle \phi_i \phi_i x, x \rangle = \langle e_i, e_i \rangle \langle x, x \rangle.$$

Assertion (2) now follows. $\qquad\qquad\qquad\qquad\qquad\qquad\qquad\qquad\square$

We now establish Theorem 4.4.1. Let $\psi \in \mathfrak{so}(\mathbb{R}^{(\ell,\ell)})$ be skew-adjoint. We adopt the notation of Eq. (1.3.b) and define:

$$A_\psi(x, y)z := (\psi y, z)\psi x - (\psi x, z)\psi y - 2(\psi x, y)\psi z.$$

We then have

$$\mathcal{J}_{A_\psi}(x)y = 3\langle y, \psi x \rangle \psi x. \tag{4.4.a}$$

Let $J \in \text{End}(\mathbb{R}^n)$. Apply Lemma 1.5.4 to choose an inner product of signature (p, q) on $\mathbb{R}^n$ so J is self-adjoint with respect to this inner product. By choosing an orthonormal basis, we may identify $\mathbb{R}^n$ with $\mathbb{R}^{(p,q)}$. Apply Lemma 4.4.1 to choose a $\text{Clif}(\mathbb{R}^{(p,q)})$ module structure ϕ on $\mathbb{R}^{(\ell,\ell)}$.

Let $\{e_i\}$ be the standard orthonormal basis for $\mathbb{R}^{(p,q)}$. Expand

$$Je_i = \sum_j J_{ij} e_j\,.$$

Let c_i and $c_{ij} = c_{ji}$ be constants to be determined presently. Let $\phi_i := \phi(e_i)$ and let

$$A := \sum_i c_i A_{\phi_i} + \tfrac{1}{2} \sum_{i \neq j} c_{ij} A_{\phi_i + \phi_j}$$

be an algebraic curvature tensor on $\mathbb{R}^{(\ell,\ell)}$. Let $\mathcal{J}_i$ be the Jacobi operator defined by A_{ϕ_i} and let $\mathcal{J}_{ij}$ be the Jacobi operator defined by $A_{\phi_i + \phi_j}$.

Let $x \in S^{\pm}(\mathbb{R}^{(\ell,\ell)})$. We will show $\mathcal{J}_A$ is conjugate to $\pm J \oplus 0$; this will complete the proof of Theorem 4.4.1. Let $f_i := \phi_i x$ and let

$$\pi(x) := \mathrm{Span}\{f_1, ..., f_{p+q}\} \in \mathrm{Gr}_{p,q}(\mathbb{R}^{(\ell,\ell)}).$$

By Lemma 4.4.1, $\{f_1, ..., f_{p+q}\}$ is an orthonormal basis for $\pi(x)$. By Eq. (4.4.a), Range $\mathcal{J}_A(x) \subset \pi(x)$ and $\mathcal{J}_A(x) = 0$ on $U := \pi(x)^{\perp}$. Expand

$$\mathcal{J}_A(x) f_i = \sum_j \tilde{J}_{ij} f_j.$$

Let i and j be distinct indices. Let $\epsilon_k := \langle e_k, e_k \rangle = \pm 1$. By Eq. (4.4.a):

$$\mathcal{J}_i f_k = \begin{cases} 3(x,x)\varepsilon_k f_i & \text{if } k = i, \\ 0 & \text{if } k \neq i, \end{cases}$$

$$\mathcal{J}_{ij} f_k = \begin{cases} 3(x,x)\varepsilon_k (f_i + f_j) & \text{if } k = i, j, \\ 0 & \text{if } k \neq i, j, \end{cases}$$

$$\tilde{J}_{ij} = \begin{cases} 3(x,x)\varepsilon_i (c_i + \sum_{k \neq i} c_{ik}) & \text{if } i = j, \\ 3(x,x)\varepsilon_i c_{ij} & \text{if } i \neq j. \end{cases}$$

We solve the relations $\tilde{J}_{ij} = (x,x) J_{ij}$ for $1 \leq i, j \leq p+q$ to see that if $i \neq j$, then:

$$c_{ij} := \tfrac{1}{3}\varepsilon_i J_{ij} \quad \text{and} \quad c_i := \tfrac{1}{3}\varepsilon_i \left(J_{ii} - \sum_{j \neq i} J_{ij} \right).$$

We have $\varepsilon_i = \pm 1$. Since J is self-adjoint we have that $\varepsilon_j J_{ij} = \varepsilon_i J_{ji}$ so $\tfrac{1}{3}\varepsilon_i J_{ij} = \tfrac{1}{3}\varepsilon_j J_{ji}$. Consequently, $c_{ij} = c_{ji}$ as required. It is now clear that $\mathcal{J}_A(x)$ is conjugate to $(x,x) J \oplus 0$. This completes the proof of Theorem 4.4.1. $\qquad\square$

4.4.2 *Rigidity results for Jordan Osserman tensors*

Before proving Theorem 4.4.2, we must recall some results from linear algebra.

Let $\Re(\lambda)$ and $\Im(\lambda)$ be the real and imaginary parts of a complex number λ. If J is a linear map of a real vector space V of dimension m, then let J_λ be the real operator on V defined by:

$$J_\lambda := \begin{cases} J - \lambda \cdot \mathrm{Id} & \text{if} \quad \lambda \in \mathbb{R}, \\ (J - \lambda \cdot \mathrm{Id})(J - \bar{\lambda} \cdot \mathrm{Id}) & \text{if} \quad \lambda \in \mathbb{C} - \mathbb{R}. \end{cases} \tag{4.4.b}$$

We define the generalized eigenspaces by setting

$$E_\lambda := \ker\{J_\lambda^m\}. \tag{4.4.c}$$

Lemma 4.4.2 *Let V be a vector space of signature (p,q) and let J be a self-adjoint linear map of V. Then V can be decomposed as an orthogonal direct sum $V = \oplus_{\Im(\lambda)\geq 0} E_\lambda$. Furthermore, the induced metrics on the generalized eigenspaces E_λ are non-degenerate.*

Proof. Let λ and μ be complex numbers with $\lambda \neq \mu$ and $\lambda \neq \bar{\mu}$. Since J_λ^m is self-adjoint and vanishes on E_λ, we have

$$0 = (J_\lambda^m x_\lambda, x_\mu) = (x_\lambda, J_\lambda^m x_\mu) \text{ for } x_\lambda \in E_\lambda \text{ and } x_\mu \in E_\mu.$$

Since J commutes with J_μ, J preserves E_μ. Since the eigenvalues of J on E_μ are μ and $\bar{\mu}$, the linear maps $J - \lambda \cdot \mathrm{Id}$, $J - \bar{\lambda} \cdot \mathrm{Id}$, and hence J_λ are isomorphisms of E_μ; thus $J_\lambda^m(E_\mu) = E_\mu$. It now follows that

$$E_\lambda \perp E_\mu \text{ and } E_\lambda \cap E_\mu = \{0\}. \tag{4.4.d}$$

Let $V^{\mathbb{C}} := V \otimes \mathbb{C}$ be the complexification of V. We extend J to $V^{\mathbb{C}}$ to be complex linear and set $E_\lambda^{\mathbb{C}} := \ker\{(J - \lambda)^m\}$. A complex vector space may be decomposed as the direct sum of the generalized complex eigenspaces defined by a linear transformation. Consequently,

$$V^{\mathbb{C}} = \oplus_\lambda E_\lambda^{\mathbb{C}}. \tag{4.4.e}$$

As $E_\lambda^{\mathbb{C}} \oplus E_{\bar{\lambda}}^{\mathbb{C}} = E_\lambda \otimes \mathbb{C}$, $V = \oplus_{\Im(\lambda)\geq 0} E_\lambda$. By Eq. (4.4.d), the direct sum given in Eq. (4.4.e) is orthogonal; thus, the induced metric on each E_λ is non-degenerate. $\qquad\square$

We continue our preparation for the proof of Theorem 4.4.2 by studying vector bundles equipped with non-degenerate fiber metrics and self-adjoint bundle morphisms which have constant Jordan normal form:

Lemma 4.4.3 *Let E be a vector bundle over a smooth manifold M which is equipped with a non-degenerate fiber metric. Let J be a self-adjoint bundle morphism of E which has constant Jordan normal form. Let λ be an eigenvalue of J. If $J_\lambda \neq 0$ on E_λ, choose $i \geq 1$ maximal so $J_\lambda^i(E_\lambda) \neq 0$. Then $J_\lambda^i(E_\lambda)$ is a totally isotropic subbundle of E of non-zero rank.*

Proof. Assume that E and J satisfy the hypothesis of the Lemma. Set:

$$E_{\lambda,i} := J_\lambda^i(E_\lambda).$$

Since J has constant Jordan normal form, $E_{\lambda,i}$ is a smooth vector bundle over M. Fix a point P of M and let v_1 and v_2 be vectors in the fiber $E_{\lambda,i}(P)$. There exist vectors $w_1, w_2 \in E_{\lambda,i}(P)$ so $v_1 = J_\lambda^i w_1$ and $v_2 = J_\lambda^i w_2$. Note that $2i \geq i + 1$, that J_λ is self-adjoint, and that $J_\lambda^{2i} = 0$ on E_λ. We demonstrate that $E_{\lambda,i}$ is totally isotropic by computing:

$$(v_1, v_2) = (J_\lambda^i w_1, J_\lambda^i w_2) = (J_\lambda^{2i} w_1, w_2) = 0.$$

The Lemma now follows. $\qquad\square$

We use Lemma 4.2.2 to establish the following Lemma. We adopt the following notational conventions. Let V be a vector space of signature (p, q). Decompose $V = V^+ \oplus V^-$ as an orthogonal direct sum where V^+ is a maximal spacelike subspace of dimension q and where V^- is the complementary maximal timelike subspace of dimension p. Let $\mathbb{RP}(V^+)$ be the associated projective space of lines through the origin in V^+. Let

$$\mathbb{V}^\pm := \mathbb{RP}(V^+) \times V^\pm$$

be the associated trivial vector bundles and let

$$\mathbb{V} := \mathbb{V}^+ \oplus \mathbb{V}^- = \mathbb{RP}(V^+) \times V.$$

Let $\mathbb{L}^\perp \subset \mathbb{V}^+$ be the orthogonal complement of the tautological line bundle.

Lemma 4.4.4 *Let $V = V^+ \oplus V^-$ be a vector space of signature (p, q) for $p < q$. There is no totally isotropic positive rank subbundle of $\mathbb{L}^\perp \oplus \mathbb{V}^-$.*

Proof. We suppose, to the contrary, that there exists a totally isotropic positive rank subbundle E of $\mathbb{L}^\perp \oplus \mathbb{V}^-$. Let π^+ be orthogonal projection on $\mathbb{L}^\perp$ and let π^- be orthogonal projection on $\mathbb{V}^-$. Set

$$E^+ := \pi^+(E) \subset \mathbb{L}^\perp \quad \text{and} \quad E^- := \pi^-(E) \subset \mathbb{V}^-.$$

Note that $\ker \pi^+ = \mathbb{V}^-$ and $\ker \pi^- = \mathbb{L}^\perp$. As E is totally isotropic, every vector in E is null. Thus

$$E \cap \mathbb{V}^- = E \cap \mathbb{L}^\perp = \{0\}.$$

Consequently, the projections $\pi^\pm$ define isomorphisms between E and $E^\pm$. Thus E^+, which is a positive rank subbundle of $\mathbb{L}^\perp$, is isomorphic to E^-, which is subbundle of $\mathbb{V}^-$. This contradicts Lemma 4.2.2. $\square$

Proof of Theorem 4.4.2. Let $V = V^+ \oplus V^-$ be a vector space of signature (p, q), where $p < q$, and where V^+ and V^- are orthogonal maximal spacelike and timelike subspaces of V, respectively. Let A be a spacelike Jordan Osserman algebraic curvature tensor on V. Let $x \in S^+(V, \langle \cdot, \cdot \rangle)$. Since $\mathcal{J}_A(x)$ is self-adjoint and since $\mathcal{J}_A(x)x = 0$, $\mathcal{J}_A(x)$ preserves the orthogonal complement $x^\perp$; we let $\tilde{\mathcal{J}}_A(x)$ denote the restriction of $\mathcal{J}_A(x)$ to $x^\perp$; this is often called the *reduced Jacobi operator*. The Jacobi operator can be represented in the form

$$\mathcal{J}_A(x) = \begin{pmatrix} 0 & 0 \\ 0 & \tilde{\mathcal{J}}_A(x) \end{pmatrix} \quad \text{on} \quad x \oplus x^\perp.$$

Thus to prove Theorem 4.4.2, it suffices to show that $\tilde{\mathcal{J}}_A(x)$ is diagonalizable.

If $\lambda \in \mathbb{C}$, then let $\tilde{J}_\lambda$ and E_λ be defined by $\tilde{\mathcal{J}}_A$ using Eqs. (4.4.b) and (4.4.c), respectively. We may then use Lemma 4.4.2 to decompose

$$\mathbb{L}^\perp \oplus \mathbb{V}^- = \oplus_{\Im(\lambda) \geq 0} E_\lambda \text{ over } \mathbb{P}(V^+),$$

where the induced metric on each eigenbundle E_λ is non-degenerate. By Lemma 4.4.4, E_λ does not contain a totally isotropic subbundle. Thus, by Lemma 4.4.3, $\tilde{\mathcal{J}}_\lambda = 0$ on E_λ. Consequently $\tilde{\mathcal{J}}_A$ is diagonalizable on E_λ if $\lambda \in \mathbb{R}$.

To complete the proof, we must show that all the eigenvalues are real. Suppose, to the contrary, that there exists an eigenvalue λ of $\tilde{\mathcal{J}}_A$ so that $\Im(\lambda) \neq 0$; we argue for a contradiction. By Lemma 4.2.1, $E_\lambda = E_\lambda^+ \oplus E_\lambda^-$ decomposes as the orthogonal direct sum of maximal spacelike and timelike subbundles. We define a bundle map $\mathcal{I}$ of E_λ by setting:

$$\mathcal{I} := \frac{\tilde{\mathcal{J}}_A - \Re(\lambda)\,\text{Id}}{\Im(\lambda)}.$$

The definition of $\tilde{\mathcal{J}}_\lambda$ given in Eq. (4.4.b) and the fact that $\tilde{\mathcal{J}}_\lambda = 0$ on E_λ then imply that $\mathcal{I}^2 = -\text{id}$ on E_λ. Since $\mathcal{I}$ is self-adjoint, $\mathcal{I}$ is a para-isometry of

E_λ that interchanges the roles of spacelike and timelike vectors. Thus, $\mathcal{I}$ defines an isomorphism between E_λ^+ and E_λ^-.

Let π^+ and π^- be orthogonal projections on $\mathbb{L}^\perp$ and $\mathbb{V}^-$, respectively. Since E_λ^+ contains no timelike vectors and since $\ker(\pi^+) = \mathbb{V}^-$ is timelike, $\ker \pi^+ \cap E_\lambda^+ = \{0\}$ and π^+ is an isomorphism from E_λ^+ to $\pi^+(E_\lambda)$. Similarly, π^- is an isomorphism from E_λ^- to $\pi^-(E_\lambda)$. Thus $\pi^+(E_\lambda)$, which is a positive rank subbundle of $\mathbb{L}^\perp$, is isomorphic to $\pi^-(E_\lambda^-)$, that is a subbundle of $\mathbb{V}^-$. This contradicts Lemma 4.2.2. $\qquad\square$

4.5   The Szabó Operator

Let $\mathfrak{M}_1 = (V, \langle \cdot, \cdot \rangle, A, A_1)$ be a 1-model. The *Szabó operator $\mathcal{S}$* is characterized by the identity:

$$\langle \mathcal{S}(x)y, z \rangle = A_1(y, x, x, z; x) \,.$$

The algebraic curvature tensor A plays no role and one can take $A = 0$ without loss of generality. One says that $\mathfrak{M}_1$ is *Szabó* if the eigenvalues of $\mathcal{S}$ are constant on $S^+(V, \langle \cdot, \cdot \rangle)$ if $q > 0$ or, equivalently by Theorem 1.9.1, on $S^-(V, \langle \cdot, \cdot \rangle)$ if $p > 0$.

We have the following result of Szabó (1991):

Theorem 4.5.1   *Let $\mathfrak{M}_1 = (V, \langle \cdot, \cdot \rangle, A, A_1)$ be a Riemannian Szabó 1-model. Then $A_1 = 0$.*

Szabó used Theorem 4.5.1 to give an elementary proof that any local 2-point homogeneous Riemannian manifold is locally symmetric; of course more is true as (M, g) is either a local rank 1 symmetric space or is flat in this setting. This result motivates the study of this operator in the higher signature setting; a proof is given in Section 4.5.1. In Section 4.5.1, we shall also prove the following results of Gilkey, Ivanova, and Stavrov (2003) and of Gilkey and Stavrov (2002):

Theorem 4.5.2   *Let $\mathfrak{M}_1 = (V, \langle \cdot, \cdot \rangle, A, A_1)$ be a Szabó 1-model. Then:*

(1) $\mathcal{S}$ is nilpotent on the null cone $\mathcal{N}$.
(2) If $\mathfrak{M}_1$ is Lorentzian, then $A_1 = 0$.

Theorem 4.5.1 and Theorem 4.5.2 show that any Riemannian or Lorentzian Szabó manifold is necessarily locally symmetric. In Section 4.5.2, we give examples of Szabó neutral signature pseudo-Riemannian manifolds which are not locally symmetric.

4.5.1   *Szabó 1-models*

Proof of Theorem 4.5.1. Since $\mathcal{S}(Z) = -\mathcal{S}(-Z)$, since $\mathcal{S}(Z)$ is self-adjoint, and since $\mathcal{S}(Z)Z = 0$, we may apply Theorem 4.2.4 to see $\mathcal{S} \equiv 0$. By Lemma 1.8.1, $A_1 = 0$. $\qquad\qquad\square$

Proof of Theorem 4.5.2 (1). Let $\mathfrak{M}_1 = (V, \langle \cdot, \cdot \rangle, A, A_1)$ be a Szabó 1-model. To show that the Szabó operator $\mathcal{S}$ is nilpotent on the null cone $\mathcal{N}$, it suffices to show that $\mathrm{Tr}\{\mathcal{S}(n)^k\} = 0$ for any $n \in \mathcal{N}$.

Let $k \in \mathbb{N}$. Since $\mathfrak{M}_1$ is Szabó, there is a constant c_k so $\mathrm{Tr}\{\mathcal{S}(x)^k\} = c_k$ for $x \in S^+(V, \langle \cdot, \cdot \rangle)$. We rescale to see that

$$\left\{ \mathrm{Tr}\{\mathcal{S}(x)^k\} \right\}^2 = c_k^2 \langle x, x \rangle^{3k}$$

if x is spacelike. Since this polynomial identity holds on an open subset of V, it holds on all V. Thus if n is a null vector, $\mathrm{Tr}\{\mathcal{S}(x)^k\}^2 = 0$ and hence $\mathrm{Tr}\{\mathcal{S}(x)^k\} = 0$ as desired. $\qquad\qquad\square$

Theorem 4.5.2 (2) will follow from the following more general result:

Lemma 4.5.1   *Let $\mathfrak{M}_1 = (V, \langle \cdot, \cdot \rangle, A, A_1)$ be a Lorentzian 1-model. If $\mathrm{Tr}\{\mathcal{S}(\cdot)^2\}$ is constant on $S^-(V, \langle \cdot, \cdot \rangle)$, then $A_1 = 0$.*

Proof.   We apply an argument similar to that used in Section 1.9.3 when we discussed natural operators with bounded spectrum. Choose an orthonormal basis $\mathcal{B} := \{e_0, e_1, ..., e_q\}$ for V, where e_0 is timelike and e_i is spacelike for $i > 0$. Let θ be a real parameter. We define a new orthonormal basis $\mathcal{B}(\theta)$ by:

$$e_0(\theta) := \cosh\theta \cdot e_0 + \sinh\theta \cdot e_1, \ e_1(\theta) := \sinh\theta \cdot e_0 + \cosh\theta \cdot e_1,$$
$$e_i(\theta) := e_i \text{ for } i \geq 2.$$

By assumption, there is a constant C so $C = \mathrm{Tr}\{\mathcal{S}(e_0(\theta))^2\}$. Since $p = 1$, one may express:

$$C = \sum_{i=1}^{q} \sum_{j=1}^{q} A_1(e_i(\theta), e_0(\theta), e_0(\theta), e_j(\theta); e_0(\theta))^2.$$

As $\cosh\theta = \frac{1}{2}(e^\theta + e^{-\theta})$ and $\sinh\theta = \frac{1}{2}(e^\theta - e^{-\theta})$, we may expand:

$$A_1(e_i(\theta), e_0(\theta), e_0(\theta), e_j(\theta); e_0(\theta)) = \sum_{\nu=-5}^{5} a_{ij,\nu} e^{\nu\theta},$$

$$C = \sum_{i=1}^{q}\sum_{j=1}^{q}\{a_{ij,5}\}^2 e^{10\theta} + O(e^{9\theta}),$$

$$Ce^{-10\theta} = \sum_{i=1}^{q}\sum_{j=1}^{q}\{a_{ij,5}\}^2 + O(e^{-\theta}).$$

We take the limit as $\theta \to \infty$ to see $\sum_{ij}\{a_{ij,5}\}^2 = 0$ and hence $a_{ij,5} = 0$ for all i, j. An analogous argument shows that $a_{ij,-5} = 0$. We proceed in this fashion to show that $a_{ij,\nu} = 0$ for $\nu \neq 0$. Consequently

$$A_1(e_2, e_0(\theta), e_0(\theta), e_2; e_0(\theta)) = a_{22,0}$$

is independent of θ. On the other hand, as there are three terms involving θ, odd powers of e^θ appear in this expression. Thus $a_{22,0} = 0$ so

$$A_1(e_2, e_0, e_0, e_2; e_0) = 0.$$

Similarly we conclude $A_1(e_i, e_0, e_0, e_i; e_0) = 0$ for any $i \geq 1$. We polarize to see $A_1(e_i, e_0, e_0, e_j; e_0) = 0$ for any i, j; the vanishing being automatic if $i = 0$ or $j = 0$. Thus $\mathcal{S}(e_0) = 0$. As e_0 was arbitrary, $\mathcal{S}(\cdot) = 0$ on $S^-(V, \langle\cdot,\cdot\rangle)$. Rescaling and analytic continuation then imply $\mathcal{S}(\cdot) = 0$ on V. By Lemma 1.8.1, $A_1 = 0$. $\qquad\square$

4.5.2 *Balanced Szabó pseudo-Riemannian manifolds*

We recall the notation of Definition 2.5.1:

Definition 4.5.1 Introduce coordinates $\{x_1, ..., x_p, \tilde{x}_1, ..., \tilde{x}_p\}$ on $\mathbb{R}^{2p}$. Let indices i, j, k range from 1 through p. Let $\psi_{ij} = \psi_{ji}$ be a symmetric 2-tensor field where $\psi_{ij} = \psi_{ij}(x_1, ..., x_p)$ only depends on the first p coordinates. Let $\mathcal{M} := (\mathbb{R}^{2p}, g)$ be the pseudo-Riemannian metric g of signature (p, p) where

$$g(\partial_{x_i}, \partial_{x_j}) := \psi_{ij}(x_1, ..., x_p) \quad \text{and} \quad g(\partial_{x_i}, \partial_{\tilde{x}_j}) := \delta_{ij}.$$

By Lemma 2.5.1, this manifold is a generalized plane wave manifold. Furthermore, the non-zero components of ∇R are given by:

$$\nabla R(\partial_{x_i}, \partial_{x_j}, \partial_{x_k}, \partial_{x_l}; \partial_{x_n})$$
$$= -\tfrac{1}{2}\partial_{x_n}(\psi_{il/jk} + \psi_{jk/il} - \psi_{ik/jl} - \psi_{jl/ik}).$$

(4.5.a)

Theorem 4.5.3 *Let $\mathcal{M}$ be as in Definition 4.5.1. Then $\mathcal{M}$ is Szabó. Furthermore, if $\mathcal{M}$ is not symmetric, then $\mathcal{M}$ is neither spacelike Jordan Szabó nor timelike Jordan Szabó.*

Proof.   Let $\mathcal{X} := \operatorname{Span}\{\partial_{x_i}\}$ and let $\tilde{\mathcal{X}} := \operatorname{Span}\{\partial_{\tilde{x}_i}\}$. Decompose a tangent vector $\xi \in T\mathbb{R}^m$ in the form $\xi = \xi_X + \xi_{\tilde{X}}$ where $\xi_X \in \mathcal{X}$ and $\xi_{\tilde{X}} \in \tilde{\mathcal{X}}$. By Eq. (4.5.a),

$$\mathcal{S}(\xi) = \mathcal{S}(\xi_X) : \mathcal{X} \to \tilde{\mathcal{X}} \to 0.$$

Thus $\mathcal{S}$ is nilpotent so 0 is the only eigenvalue of $\mathcal{S}$ and $\mathcal{M}$ is Szabó.

Let $\mathfrak{M}_1(\mathcal{M}, P) := (T_P\mathbb{R}^m, g_P, R_P, \nabla R_P)$ be the 1-model of $\mathcal{M}$ at a point P of $\mathbb{R}^{2p}$. Suppose that $\mathfrak{M}_1$ is spacelike Jordan Szabó. We will show this implies that $\nabla R_P = 0$. A similar argument can be used to prove that if $\mathfrak{M}_1$ is timelike Jordan Szabó, then $\nabla R_P = 0$. This will complete the proof of the Theorem.

If $\mathfrak{M}_1(\mathcal{M}, P)$ is spacelike Jordan Szabó, then $\operatorname{Rank}\{\mathcal{S}(\cdot)\} = r > 0$ is constant on $S^+(T_PM)$. Let $\mathcal{V}^+$ be any maximal spacelike subspace of T_PM and let $\mathcal{V}^- := (\mathcal{V}^+)^\perp$ be the complementary timelike subspace. Let $\rho^\pm$ be orthogonal projection on $\mathcal{V}^\pm$. If $Z \in S^+(\mathcal{V}^+)$, then we define:

$$\check{\mathcal{S}}(Z) := \rho^+ \mathcal{S}(Z)\rho^+.$$

We wish to show that $\operatorname{Rank}\{\check{\mathcal{S}}(Z)\} = r$. Let $\{Z_1, ..., Z_r\}$ be tangent vectors at P so $\{\mathcal{S}(Z)Z_1, ..., \mathcal{S}(Z)Z_r\}$ is a basis for $\operatorname{Range}(\mathcal{S}(Z))$. Use the decomposition $T_PM = \mathcal{V}^+ + \tilde{\mathcal{X}}$ to decompose $Z_i = V_i^+ + \tilde{X}_i$, where $V_i^+ \in \mathcal{V}^+$ and $\tilde{X}_i \in \tilde{\mathcal{X}}$. Since $\tilde{\mathcal{X}} \subset \ker \mathcal{S}(Z)$, $\mathcal{S}(Z)Z_i = \mathcal{S}(Z)V_i^+$ and thus $\{\mathcal{S}(Z)V_1^+, ..., \mathcal{S}(Z)V_r^+\}$ is a basis for $\operatorname{Range}(\mathcal{S}(Z))$. As $\ker \rho^+ = \mathcal{V}^-$ is timelike, as $\tilde{\mathcal{X}}$ is totally isotropic, and as $\operatorname{Range}(\mathcal{S}(Z)) \subset \tilde{\mathcal{X}}$, the vectors

$$\{\rho^+ \mathcal{S}(Z)\rho^+ V_1^+, ..., \rho^+ \mathcal{S}(Z)\rho^+ V_r^+\}$$

are linearly independent. Consequently, $\operatorname{Rank}\{\check{\mathcal{S}}(Z)\} \geq r$. Since the reverse inequality is immediate, we have as desired that

$$\operatorname{Rank}\{\check{\mathcal{S}}(Z)\} = r \quad \text{for} \quad Z \in S^{p-1} := S^+(\mathcal{V}^+).$$

Since $\mathcal{S}$ is self-adjoint and ρ^+ is self-adjoint, $\check{\mathcal{S}}(Z)$ is a self-adjoint map of $\mathcal{V}^+$. $\check{\mathcal{S}}(-Z) = -\check{\mathcal{S}}(Z)$, and $\check{\mathcal{S}}(Z)Z = 0$. Thus by Theorem 4.2.4, $\check{\mathcal{S}} = 0$. This shows that $r = 0$. Hence, $\mathcal{S}(\cdot)$ vanishes identically on $S^+(T_PM)$. Thus by analytic continuation, $\mathcal{S} = 0$. Lemma 1.8.1 now shows $\nabla R_P = 0$ as desired. $\qquad\Box$

4.6   Conformal Geometry

In Section 4.6, we discuss the spectral geometry of the conformal Jacobi operator and of the conformal skew-symmetric curvature operator. Section 4.6.1 contains a brief review of conformal geometry. In Section 4.6.2, we discuss the spectral geometry of the conformal Jacobi operator. In Section 4.6.3, we discuss 4-dimensional conformal Osserman manifolds. In Section 4.6.4, we discuss the spectral geometry of the conformal curvature operator.

4.6.1   *The Weyl model*

Let $\mathfrak{M} = (V, \langle \cdot, \cdot \rangle, A)$ be a 0-model, let $\{e_i\}$ be an orthonormal basis for V, and let $\varepsilon_i := \langle e_i, e_i \rangle$. The Ricci tensor $\rho = \rho_A$, the scalar curvature $\tau = \tau_A$, and the Weyl conformal curvature tensor $W = W_A$ are given by:

$$\rho(x,y) := \sum_{i=1}^{m} \varepsilon_i A(x, e_i, e_i, y), \quad \tau := \sum_{i=1}^{m} \varepsilon_i \rho(e_i, e_i),$$

$$\begin{aligned} W(x,y,z,w) := \; & A(x,y,z,w) - \tfrac{1}{m-2}\{\rho(x,w)\langle y,z\rangle + \langle x,w\rangle\rho(y,z)\} \\ & + \tfrac{1}{m-2}\{\rho(x,z)\langle y,w\rangle + \langle x,z\rangle\rho(y,w)\} \\ & + \tfrac{1}{(m-1)(m-2)}\tau\{\langle x,w\rangle\langle y,z\rangle - \langle x,z\rangle\langle y,w\rangle\}. \end{aligned}$$

Note that $\rho_W = 0$. Since W is an algebraic curvature tensor, we can consider the associated *Weyl 0-model*

$$\mathfrak{M}_W := (V, \langle \cdot, \cdot \rangle, W).$$

We say that $\mathfrak{M}$ is conformally Osserman if and only if $\mathfrak{M}_W$ is Osserman; the other conformal properties are defined analogously. Note that by definition, $\mathfrak{M}_W$ is Ricci flat.

If $\mathcal{M}$ is a pseudo-Riemannian manifold, let

$$\mathfrak{M}_W(\mathcal{M}, P) := (T_PM, g_P, W_P)$$

be the associated Weyl model. We say that $\mathcal{M}$ is *conformally Osserman* if $\mathfrak{M}_W$ is Osserman for every point $P \in M$; the notions *conformally spacelike/timelike Jordan Osserman*, *conformally Ivanov–Petrova*, and *conformally spacelike/timelike/mixed Jordan Ivanov–Petrova* are defined similarly. These are conformal notions by Theorem 1.9.9; if $\tilde{\mathcal{M}} = (M, e^f g)$ is conformally equivalent to $\mathcal{M}$, then $\mathcal{M}$ has one of these properties if and only if $\tilde{\mathcal{M}}$ has the corresponding property.

4.6.2 *Conformally Jordan Osserman* 0-*models*

The following Theorem of Blažić, Gilkey, Nikčević, and Simon (2005a) deals with conformal geometry.

Theorem 4.6.1 *If $\mathfrak{M}$ is an Einstein* 0-*model, then $\mathfrak{M}$ is conformally Osserman (respectively conformally spacelike Jordan Osserman or conformally timelike Jordan Osserman) if and only if $\mathfrak{M}$ is Osserman (respectively spacelike Jordan Osserman or timelike Jordan Osserman).*

Proof. If $\mathfrak{M}$ is Einstein, then there is a scalar $\lambda = \lambda(m, \tau)$ so that

$$W(x, y, z, w) = A(x, y, z, w) + \lambda\{\langle x, w\rangle\langle y, z\rangle - \langle x, z\rangle\langle y, w\rangle\},$$

$$\mathcal{J}_W(x)y = \begin{cases} 0 & \text{if} \quad y = x, \\ \{\mathcal{J}_A(x) + \lambda\langle x, x\rangle \operatorname{Id}\}y & \text{if} \quad y \perp x. \end{cases}$$

Thus apart from the trivial eigenvalue 0, the Jordan normal form of $\mathcal{J}_W(x)$ and $\mathcal{J}_A(x)$ are simply shifted by adding a scalar multiple of the identity if x is not a null vector. Theorem 4.6.1 is now immediate. $\square$

The classification is complete in certain settings:

Theorem 4.6.2 *Let $\mathfrak{M} = (V, \langle \cdot, \cdot \rangle, A)$ be a* 0-*model which either is odd-dimensional and Riemannian or Lorentzian. Then $\mathfrak{M}$ is conformally spacelike Jordan Osserman if and only if $\mathfrak{M}$ is conformally flat.*

Proof. If the associated Weyl model $\mathfrak{M}_W = (V, \langle \cdot, \cdot \rangle, W)$ is Osserman and if m is odd, work of Chi (1988) shows that $\mathfrak{M}_W$ has constant sectional curvature λ. Similarly if $\mathfrak{M}_W$ is Lorentzian, then $\mathfrak{M}_W$ has constant sectional curvature λ by Blažić, Bokan, and Gilkey (1997a) and García-Río, Kupeli, and Vázquez-Abal (1997). Since $\mathfrak{M}_W$ has vanishing Ricci tensor, $\lambda = 0$ and $W = 0$. This implies $\mathfrak{M}$ is conformally flat. $\square$

The classification is almost complete in Riemannian setting if $m \equiv 2$ mod 4 as well. We refer to Blažić and Gilkey (2004) for the proof of the following result and omit details in the interests of brevity:

Theorem 4.6.3 *Let (M, g) be a conformally Osserman Riemannian manifold of dimension $m = 4k + 2 \geq 10$. Let P be a point of M where $W_P \neq 0$. Then there is an open neighborhood of P in M which is conformally equivalent to an open subset of either complex projective space or its negative curvature dual.*

Any local rank 1 Riemannian symmetric space is necessarily conformally Osserman since the group of local isometries acts transitively on the unit sphere bundle. Based upon Theorems 4.6.2 and 4.6.3, we conjecture that the converse holds; this is a corresponding counterpart to the Osserman conjecture in this setting:

Conjecture 4.6.1 A connected Riemannian manifold $\mathcal{M}$ is conformally Osserman if and only if $\mathcal{M}$ is locally conformally equivalent to a rank 1 symmetric space.

Theorem 2.5.1 and Theorem 2.7.3 show Conjecture 4.6.1 fails in the higher signature setting.

4.6.3 *Conformally Osserman 4-dimensional manifolds*

One can say a bit more in the 4-dimensional setting. We follow the discussion in Blažić and Gilkey (2005).

Theorem 4.6.4 *Let $\mathfrak{M}$ be a 4-dimensional Riemannian 0-model. The following conditions are equivalent:*

(1) $\mathfrak{M}$ is conformally Osserman.
(2) $\mathfrak{M}$ is self-dual or anti self-dual.

Let Φ be a skew-symmetric endomorphism of V with $\Phi^2 = -1$. Following Eq. (1.3.a), define an algebraic curvature tensor A_Φ with associated curvature operator

$$\mathcal{A}_\Phi(x, y)z := \langle \Phi y, z \rangle \Phi x - \langle \Phi x, z \rangle \Phi y - 2\langle \Phi x, y \rangle \Phi z .$$

We say that $\{\Phi_1, \Phi_2, \Phi_3\}$ is a *unitary quaternion* structure on V if the Φ_i are skew-adjoint and if the usual structure equations are satisfied:

$$\Phi_i \Phi_j + \Phi_j \Phi_i = -2\delta_{ij}\,\mathrm{id} \quad \text{and} \quad \Phi_1 \Phi_2 \Phi_3 = -\,\mathrm{id} .$$

Theorem 4.6.3 is a consequence of the following purely algebraic fact:

Lemma 4.6.1   *Let $\mathfrak{M} = (\mathbb{R}^4, \langle \cdot, \cdot \rangle, A)$ be a Riemannian 0-model. The following assertions are equivalent:*

(1) A is conformally Osserman.
(2) A is self-dual or anti self-dual.
(3) There exists a unitary quaternion structure on $\mathbb{R}^4$ so that
   $W = \lambda_1 A_{\Phi_1} + \lambda_2 A_{\Phi_2} + \lambda_3 A_{\Phi_3}$ *where* $\lambda_1 + \lambda_2 + \lambda_3 = 0$.

Proof.   We begin by showing that Assertion (1) implies Assertion (2) in Lemma 4.6.1. Suppose $\mathfrak{M}$ is conformally Osserman. Zero is always an eigenvalue of $\mathcal{J}_W$ since $\mathcal{J}_W(x)x = 0$. Let e_1 be a unit vector. Since $\mathcal{J}_W(\cdot)$ is symmetric, it has an orthonormal basis of real eigenvectors. Thus we may extend e_1 to an orthonormal basis $\{e_1, e_2, e_3, e_4\}$ so that

$$\mathcal{J}_W(e_1)e_2 = ae_2, \quad \mathcal{J}_W(e_1)e_3 = be_3, \quad \mathcal{J}_W(e_1)e_4 = ce_4.$$

Since $\operatorname{Tr}(\mathcal{J}_W) = 0$, we have $a + b + c = 0$.

The argument given by Chi (1988) in his analysis of the 4-dimensional setting was in part purely algebraic. This algebraic argument extends without change to this setting to show that, after possibly replacing $\{e_2, e_3, e_4\}$ by $\{-e_2, -e_3, -e_4\}$ that the non-vanishing components of the Weyl curvature on this basis are given by:

$$\begin{aligned}
W_{1221} &= W_{3443} = -W_{1234} = a, \\
W_{1331} &= W_{2442} = -W_{1342} = b, \\
W_{1441} &= W_{2332} = -W_{1423} = c.
\end{aligned} \qquad (4.6.\text{a})$$

Let $e^{ij} := e^i \wedge e^j$ where $\{e^i\}$ is the dual basis for $(\mathbb{R}^4)^*$. We consider the following bases for $\Lambda_2^{\pm}(\mathbb{R}^4)$:

$$f_1^{\pm} = e^{12} \pm e^{34}, \ f_2^{\pm} = e^{13} \mp e^{24}, \ f_3^{\pm} = e^{14} \pm e^{23}.$$

Since $\mathcal{W}(e^{pq}) = \frac{1}{2} W_{pqij} e^{ij}$,

$$\begin{aligned}
\mathcal{W}(f_1^-) &= 0, & \mathcal{W}(f_2^-) &= 0, & \mathcal{W}(f_3^-) &= 0, \\
\mathcal{W}(f_1^+) &= -2a f_1^+, & \mathcal{W}(f_2^+) &= -2b f_2^+, & \mathcal{W}(f_3^+) &= -2c f_3^+.
\end{aligned}$$

Thus conformally Osserman algebraic curvature tensors are self-dual.

Next we show that Assertion (2) implies Assertion (3) in Lemma 4.6.1. Suppose that A is a self-dual algebraic curvature tensor on $\mathbb{R}^4$. Let e_1 be a unit vector. Choose an orthonormal basis $\{e_1, e_2, e_3, e_4\}$ for $\mathbb{R}^4$ so that

$$\mathcal{J}_W(e_1)e_2 = ae_2, \quad \mathcal{J}_W(e_1)e_3 = be_3, \quad \mathcal{J}_W(e_1)e_4 = ce_4.$$

We then have

$$W_{1221} = a, \; W_{1231} = 0, \; W_{1241} = 0,$$
$$W_{1321} = 0, \; W_{1331} = b, \; W_{1341} = 0, \qquad\qquad (4.6.b)$$
$$W_{1421} = 0, \; W_{1431} = 0, \; W_{1441} = c.$$

By replacing $\{e_2, e_3, e_4\}$ by $\{-e_2, -e_3, -e_4\}$ if necessary, we can assume that $\{e_1, e_2, e_3, e_4\}$ is an oriented orthonormal basis. We have

$$\begin{aligned}
\mathcal{W}(e^{12} - e^{34}) = {} & (W_{1212} - W_{3412})e^{12} + (W_{1234} - W_{3434})e^{34} \\
& + (W_{1213} - W_{3413})e^{13} + (W_{1214} - W_{3414})e^{14} \\
& + (W_{1223} - W_{3423})e^{23} + (W_{1224} - W_{3424})e^{24}.
\end{aligned}$$

Since W is self-dual, $\mathcal{W}(e^{12} - e^{34}) = 0$. One may use Eq. (4.6.b) to see that

$$W_{3412} = W_{1212} = -a, \; W_{3434} = W_{1234} = -a, \; W_{3413} = W_{1213} = 0,$$
$$W_{3414} = W_{1214} = 0, \quad W_{3423} = W_{1223} = 0, \quad W_{3424} = W_{1224} = 0.$$

We argue similarly using $e^{13} + e^{24}$ and $e^{14} - e^{23}$ to see that the formulas of Eq. (4.6.a) hold. We define a unitary quaternion structure by defining $\Phi_3 := \Phi_1 \Phi_2$ where

$$\Phi_1 : e_1 \to e_2, \; \Phi_1 : e_2 \to -e_1, \; \Phi_1 : e_3 \to e_4, \quad \Phi_1 : e_4 \to -e_3,$$
$$\Phi_2 : e_1 \to e_3, \; \Phi_2 : e_3 \to -e_1, \; \Phi_2 : e_4 \to -e_2, \; \Phi_2 : e_2 \to e_4.$$

It is then immediate that the formulas of Eq. (4.6.a) hold for

$$\tilde{W} := a W_{\Phi_1} + b W_{\Phi_2} + c W_{\Phi_3}$$

and thus $W = \tilde{W}$. Thus Assertion (2) of the Lemma implies Assertion (3).

Finally, if W is given by a unitary quaternion structure, then the discussion of Gilkey (1994) shows that W is Osserman. $\qquad\square$

4.6.4 *Conformally Jordan Ivanov–Petrova 0-models*

The classification of Riemannian conformally Ivanov–Petrova 0-models is complete if $m \neq 3$.

Theorem 4.6.5 *Let $\mathfrak{M} = (V, \langle \cdot, \cdot \rangle, A)$ be a conformally Ivanov–Petrova Riemannian 0-model of dimension $m \neq 3$. Then $\mathfrak{M}$ is conformally flat.*

Proof. If $m = 2$, then $\mathfrak{M}$ is conformally flat. Suppose first $m \geq 5$. We apply Theorem 1.9.11 to see that there exists a self-adjoint isometry ϕ of V with $\phi^2 = \mathrm{Id}$ so that $W = \lambda A_\phi$ where A_ϕ is the canonical curvature tensor given in Eq. (1.3.b) and whose associated curvature operator and Jacobi operators are:

$$\mathcal{A}_\phi(x,y)z := \langle \phi y, z \rangle \phi x - \langle \phi x, z \rangle \phi y,$$
$$\mathcal{J}_\phi(x)y = \langle \phi x, x \rangle \phi y - \langle \phi x, y \rangle \phi x. \tag{4.6.c}$$

Decompose $V = V^+ \oplus V^-$ into the ± 1 eigenspaces of ϕ. Since W has constant sectional curvature if $\phi = \pm \operatorname{Id}$, the argument given to establish Theorem 4.6.2 shows $W = 0$ if $\phi = \pm \operatorname{Id}$. Thus we may assume that $a^\pm := \dim\{V^\pm\} \geq 1$. Let $e^\pm \in S(V^\pm)$ and let $x \in S(V)$. One has:

$$\mathcal{J}_W(x)y = \lambda \langle \phi x, x \rangle \phi y \quad \text{if} \quad y \perp \phi x,$$
$$\operatorname{Tr}\{\mathcal{J}_W(e^+)\} = \lambda(a^+ - 1 - a^-), \tag{4.6.d}$$
$$\operatorname{Tr}\{\mathcal{J}_W(e^-)\} = \lambda(a^- - 1 - a^+).$$

As W is Ricci flat, $\operatorname{Tr}\{\mathcal{J}_W(x)\} = 0$ for any $x \in V$. Thus by Eq. (4.6.d),

$$(a^+ - a^- - 1)\lambda = 0 \quad \text{and} \quad (a^- - a^+ - 1)\lambda = 0.$$

Adding these two equations implies $-2\lambda = 0$ and hence $W = 0$. This establishes the Lemma except when $m = 4$.

We complete the proof of the Lemma by dealing with the exceptional case $m = 4$. We follow the discussion in Ivanov and Petrova (1998) to see that either W has the form given in Eq. (4.6.c), in which case the argument given above shows $W = 0$, or that there exists an orthonormal basis $\{e_1, e_2, e_3, e_4\}$ for V so that the non-zero components of W are:

$$W_{1212} = a_1, \ W_{1234} = \ \ a_2, \ W_{1313} = a_2, \ W_{1324} = -a_1,$$
$$W_{1414} = a_2, \ W_{1423} = \ \ a_1, \ W_{2323} = a_2, \ W_{2314} = \ \ a_1,$$
$$W_{2424} = a_2, \ W_{2413} = -a_1, \ W_{3434} = a_1, \ W_{3412} = \ \ a_2,$$

where $a_2 + 2a_1 = 0$. Since W is Ricci flat, $\rho_W(e_1, e_1) = -2a_2 - a_1 = 0$. Consequently $a_1 = a_2 = 0$, which once again implies $W = 0$. $\qquad\square$

There are analogous results in the higher signature setting, although with slightly more restrictive hypotheses.

Theorem 4.6.6   *Let $\mathfrak{M}$ be a 0-model of signature (p,q) which is conformally spacelike Jordan Ivanov–Petrova. Assume that one of the following conditions holds:*

(1) $p = 1$ and $q \geq 9$.
(2) $p = 2$, $q \geq 10$, and neither q nor $q + 2$ are powers of 2.
(3) $3 \leq p \leq \frac{1}{4}q - 6$ and $\{q, q+1, ..., q+p\}$ does not contain a power of 2.

Then either $\mathfrak{M}$ *is conformally flat or the skew-symmetric curvature operator* $\mathcal{A}$ *is nilpotent.*

Proof.   Theorem 1.9.15 shows that $\mathrm{Rank}\{\mathcal{A}(\pi)\} = 2$ for any spacelike 2-plane π. By Theorem 1.9.13, there exists a normalizing constant λ so that $W = \lambda A_\phi$ where A_ϕ is given by Eq. (4.6.c) and where one of the following conditions holds:

(1) $\phi^2 = \mathrm{Id}$ and ϕ is a self-adjoint isometry of V.
(2) $\phi^2 = -\mathrm{Id}$ and ϕ is a self-adjoint para-isometry of V.
(3) $\phi^2 = 0$.

If $\phi^2 = 0$, then $W(\pi)$ is always nilpotent. We complete the proof by showing that either (1) or (2) imply $\lambda = 0$.

Suppose ϕ is a self-adjoint isometry of V with $\phi^2 = \mathrm{Id}$. As in the proof of Theorem 4.6.5, we decompose $V = V^+ \oplus V^-$ into the ± 1 eigenspaces of ϕ. Again, set $a^\pm = \dim\{V^\pm\}$ where $a^+ \geq 1$ and $a^- \geq 1$. These eigenspaces are orthogonal with respect to the metric $\langle \cdot, \cdot \rangle$ and thus the restriction of the metric to each eigenspace is non-degenerate. Choose unit vectors $e^\pm$ in $V^\pm$ and let $\varepsilon^\pm := \langle e^\pm, e^\pm \rangle \neq 0$. One may extend Eq. (4.6.d) to have

$$\mathrm{Tr}\{\mathcal{J}_W(e^+)\} = \varepsilon^+ \lambda(a^+ - 1 - a^-), \text{ and}$$
$$\mathrm{Tr}\{\mathcal{J}_W(e^-)\} = \varepsilon^- \lambda(a^- - 1 - a^+).$$

We argue as in the proof of Theorem 4.6.5 to see that this implies $\lambda = 0$.

If ϕ is a para-isometry, we complexify. Replacing ϕ by $\tilde{\phi} := \sqrt{-1}\phi$ and applying the argument given above to the self-adjoint (complex) isometry $\tilde{\phi}$ to see that $\sqrt{-1}\lambda = 0$ and thus, again, $W = 0$. $\square$

Remark 4.6.1   Examples of such manifolds are given in Theorem 2.5.1. On the other hand, the manifolds of Theorem 2.7.3 are Ricci flat and provide examples of conformally Ivanova–Petrova manifolds which are not of this type.

4.7   Stanilov Models

We follow the notation introduced in Section 1.8.3. Let $\mathfrak{M} = (V, \langle \cdot, \cdot \rangle, A)$ be a 0-model of signature (p, q). Let π be a spacelike or a timelike k-plane. Let $\{e_1, ..., e_k\}$ be an orthonormal basis for π. The *Stanilov operator* is defined by setting:

$$\Theta(\pi) := \sum_{i,j} \mathcal{A}(e_i, e_j)\mathcal{A}(e_i, e_j)\,.$$

Let $1 < k < p+q$ and let $k \le q$. We say that $\mathfrak{M}$ is a spacelike k-*Stanilov* 0-model if the eigenvalues of $\mathfrak{M}$ are constant on the Grassmannian of spacelike k-planes in V. The notions timelike k-Stanilov, spacelike Jordan k-Stanilov, and timelike k-Jordan Stanilov are defined similarly.

One says that a pseudo-Riemannian manifold $\mathcal{M}$ has one of these properties if $\mathfrak{M}(\mathcal{M}, P)$ has the appropriate property for all $P \in M$. Examples of Stanilov manifolds were given previously in Theorem 2.5.1 and in Theorem 2.7.3. In this section, we follow the treatment in Gilkey, Nikčević, and Videv (2004) and present some theoretical results concerning these manifolds. We begin by showing that Ivanov–Petrova models are k-Stanilov under certain conditions.

Theorem 4.7.1   *Let $\mathfrak{M}$ be a spacelike Jordan Ivanov-Petrova model of signature (p, q). Assume that one of the following conditions holds:*

(1) $p = 0$ and $q \ge 5$.
(2) $p = 1$ and $q \ge 9$.
(3) $p = 2$, $q \ge 10$, and neither q nor $q + 2$ are powers of 2.
(4) $3 \le p \le \frac{1}{4}q - 6$ and $\{q, q+1, ..., q+p\}$ does not contain a power of 2.

Let $2 \le k \le p+q-1$. If $k \le q$, then $\mathfrak{M}$ is k-spacelike Jordan Stanilov. If $k \le p$, then $\mathfrak{M}$ is k-timelike Jordan Stanilov.

Proof.   The hypotheses of the Theorem permit the use of Theorem 1.9.13 to see that if π is a spacelike 2-plane, then $\mathrm{Rank}\{\mathcal{A}(\pi)\} = 2$. Thus by Theorem 1.9.15, there exists a self-adjoint map ϕ so that $A = cA_\phi$, where A_ϕ is as described in Section 1.6.3, and where $\phi^2 = \mathrm{Id}$, $\phi^2 = -\mathrm{Id}$, or $\phi^2 = 0$. If $\phi^2 = 0$, then $\mathcal{A}(\pi)^2 = 0$ and hence $\Theta = 0$. Thus we suppose $\phi^2 = \varepsilon\,\mathrm{Id}$. If $\{e_1, e_2\}$ is an orthonormal basis for an oriented spacelike 2-plane π, then

$$\mathcal{A}(e_1, e_2)x = c\{\langle \phi e_2, x\rangle \phi e_1 - \langle \phi e_1, x\rangle \phi e_2\}\,.$$

Let ρ_π be orthogonal projection on π and let $x \perp \phi\pi$. Then

$$\mathcal{A}(\pi) : \phi e_2 \to c\varepsilon \phi e_1, \quad \mathcal{A}(\pi) : \phi e_1 \to -c\varepsilon \phi e_2, \quad \mathcal{A}(\pi) : x \to 0,$$
$$\mathcal{A}(\pi)^2 : \phi e_2 \to -c^2 \phi e_2, \ \mathcal{A}(\pi)^2 : \phi e_1 \to -c^2 \phi e_1, \ \mathcal{A}(\pi)^2 : x \to 0\,.$$
$$\mathcal{A}(\pi)^2 = -c^2 \rho_{\phi\pi}\,.$$

If π_k is a spacelike k-plane, then

$$\Theta(\pi_k) = -2(k-1)c^2\rho_{\phi\pi_k}\,.$$

This shows $\Theta(\pi_k)$ has constant Jordan normal form and hence $\mathfrak{M}$ is k-spacelike Jordan Stanilov. The argument that $\mathfrak{M}$ is k-timelike Jordan Stanilov is essentially the same modulo an appropriate change of signs and thus is omitted. $\qquad\square$

The situation is particularly simple if $k = 2$:

Theorem 4.7.2 *Any 2-Stanilov 0-model is Ivanov–Petrova.*

Proof. There is nothing to prove if $m = 2$ so we assume $m \geq 3$. Let $\mathfrak{M}$ be 2-Stanilov. Let $\{\lambda_i(\pi)\}$ be the eigenvalues of $\mathcal{A}(\pi)$ for an oriented 2-plane π. Then $\{\lambda_i^2(\pi)\}$ are the eigenvalues of $\Theta(\pi)$. Since these eigenvalues are independent of π, since the Grassmannian of oriented 2-planes is connected, and since the eigenvalues vary continuously, we may conclude that $\mathcal{A}(\pi)$ also has constant eigenvalues so $\mathfrak{M}$ is Ivanov–Petrova. $\qquad\square$

4.8 Complex Geometry

We adopt the notation of Section 1.6.4. Let $\mathfrak{M} := \{V, \langle\cdot,\cdot\rangle, J, A\}$ be a Riemannian complex 0-model. This means that $(V, \langle\cdot,\cdot\rangle, A)$ is a Riemannian 0-model and that J is a unitary almost complex structure on V. We assume that A and J are compatible; this means, by Lemma 1.6.6, that

$$JA(x, Jx) = A(x, Jx)J \quad \text{for all} \quad x \in V\,.$$

Thus this operator is complex linear and self-adjoint.

We say that $\mathfrak{M}$ is *complex Ivanov–Petrova* if the eigenvalues of the operator $JA(\cdot)$ are constant on the associated complex projective space or, equivalently, if the eigenvalues of the operator $JA(x, Jx)$ are constant on $S(V)$. As this operator is self-adjoint, the eigenvalues are necessarily real. Let $\{\lambda_i, \mu_i\}$ be the eigenvalues and multiplicities of $JA(\pi)$ where we order the multiplicities so $\mu_0 \geq ... \geq \mu_\ell > 0$. Since $JA(\pi)$ is J-linear, the μ_i are even.

Theorem 4.8.1 *Let $\mathfrak{M} = (V, \langle\cdot,\cdot\rangle, J, A)$ be a complex Ivanov–Petrova Riemannian 0-model with eigenvalue structure $\{\lambda_i, \mu_i\}$. Assume $\ell > 0$. If $m \equiv 2 \bmod 4$, then $\ell = 1$ and $\mu_1 = 2$. If $m \equiv 0 \bmod 4$, then either $\ell = 1$ and $\mu_1 = 2, 4$ or $\ell = 2$ and $\mu_1 = \mu_2 = 2$.*

Proof. We use methods of algebraic topology and follow the discussion in Gilkey (2001a) to prove Assertion (1). Let $m = 2n$. Choose a unitary basis for V to identify $V = \mathbb{C}^n$ and to identify $\langle \cdot, \cdot \rangle$ with the standard Euclidean inner product. Since A and J are compatible, the operator $J\mathcal{A}(x, Jx)$ is both self-adjoint and complex. Let

$$E_\lambda(x) := \{\xi : J\mathcal{A}(x, JX)\xi = \lambda\xi\}$$

be the associated eigenspaces. Since $\dim E_\lambda(x)$ is constant on S^{m-1}, the $E_\lambda(x)$ patch together to define smooth real vector bundles over S^{m-1}. They are in fact complex vector bundles which are well defined over $\mathbb{CP}^{n-1}$. They define a decomposition of

$$\mathbb{1}^n = \oplus_\lambda E_\lambda \quad \text{over} \quad \mathbb{CP}^{n-1}.$$

The desired conclusion will follow from Theorem 4.2.5 if we can show that any of the E_λ is geometrically symmetric.

Let λ be the maximal real eigenvalue. Let $\xi \in E_\lambda(x)$ be a unit eigenvector. We then have $\lambda\xi = J\mathcal{A}(x, JX)\xi$ and thus

$$\begin{aligned}
\lambda &= \langle J\mathcal{A}(x, Jx)\xi, \xi \rangle = -\langle \mathcal{A}(x, Jx)\xi, J\xi \rangle \\
&= A(x, Jx, J\xi, \xi) = A(\xi, J\xi, Jx, x) \\
&= \langle J\mathcal{A}(\xi, J\xi)x, x \rangle.
\end{aligned}$$

Since λ is maximal, we may apply Lemma 1.5.1 to conclude that

$$\mathcal{A}(\xi, J\xi)x = \lambda x.$$

Consequently, $x \in E_\lambda(\xi)$. This shows that E_λ is a geometrically symmetric vector bundle as desired. $\square$

These results are sharp as the following examples show. Let ϕ be a skew-adjoint linear map. Let the associated skew-symmetric bilinear form $\Phi \in \Lambda^2(V^*)$ be defined by $\Phi(v, w) := \langle \phi v, w \rangle$. Use Eq. (1.3.a) to define:

$$\begin{aligned}
A_\phi(x, y, z, w) &:= \Phi(x, w)\Phi(y, z) - \Phi(x, z)\Phi(y, w) \\
&\quad -2\Phi(x, y)\Phi(z, w).
\end{aligned}$$

This is an algebraic curvature tensor by Lemma 1.6.3; by Eq. (1.3.b), the associated curvature operator is given by:

$$\mathcal{A}_\phi(x, y)z := \langle \phi y, z \rangle \phi x - \langle \phi x, z \rangle \phi y - 2\langle \phi x, y \rangle \phi z.$$

Example 4.8.1 Let $V = \mathbb{C}^n$ with the usual Hermitian almost complex structure J. Let $A = A_J$ and let $\mathfrak{M} := \{\mathbb{R}^{2n}, \langle \cdot, \cdot \rangle, J, A_J\}$. Let $x \in S(V)$. Then:

$$\mathcal{A}(x, Jx)z = \langle JJx, z \rangle Jx - \langle Jx, z \rangle JJx - 2\langle Jx, Jx \rangle Jz,$$

$$J\mathcal{A}(x, Jx)z = \langle x, z \rangle x + \langle Jx, z \rangle Jx + 2z,$$

$$= \begin{cases} 3z & \text{if } z \in \operatorname{Span}\{x, Jx\}, \\ 2z & \text{if } z \perp \operatorname{Span}\{x, Jx\}. \end{cases}$$

This shows $\mathfrak{M}$ is complex Ivanov–Petrova. There are two distinct eigenvalues; one has multiplicity 2 and one has multplicity $m - 2$.

Example 4.8.2 Let $V = \mathbb{H}^n$ with the usual Hermitian quaternion structure $\{J_1, J_2, J_3\}$ where $J_3 = J_1 J_2$ and $J_1^2 = J_2^2 = -\operatorname{Id}$. Let $\mathcal{A} = c_1 \mathcal{A}_{J_1} + c_2 \mathcal{A}_{J_2}$. We show that $\mathfrak{M} := \{\mathbb{R}^{2n}, \langle \cdot, \cdot \rangle, J_1, A\}$ is complex Ivanov–Petrova by computing:

$$\mathcal{A}(x, J_1 x)z = c_1\{\langle J_1 J_1 x, z \rangle J_1 x - \langle J_1 x, z \rangle J_1 J_1 x - 2\langle J_1 x, J_1 x \rangle J_1 z\}$$
$$+ c_2\{\langle J_2 J_1 x, z \rangle J_2 x - \langle J_2 x, z \rangle J_2 J_1 x - 2\langle J_2 x, J_1 x \rangle J_2 z\},$$

$$J_1 \mathcal{A}(x, J_1 x)z = c_1\{\langle x, z \rangle x + \langle J_1 x, z \rangle J_1 x + 2z\}$$
$$- c_2\{\langle J_1 J_2 x, z \rangle J_1 J_2 x + \langle J_2 x, z \rangle J_2 x\}$$

$$= \begin{cases} 3c_1 z & \text{if } z \in \operatorname{Span}\{x, J_1 x\}, \\ -2c_2 z & \text{if } z \in \operatorname{Span}\{J_2 x, J_3 x\}, \\ 2c_1 z & \text{if } z \perp \operatorname{Span}\{x, J_1 x, J_2 x, J_3 x\}. \end{cases}$$

This shows that $\mathfrak{M}$ is complex Ivanov–Petrova. Furthermore, the possible eigenvalue structures of Theorem 4.8.1 may be illustrated by choosing c_1 and c_2 appropriately.

Remark 4.8.1 In Chapter 5, we will show that the 0-models in these two examples are also complex Osserman. These examples can be modified by adding the algebraic curvature of constant sectional curvature c to shift the eigenvalues. If we set

$$A = c_0 R_0 + c_1 R_J$$

where $c_1 = \frac{1}{2}\lambda_1$ and $c_0 = \lambda_0 - 3c_1$, then

$$J\mathcal{A}(\pi_x) = \begin{cases} (c_0 + 3c_1)z & \text{if } z \in \operatorname{Span}\{x, Jx\}, \\ 2c_0 z & \text{if } z \perp \operatorname{Span}\{x, Jx\}. \end{cases}$$

$$= \begin{cases} \lambda_1 z & \text{if } z \in \operatorname{Span}\{x, Jx\}, \\ \lambda_0 z & \text{if } z \perp \operatorname{Span}\{x, Jx\}. \end{cases}$$

This has eigenvalue structure $\{(\lambda_0, m - 2), (\lambda_1, 2)\}$. We can also modify Example 4.8.2 similarly.

Chapter 5

Complex Osserman Algebraic Curvature Tensors

5.1 Introduction

In Chapter 5, we give a partial classification of complex Osserman algebraic curvature tensors which are given by Clifford families. The material of Chapter 5 is joint work with and as well is coauthored by M. Brozos-Vázquez.

Section 5.1 is both a statement of results and also serves as an outline to the remaining sections in Chapter 5. In Section 5.1.1, we review material concerning Clifford families. In Section 5.1.2, we define complex Osserman models. Section 5.1.3 summarizes our classification results in the algebraic context. Section 5.1.4 presents some geometric examples. Section 5.1.5 outlines the remainder of Chapter 5.

5.1.1 *Clifford families*

We shall work exclusively in the Riemannian setting in Chapter 5. We recall some notation established previously. Fix a positive definite inner product $\langle \cdot, \cdot \rangle$ on a real vector space V of dimension m. We say that a linear map J is a *Hermitian almost complex structure* on V if J is an isometry of V with $J^2 = -\,\mathrm{id}$. The *canonical curvature tensors* discussed in Section 1.3.2 are given by:

$$A_{\langle \cdot, \cdot \rangle}(x, y, z, w) := \langle x, w \rangle \langle y, z \rangle - \langle x, z \rangle \langle y, w \rangle,$$

and

$$A_J(x, y, z, w) := \langle x, Jw \rangle \langle y, Jz \rangle - \langle x, Jz \rangle \langle y, Jw \rangle - 2\langle x, Jy \rangle \langle z, Jw \rangle .$$

It now follows that the associated Jacobi operators are given by:

257

$$\mathcal{J}_{A_{\langle\cdot,\cdot\rangle}}(x)y = \left\{ \begin{array}{l} 0 \text{ if } y = x, \\ y \text{ if } y \perp x, \end{array} \right.$$
$$\mathcal{J}_{A_J}(x)y = \left\{ \begin{array}{l} 0 \text{ if } y \perp Jx, \\ 3y \text{ if } y = Jx. \end{array} \right. \tag{5.1.a}$$

We say that $\mathcal{F} = \{J_1, ..., J_\ell\}$ is a Clifford family *of rank ℓ* if each J_i is a Hermitian almost complex structure and if the *Clifford commutation relations*

$$J_i J_j + J_j J_i = -2\delta_{ij} \operatorname{Id}$$

are satisfied. If $\Psi = (\psi_{ij})$ belongs to the orthogonal group $O(\ell)$, let

$$\tilde{J}_i := \psi_{i1} J_1 + ... + \psi_{il} J_l \quad \text{for} \quad 1 \le i \le \ell$$

define a new Clifford family $\tilde{\mathcal{F}} := \{\tilde{J}_1, ..., \tilde{J}_\ell\}$ which will be said to be a *reparametrization* of $\mathcal{F}$. If $\ell = 3$, then $\mathcal{F} = \{J_1, J_2, J_3\}$ is said to be a *quaternion structure* on V if $J_3 = J_1 J_2$; such structures exist if and only if the dimension of V is divisible by 4. If

$$A = c_0 A_{\langle\cdot,\cdot\rangle} + c_1 A_{J_1} + ... + c_\ell A_{J_\ell}$$

with $c_1 \ne 0$, ..., $c_\ell \ne 0$, then A is said to be *given by a Clifford family of rank ℓ*. Let $(V, \langle\cdot,\cdot\rangle, A)$ be a 0-model. We suppose $m \ne 16$. Work of Chi and Nikolayevsky as discussed in Theorem 1.9.6 shows that $\mathfrak{M}$ is Osserman if and only if A is given by a Clifford family. Thus these tensors form a very natural family of examples.

5.1.2   *Complex Osserman tensors*

We first review some notation that we have established previously. Let $\mathbb{CP}(V) = \mathbb{CP}(V, \langle\cdot,\cdot\rangle, J)$ be the projective space of complex lines in V. Let $\pi_x := \operatorname{Span}\{x, Jx\} \in \mathbb{CP}(V)$ for $x \in S(V)$. The map $x \to \pi_x$ defines the *Hopf fibration* $S(V) \to \mathbb{CP}(V)$. The *higher order Jacobi operator* $\mathcal{J}(\pi_x)$ is given by setting

$$\mathcal{J}(\pi_x) := \mathcal{J}(x) + \mathcal{J}(Jx).$$

We say that $\mathfrak{M} = (V, \langle\cdot,\cdot\rangle, J, A)$ is a *complex 0-model* if J is a Hermitian almost complex structure on $(V, \langle\cdot,\cdot\rangle)$ and if A is an algebraic curvature tensor. We say that J and A are *compatible* and that $\mathfrak{M}$ is a *compatible complex model* if any of the following conditions, which are equivalent by Lemma 1.6.6, is satisfied:

(1) $J^*A = A$.
(2) $\mathcal{J}(\pi)$ is complex linear for every π in $\mathbb{CP}(V)$.
(3) $\mathcal{R}(\pi)$ is complex linear for every π in $\mathbb{CP}(V)$.

We say that a complex 0-model $\mathfrak{M}$ is *complex Osserman* if J and A are compatible and if the eigenvalues of $\mathcal{J}$ are constant on $\mathbb{CP}(V)$. Since the eigenvalues are constant, the multiplicities are constant as well. If $m = 2$ and if $\mathfrak{M}$ is a complex 0-model, then necessarily $\mathfrak{M}$ is complex Osserman since $\mathbb{CP}(V)$ consists of a single point. We shall therefore assume $m \geq 4$ for the remainder of Chapter 5.

5.1.3 *Classification results in the algebraic setting*

We have a complete classification if $\text{Rank}(\mathcal{F}) \leq 3$ except in a few cases.

Theorem 5.1.1 *Let $\mathfrak{M} = (V, \langle \cdot, \cdot \rangle, J, A)$ be a complex 0-model. Assume that $A = c_0 A_{\langle \cdot, \cdot \rangle} + c_1 A_{J_1} + ... + c_\ell A_{J_\ell}$ is given by a Clifford family of rank ℓ. Assume $c_i \neq 0$ for $1 \leq i \leq \ell$.*

(1) Let $\ell = 0$. Then $A = c_0 A_{\langle \cdot, \cdot \rangle}$ and $\mathfrak{M}$ is complex Osserman.

(2) Let $\ell = 1$. Then:

    (a) If $c_0 = 0$, then $\mathfrak{M}$ is complex Osserman if and only if $JJ_1 = \pm J_1 J$.

    (b) If $c_0 \neq 0$, then $\mathfrak{M}$ is complex Osserman if and only if $J = \pm J_1$ or $JJ_1 = -J_1 J$.

(3) Let $\ell = 2$. If $c_0 \neq 0$, assume that $\dim(V) \geq 12$. Then $\mathfrak{M}$ is complex Osserman if and only if there is a reparametrization $\{\tilde{J}_i\}$ of $\mathcal{F}$ with $A = c_0 A_{\langle \cdot, \cdot \rangle} + \tilde{c}_1 A_{\tilde{J}_1} + \tilde{c}_2 A_{\tilde{J}_2}$ so at least one of the following holds:

    (a) $J = \tilde{J}_1 \tilde{J}_2$.

    (b) $J = \tilde{J}_1$.

    (c) $c_0 = 0$, $J\tilde{J}_1 = \tilde{J}_1 J$, and $J\tilde{J}_2 = -\tilde{J}_2 J$.

(4) Let $\ell = 3$. Then:

    (a) If $c_0 = 0$ and if $\dim(V) \geq 12$, then $\mathfrak{M}$ is complex Osserman if and only if there is a reparametrization $\{\tilde{J}_i\}$ of $\mathcal{F}$ so $J = \tilde{J}_2 \tilde{J}_3$ and so $A = \tilde{c}_1 A_{\tilde{J}_1} + \tilde{c}_2 A_{\tilde{J}_2} + \tilde{c}_3 A_{\tilde{J}_3}$.

    (b) If $c_0 \neq 0$ and if $\dim(V) \geq 16$, then $\mathfrak{M}$ is complex Osserman if and only if there is a reparametrization $\{\tilde{J}_i\}$ of $\mathcal{F}$ so $J = \tilde{J}_1$, so $\tilde{J}_1 \tilde{J}_2 \tilde{J}_3 = \text{Id}$, and so $A = c_0 A_{\langle \cdot, \cdot \rangle} + \tilde{c}_1 A_{\tilde{J}_1} + \tilde{c}_2 A_{\tilde{J}_2} + \tilde{c}_3 A_{\tilde{J}_3}$.

The work of Chi and of Nikolayevsky which was discussed in Theorem 1.9.6 also shows that the higher rank Clifford module structures do not arise in the geometric setting. Still, it is worth examining the algebraic context. We will prove the following results in the higher rank setting:

Theorem 5.1.2   *Let* $\mathfrak{M} = (V, \langle \cdot, \cdot \rangle, J, A)$ *be a complex 0-model.*

(1) *Let* $A = \sum_i c_i A_{J_i}$ *be given by a Clifford family of rank* $\ell \geq 4$ *where* $c_i \neq 0$ *for* $1 \leq i \leq \ell$. *If* $\ell = 4, 5$, *assume that* $m \geq 2^\ell$. *If* $m \geq 6$, *assume that* $m \geq \ell(\ell - 1)$. *Then* $\mathfrak{M}$ *is not complex Osserman.*
(2) *Let* $A = c_0 A_{\langle \cdot, \cdot \rangle} + \sum_i c_i A_{J_i}$ *be given by a Clifford family of rank* $\ell \geq 4$ *where* $c_i \neq 0$ *for* $0 \leq i \leq \ell$. *If* $\ell = 4$, *assume* $m \geq 32$. *If* $\ell = 5, 6, 7$, *assume that* $m \geq 2^\ell$. *If* $\ell \geq 8$, *assume that* $m \geq \ell(\ell - 1)$. *Then* $\mathfrak{M}$ *is not complex Osserman.*

Remark 5.1.1   It follows from Lemma 4.2.3 that if $\ell \geq 16$, then necessarily $m \geq \ell(\ell - 1)$. Thus this condition plays no role if $\ell \geq 16$.

5.1.4   *Geometric examples*

There are examples of complex Osserman Riemannian manifolds. Recall that $\mathfrak{M}$ is a complex space form if $\mathfrak{M}$ is an open subset of m-dimensional complex projective space with the Fubini–Study metric or the negative curvature dual.

Theorem 5.1.3

(1) *Let* (M, g) *be an even-dimensional contractible manifold of constant sectional curvature* c. *Let* J *be any Hermitian almost complex structure on* M. *Then* (M, g, J) *is complex Osserman.*
(2) *Let* (M, g) *be a complex space form. Let* J *be the canonical almost complex structure.*

    (a) (M, g, J) *is complex Osserman.*
    (b) *Suppose that* $m \equiv 0 \bmod 4$ *and* M *is contractible. Let* K *be a second Hermitian almost complex structure on* (M, g) *with* $JK = -KJ$. *Then* (M, g, K) *is complex Osserman.*

Proof.   Let (M, g) be a contractible manifold of constant sectional curvature c_0. Then $R = c_0 R_g$ and the tangent bundle $T(M)$ is trivial. Thus we can choose a global orthonormal frame $\{e_1, ..., e_m\}$. As m is even, we can define a Hermitian almost complex structure J by

$$Je_{2i} = e_{2i-1} \quad \text{and} \quad Je_{2i-1} = -e_{2i}.$$

Thus Hermitian almost complex structures exist. The desired conclusion then follows from Theorem 5.1.1 (1).

If (M, g, J) is a complex space form, then $R = c_0\{R_g + R_J\}$; see, for example, Lemma 1.15.1 of Gilkey (2001a). Theorem 5.1.1 shows $\mathcal{M}$ is complex Osserman. If M is contractible, we can choose a global orthonormal frame $\{e_1, Je_1, ..., e_{\bar{m}}, Je_{\bar{m}}\}$ for $T(M)$. If $m \equiv 0 \bmod 4$, then $\bar{m}$ is even and we can define K by setting

$$Ke_{2i} = e_{2i-1}, \qquad Ke_{2i-1} = -e_{2i},$$
$$KJe_{2i} = -Je_{2i-1}, \quad KJe_{2i-1} = Je_{2i}.$$

We set $\tilde{J}_1 = K$, $c_1 = 0$, $\tilde{J}_2 = J$, $c_2 = c_0$, $\tilde{J}_3 = \tilde{J}_1\tilde{J}_2$, and $c_3 = 0$. Theorem 5.1.1 (2) then implies that (M, g, K) is complex Osserman. $\qquad\square$

5.1.5 *Chapter outline*

The remainder of Chapter 5 is devoted to the proof of Theorem 5.1.1 and of Theorem 5.1.2. In Section 5.2, we establish some technical preliminaries. We give criteria to ensure that a 0-model is complex Osserman, we examine the eigenvalue structure of a complex Osserman 0-model, we give some examples of complex Osserman 0-models, and we establish some additional useful observations of a technical nature. We then consider Clifford families of low rank. Assertions (1) and (2) of Theorem 5.1.1 are proved in Section 5.3, Assertion (3) of Theorem 5.1.1 is proved in Section 5.4, and we complete the proof of Theorem 5.1.1 in Section 5.5 by establishing Assertion (4). In Section 5.6 and in Section 5.7 we establish Theorem 5.1.3 which deals with the higher rank case. Section 5.6 deals with the case $A = \sum_i c_i A_{J_i}$ and Section 5.7 deals with the case $A = c_0 A_{\langle \cdot, \cdot \rangle} + \sum_i c_i A_{J_i}$.

5.2 Technical Preliminaries

In this section, we present some results we shall need subsequently. In Section 5.2.1, we give necessary and sufficient conditions that a complex model $\mathfrak{M} = (V, \langle \cdot, \cdot \rangle, J, A)$ is complex Osserman in terms of the eigenspaces of the complex Jacobi operator. In Section 5.2.2, we use methods from algebraic topology to control the eigenvalue structure. In Section 5.2.3, we give examples of complex Osserman 0-models that show the results of Section 5.2.2 are sharp. In Section 5.2.4, we discuss reparametrization of a Clifford family. In Section 5.2.5, we examine the dual Clifford family associated to a rank 3 Clifford family. In Section 5.2.6 we study complex 0-models given by Clifford families where J and A are compatible and where certain additional hypotheses are imposed on J. In Section 5.2.7, we present some technical results which deal with linear independence of endomorphisms defined by a Clifford family. We conclude in Section 5.2.8 by establishing some additional technical results concerning Clifford families.

5.2.1 *Criteria for complex Osserman models*

Let $\mathrm{Spec}\{\mathcal{J}(\pi_x)\} \subset \mathbb{R}$ be the set of eigenvalues of the complex Jacobi operator:

$$\mathrm{Spec}\{\mathcal{J}(\pi_x)\} := \{\lambda \in \mathbb{R} : \det\{\mathcal{J}(\pi_x) - \lambda\,\mathrm{Id}\} = 0\}\,.$$

Let $E_\lambda(\pi_x)$ be the associated eigenspaces:

$$E_\lambda(\pi_x) := \{v \in V : \mathcal{J}(\pi_\lambda)v = \lambda v\}\,.$$

We then have an orthogonal direct sum decomposition for any $\pi_x \in \mathbb{CP}(V)$:

$$V = \oplus_{\lambda \in \mathrm{Spec}\{\mathcal{J}(\pi_x)\}} E_\lambda(\pi_x)\,.$$

The following result will be central to our discussion.

Lemma 5.2.1   *A complex 0-model $\mathfrak{M} = (V, \langle\cdot,\cdot\rangle, J, A)$ is complex Osserman if and only if the following two conditions are satisfied:*

(1) $JE_\lambda(\pi_x) = E_\lambda(\pi_x)$ for all $\pi_x \in \mathbb{CP}(V)$ and for all $\lambda \in \mathrm{Spec}\{\mathcal{J}(\pi_x)\}$.
(2) $\mathrm{Spec}\{\mathcal{J}(\pi_x)\} = \mathrm{Spec}\{\mathcal{J}(\pi_y)\}$ for all $\pi_x, \pi_y \in \mathbb{CP}(V)$.

Proof.   If $\mathfrak{M}$ is complex Osserman, then A and J are compatible and Condition (1) holds. Furthermore, the eigenvalue structure is independent of π and Condition (2) holds. This establishes one implication of the Lemma. Conversely, if Condition (1) holds, then all the eigenspaces E_λ are preserved by J. This implies that $J\mathcal{J}(\pi_x) = \mathcal{J}(\pi_x)J$ and consequently by Lemma 1.6.6, J and A are compatible. Assumption (2) then implies $\mathfrak{M}$ is complex Osserman. $\square$

5.2.2 *Controlling the eigenvalue structure*

If $\mathfrak{M}$ is complex Osserman, let $\vec{\lambda} = (\lambda_0, ..., \lambda_\kappa)$ and $\vec{\mu} = (\mu_0, ..., \mu_\kappa)$ be the associated eigenvalues λ_i and eigenvalue multiplicities $\mu_i := \dim(E_{\lambda_i})$ where the eigenvalues are ordered so that $\mu_0 \geq \mu_1 \geq ... \geq \mu_\kappa > 0$. The following result is based on Theorem 4.2.5.

Theorem 5.2.1 *Let $\mathfrak{M} = (V, \langle \cdot, \cdot \rangle, J, A)$ be a complex Osserman 0-model. Assume that $\kappa \geq 1$ so there are at least 2 distinct eigenvalues. Then:*

(1) If $m \equiv 2 \bmod 4$, then $\kappa = 1$ and $\mu_1 = 2$.
(2) If $m \equiv 0 \bmod 4$, then one of the following alternatives holds:

 (a) $\kappa = 1$ and $\mu_1 = 2$.
 (b) $\kappa = 1$ and $\mu_1 = 4$.
 (c) $\kappa = 2$ and $\mu_1 = \mu_2 = 2$.

Proof. Recall that a sub-bundle E of the trivial bundle $\mathbb{V} := \mathbb{CP}(V) \times V$ over $\mathbb{CP}(V)$ is said to be a *geometrically symmetric vector bundle* if for all $\sigma, \tau \in \mathbb{CP}(V)$, $\tau \subset E(\sigma)$ implies that $\sigma \subset E(\tau)$.

The proof we shall give is similar to the proof of Theorem 4.8.1. By Lemma 5.2.1, the eigenspaces $E_{\lambda_i}(\pi)$ have constant rank and patch together to define smooth vector bundles E_{λ_i} over V which give an orthogonal direct sum decomposition

$$\mathbb{V} = E_{\lambda_0} \oplus ... \oplus E_{\lambda_\kappa}.$$

Furthermore, since the eigenbundles are invariant under J, they inherit natural complex structures so the decomposition given above is in the category of complex vector bundles. The desired conclusion will follow from Theorem 4.2.5 if we can show that one of the eigenbundles E_{λ_i} is geometrically symmetric.

Let λ be the maximal eigenvalue. Let $x, y \in S(V)$. We must show $y \in E_\lambda(\pi_x)$ implies $x \in E_\lambda(\pi_y)$. We have $\mathcal{J}(\pi_x) = \mathcal{J}(x) + \mathcal{J}(Jx)$ and $\mathcal{J}(\pi_y) = \mathcal{J}(y) + \mathcal{J}(Jy)$. Since A and J are compatible,

$$\lambda = \langle \mathcal{J}(\pi_x)y, y \rangle = A(y, x, x, y) + A(y, Jx, Jx, y)$$
$$= A(x, y, y, x) + A(x, Jy, Jy, x) = \langle \mathcal{J}(\pi_y)x, x \rangle .$$

Since λ is the maximal eigenvalue, Lemma 1.5.1 shows $\mathcal{J}(\pi_y)x = \lambda x$. $\qquad \square$

If $A = c_0 A_0 + c_1 A_{J_1} + ... + c_\ell A_{J_\ell}$ is given by a Clifford family, we introduce the *reduced complex Jacobi operator*

$$\tilde{\mathcal{J}}(\pi_x) := \mathcal{J}(\pi_x) - 2c_0 \operatorname{Id} .$$

Corollary 5.2.1 *Let $\mathfrak{M} = (V, \langle \cdot, \cdot \rangle, J, A)$ be a complex Osserman model which is given by a Clifford family of rank ℓ. Assume that $m > 2\ell + 6$. Then* $\operatorname{Rank}\{\tilde{\mathcal{J}}(\pi_x)\} \leq 4$.

Proof. We compute that

$$\mathcal{J}(\pi_x)y = 2c_0 y + \sum_{i=1}^{\ell} 3c_i\{\langle y, J_i x\rangle J_i x + \langle y, J_i J x\rangle J_i J x\}$$
$$- c_0\{\langle y, x\rangle x + \langle y, J x\rangle J x\} .$$

Consequently

$$\mathcal{J}(\pi_x)y = 2c_0 y \text{ for } y \perp \operatorname{Span}\{x, Jx, J_1 x, ..., J_\ell x, J_1 J x, ..., J_\ell J x\} .$$

This implies that $2c_0$ is an eigenvalue of multiplicity at least $m - 2\ell - 2$. Thus if $m > 2\ell + 6$, then $2c_0$ is the dominant eigenvalue and the other eigenvalues have multiplicity at most 4. The Corollary now follows since $\tilde{\mathcal{J}}(\pi_x)$ is defined by shifting the spectrum appropriately. $\square$

5.2.3 *Examples of complex Osserman 0-models*

If $m \equiv 0 \bmod 2$, then there exists a Hermitian almost complex structure J_1 on V. If $m \equiv 0 \bmod 4$, then there exists a quaternion structure on V; this means that there is a Clifford family $\{J_1, J_2, J_3\}$ so that $J_3 = J_1 J_2$. The following result shows that Theorem 5.2.1 is sharp.

Theorem 5.2.2 *Let $A = c_0 A_{\langle \cdot, \cdot\rangle} + c_1 A_{J_1} + ... + c_\ell A_{J_\ell}$ where $\mathcal{F} = \{J_i\}$ is a Clifford family. Let $\mathfrak{M} := (V, \langle \cdot, \cdot\rangle, J_1, A)$ be a complex 0-model. Not all the c_i need be non-zero.*

(1) If $\ell = 1$, then $\mathfrak{M}$ is complex Osserman. Furthermore:

    (a) If $(c_0, c_1) = (3, 1)$, then $\kappa = 0$ and $\vec{\mu} = (m)$.
    (b) If $(c_0, c_1) = (0, 1)$, then $\kappa = 1$ and $\vec{\mu} = (m - 2, 2)$.

(2) If $\ell = 3$ and if $J_3 = J_1 J_2$, then $\mathfrak{M}$ is complex Osserman. Furthermore:

    (a) If $(c_0, c_1, c_2, c_3) = (3, 1, 0, 0)$, then $\kappa = 0$ and $\vec{\mu} = (m)$.
    (b) If $(c_0, c_1, c_2, c_3) = (0, 1, 0, 0)$, then $\kappa = 1$ and $\vec{\mu} = (m - 2, 2)$.
    (c) If $(c_0, c_1, c_2, c_3) = (0, 2, 1, 1)$, then $\kappa = 1$ and $\vec{\mu} = (m - 4, 4)$.

(d) If $(c_0, c_1, c_2, c_3) = (0, 2, 2, 2)$, then $\kappa = 2$ and $\vec{\mu} = (m - 4, 2, 2)$.

Proof. Suppose that $A = c_0 A_{\langle \cdot, \cdot \rangle} + c_1 A_{J_1}$ and that $J = J_1$. By Eq. (5.1.a),

$$\mathcal{J}(\pi_x)y = \begin{cases} 2c_0 y & \text{if } y \perp \pi_x, \\ (c_0 + 3c_1)y & \text{if } y \in \pi_x. \end{cases}$$

Lemma 5.2.1 implies $\mathfrak{M}$ is complex Osserman. The eigenvalues are distinct if and only if $3c_1 \neq c_0$; Assertion (1) follows.

Next let $A = c_0 A_{\langle \cdot, \cdot \rangle} + c_1 A_{J_1} + c_2 A_{J_2} + c_3 A_{J_3}$, let $\tilde{\pi}_x := \mathrm{Span}\{J_2 x, J_3 x\}$, and let $J = J_1$. Assume that $J_1 J_2 = J_3$. By Eq. (5.1.a),

$$\mathcal{J}(\pi_x)y = \begin{cases} 2c_0 y & \text{if } y \perp \pi_x, y \perp \tilde{\pi}_x, \\ (c_0 + 3c_1)y & \text{if } y \in \pi_x, \\ (2c_0 + 3c_2 + 3c_3)y & \text{if } y \in \tilde{\pi}_x. \end{cases}$$

Assertion (2) now follows similarly. $\qquad\square$

5.2.4 *Reparametrization of a Clifford family*

Let $\mathcal{F} = \{J_1, ..., J_\ell\}$ be a Clifford family. If $\Psi = (\psi_{ij})$ belongs to the orthogonal group $O(\ell)$, we defined the *reparametrization* $\tilde{\mathcal{F}} = \{\tilde{J}_1, ..., \tilde{J}_\ell\}$ of $\mathcal{F}$ by setting $\tilde{J}_i := \psi_{i1} J_1 + ... + \psi_{il} J_l$ for $1 \leq i \leq \ell$.

Lemma 5.2.2 *Suppose that $\tilde{\mathcal{F}}$ is a reparametrization of the Clifford family $\mathcal{F}$. Then $A_{J_1} + ... + A_{J_\ell} = A_{\tilde{J}_1} + ... + A_{\tilde{J}_\ell}$.*

Proof. Let $A = A_{J_1} + ... + A_{J_\ell}$ and let $\tilde{A} = A_{\tilde{J}_1} + ... + A_{\tilde{J}_\ell}$. We use Eq. (5.1.a) to see that

$$\mathcal{J}_A(y) = \begin{cases} 0 & \text{if } y \perp \mathrm{Span}\{J_1 x, ..., J_\ell x\}, \\ 3y & \text{if } y \in \mathrm{Span}\{J_1 x, ..., J_\ell x\}. \end{cases}$$

This shows that $\mathcal{J}_A = \mathcal{J}_{\tilde{A}}$ and hence $A = \tilde{A}$ by Lemma 1.7.1. $\qquad\square$

5.2.5 *The dual Clifford family*

Let $\mathcal{F} = \{J_1, J_2, J_3\}$ be a Clifford family on V. The *dual Clifford family* $\mathcal{F}^*$ is given by:

$$\mathcal{F}^* := \left\{ J_1^* := J_2 J_3, J_2^* := J_3 J_1, J_3^* = J_1 J_2 \right\}.$$

The dual structure is a skew-quaternion structure since $J_1^* J_2^* J_3^* = -\operatorname{Id}$; $\mathcal{F}^*$ is a reparametrization of $\mathcal{F}$ if and only if $J_1 J_2 = \pm J_3$. Set

$$\Theta := J_1 J_2 J_3, \quad V_{\pm} := \{ x \in V : \Theta x = \mp x \}, \text{ and}$$
$$\mathcal{N} := \left\{ \tfrac{1}{\sqrt{2}}(x_+ + x_-) : x_{\pm} \in S(V_{\pm}) \right\}. \tag{5.2.a}$$

Lemma 5.2.3 *Let $\mathcal{F} = \{ J_1, J_2, J_3 \}$ and let $\mathcal{F}^* = \{ J_1^*, J_2^*, J_3^* \}$ be a Clifford family and the associated dual Clifford family. Assume $J_1 J_2 \neq \pm J_3$ so $\mathcal{N}$ is non-empty.*

(1) If $x \in \mathcal{N}$, then $\{ x, J_1 x, J_2 x, J_3 x, J_1^* x, J_2^* x, J_3^* x, \Theta x \}$ is an orthonormal set.

(2) If $\tilde{\mathcal{F}}$ is a reparametrization of $\mathcal{F}$, then $\tilde{\mathcal{F}}^*$ is a reparametrization of $\mathcal{F}^*$.

Proof. Suppose that $J_1 J_2 \neq \pm J_3$ so $\mathcal{N}$ is non-empty. Let $x \in \mathcal{N}$. Assertion (1) follows from the decomposition:

$$\begin{aligned}
x &= \tfrac{1}{\sqrt{2}}(x_+ + x_-), & \Theta x &= \tfrac{1}{\sqrt{2}}(x_+ - x_-), \\
J_1 x &= \tfrac{1}{\sqrt{2}}(J_1 x_+ + J_1 x_-), & J_1^* x &= \tfrac{1}{\sqrt{2}}(J_1 x_+ - J_1 x_-), \\
J_2 x &= \tfrac{1}{\sqrt{2}}(J_2 x_+ + J_2 x_-), & J_2^* x &= \tfrac{1}{\sqrt{2}}(J_2 x_+ - J_2 x_-), \\
J_3 x &= \tfrac{1}{\sqrt{2}}(J_3 x_+ + J_3 x_-), & J_3^* x &= \tfrac{1}{\sqrt{2}}(J_3 x_+ - J_3 x_-).
\end{aligned}$$

Let the orthonormal basis $\{ \vec{v}_1, \vec{v}_2, \vec{v}_3 \}$ for $\mathbb{R}^3$ define a reparametrization

$$\tilde{J}_i := \vec{v}_i \cdot (J_1, J_2, J_3) = v_{i1} J_1 + v_{i2} J_2 + v_{i3} J_3.$$

Let $v_i \times v_j$ denote the cross product on $\mathbb{R}^3$. Then

$$\begin{aligned}
\tilde{J}_1^* &= (\vec{v}_2 \times \vec{v}_3) \cdot (J_1^*, J_2^*, J_3^*), \\
\tilde{J}_2^* &= (\vec{v}_3 \times \vec{v}_1) \cdot (J_1^*, J_2^*, J_3^*), \\
\tilde{J}_3^* &= (\vec{v}_1 \times \vec{v}_2) \cdot (J_1^*, J_2^*, J_3^*).
\end{aligned}$$

Since $\{ \vec{v}_2 \times \vec{v}_3, \vec{v}_3 \times \vec{v}_1, \vec{v}_1 \times \vec{v}_2 \}$ is also an orthonormal basis for $\mathbb{R}^3$, $\{ \tilde{J}_1^*, \tilde{J}_2^*, \tilde{J}_3^* \}$ is a reparametrization of $\mathcal{F}^*$. $\qquad\square$

5.2.6 *Compatible complex models given by Clifford families*

Lemma 5.2.4 *Let $\mathfrak{M} := (V, \langle \cdot, \cdot \rangle, J, A := c_0 A_{\langle \cdot, \cdot \rangle} + c_1 A_{J_1} + c_2 A_{J_2} + c_3 A_{J_3})$ be a complex Osserman 0 model which is given by a Clifford family of rank at most 3; we permit some of the c_i to be zero. Assume J and A are compatible.*

(1) *Suppose that $Jx = (a_1 J_1 + a_2 J_2 + a_3 J_3)x$ for some $x \in S(V)$. Then $(c_i - c_j)a_i a_j = 0$ for $i < j$.*

(2) *Suppose that $J_1 J_2 \neq \pm J_3$ so $\mathcal{N}$ is non-empty. Also suppose that Jx belongs to $\mathrm{Span}\{J_1 x, J_2 x, J_3 x, J_1^* x, J_2^* x, J_3^* x\}$ for all $x \in \mathcal{N}$. Then either $J \in \mathrm{Span}\{J_1^*, J_2^*, J_3^*\}$ or $J \in \mathrm{Span}\{J_1, J_2, J_3\}$.*

Proof. Let $Jx = (a_1 J_1 + a_2 J_2 + a_3 J_3)x$ for some $x \in S(V)$. One has:

$$JA(x, Jx)x = -JJ(x)Jx$$
$$= -J(c_0 Jx + 3c_1\langle Jx, J_1 x\rangle J_1 x + 3c_2\langle Jx, J_2 x\rangle J_2 x + 3c_3\langle Jx, J_3 x\rangle J_3 x)$$
$$= c_0 x - 3c_1 a_1 J J_1 x - 3c_2 a_2 J J_2 x - 3c_3 a_3 J J_3 x,$$
$$A(x, Jx)Jx = J(Jx)x$$
$$= c_0 x + 3c_1\langle x, J_1 Jx\rangle J_1 Jx + 3c_2\langle x, J_2 Jx\rangle J_2 Jx + 3c_3\langle x, J_3 Jx\rangle J_3 Jx$$
$$= c_0 x - 3c_1 a_1 J_1 Jx - 3c_2 a_2 J_2 Jx - 3c_3 a_3 J_3 Jx.$$

Since J and A are compatible, $JA(x, Jx)x = A(x, Jx)Jx$ so

$$0 = (c_1 - c_2)a_1 a_2 J_1 J_2 x + (c_1 - c_3)a_1 a_3 J_1 J_3 x + (c_2 - c_3)a_2 a_3 J_2 J_3 x.$$

Since $\{J_1 J_2 x, J_1 J_3 x, J_2 J_3 x\}$ is an orthonormal set, Assertion (1) follows. Let $J_1 J_2 \neq J_3$ so $\mathcal{N}$ is non-empty. Assume we may decompose

$$Jx = \sum_{i=1}^{3}(a_i(x)J_i + a_i^*(x)J_i^*)x$$

for every $x \in \mathcal{N}$; the coefficients are uniquely determined and vary continuously by Lemma 5.2.3 (1). We generalize the discussion given to establish Assertion (1). Let $x \in \mathcal{N}$. We compute:

$$JA(x, Jx)x = -JJ(x)Jx = -J\left\{c_0 Jx + 3\sum_{i=1}^{3} c_i\langle Jx, J_i x\rangle J_i x\right\},$$
$$A(x, Jx)Jx = J(Jx)x = c_0 x + 3\sum_{i=1}^{3} c_i\langle x, J_i Jx\rangle J_i Jx.$$

Since J and A are compatible, $JA(x, Jx)x = A(x, Jx)Jx$ so

$$0 = \sum_{i=1}^{3} c_i\langle Jx, J_i x\rangle(JJ_i - J_i J)x$$
$$= c_1 a_1(x)\{a_2(x)(J_2 J_1 - J_1 J_2) + a_3(x)(J_3 J_1 - J_1 J_3)$$
$$+ a_2^*(x)(J_3 J_1 J_1 - J_1 J_3 J_1) + a_3^*(x)(J_1 J_2 J_1 - J_1 J_2 J_1)\}x$$

$$+c_2 a_2(x)\{a_1(x)(J_1 J_2 - J_2 J_1) + a_3(x)(J_3 J_2 - J_2 J_3)$$

$$+a_1^*(x)(J_2 J_3 J_2 - J_2 J_2 J_3) + a_3^*(x)(J_1 J_2 J_2 - J_2 J_1 J_2)\}x$$

$$+c_3 a_3(x)\{a_1(x)(J_1 J_3 - J_3 J_1) + a_2(x)(J_2 J_3 - J_3 J_2)$$

$$+a_1^*(x)(J_2 J_3 J_3 - J_3 J_2 J_3) + a_2^*(x)(J_3 J_1 J_3 - J_3 J_3 J_1)\}x.$$

By Lemma 5.2.3 (1), $\{J_1 x, J_2 x, J_3 x, J_1^* x, J_2^* x, J_3^* x\}$ is an orthonormal set. Consequently the relations derived above yield:

$$
\begin{aligned}
(c_1 - c_2)a_1(x)a_2(x) = 0, \quad & c_1 a_1(x)a_2^*(x) - c_2 a_2(x)a_1^*(x) = 0, \\
(c_2 - c_3)a_2(x)a_3(x) = 0, \quad & c_2 a_2(x)a_3^*(x) - c_3 a_3(x)a_2^*(x) = 0, \\
(c_3 - c_1)a_3(x)a_1(x) = 0, \quad & c_3 a_3(x)a_1^*(x) - c_1 a_1(x)a_3^*(x) = 0 .
\end{aligned}
\qquad (5.2.\text{b})
$$

Our first step is to establish the dichotomy that

$$
\begin{aligned}
Jx \in \mathrm{Span}\{J_1 x, J_2 x, J_3 x\} \quad & \text{or} \\
Jx \in \mathrm{Span}\{J_1^* x, J_2^* x, J_3^* x\} \quad & \text{if} \quad x \in \mathcal{N} .
\end{aligned}
\qquad (5.2.\text{c})
$$

If $Jx \notin \mathrm{Span}\{J_1^* x, J_2^* x, J_3^* x\}$ we must show $Jx \in \mathrm{Span}\{J_1 x, J_2 x, J_3 x\}$; this is independent of reparametrization. Since $Jx \notin \mathrm{Span}\{J_1^* x, J_2^* x, J_3^* x\}$, $a_i(x) \neq 0$ for some i; we may suppose without loss of generality $a_1(x) \neq 0$. We apply Eq. (5.2.b). If $c_1 \neq c_3$ and $c_2 \neq c_3$, then we have $a_2(x) = 0$ and $a_3(x) = 0$. If $c_1 = c_2$ but $c_1 \neq c_3$, then $a_3(x) = 0$ and we can reparametrize $\mathcal{F}$ (and thereby $\mathcal{F}^*$ as well by Lemma 5.2.3) to have that $a_2(x) = a_3(x) = 0$. Finally, if $c_1 = c_2 = c_3$, we can reparametrize $\mathcal{F}$ and $\mathcal{F}^*$ so $a_2(x) = a_3(x) = 0$. Thus after a suitable reparametrization of $\mathcal{F}$, we may assume $a_2(x) = a_3(x) = 0$. By Eq. (5.2.b) we then have $a_2^*(x) = a_3^*(x) = 0$ and thus

$$Jx = a_1(x)J_1 x + a_1^*(x)J_1^* x .$$

As Jx is a unit vector, $a_1(x)^2 + a_1^*(x)^2 = 1$. We expand

$$
\begin{aligned}
Jx &= a_1(x)\tfrac{1}{\sqrt{2}}(J_1 x_+ + J_1 x_-) + a_1^*(x)\tfrac{1}{\sqrt{2}}(J_1 x_+ - J_1 x_-) \\
&= \tfrac{1}{\sqrt{2}}(Jx_+ + Jx_-) .
\end{aligned}
$$

This shows $Jx_+ = (a_1(x) + a_1^*(x))J_1 x_+$ so $(a_1(x) + a_1^*(x))^2 = 1$ as well. This implies $a_1(x)a_1^*(x) = 0$ and thus $a_1(x) = \pm 1$ and hence $Jx \in \mathrm{Span}\{J_1 x, J_2 x, J_3 x\}$. This establishes Eq. (5.2.c).

Since Jx is a continuous function of x, since $\mathcal{N}$ is connected, and since $\mathrm{Span}\{J_1 x, J_2 x, J_3 x\} \perp \mathrm{Span}\{J_1^* x, J_2^* x, J_3^* x\}$, Eq. (5.2.c) implies

$$\text{either} \quad Jx \in \mathrm{Span}\{J_1 x, J_2 x, J_3 x\} \ \forall x \in \mathcal{N}$$
$$\text{or} \quad Jx \in \mathrm{Span}\{J_1^* x, J_2^* x, J_3^* x\} \ \forall x \in \mathcal{N}.$$

Let $\mathcal{G} := \{G_1, G_2, G_3\}$ where $\mathcal{G} = \mathcal{F}$ or where $\mathcal{G} = \mathcal{F}^*$ is the Clifford family of rank 3 so that $Jx \in \mathrm{Span}\{G_1 x, G_2 x, G_3 x\}$ for all $x \in \mathcal{N}$. Expand

$$\tfrac{1}{\sqrt{2}} J(x_+ + x_-) = \tfrac{1}{\sqrt{2}} \sum_{i=1}^{3} \hat{a}_i (\tfrac{1}{\sqrt{2}}(x_+ + x_-)) \{G_i x_+ + G_i x_-\}.$$

This implies

$$Jx_+ = \sum_{i=1}^{3} \hat{a}_i (\tfrac{1}{\sqrt{2}}(x_+ + x_-)) G_i x_+, \qquad (5.2.\text{d})$$

$$Jx_- = \sum_{i=1}^{3} \hat{a}_i (\tfrac{1}{\sqrt{2}}(x_+ + x_-)) G_i x_- . \qquad (5.2.\text{e})$$

By Eq. (5.2.d), $\hat{a}_i(\tfrac{1}{\sqrt{2}}(x_+ + x_-))$ is independent of x_-; by Eq. (5.2.e), $\hat{a}_i(\tfrac{1}{\sqrt{2}}(x_+ + x_-))$ is independent of x_+. Thus $\hat{a}_i$ is a constant and

$$Jx_\pm = \sum_{i=1}^{3} \hat{a}_i G_i x_\pm \quad \forall x_\pm \in V_\pm .$$

Since $V = V_+ \oplus V_-$, we conclude $Jv = \sum_i \hat{a}_i G_i v$ for all $v \in V$. $\qquad \square$

5.2.7 *Linearly independent endomorphisms*

If $\{T_1, ..., T_\kappa\}$ are linearly independent linear maps from a vector space V to another vector space W, it does not necessarily follow that there exists $x \in V$ so that $\{T_1 x, ..., T_\kappa x\}$ are linearly independent vectors in W; it is clearly necessary that $\dim(W) \geq \kappa$ for this to be possible. The following Lemma gives conditions under which the condition of linear independence as linear transformations can manifest itself as linear independence of vectors.

Lemma 5.2.5 *Let* $\mathcal{T} := \{T_1, ..., T_\kappa\}$ *be a collection of linear maps from a vector space V of dimension m to a vector space W. Assume there exists $N \in \mathbb{N}$ so that if $(a_1, ..., a_\kappa)$ is any collection of real constants, not all of which are zero, $\dim\{\mathrm{Range}(a_1 T_1 + ... + a_\kappa T_\kappa)\} \geq N$. Then:*

(1) If $N \geq \kappa$, there exists x in V so $\{T_1 x, ..., T_\kappa x\}$ is a set of κ linearly independent vectors.

(2) If $N \geq 2\kappa$, there exist x and y in V so $\{T_1 x, ..., T_\kappa x, T_1 y, ..., T_\kappa y\}$ is a set of 2κ linearly independent vectors.

(3) Suppose that $Tz \in \mathrm{Span}\{T_1 z, ..., T_\kappa z\}$ for all z in V and that there exist $x, y \in V$ so $\{T_1 x, ..., T_\kappa x, T_1 y, ..., T_\kappa y\}$ is a set of 2κ linearly independent vectors. Then T belongs to $\mathrm{Span}\{T_1, ..., T_\kappa\}$.

Proof.   Suppose $N \geq \kappa$. For each x in V, choose $r(x)$ minimal so $\{T_1 x, ..., T_r x\}$ is a set of r linearly independent vectors but $\{T_1 x, ..., T_{r+1} x\}$ is a linearly dependent set; if $T_1 x = 0$, we set $r(x) = 0$. Choose x in V so $r(x)$ is maximal. If $r(x) = \kappa$, then Assertion (1) holds. We therefore assume $r = r(x) < \kappa$ and argue for a contradiction. Note that since $T_1 \neq 0$ we have $r(x) \geq 1$. Let

$$S := a_1 T_1 + ... + a_r T_r + T_{r+1}$$

where the constants $(a_1, ..., a_r)$ are chosen so $Sx = 0$. By hypothesis, we have that $\dim\{\mathrm{Range}(S)\} \geq N \geq \kappa$. Since $r < \kappa$, we can find z in V so $\{T_1 x, ..., T_r x, Sz\}$ is a linearly independent set of $r + 1$ vectors. By continuity, there exists $\varepsilon > 0$ small so that $\{T_1(x + \varepsilon z), ..., T_r(x + \varepsilon z), Sz\}$ is a linearly independent set. Since

$$\begin{aligned}
T_{r+1}(x + \varepsilon z) &= S(x + \varepsilon z) - a_1 T_1(x + \varepsilon z) - ... - a_r T_r(x + \varepsilon z) \\
&= \varepsilon S(z) - a_1 T_1(x + \varepsilon z) - ... - a_r T_r(x + \varepsilon z),
\end{aligned}$$

one has that $\{T_1(x + \varepsilon z), ..., T_r(x + \varepsilon z), T_{r+1}(x + \varepsilon z)\}$ is a linearly independent set of $r + 1$ vectors. Thus $r(x + \varepsilon z) \geq r + 1$ which contradicts the choice of x. This contradiction establishes Assertion (1).

Suppose $N \geq 2\kappa$. By Assertion (1), choose x in V so $\{T_1 x, ..., T_\kappa x\}$ is a linearly independent set of κ vectors. Let $W_0 := \mathrm{Span}\{T_1 x, ..., T_\kappa x\}$ and let $\pi : W \to W/W_0$ be the natural projection. Let $\bar{T}_i := \pi T_i : V \to W/W_0$. If $(a_1, ..., a_\kappa)$ are constants not all of which are zero, then

$$\dim\left\{\mathrm{Range}\{a_1 \bar{T}_1 + ... + a_\kappa \bar{T}_\kappa\}\right\} \geq N - \dim W_0 = N - \kappa.$$

Since $N - \kappa \geq \kappa$, we can apply Assertion (1) to the family $\{\bar{T}_1, ..., \bar{T}_k\}$ to choose y in V so that $\{\bar{T}_1 y, ..., \bar{T}_k y\}$ is a linearly independent set of κ vectors. The pair $\{x, y\}$ now satisfies the conclusions of Assertion (2).

Suppose the hypotheses of Assertion (3) hold. Choose x and y in V so $\{T_1 x, ..., T_\kappa x, T_1 y, ..., T_\kappa y\}$ is a set of 2κ linearly independent vectors. There then exist small neighborhoods $\mathcal{O}_x$ and $\mathcal{O}_y$ of x and y, respectively, so that if $x_1 \in \mathcal{O}_x$ and $y_1 \in \mathcal{O}_y$, then $\{T_1 x_1, ..., T_\kappa x_1, T_1 y_1, ..., T_\kappa y_1\}$ is a linearly independent set. We may expand

$$Tx_1 = \sum_{i=1}^{\kappa} a_i(x_1) T_i x_1, \quad Ty_1 = \sum_{i=1}^{\kappa} a_i(y_1) T_i y_1,$$

$$T(x_1 + y_1) = \sum_{i=1}^{\kappa} a_i(x_1 + y_1) T_i(x_1 + y_1).$$

The condition of linear independence permits to equate coefficients and to conclude

$$a_i(x_1) = a_i(x_1 + y_1) = a_i(y_1).$$

Consequently $a_i := a_i(x_1) = a_i(y_1)$ is independent of the choice of $x_1 \in \mathcal{O}_x$ or of $y_1 \in \mathcal{O}_y$. We therefore have $Tx_1 = \sum_i a_i T_i x_1$ on $\mathcal{O}_x$. Since this polynomial identity holds on an open subset of V, it holds on all of V so $T = \sum_i a_i T_i$. $\square$

We apply Lemma 5.2.5 in three different situations. We begin with:

Lemma 5.2.6 *Let $\mathcal{F} := \{J_1, ..., J_\ell\}$ be a Clifford family on a vector space V of dimension m.*

(1) Let $T = \sum_i a_i J_i$ where at least one of the coefficients is non-zero. Then $\dim \{\mathrm{Range}\{T\}\} = m$.

(2) If $m \geq \ell$, then there exists x in V so $\{J_i x\}$ is a linearly independent set of ℓ vectors.

(3) If $m \geq 2\ell$, then there exist x and y in V so $\{J_i x, J_i y\}$ is a linearly independent set of 2ℓ vectors.

(4) If $m \geq 2\ell$ and if $Jx \in \mathrm{Span}\{J_i x\}$ for all $x \in V$, then $J \in \mathrm{Span}\{J_i\}$.

Proof. Let $T = \sum_i a_i J_i$ where not all the coefficients a_i vanish. We have $T^2 = -\sum_i a_i^2 \, \mathrm{id}$. Thus T is invertible and (1) follows. We take $N = m$. Assertions (2), (3), and (4) then follow from Lemma 5.2.5. $\square$

Next we apply Lemma 5.2.5 to the family $\{J_i J_j\}_{i<j}$:

Lemma 5.2.7 *Let $\mathcal{F} := \{J_1, ..., J_\ell\}$ be a Clifford family on a vector space V of dimension m.*

(1) Let $T = \sum_{j,k} a_{jk} J_j J_k$ where $a_{jk} = -a_{kj}$ and where at least one of the coefficients is non-zero. Then $\dim \{\mathrm{Range}\{T\}\} \geq \frac{1}{2}m$.

(2) If $m \geq \ell(\ell-1)$, then there exists x in V so $\{J_j J_k x\}$ is a set of $\frac{1}{2}\ell(\ell-1)$ linearly independent vectors.

(3) If $m \geq 2\ell(\ell - 1)$, then there exist x and y in V so $\{J_j J_k x, J_j J_k y\}$ is a set of $\ell(\ell - 1)$ linearly independent vectors.

(4) If $m \geq 2\ell(\ell - 1)$ and if $Jx \in \mathrm{Span}\{J_i J_j x\}_{i<j}$ for all $x \in V$, then $J \in \mathrm{Span}\{J_i J_j\}$.

Proof. To prove Assertion (1), we suppose to the contrary $\dim\{\mathrm{Range}\{T\}\} < \frac{1}{2}m$. Let $K := \ker\{T\}$. Then $\dim\{K\} > \frac{1}{2}m$. Choose indexes u and v so $a_{uv} \neq 0$. Since $2\dim\{K\} > m$, we may choose $x \in S(K \cap J_u K)$. We establish the desired contradiction by noting:

$$
\begin{aligned}
0 &= \langle Tx, J_v J_u x\rangle + \langle J_u T J_u x, J_v J_u x\rangle \\
&= \sum_{jk} a_{jk}\langle (J_j J_k + J_u J_j J_k J_u)x, J_v J_u x\rangle \\
&= 4\sum_{i\neq u} a_{iu}\langle J_i J_u x, J_v J_u x\rangle = 4a_{vu} \neq 0\,.
\end{aligned}
$$

This contradiction establishes Assertion (1). Assertions (2), (3), and (4) then follow as in the proof of Lemma 5.2.6 from Lemma 5.2.5. $\square$

Remark 5.2.1   The estimate of Lemma 5.2.7 is sharp. Let $\ell \geq 4$ and let $\alpha := J_1 J_2 J_3 J_4$. Then $\alpha^2 = \mathrm{Id}$ so we can decompose $V = V_+ \oplus V_-$ into the ± 1 eigenspaces of α. Since J_1 anti-commutes with α, J_1 intertwines V_+ and V_- so $\dim V_+ = \dim V_- = \frac{1}{2}m$. Thus

$$
\begin{aligned}
\dim \ker\{J_1 J_2 \pm J_3 J_4\} &= \dim \ker\{J_3 J_4 (J_1 J_2 \pm J_3 J_4)\} \\
&= \dim \ker\{\alpha \mp \mathrm{Id}\} = \dim V_\pm = \tfrac{1}{2}m, \\
\dim \mathrm{Range}\{J_1 J_2 \pm J_3 J_4\} &= m - \tfrac{1}{2}m = \tfrac{1}{2}m\,.
\end{aligned}
$$

5.2.8   *Technical results concerning Clifford algebras*

Let $\mathcal{F} = \{J_1, ..., J_\ell\}$ be a Clifford family of rank ℓ on a vector space V of dimension m. We set $J_{ij} := J_i J_j$, $J_{ijk} := J_i J_j J_k$, and $J_{ijkn} := J_i J_j J_k J_n$. In Section 5.2.8, we study the cases $\ell = 4$, $\ell = 5$, $\ell = 6$, and $\ell = 7$.

Lemma 5.2.8   *Let $\mathcal{F} := \{J_1, ..., J_4\}$ be a Clifford family of rank 4 on a vector space V of dimension m. Then:*

(1) If $m \geq 8$, there exists $x \in S(V)$ so that $\langle J_{123}y, y\rangle = 1$, so that $\langle J_{ijk}y, y\rangle = 0$ if $\{i, j, k\}$ is not a permutation of $\{1, 2, 3\}$, and so that $\{J_{ij}x\}_{i<j}$ is an orthonormal set of 6 vectors.

(2) If $m \geq 16$, there exist $x, y \in S(V)$ so that $\{J_{ij}x, J_{ij}y\}_{i<j}$ is an orthonormal set of 12 vectors.

(3) If $m \geq 16$, there exists $z \in S(V)$ so that $\{J_I\}$ is an orthonormal set of 16 vectors where I ranges over all possible multi-indices.

(4) If $m \geq 32$, there exists $x, y \in S(V)$ so that $\{J_i x, J_{jk} x, J_i y, J_{jk} y\}_{j<k}$ is an orthonormal set of 20 vectors.

Assertion (1) is a bit technical. It, and a similar result for rank 5 Clifford structures to be established presently in Lemma 5.2.9, will be used in the proof of Lemma 5.6.2 when we generalize Lemma 5.2.4 (1) to the higher rank setting.

Proof. In studying $\langle J_{ijk} x, x \rangle$, we may assume all 3 indices are distinct since if at least 2 indices agree, $J_{ijk} = \pm J_a$ and $x \perp J_a x$. Since J_{123} is self-adjoint and since $J_{123}^2 = \mathrm{Id}$, there is an orthonormal decomposition of $V = V_+ \oplus V_-$ into the $\pm$ eigenvalues of J_{123}. The eigenspaces $V_{\pm}$ are preserved by J_1, J_2, and J_3 since these commute with J_{123}. However, since J_4 anti-commutes with J_{123}, J_4 intertwines V_+ and V_-. Thus

$$\dim(V_+) = \dim(V_-) = \tfrac{1}{2} m.$$

Choose $x_+ \in S(V_+)$. Then $\langle J_{123} x_+, x_+ \rangle = 1$. If $\{i, j, k\}$ are distinct indices which do not form a permutation of $\{1, 2, 3\}$, then $J_{ijk} x_+ \in \Delta_-$ so $\langle J_{ijk} x_+, x_+ \rangle = 0$.

Let $i < j$ and $k < n$. Then $\langle J_{ij} x_+, J_{kn} x_+ \rangle = -\langle J_{ijkn} x_+, x_+ \rangle$. If exactly 2 indices agree, this vanishes since $\langle J_{ab} x_+, x_+ \rangle = 0$ for $a < b$. If all 4 indices are different, then $J_{ijkn} = \pm J_{1234}$ so $\langle J_{1234} x_+, x_+ \rangle = 0$ as $J_{1234} x_+ \in V_-$. Assertion (1) now follows.

We have $\{J_{14} x_+, J_{24} x_+, J_{34} x_+\} \subset V_-$. Since $m \geq 16$, $\dim(V_+) \geq 8$ so we may choose $y_+ \in V_+$ with $y_+ \perp \{x_+, J_{12} x_+, J_{13} x_+, J_{23} x_+\}$; $\{x_+, y_+\}$ then satisfy the conditions of Assertion (2). Finally $z = (x_+ + J_4 y_+)/\sqrt{2}$ satisfies the conditions of Assertion (3). Doubling the dimension permits to derive Assertion (4) from Assertion (3). $\qquad\square$

Next we study rank 5 Clifford families.

Lemma 5.2.9 *Let $\mathcal{F} := \{J_1, ..., J_5\}$ be a Clifford family of rank 5 on a vector space of dimension m.*

(1) If $m \geq 16$, there exists $x \in S(V)$ so $x \perp J_{ijk} x$ for all i, j, k and so $J_{12} x \perp \mathrm{Span}\{J_{ij} x\}_{i<j, (i,j) \neq (1,2)}$.

(2) If $m \geq 16$, there exists $x \in S(V)$ so $\langle x, J_{123} x \rangle = 1$ and so $x \perp J_i J_j J_k x$ if $\{i, j, k\}$ is not a permutation of $\{1, 2, 3\}$.

(3) If $m \geq 16$, there exists $x \in S(V)$ so $\{J_{ij} x\}_{i<j}$ is an orthonormal set.

(4) If $m \geq 32$, there exists $x, y \in S(V)$ so $\{J_{ij}x, J_{ij}y\}_{i<j}$ is an orthonormal set.

Proof. Since $x \perp J_i x$, we may assume $1 \leq i < j < k \leq 5$ in proving Assertion (1). Because $J_{123}^2 = J_{145}^2 = \mathrm{Id}$ and because $J_{123}J_{145} = J_{145}J_{123}$, we can decompose

$$V = V_{++} \oplus V_{+-} \oplus V_{-+} \oplus V_{--}$$

into the simultaneous eigenspaces of J_{123} and J_{145}. Since J_2 anti-commutes with J_{145} and commutes with J_{123} and since J_4 anti-commutes with J_{123} and commutes with J_{145}, they intertwine these spaces so all these spaces have the same dimension; since $m \geq 16$, $\dim\{V_{\vec{\varepsilon}}\} \geq 4$ for any choice of signs $\vec{\varepsilon} = (\varepsilon_1, \varepsilon_2)$. Let $x \in V_{\vec{\varepsilon}}$. We have

$$J_{124}x = \pm J_{124}J_{123}J_{145}x = \pm J_{135}x \in V_{-\varepsilon_1, -\varepsilon_2},$$
$$J_{125}x = \pm J_{125}J_{123}J_{145}x = \pm J_{134}x \in V_{-\varepsilon_1, -\varepsilon_2},$$
$$\{J_{234}x, J_{235}x\} \subset V_{-\varepsilon_1, \varepsilon_2},$$
$$\{J_{245}x, J_{345}x\} \subset V_{\varepsilon_1, -\varepsilon_2},$$
$$\{J_{123}x, J_{145}x\} \subset V_{\varepsilon_1, \varepsilon_2}.$$

Since $\dim V_{\vec{\varepsilon}} \geq 4$, we may choose $x_{++} \in S(V_{++})$ and $x_{--} \in S(V_{--})$ so that $J_{ijk}x_{++} \perp x_{--}$ for i, j, k distinct. Let $x = (x_{++} + x_{--})/\sqrt{2} \in S(V)$. Since

$$\{J_{1234}x, J_{1235}x, J_{1245}x\} \subset V_{+-} \oplus V_{-+}$$

we have $x \perp J_{1234}x$, $x \perp J_{1235}x$, and $x \perp J_{1245}x$. Since $x \perp J_{ij}x$ for $i \neq j$, we may conclude

$$J_{12}x \perp \mathrm{Span}\{J_{ij}x\}_{i<j, (i,j) \neq (1,2)}.$$

Furthermore $J_{ijk}x \perp x$ except possibly for J_{123} and J_{145}. We complete the proof of Assertion (1) by checking:

$$\langle x, J_{123}x \rangle = \langle x, J_{145}x \rangle = \tfrac{1}{2} - \tfrac{1}{2} = 0.$$

To prove Assertion (2), we choose x_{++} in $S(V_{++})$ and x_{+-} in $S(V_{+-})$ so that $J_{ijk}x_{++} \perp x_{+-}$ for i, j, k distinct. Let $x = (x_{++} + x_{+-})/\sqrt{2} \in S(V)$. As $J_{ijk}x \perp x$ except possibly for J_{123} and J_{145}, Assertion (2) follows since:

$$\langle x, J_{123}x \rangle = \tfrac{1}{2} + \tfrac{1}{2} = 1, \quad \text{and} \quad \langle x, J_{145}x \rangle = \tfrac{1}{2} - \tfrac{1}{2} = 0.$$

Let $i < j$ and $k < l$. To prove Assertion (3), we must choose x so $x \perp J_{ijkl}x$ unless $(i,j) = (k,l)$. Since $x \perp J_{uv}x$ for $u < v$, we may assume i, j, k, l distinct. Let $x \in V_{\varepsilon_1, \varepsilon_2}$. Then

$$J_{1234}x \in V_{-\varepsilon_1, \varepsilon_2}, \qquad J_{1235}x \in V_{-\varepsilon_1, \varepsilon_2},$$
$$J_{1245}x \in V_{\varepsilon_1, -\varepsilon_2}, \qquad J_{1345}x \in V_{\varepsilon_1, -\varepsilon_2},$$
$$J_{2345}x = -\varepsilon_1 \varepsilon_2 x \in V_{\varepsilon_1, \varepsilon_2}.$$

Choose $x_{++} \in V_{++}$ and $x_{-+} \in V_{-+}$ so

$$J_{ijkn}x_{++} \perp x_{-+} \quad \text{for all} \quad i, j, k, n.$$

Consequently

$$x := (x_{++} + x_{-+})/\sqrt{2}$$

satisfies the conditions of Assertion (3). If $\dim(V) \geq 32$, we can also choose $y_{++} \in V_{++}$ and $y_{-+} \in V_{-+}$ so $J_{ijkn}y_{\bar{\varepsilon}} \perp x_{\bar{\varepsilon}}$ and setting

$$y := (y_{++} + y_{+-})/\sqrt{2}$$

then yields a pair $\{x, y\}$ which satisfies the conditions of Assertion (4). $\quad\square$

We will use the following Lemma for rank 6 and rank 7 Clifford algebras; the estimate $m \geq 2^\ell$ is significantly worse than the polynomial estimates obtained in Section 5.2.7.

Lemma 5.2.10 *Let $\mathcal{F} = \{J_1, ..., J_\ell\}$ be a Clifford family of rank $\ell = 6$ or $\ell = 7$ on a vector space of dimension $m \geq 2^\ell$. Then there exists $x \in S(V)$ so that $x \perp J_{ijk}x$ for any i, j, k and so that $\{J_{ij}x\}_{i<j}$ is an orthonormal set of $\frac{1}{2}\ell(\ell-1)$ elements.*

Proof. Suppose first that $\ell = 6$. Let $\mathcal{A} := \text{Clif}(\mathbb{R}^6)$. By Theorem 4.2.1, $\mathcal{A} = M_8(\mathbb{R})$. Thus there is only one irreducible $\mathcal{A}$ module of dimension 8 and any two $\mathcal{A}$ modules of dimension m are isomorphic. If $\xi \in \mathbb{R}^6$, let $\text{ext}(\xi)$ denote exterior multiplication and let $\text{int}(\xi)$ denote the dual operator interior multiplication on the exterior algebra $\Lambda(\mathbb{R}^6)$. We set $c(\xi) := \text{ext}(\xi) - \text{int}(\xi)$ to define a Clifford module structure on $\Lambda(\mathbb{R}^6)$. Let $\{e_1, ..., e_6\}$ be the standard orthonormal basis for $\mathbb{R}^6$ and set $J_i := c(e_i)$. If $1 \leq i_1 < ... < i_p \leq 6$, then

$$J_{i_1}...J_{i_p}1 = e_{i_1} \wedge ... \wedge e_{i_p}$$

so $\{J_I 1\}$ forms an orthonormal family. Since $\dim\{\Lambda(\mathbb{R}^6)\} = 64$, $(V, \mathcal{F})$ contains $(\Lambda(\mathbb{R}^6), \mathcal{F})$ as a submodule. The Lemma now follows if $\ell = 6$.

If $\ell = 7$, let $\alpha := J_1...J_7$. Because $\alpha^2 = \mathrm{Id}$, we may decompose

$$V = V_+ \oplus V_-$$

into the ± 1 eigenspaces of α. Since the J_i commute with α, the spaces $V_\pm$ are $\mathcal{A}$ modules. Since $\dim(V) \geq 128$, either $\dim(V_+)$ or $\dim(V_-)$ is at least 64. By replacing J_1 by $-J_1$, we may assume without loss of generality that $\dim(V_+) \geq 64$. This means that $J_7 = -J_1...J_6$. The argument given above shows we may find x so $\{J_I x\}$ forms an orthonormal set as I ranges over all collections of indices less than 7. Thus

$$\{J_I x, J_K J_7 x\}_{|I|\leq 3, |K|\leq 2}$$

is an orthonormal set where I and J consist of indices less than 7 since $J_K J_7 = J_{\bar{K}}$ where $\bar{K}$ consists of the complementary set of at least 5 indices less than 7. $\qquad\square$

5.3 Clifford Families of Rank 1

Let $\mathfrak{M} = (V, \langle \cdot, \cdot \rangle, J, A := c_0 A_{\langle \cdot, \cdot \rangle} + c_1 A_{J_1})$ be a complex 0-model. If $c_1 = 0$, then $A = c_0 A_{\langle \cdot, \cdot \rangle}$ and $\mathfrak{M}$ is complex Osserman by Theorem 5.2.2. This establishes Assertion (1) of Theorem 5.1.1 when the Clifford family has rank 0. We therefore suppose $c_1 \neq 0$ so the rank is 1. We have the following examples:

Lemma 5.3.1   *Let $\mathfrak{M} = (V, \langle \cdot, \cdot \rangle, J, A := c_0 A_{\langle \cdot, \cdot \rangle} + c_1 A_{J_1})$ be a complex 0-model which is given by a Clifford family of rank 1 with $c_1 \neq 0$.*

(1) If $c_0 = 0$ and if $JJ_1 = JJ_1$ or $JJ_1 = -J_1 J$, $\mathfrak{M}$ is complex Osserman.
(2) If $c_0 \neq 0$ and if $J = \pm J_1$ or if $JJ_1 = -J_1 J$, $\mathfrak{M}$ is complex Osserman.

Proof.   Suppose that $c_0 = 0$ and $JJ_1 = \pm J_1 J$. If $x \in S(V)$, then we have that $J_1 x \perp J_1 J x$. By Eq. (5.1.a),

$$c_1 \mathcal{J}_A(\pi_x) y = \begin{cases} 0 & \text{if } y \perp \mathrm{Span}\{J_1 x, J_1 J x\}, \\ 3 c_1 y & \text{if } y \in \mathrm{Span}\{J_1 x, J_1 J x\}. \end{cases}$$

Thus $\mathrm{Spec}\{\mathcal{J}_A(\pi_x)\} = \{0, 3c_1\}$ is independent of $x \in S(V)$. We have

$$E_{3c_1}(\pi_x) = J_1 \pi_x \quad \text{and} \quad E_0(\pi_x) = J_1 \pi_x^\perp .$$

Since $JJ_1 = \pm J_1 J$, $JJ_1\pi_x = J_1\pi_x$ and $JJ_1\pi_x^\perp = J_1\pi_x^\perp$. Thus the eigenspaces are J-invariant. Lemma 5.2.1 now implies $\mathfrak{M}$ is complex Osserman.

Suppose $c_0 \neq 0$. If $J = \pm J_1$, then we may use Theorem 5.2.2 (1) to see $\mathfrak{M}$ is complex Osserman. If $JJ_1 = -J_1 J$, then $\{J, J_1\}$ is a Clifford family. Furthermore, $A = c_0 R_0 + 0 R_J + c_1 R_{J_1}$. Theorem 5.2.2 (2) then shows $\mathfrak{M}$ is complex Osserman. $\qquad\qquad\square$

We complete our study of the case $\ell = 1$ and establish Assertion (2) of Theorem 5.1.1 by showing:

Lemma 5.3.2 *Let* $\mathfrak{M} = (V, \langle\cdot,\cdot\rangle, J, A := c_0 A_{\langle\cdot,\cdot\rangle} + c_1 A_{J_1})$ *be a complex Osserman 0-model which is given by a Clifford family of rank 1 with* $c_1 \neq 0$.

(1) If $c_0 = 0$, *then* $JJ_1 = JJ_1$ *or* $JJ_1 = -J_1 J$.
(2) If $c_0 \neq 0$, *then* $J = \pm J_1$ *or* $JJ_1 = -J_1 J$.

Proof. Since $A_{\langle\cdot,\cdot\rangle}$ and J are compatible and since A and J are compatible, A_{J_1} and J are compatible. If $x \in S(V)$, then $\{J_1 x, J_1 Jx\}$ is an orthonormal set. By Eq. (5.1.a),

$$
\begin{aligned}
\mathcal{J}_{A_{\langle\cdot,\cdot\rangle}}(\pi_x)y &= 2y - \langle y, x\rangle x - \langle y, Jx\rangle Jx, \\
\mathcal{J}_{A_{J_1}}(\pi_x)y &= 3\langle y, J_1 x\rangle J_1 x + 3\langle y, J_1 Jx\rangle J_1 Jx .
\end{aligned}
\tag{5.3.a}
$$

Thus $\mathrm{Range}\{\mathcal{J}_{A_{J_1}}(\pi)\} = \mathrm{Span}\{J_1 x, J_1 Jx\}$. As both J and J_1 are Hermitian almost complex structures on this 2-dimensional space,

$$
JJ_1 x = \varepsilon_x J_1 Jx \quad \text{for} \quad \varepsilon_x = \langle JJ_1 JJ_1 x, x\rangle = \pm 1 .
$$

Since the map $x \to \varepsilon_x$ is continuous, since $S(V)$ is connected, and since $\varepsilon_x = \pm 1$, ε_x is constant. Therefore either $J_1 J = JJ_1$ or $J_1 J = -JJ_1$. This proves Assertion (1).

Suppose $c_0 \neq 0$. We argue as above to see $J_1 J = \pm JJ_1$. To complete the proof of Assertion (2), we must deal with the case $JJ_1 = J_1 J$. We suppose $J_1 \neq \pm J$ and argue for a contradiction. Since $(JJ_1)^2 = \mathrm{Id}$, there is a non-trivial decomposition of V of the form

$$
V = V_+ \oplus V_- \quad \text{where} \quad J_1 = \pm J \quad \text{on} \quad V_\pm .
$$

Let $x_\pm \in S(V_\pm)$. Since $Jx_\pm = \pm J_1 x_\pm$, Eq. (5.3.a) yields:

$$
\mathcal{J}_R(\pi_{x_\pm})y = \begin{cases} (c_0 + 3c_1)y & \text{if } y \in \pi_{x_\pm}, \\ 2c_0 y & \text{if } y \perp \pi_{x_\pm} . \end{cases}
$$

This shows that:

$$\begin{aligned}
\text{if } c_0 \neq 3c_1 \text{ then } \vec{\lambda} &= (2c_0, c_0 + 3c_1), &&\text{and } \vec{\mu} = (m - 2, 2),\\
\text{if } c_0 = 3c_1 \text{ then } \vec{\lambda} &= (2c_0), &&\text{and } \vec{\mu} = (m).
\end{aligned} \qquad (5.3.\text{b})$$

Let $x = (x_+ + x_-)/\sqrt{2} \in S(V)$. The following vectors

$$\begin{aligned}
x &= (x_+ + x_-)/\sqrt{2}, & Jx &= (Jx_+ + Jx_-)/\sqrt{2},\\
J_1 x &= (Jx_+ - Jx_-)/\sqrt{2}, & J_1 Jx &= (-x_+ + x_-)/\sqrt{2},
\end{aligned}$$

form an orthonormal set. We use Eq. (5.3.a) to see:

$$\mathcal{J}(\pi_x)y = \begin{cases}
2c_0 y & \text{if } y \perp \{x, J_1 x, Jx, J_1 Jx\},\\
c_0 y & \text{if } y \in \mathrm{Span}\{x, Jx\},\\
(2c_0 + 3c_1)y & \text{if } y \in \mathrm{Span}\{J_1 x, J_1 Jx\}.
\end{cases}$$

This shows that:

$$\begin{aligned}
\text{if } c_0 \neq -3c_1 \text{ then } \vec{\lambda} &= (2c_0, c_0, 2c_0 + 3c_1), &&\text{and } \vec{\mu} = (m - 4, 2, 2),\\
\text{if } c_0 = -3c_1 \text{ then } \vec{\lambda} &= (2c_0, c_0), &&\text{and } \vec{\mu} = (m - 4, 4).
\end{aligned} \qquad (5.3.\text{c})$$

It follows from Eqs. (5.3.b) and (5.3.c) that $\mathcal{J}(\pi_{x_\pm})$ and $\mathcal{J}(\pi_x)$ do not have the same eigenvalues and eigenvalue multiplicities. This contradicts the assumption that $\mathfrak{M}$ is complex Osserman and establishes Assertion(2). $\square$

5.4   Clifford Families of Rank 2

Let $\mathfrak{M} = (V, \langle \cdot, \cdot \rangle, J, A := c_0 A_{\langle \cdot, \cdot \rangle} + c_1 A_{J_1} + c_2 A_{J_2})$ be a complex 0-model which is given by a Clifford family of rank 2 for $c_1 \neq 0$ and $c_2 \neq 0$. Assertion (3) of Theorem 5.1.1 will follow from the results of this section. By Eq. (5.1.a),

$$\begin{aligned}
\mathcal{J}_A(\pi_x)y = c_0\{2y - \langle y, x \rangle x - \langle y, Jx \rangle Jx\}\\
+ 3c_1 \langle y, J_1 x \rangle J_1 x + 3c_1 \langle y, J_1 Jx \rangle J_1 Jx\\
+ 3c_2 \langle y, J_2 x \rangle J_2 x + 3c_2 \langle y, J_2 Jx \rangle J_2 Jx.
\end{aligned} \qquad (5.4.\text{a})$$

One implication of Theorem 5.1.1 (3) is provided by the following result:

Lemma 5.4.1   *Let $\mathfrak{M} = (V, \langle \cdot, \cdot \rangle, J, A := c_0 A_{\langle \cdot, \cdot \rangle} + c_1 A_{J_1} + c_2 A_{J_2})$ be a complex 0-model which is given by a Clifford family of rank 2 where $c_1 \neq 0$ and $c_2 \neq 0$. Then $\mathfrak{M}$ is complex Osserman if any of the following holds:*

(1) $J = J_1 J_2$.

(2) $J = J_1$.

(3) $c_0 = 0$, $JJ_1 = J_1 J$, and $JJ_2 = -J_2 J$.

Proof. Let $J_3 := J_1 J_2$. Then $\{J_1, J_2, J_3\}$ is a Clifford family of rank 3. By Theorem 5.2.2 (2),

$$\mathfrak{M} := (V, \langle \cdot, \cdot \rangle, J_i, c_0 A_{\langle \cdot, \cdot \rangle} + c_1 A_{J_1} + c_2 A_{J_2} + c_3 A_{J_3})$$

is complex Osserman for $i = 1, 2, 3$. Assertions (1) and (2) follow by setting $c_3 = 0$ and by choosing $i = 3$ or $i = 1$.

Suppose that $c_0 = 0$, that $JJ_1 = J_1 J$, and that $JJ_2 = -J_2 J$. Let $x \in S(V)$; the vectors $\{J_1 x, J_1 J x, J_2 x, J_2 J x\}$ form an orthonormal set. Thus by Eq. (5.4.a),

$$\mathcal{J}_A(\pi_x) y = \begin{cases} 3c_1 y & \text{if } y \in \mathrm{Span}\{J_1 x, J_1 J x\}, \\ 3c_2 y & \text{if } y \in \mathrm{Span}\{J_2 x, J_2 J x\}, \\ 0 & \text{if } y \perp \{J_1 x, J_1 J x, J_2 x, J_2 J x\}. \end{cases}$$

This shows that the spectrum of $\mathcal{J}_A(\pi_x)$ is independent of x. Furthermore the eigenspaces are invariant under J as

$$\mathrm{Span}\{J_i x, J_i J x\} = \mathrm{Span}\{J_i x, J J_i x\} \quad \text{for} \quad i = 1, 2.$$

Consequently by Lemma 5.2.1, $\mathfrak{M}$ is complex Osserman. $\qquad\square$

5.4.1 *The tensor* $c_1 A_{J_1} + c_2 A_{J_2}$

We begin our study of the case $c_0 = 0$ with the following technical Lemma. Let $r := \mathrm{Rank}\{\mathcal{J}_A(\pi)\}$; this is independent of the particular $\pi \in \mathbb{CP}(V)$ if $\mathfrak{M}$ is complex Osserman.

Lemma 5.4.2 *Let* $\mathfrak{M} := (V, \langle \cdot, \cdot \rangle, J, A := c_1 A_{J_1} + c_2 A_{J_2})$ *be a complex Osserman* 0*-model which is given by a Clifford family of rank* 2 *where* $c_1 \neq 0$ *and* $c_2 \neq 0$. *Define* $\alpha_x := \langle J_1 J_2 x, J x \rangle \in [-1, 1]$. *Then*

(1) The function α_x *is constant on* $S(V)$.

(2) If $r = 4$, *then* $\alpha \neq \pm 1$.

(3) If $r < 4$, *then* $\alpha = \pm 1$.

(4) If $\alpha = 0$, *then* $J_1 J_2 J + J J_1 J_2 = 0$.

(5) If $\alpha = \pm 1$, *then* $J = \pm J_1 J_2$.

Proof. We have by definition that:

$$\alpha_x = \langle J_1 J_2 x, J x \rangle = \langle J_1 x, J_2 J x \rangle = -\langle J_2 x, J_1 J x \rangle.$$

By Eq. (5.4.a) one has that:

$$\mathrm{Range}\{\mathcal{J}_A(\pi_x)\} \subset \mathrm{Span}\{J_1 x, J_2 Jx, J_2 x, J_1 Jx\},$$
$$\mathcal{J}_A(\pi_x) J_1 x = 3c_1 J_1 x + 3\alpha_x c_2 J_2 Jx,$$
$$\mathcal{J}_A(\pi_x) J_2 Jx = 3\alpha_x c_1 J_1 x + 3c_2 J_2 Jx,$$
$$\mathcal{J}_A(\pi_x) J_1 Jx = 3c_1 J_1 Jx + 3\alpha_x c_2 (-J_2 x),$$
$$\mathcal{J}_A(\pi_x)(-J_2 x) = 3\alpha_x c_1 J_1 Jx + 3c_2 (-J_2 x).$$

This shows that the spaces

$$V_1(x) := \mathrm{Span}\{J_1 x, J_2 Jx\} \quad \text{and}$$
$$V_2(x) := V_1(Jx) = \mathrm{Span}\{J_1 Jx, -J_2 x\}$$

are invariant under $\mathcal{J}_A(\pi_x)$. Clearly

$$\mathrm{Range}\{\mathcal{J}_A(\pi_x)\} \subset V_1(x) \oplus V_2(x) \quad \text{and}$$
$$V_1(x) \perp V_2(x).$$

Suppose that $r = 4$. This implies that:

$$\dim\{V_1(x)\} = 2, \quad \dim\{V_2(x)\} = 2, \quad \text{and}$$
$$\mathrm{Range}\{\mathcal{J}_A(\pi_x)\} = V_1(x) \oplus V_2(x).$$

Furthermore, relative to the given bases we have

$$\mathcal{J}_A(\pi_x)|_{V_1(x)} = \mathcal{J}_A(\pi_x)|_{V_2(x)} = \begin{pmatrix} 3c_1 & 3\alpha_x c_1 \\ 3\alpha_x c_2 & 3c_2 \end{pmatrix} \tag{5.4.b}$$

$$\det\left\{\mathcal{J}_A(\pi_x)|_{\mathrm{Range}\{\mathcal{J}_A(\pi_x)\}}\right\} = \left\{9c_1 c_2 (1 - \alpha_x^2)\right\}^2.$$

Since the eigenvalues of $\mathcal{J}_A(\pi)$ are constant, $\det\{\mathcal{J}_A(\pi)\}$ is constant. This implies that $\alpha := \alpha_x$ is independent of x. Furthermore, $r = 4$ implies $\alpha^2 \neq 1$. This establishes Assertions (1) and (2) if $r = 4$.

Suppose next $r = 2$. We note $J_1 J_2 : V_1(x) \leftrightarrow V_2(x)$ and hence

$$\dim\{V_1(x)\} = \dim\{V_2(x)\}.$$

If $\dim\{V_1(x)\} = \dim\{V_2(x)\} = 2$, then $\alpha_x^2 = 1$ by Eq. (5.4.b). On the other hand, if

$$\dim\{V_1(x)\} = \dim\{V_2(x)\} = 1,$$

then $J_1 x$ is a multiple of $J_2 Jx$. As these are unit vectors, $J_1 x = \pm J_2 Jx$ and again $\alpha_x^2 = 1$. Suppose finally $r = 0$. This is not possible if $\dim\{V_1(x)\} = 2$

by Eq. (5.4.b) since $c_1 \neq 0$. Thus

$$\dim\{V_1(x)\} = \dim\{V_2(x)\} = 1$$

so again $\alpha_x^2 = 1$. As α_x varies continuously, $\alpha = \alpha_x$ is constant and takes the value ± 1. This completes the proof of Assertion (1) and also completes the proof of Assertion (3).

Suppose $\alpha = 0$. Let $\Theta := J J_1 J_2$. As $\alpha = 0$, $(\Theta x, x) = 0$ for all x. We polarize to see

$$(\Theta \xi_1, \xi_2) + (\Theta \xi_2, \xi_1) = 0 \quad \text{for all} \quad \xi_i \in V$$

so $\Theta + \Theta^* = 0$. Assertion (4) follows since

$$\Theta^* = -J_2 J_1 J = J_1 J_2 J.$$

Since $J_1 J_2 x$ and Jx are unit vectors, $\alpha = \pm 1$ implies $J_1 J_2 x = \pm Jx$. This proves Assertion (5). $\square$

The case $J = \pm J_1 J_2$ leads to Case (3a) of Theorem 5.1.1. We therefore assume $\alpha \neq \pm 1$ and continue our analysis. If $\alpha \neq 0$ or $c_1 \neq c_2$, then the following Lemma shows that we obtain Case (3c) of Theorem 5.1.1:

Lemma 5.4.3 *Let $\mathfrak{M} := (V, \langle \cdot, \cdot \rangle, J, c_1 A_{J_1} + c_2 A_{J_2})$ be a complex Osserman 0-model which is given by a Clifford family of rank 2 where $c_1 \neq 0$ and $c_2 \neq 0$. Let $\alpha \neq \pm 1$. Assume either $\alpha \neq 0$ or $c_1 \neq c_2$. Then either $J_1 J = J J_1$ and $J_2 J = -J J_2$ or $J_1 J = -J J_1$ and $J_2 J = J J_2$.*

Proof. The operator $\mathcal{J}_A(\pi)$ is self-adjoint and hence diagonalizable. It also has rank 4. Since either $\alpha \neq 0$ or $c_1 \neq c_2$, we may apply Eq. (5.4.b) to see that $\mathcal{J}_A$ is not a scalar multiple of the identity on $V_1(x)$. Thus the eigenvalues of $\mathcal{J}_A$ are not equal on $V_1(x)$. As J defines an almost complex structure and preserves the eigenspaces of $\mathcal{J}_A(\pi_x)$, J maps $V_1(x)$ to $V_2(x)$. Let

$$e_1(x) := J_1 x, \quad e_2(x) := (\sqrt{1-\alpha^2})^{-1}\{J_2 J x - \alpha J_1 x\},$$
$$f_1(x) := J_1 J x, \; f_2(x) := (\sqrt{1-\alpha^2})^{-1}\{-J_2 x - \alpha J_1 J x\}$$

define, respectively, an orthonormal basis for $V_1(x)$ and an orthonormal basis for $V_2(x) = V_1(Jx)$. Note that $f_1(x) = e_1(Jx)$ and $f_2(x) = e_2(Jx)$. Relative to these bases, the action of $\mathcal{J}_A(\pi_x)$ on $V_i(x)$ is given by the matrix

$$\Theta_{\mathcal{J}} := \begin{pmatrix} 3c_1 + 3\alpha^2 c_2 & 3\alpha\sqrt{1-\alpha^2}\, c_2 \\ 3\alpha\sqrt{1-\alpha^2}\, c_2 & 3(1-\alpha^2)c_2 \end{pmatrix}.$$

Furthermore, using this identification of $V_1(x)$ and $V_2(x)$ with $\mathbb{R}^2$, we may express J as an orthogonal matrix Ψ_x which commutes with the matrix Θ_J. There exists an θ_x and a sign $\delta_x = \pm 1$ so that:

$$\begin{aligned}
\Psi_x &:= \begin{pmatrix} \langle Je_1(x), f_1(x) \rangle & \langle Je_2(x), f_1(x) \rangle \\ \langle Je_1(x), f_2(x) \rangle & \langle Je_2(x), f_2(x) \rangle \end{pmatrix} \\
&= \begin{pmatrix} \cos\theta_x & \sin\theta_x \\ -\delta_x \sin\theta_x & \delta_x \cos\theta_x \end{pmatrix} .
\end{aligned} \tag{5.4.c}$$

Let $\{v_1, v_2\}$ be unit eigenvectors for Θ_J corresponding to the two distinct eigenvalues. Then $\Psi_x v_i = \delta_i(x) v_i$ where $\delta_i(x) = \pm 1$. As $\delta_i(x)$ varies continuously, δ_i is constant; thus $\Psi = \Psi_x$, $\theta = \theta_x$, and $\delta = \delta_1 \delta_2$ are independent of x. Furthermore, the inner products $\langle Je_i(x), f_j(x) \rangle$ are constant and

$$\langle Je_1(x), f_1(x) \rangle = \det\{\Psi\} \langle Je_2(x), f_2(x) \rangle$$
$$\text{where} \quad \delta = \det\{\Psi\} = \pm 1 .$$

Suppose that $\det(\Psi) = 1$. Then $\delta_1 = \delta_2 = \pm 1$ so $\Psi = \pm \operatorname{Id}$. This implies

$$J J_i x = \delta_i J_i J x \quad \text{so} \quad J J_1 J_2 x = J_1 J_2 J x .$$

Since $(J J_1 J_2)^2 = \operatorname{Id}$, we can find x so $J J_1 J_2 x = \pm x$ and consequently $\alpha = \alpha_x = \pm 1$. This is false as we assumed $\alpha \neq \pm 1$. Consequently $\delta_1 \delta_2 = -1$ so

$$\langle Je_1(x), f_1(x) \rangle = -\langle Je_2(x), f_2(x) \rangle . \tag{5.4.d}$$

Since α is constant, we replace x by $J_2 x$, $J_1 x$, and by $J_1 J x$ to see:

$$\begin{aligned}
\alpha &= \langle J_1 J_2 x, J x \rangle = -\langle J_2 x, J_1 J x \rangle = \langle J_1 x, J_2 J x \rangle, \\
\alpha &= \langle J_1 J_2 J_2 x, J J_2 x \rangle = -\langle J_1 x, J J_2 x \rangle, \\
\alpha &= \langle J_1 J_2 J_1 x, J J_1 x \rangle = \langle J_2 x, J J_1 x \rangle, \\
\alpha &= \langle J_2 J x, J J_1 J x \rangle = -\langle J J_2 J x, J_1 J x \rangle .
\end{aligned} \tag{5.4.e}$$

Let $\{J_2 x, \xi_x\}$ be an orthonormal basis for $V_2(x)$. We may use Eq. (5.4.e) to choose

$$\xi_x := (1 - \alpha^2)^{-1/2} \{J_1 J x + \alpha J_2 x\} .$$

Since $\langle JJ_1 x, J_2 x \rangle = \alpha$, there is a sign $\varepsilon_1 = \pm 1$, which by continuity is independent of x, so $\langle JJ_1 x, \xi_x \rangle = \varepsilon_1 \sqrt{1 - \alpha^2}$. Thus:

$$\langle JJ_1 x, J_1 Jx \rangle = \langle JJ_1 x, (1 - \alpha^2)^{1/2} \xi_x - \alpha J_2 x \rangle$$
$$= \varepsilon_1 (1 - \alpha^2) - \alpha^2 \,.$$

If $\varepsilon_1 = -1$, then $\langle JJ_1 x, J_1 Jx \rangle = -1$. Consequently

$$JJ_1 = -J_1 J \quad \text{and} \quad \cos \theta_x = -1$$

in Eq. (5.4.c). Thus $\Psi = \mathrm{diag}(-1, 1)$. Since $J\mathcal{J}_A(\pi_x) = \mathcal{J}_A(\pi_x)J$,

$$\begin{pmatrix} -1 & 0 \\ 0 & 1 \end{pmatrix} \begin{pmatrix} 3c_1 + 3\alpha^2 c_2 & 3\alpha\sqrt{1 - \alpha^2}\,c_2 \\ 3\alpha\sqrt{1 - \alpha^2}\,c_2 & 3(1 - \alpha^2)c_2 \end{pmatrix}$$
$$= \begin{pmatrix} 3c_1 + 3\alpha^2 c_2 & 3\alpha\sqrt{1 - \alpha^2}\,c_2 \\ 3\alpha\sqrt{1 - \alpha^2}\,c_2 & 3(1 - \alpha^2)c_2 \end{pmatrix} \begin{pmatrix} -1 & 0 \\ 0 & 1 \end{pmatrix} \,.$$

This implies

$$-3\alpha\sqrt{1 - \alpha^2}\,c_2 = 3\alpha\sqrt{1 - \alpha^2}\,c_2 \,.$$

Since $c_2 \neq 0$ and since $\alpha \neq \pm 1$, this shows that $\alpha = 0$. Thus by Lemma 5.4.2 (4), $JJ_1 J_2 = -J_1 J_2 J$. Thus $J_2 J = JJ_2$. This is one of the possibilities given in the Lemma.

After applying a similar argument to J_2, we have $\varepsilon_1 = \varepsilon_2 = 1$ so

$$\langle JJ_1 x, J_1 Jx \rangle = \langle JJ_2 x, J_2 Jx \rangle = 1 - 2\alpha^2 \,. \tag{5.4.f}$$

We argue for a contradiction. It is implied by Eqs. (5.4.e), and (5.4.f) that:

$$\langle Je_1(x), f_1(x) \rangle = \langle JJ_1 x, J_1 Jx \rangle = 1 - 2\alpha^2 ,$$
$$\langle Je_2(x), f_2(x) \rangle = (1 - \alpha^2)^{-1} \langle JJ_2 Jx - \alpha JJ_1 x, -J_2 x - \alpha J_1 Jx \rangle$$
$$= (1 - \alpha^2)^{-1} \{ \langle J_2 Jx, JJ_2 x \rangle + \alpha^2 \langle JJ_1 x, J_1 Jx \rangle$$
$$+ \alpha \langle JJ_1 x, J_2 x \rangle - \alpha \langle JJ_2 Jx, J_1 Jx \rangle \}$$
$$= (1 - \alpha^2)^{-1} \{ (1 - 2\alpha^2) + \alpha^2 (1 - 2\alpha^2) + 2\alpha^2 \} = 2\alpha^2 + 1 \,.$$

Applying Eq. (5.4.d) then yields the identity:

$$-(1 - 2\alpha^2) = 2\alpha^2 + 1 \,.$$

This is not possible; this contradiction completes the proof. $\qquad\square$

We deal with the remaining case $\alpha = 0$ and $c_1 = c_2 \neq 0$.

Lemma 5.4.4   *Let* $\mathfrak{M} := (V, \langle \cdot, \cdot \rangle, J, A := c_1 A_{J_1} + c_2 A_{J_2})$ *be a complex Osserman 0 model which is given by a Clifford family of rank 2 where* $c_1 = c_2 \neq 0$ *and* $\alpha = 0$. *Then there exists a reparametrization* $\tilde{\mathcal{F}}$ *of* $\mathcal{F}$ *so* $A = c_1(A_{\tilde{J}_1} + A_{\tilde{J}_2})$, *so* $\tilde{J}_1 J = J \tilde{J}_1$, *and so* $\tilde{J}_2 J = -J \tilde{J}_2$.

Proof.   Since $\alpha = 0$, $\mathrm{Rank}\{\mathcal{J}_A(\pi_x)\} = 4$ and $\{J_1 x, J_1 J x, J_2 x, J_2 J x\}$ is an orthonormal set. Lemma 5.4.2 (4) implies $J J_1 J_2 = -J_1 J_2 J$. Set

$$\Theta_1 := J J_1 \quad \text{and} \quad \Theta_2 = J J_2 \, .$$

Then

$$\Theta_1 \Theta_1^* = J J_1 J_1 J = \mathrm{id}, \quad \Theta_2 \Theta_2^* = J J_2 J_2 J = \mathrm{id},$$
$$\Theta_1 \Theta_2^* + \Theta_2 \Theta_1^* = J J_1 J_2 J + J J_2 J_1 J = 0, \tag{5.4.g}$$
$$\Theta_1 \Theta_2 = J J_1 J J_2 = J J_1 J J_1 J_2 J_1 = -J J_1 J_1 J_2 J J_1 = \Theta_2 \Theta_1 \, .$$

We have that

$$\mathrm{Range}\{\mathcal{J}_A(\pi_x)\} = \mathrm{Span}\{J_1 x, J_2 x, J_1 J x, J_2 J x\} \, .$$

This subspace is J invariant. Since $\alpha = 0$, Eq. (5.4.e) yields $J J_1 x \perp J_2 x$ and $J J_1 x \perp J_1 x$. Consequently:

$$J J_1 x = \langle J J_1 x, J_1 J x \rangle J_1 J x + \langle J J_1 x, J_2 J x \rangle J_2 J x,$$
$$1 = \langle J J_1 x, J J_1 x \rangle = \langle \Theta_1^2 x, x \rangle^2 + \langle \Theta_2 \Theta_1 x, x \rangle^2 \, . \tag{5.4.h}$$

By Eq. (5.4.g), Θ_1 and Θ_2 are commuting orthogonal maps. Let

$$V = V_+ \oplus V_- \oplus V_1 \oplus \ldots \oplus V_k$$

be a decomposition of Θ_1 where $\Theta_1 = \pm \mathrm{Id}$ on $V_\pm$ and where Θ_1 is a rotation through an angle between 0 and π on each V_i.

Suppose $k \geq 1$ so that Θ has a non-trivial rotation angle $0 < \theta_1 < \pi$. The maps Θ_1 and Θ_2 commute. We can simultaneously diagonalize the complexification of Θ_1 and Θ_2 on $V_1 \otimes \mathbb{C}$. Over the reals, this means we can find a 2-dimensional subspace of V_1 which is spanned by an orthonormal set $\{x_1, x_2\}$ which is invariant under Θ_1 and Θ_2. We may choose $\{x_1, x_2\}$ so that

$$\Theta_1 x_1 = \cos(\theta_1) x_1 + \sin(\theta_1) x_2,$$
$$\Theta_1 x_2 = -\sin(\theta_1) x_1 + \cos(\theta_1) x_2 \, .$$

If Θ_2 is a reflection on this subspace, we could further normalize the basis so that $\Theta_2 x_1 = x_1$ and $\Theta_2 x_2 = -x_2$. The identity $\Theta_2 \Theta_1 x_1 = \Theta_1 \Theta_2 x_1$ would then imply $\sin(\theta_1) = 0$ which is false. Thus Θ_2 is a rotation on this subspace and hence there exists $\theta_2 \in [0, 2\pi]$ so that:

$$\Theta_2 x_1 = \cos(\theta_2) x_1 + \sin(\theta_2) x_2,$$
$$\Theta_2 x_2 = -\sin(\theta_2) x_1 + \cos(\theta_2) x_2 \,.$$

We compute:

$$\Theta_1 \Theta_2^* x_1 = \cos(\theta_1 - \theta_2) x_1 + \sin(\theta_1 - \theta_2) x_2,$$
$$\Theta_2 \Theta_1^* x_1 = \cos(\theta_2 - \theta_1) x_1 + \sin(\theta_2 - \theta_1) x_2 \,.$$

As $\Theta_1 \Theta_2^* + \Theta_2 \Theta_1^* = 0$ by Eq. (5.4.g), $\cos(\theta_1 - \theta_2) = 0$. Thus $\theta_2 = \theta_1 \pm \frac{\pi}{2}$ so

$$J J_2 x_1 = \Theta_2 x_1 = \mp \sin \theta_1 x_1 \pm \cos \theta_1 x_2,$$
$$J J_2 x_2 = \Theta_2 x_2 = \mp \cos \theta_1 x_1 \mp \sin \theta_1 x_2 \,.$$

Set

$$\tilde{J}_1 := \cos \theta_1 J_1 \mp \sin \theta_1 J_2 \quad \text{and} \quad \tilde{J}_2 = \pm \sin \theta_1 J_1 + \cos \theta_1 J_2 \,.$$

This yields a reparametrization of $\mathcal{F}$. The curvature tensor is unchanged; by Lemma 5.2.2

$$A = c_1 \{A_{J_1} + A_{J_2}\} = c_1 \{A_{\tilde{J}_1} + A_{\tilde{J}_2}\} \,.$$

Furthermore,

$$\begin{aligned}
\tilde{\Theta}_1 x_1 &= \cos \theta_1 \Theta_1 x_1 \mp \sin \theta_1 \Theta_2 x_1 \\
&= \cos(\theta_1) \{\cos(\theta_1) x_1 + \sin(\theta_1) x_2\} \\
&\quad \mp \sin(\theta_1) \{\mp \sin(\theta_1) x_1 \pm \cos(\theta_1) x_2\} \\
&= x_1 \,.
\end{aligned}$$

Consequently, $\tilde{\Theta}_1$ has the eigenvalue $+1$. We note that if -1 is an eigenvalue of Θ_1, then setting $\tilde{J}_1 = -J_1$ and $\tilde{J}_2 = J_2$ yields a reparametrization so 1 is an eigenvalue of $\tilde{\Theta}_1$.

The argument given above shows that after a possible reparametrization, that we may assume $\dim(V_+) > 0$; this means that $+1$ is an eigenvalue of Θ_1. Suppose Θ_1 has an additional non-trivial rotation angle $0 < \theta_1 < \pi$ on V_1. Choose x_+ so $\Theta_1 x_+ = x_+$ and choose x_1 in the rotation block V_1 corresponding to θ_1. Let

$$x = (x_+ + x_1)/\sqrt{2} \,.$$

Since $\Theta_1 = \mathrm{Id}$ on V_+, $\Theta_1^* = \Theta_1$ so $\Theta_2^* = -\Theta_2$ on V_+. Furthermore Θ_1 is given by a rotation of θ_1 and Θ_2 is given by a rotation of $\theta_1 \pm \frac{\pi}{2}$ on V_1. Thus

$$\langle \Theta_1^2 x_+, x_+ \rangle = 1, \qquad \langle \Theta_1 \Theta_2 x_+, x_+ \rangle = 0,$$
$$\langle \Theta_1^2 x_1, x_1 \rangle = \cos(2\theta_1), \quad \langle \Theta_1 \Theta_2 x_1, x_1 \rangle = \mp \sin(2\theta_1).$$

We may now apply Eq. (5.4.h) to compute:

$$\begin{aligned}
1 &= \langle \Theta_1^2 x, x \rangle^2 + \langle \Theta_1 \Theta_2 x, x \rangle^2 \\
&= \tfrac{1}{4}\{(1 + \cos(2\theta_1))^2 + (0 \mp \sin(2\theta_1))^2\} \\
&= \tfrac{1}{4}\{2 + 2\cos(2\theta_1)\}\,.
\end{aligned}$$

This implies $\cos(2\theta_1) = 1$; this is not possible for $0 < \theta_1 < \pi$. Consequently Θ_1 has only the eigenspaces ± 1. Since $J = \pm J_1$ on $V_\mp$, we have $JJ_1 = J_1 J$. Since $JJ_1 J_2 = J_1 J_2 J$, $JJ_2 = -J_2 J$. $\qquad\qquad \square$

5.4.2   The tensor $c_0 A_{\langle \cdot, \cdot \rangle} + c_1 A_{J_1} + c_2 A_{J_2}$

We complete the proof of Assertion (3) of Theorem 5.1.1 by considering the case $c_0 \neq 0$. This will complete our study of the rank 2 case.

Lemma 5.4.5   *Let $\mathfrak{M} := (V, \langle \cdot, \cdot \rangle, J, A := c_0 A_{\langle \cdot, \cdot \rangle} + c_1 A_{J_1} + c_2 A_{J_2})$ be a complex Osserman 0 model which is given by a Clifford family of rank 2 where $c_0 \neq 0$, $c_1 \neq 0$ and $c_2 \neq 0$. If $\dim\{V\} \geq 12$, then there exists a reparametrization of $\mathcal{F}$ so that $A = c_0 A_{\langle \cdot, \cdot \rangle} + \tilde{c}_1 A_{\tilde{J}_1} + \tilde{c}_2 A_{\tilde{J}_2}$ and so that either $J = \tilde{J}_1$ or $J = \tilde{J}_1 \tilde{J}_2$.*

Proof.   Let

$$W := \mathrm{Span}\{x, J_1 x, J_2 x, Jx, J_1 Jx, J_2 Jx\}\,.$$

Suppose first $\dim_{\mathbb{R}}(W) = 6$. We argue for a contradiction. Set

$$\widetilde{\mathcal{J}}(\pi_x) := \mathcal{J}_A(\pi_x) - 2c_0\,\mathrm{id}, \quad \alpha := \langle J_1 J_2 x, Jx \rangle,$$
$$\alpha_1 := \langle J_1 x, Jx \rangle, \qquad\qquad \alpha_2 := \langle J_2 x, Jx \rangle\,.$$

We change notation slightly and assume $R := c_0 A_{\langle \cdot, \cdot \rangle} + \tfrac{1}{3}c_1 A_{J_1} + \tfrac{1}{3}c_2 A_{J_2}$ to simplify the computations. We have

$$\begin{aligned}
\widetilde{\mathcal{J}}(\pi_x)x &= -c_0 x - c_1 \alpha_1 J_1 Jx - c_2 \alpha_2 J_2 Jx, \\
\widetilde{\mathcal{J}}(\pi_x)Jx &= -c_0 Jx + c_1 \alpha_1 J_1 x + c_2 \alpha_2 J_2 x, \\
\widetilde{\mathcal{J}}(\pi_x)J_1 x &= -c_0 \alpha_1 Jx + c_1 J_1 x + c_2 \alpha J_2 Jx,
\end{aligned}$$

$$\widetilde{\mathcal{J}}(\pi_x)J_2x = -c_0\alpha_2 Jx - c_1\alpha J_1 Jx + c_2 J_2 x,$$
$$\widetilde{\mathcal{J}}(\pi_x)J_1 Jx = c_0\alpha_1 x + c_1 J_1 Jx - c_2\alpha J_2 x,$$
$$\widetilde{\mathcal{J}}(\pi_x)J_2 Jx = c_0\alpha_2 x + c_1\alpha J_1 x + c_2 J_2 Jx,$$
$$\widetilde{\mathcal{J}}(\pi_x)\xi = 0 \text{ if } \xi \in W^\perp.$$

As $\{x, Jx, J_1 x, J_2 x, J_1 Jx, J_2 Jx\}$ is a basis for W; the matrix of $\widetilde{\mathcal{J}}(x)|_W$ is given relative to this basis by:

$$M = \begin{pmatrix}
-c_0 & 0 & 0 & 0 & c_0\alpha_1 & c_0\alpha_2 \\
0 & -c_0 & -c_0\alpha_1 & -c_0\alpha_2 & 0 & 0 \\
0 & c_1\alpha_1 & c_1 & 0 & 0 & c_1\alpha \\
0 & c_2\alpha_2 & 0 & c_2 & -c_2\alpha & 0 \\
-c_1\alpha_1 & 0 & 0 & -c_1\alpha & c_1 & 0 \\
-c_2\alpha_2 & 0 & c_2\alpha & 0 & 0 & c_2
\end{pmatrix}.$$

One then has

$$\det(M) = c_0^2 c_1^2 c_2^2 (-1 + \alpha^2 + \alpha_1^2 + \alpha_2^2)^2.$$

Since $\dim\{W^\perp\} \geq 6$ and since $\widetilde{\mathcal{J}} = 0$ on $W^\perp$, we may apply Corollary 5.2.1 to see 0 is an eigenvalue of multiplicity $m - 6$. Consequently 0 is an eigenvalue of $\widetilde{\mathcal{J}}$ on W and $\det(M) = 0$. Since $c_0 c_1 c_2 \neq 0$, we have $\alpha^2 + \alpha_1^2 + \alpha_2^2 = 1$. Since α, α_1, and α_2 are the Fourier coefficients of Jx with respect to an orthonormal set $\{J_1 J_2 x, J_1 x, J_2 x\}$, we have

$$Jx = \alpha J_1 J_2 x + \alpha_1 J_1 x + \alpha_2 J_2 x.$$

It now follows that $\dim(W) = 4$ which is a contradiction. Consequently $\dim\{W\} \leq 5$ so

$$\mathrm{Span}\{x, J_1 x, J_2 x\} \cap \mathrm{Span}\{Jx, J_1 Jx, J_2 Jx\} \neq \{0\}.$$

Thus we can find $\vec{\xi} = (\xi_0, \xi_1, \xi_2)$ so that $\xi_0^2 + \xi_1^2 + \xi_2^2 = 1$ with

$$(\xi_0 + \xi_1 J_1 + \xi_2 J_2)Jx \in \mathrm{Span}\{x, J_1 x, J_2 x\}. \qquad (5.4.\mathrm{i})$$

Set $J_3 := J_1 J_2$ to define a quaternion structure $\mathbb{H}$ on V. Then Eq. (5.4.i) implies $Jx \in \mathbb{H}x$. Consequently for any unit vector x, there exist constants $a_i(x)$ so that

$$Jx = a_1(x)J_1 x + a_2(x)J_2 x + a_3(x)J_3 x.$$

Since $m \geq 12$, we can apply Lemma 5.2.6 (4) to see

$$J = a_1 J_1 + a_2 J_2 + a_3 J_3.$$

Note that A and J are compatible. Since $c_1 \neq 0$, $c_2 \neq 0$, and $c_3 = 0$, Lemma 5.2.4 implies

$$(c_2 - c_3)a_2 a_3 = (c_1 - c_3)a_1 a_3 = 0 \quad \text{so} \quad a_1 a_3 = a_2 a_3 = 0 \,.$$

Thus if $a_3 \neq 0$, $a_1 = a_2 = 0$ so $J = \pm J_3$ as desired. If $a_3 = 0$, then $J = a_1 J_1 + a_2 J_2$. Furthermore, the relation $(c_1 - c_2)a_1 a_2 = 0$ shows either $a_1 = 0$, or $a_2 = 0$, or $c_1 = c_2$. If $a_1 = \pm 1$, we are done. If $a_1 = 0$, then $a_2 = \pm 1$ and we simply interchange the roles of J_1 and J_2. If $a_1 \neq 0$ and $a_2 \neq 0$, then $c_1 = c_2$ and we may reparametrize the family by taking

$$\tilde{J}_1 := a_1 J_1 + a_2 J_2 \quad \text{and} \quad \tilde{J}_2 := a_2 J_1 - a_1 J_2$$

and still have $A = c_0 A_{\langle \cdot, \cdot \rangle} + c_1 A_{\tilde{J}_1} + c_1 A_{\tilde{J}_2}$ and $J = \tilde{J}_1$. $\qquad\square$

5.5   Clifford Families of Rank 3

In this section, we complete the proof of Theorem 5.1.1 by studying a complex 0-model where the algebraic curvature tensor is given by a Clifford family of rank 3. We establish one implication of Theorem 5.1.1 (4) by giving the following cases where $\mathfrak{M}$ is complex Osserman:

Lemma 5.5.1   *Let $\mathfrak{M} = (V, \langle \cdot, \cdot \rangle, J, A := c_0 A_{\langle \cdot, \cdot \rangle} + c_1 A_{J_1} + c_2 A_{J_2} + c_3 A_{J_3})$ be a complex 0-model which is given by a Clifford family of rank 3 where $c_1 \neq 0$, $c_2 \neq 0$, and $c_3 \neq 0$.*

(1) Let $c_0 = 0$. If $J = J_2 J_3$, then $\mathfrak{M}$ is complex Osserman.
(2) If $J_1 J_2 J_3 = \mathrm{Id}$ and if $J = J_1$, then $\mathfrak{M}$ is complex Osserman.

Proof.   Suppose that $c_0 = 0$ and that $J = J_2 J_3$. By Eq. (5.1.a),

$$\mathcal{J}(\pi_x)y = \begin{cases} 3c_1 y & \text{if } y \in \mathrm{Span}\{J_1 x, J_1 J_2 J_3 x\}, \\ 3(c_2 + c_3)y & \text{if } y \in \mathrm{Span}\{J_2 x, J_3 x\}, \\ 0 & \text{if } y \perp \mathrm{Span}\{J_1 x, J_1 J_2 J_3 x, J_2 x, J_3 x\} \,. \end{cases}$$

Assertion (1) now follows from Lemma 5.2.1. Assertion (2) follows from Theorem 5.2.2 (2). $\qquad\square$

5.5.1   *Technical results*

Recall that we defined the dual family

$$\mathcal{F}^* := \{J_1^* := J_2 J_3, \; J_2^* := J_3 J_1, \; J_3^* = J_1 J_2\} \,.$$

Adopt the notation of Eq. (5.2.a) to define $V_\pm$, and $\mathcal{N}$.

Lemma 5.5.2 *Let $\mathfrak{M} = (V, \langle \cdot, \cdot \rangle, J, A := c_0 A_{\langle \cdot, \cdot \rangle} + c_1 A_{J_1} + c_2 A_{J_2} + c_3 A_{J_3})$ be a complex Osserman 0-model which is given by a Clifford family of rank 3 where $c_1 \neq 0$, $c_2 \neq 0$, and $c_3 \neq 0$.*

(1) If $J \in \mathrm{Span}\{J_1, J_2, J_3\}$, then there exists a reparametrization of $\mathcal{F}$ so $A = c_0 A_{\langle \cdot, \cdot \rangle} + \tilde{c}_1 A_{\tilde{J}_1} + \tilde{c}_2 A_{\tilde{J}_2} + \tilde{c}_3 A_{\tilde{J}_3}$ and so $J = \tilde{J}_1$.
(2) Suppose $J_1 J_2 \neq \pm J_3$. Then $J \neq J_1$. If $c_0 \neq 0$, then $J \neq J_2 J_3$.

Proof. Expand $J = a_1 J_1 + a_2 J_2 + a_3 J_3$. By reparametrizing $\mathcal{F}$, we can permute the indices. Thus all indices play the same role. We suppose $a_1 \neq 0$. Assertion (1) is immediate if $a_2 = a_3 = 0$. Suppose $a_2 \neq 0$ but $a_3 = 0$. Then $J = a_1 J_1 + a_2 J_2$ and, by Lemma 5.2.4 (1), $c_1 = c_2$. Define a reparametrization

$$\tilde{J}_1 := J = a_1 J_1 + a_2 J_2, \quad \tilde{J}_2 := -a_2 J_1 + a_1 J_2, \quad \tilde{J}_3 := J_3.$$

Assertion (1) then follows from Lemma 5.2.2. Finally, suppose $a_1 \neq 0$, $a_2 \neq 0$, and $a_3 \neq 0$. By Lemma 5.2.4, $c_1 = c_2 = c_3$. The desired conclusion will then follow from Lemma 5.2.2 after choosing any $\alpha \in O(3)$ so $\alpha J_1 = J$.

To prove Assertion (2), we suppose that $\mathfrak{M}$ is complex Osserman and that $J_1 J_2 \neq \pm J_3$. Suppose first that $J = J_1$. We argue for a contradiction. Choose $x_\pm \in V_\pm$ so $J_1 J_2 J_3 = \pm x_\pm$. Then:

$$\mathcal{J}(\pi_{x_\pm})y = \begin{cases} (c_0 + 3c_1)y & \text{if } y \in \mathrm{Span}\{x_\pm, J_1 x_\pm\}, \\ (2c_0 + 3c_2 + 3c_3)y & \text{if } y \in \mathrm{Span}\{J_2 x_\pm, J_3 x_\pm\}, \\ 2c_0 y & \text{if } y \perp \mathrm{Span}\{x_\pm, J_1 x_\pm, J_2 x_\pm, J_3 x_\pm\}. \end{cases} \quad (5.5.\text{a})$$

On the other hand, if we let $x := \frac{1}{\sqrt{2}}(x_+ + x_-) \in \mathcal{N}$, then the orthogonality relations of Lemma 5.2.3 (1) imply:

$$\mathcal{J}(\pi_x) = \begin{cases} (c_0 + 3c_1)y & \text{if } y \in \mathrm{Span}\{x, J_1 x\}, \\ (2c_0 + 3c_2)y & \text{if } y \in \mathrm{Span}\{J_1 J_2 x, J_2 x\}, \\ (2c_0 + 3c_3)y & \text{if } y \in \mathrm{Span}\{J_1 J_3 x, J_3 x\}, \\ 2c_0 y & \text{if } y \perp \mathrm{Span}\{x, J_1 x, J_2 x, J_3 x, J_1 J_2 x, J_1 J_3 x\}. \end{cases} \quad (5.5.\text{b})$$

Since $\mathrm{Spec}\{\mathcal{J}(\pi_{x_+})\} = \mathrm{Spec}\{\mathcal{J}(\pi_x)\}$, and since the multiplicities must be the same, either $2c_0 = 2c_0 + 3c_2$ and $c_2 = 0$ or $2c_0 = 2c_0 + 3c_3$ and $c_3 = 0$. This is false.

Finally, assume that $c_0 \neq 0$, and that $J = J_2 J_3$. Let $x_\pm$ and x be as above. Since $J = \pm J_1$ on $V_\pm$, Eq. (5.5.a) continues to hold. However

Eq. (5.5.b) must be replaced by

$$\mathcal{J}(\pi_{x_\pm}) = \begin{cases} c_0 y & \text{if } y \in \operatorname{Span}\{x, J_2 J_3 x\}, \\ (2c_0 + 3c_1)y & \text{if } y \in \operatorname{Span}\{J_1 x, J_1 J_2 J_3 x\}, \\ (2c_0 + 3c_2 + 3c_3)y & \text{if } y \in \operatorname{Span}\{J_2 x, J_3 x\}, \\ 2c_0 y & \text{if } y \perp \operatorname{Span}\{x, J_1 x, J_2 x, J_3 x, J_1 J_2 J_3 x, J_2 J_3 x\}. \end{cases}$$

Since $\operatorname{Spec}\{\mathcal{J}(\pi_{x_+})\} = \operatorname{Spec}\{\mathcal{J}(\pi_x)\}$, since the multiplicities must be the same, either the following unordered sets must be equal:

$$\{c_0 + 3c_1, 2c_0\} = \{c_0, 2c_0 + 3c_1\}.$$

This is not possible as $c_0 \neq 0$ and $c_1 \neq 0$. This contradiction completes the proof. $\qquad\square$

We continue our study of the rank 3 case.

Lemma 5.5.3 *Let $\mathfrak{M} = (V, \langle \cdot, \cdot \rangle, J, A := c_0 A_{\langle \cdot, \cdot \rangle} + c_1 A_{J_1} + c_2 A_{J_2} + c_3 A_{J_3})$ be a complex 0-model which is given by a Clifford family of rank 3 where $c_1 \neq 0$, $c_2 \neq 0$, and $c_3 \neq 0$. Assume that $J_1 J_2 \neq \pm J_3$ and that*

$$Jx \in \operatorname{Span}\{J_1 x, J_2 x, J_3 x, \tilde{J}_1 x, \tilde{J}_2 x, \tilde{J}_3 x\} \quad \text{if } x \in \mathcal{N}.$$

Then $c_0 = 0$ and there exists a reparametrization of $\mathcal{F}$ so that $J = J_2 J_3$ and so that $A = \tilde{c}_1 A_{\tilde{J}_1} + \tilde{c}_2 A_{\tilde{J}_2} + \tilde{c}_3 A_{\tilde{J}_3}$.

Proof. We apply Lemma 5.2.4 to see either that $J \in \operatorname{Span}\{J_1, J_2, J_3\}$ or that $J \in \operatorname{Span}\{J_1^*, J_2^*, J_3^*\}$. Suppose $J \in \operatorname{Span}\{J_1, J_2, J_3\}$. We can apply Lemma 5.5.2 (1) to reparametrize $\mathcal{F}$ and assume that $J = J_1$. We may then use Lemma 5.5.2 (2) to rule out this possibility.

Thus we may suppose that $J = a_1^* J_1^* + a_2^* J_3^* + a_3^* J_3^*$. Restrict the structures to define:

$$\mathcal{F}_\pm := \{J_1|_{V_\pm}, J_2|_{V_\pm}, J_3|_{V_\pm}\},$$
$$\mathfrak{M}_\pm := (V_\pm, \langle \cdot, \cdot \rangle|_{V_\pm}, J|_{V_\pm}, A|_{V_\pm}).$$

We then have $J = \pm\{a_1 J_1 + a_2 J_2 + a_3 J_3\}$ on $V_\pm$. The argument given to prove Lemma 5.5.2 (1) shows we can reparametrize $\mathcal{F}$ so that $J = \pm J_1$ on $V_\pm$ where the signs must differ. This ensures that $J = J_1^*$ on V. By Lemma 5.5.2 (2), $c_0 = 0$. $\qquad\square$

5.5.2 The tensor $A = c_1 A_{J_1} + c_2 A_{J_2} + c_3 A_{J_3}$

Lemma 5.5.4 *Let $\mathfrak{M} = (V, \langle \cdot, \cdot \rangle, J, A := \frac{1}{3}(c_1 A_{J_1} + c_2 A_{J_2} + c_3 A_{J_3}))$ be a complex Osserman 0-model which is given by a Clifford family of rank 3 where $c_1 \neq 0$, $c_2 \neq 0$, and $c_3 \neq 0$. Assume $m \geq 12$. Then there is a reparametrization of $\mathcal{F}$ so that either $J = \tilde{J}_1$ or $J = \tilde{J}_1^*$ and so that $A = \frac{1}{3}(\tilde{c}_1 A_{\tilde{J}_1} + \tilde{c}_2 A_{\tilde{J}_2} + \tilde{c}_3 A_{\tilde{J}_3})$.*

Proof. Set $\alpha_{ij} = \langle J_i J_j x, J x \rangle$. By Eq. (5.1.a),

$$\mathcal{J}_R(\pi_x)J_1 x = c_1 J_1 x + c_2 \alpha_{12} J_2 J x + c_3 \alpha_{13} J_3 J x,$$

$$\mathcal{J}_R(\pi_x)J_2 x = c_2 J_2 x - c_1 \alpha_{12} J_1 J x + c_3 \alpha_{23} J_3 J x,$$

$$\mathcal{J}_R(\pi_x)J_3 x = c_3 J_3 x - c_1 \alpha_{13} J_1 J x - c_2 \alpha_{23} J_2 J x,$$

$$\mathcal{J}_R(\pi_x)J_1 J x = c_1 J_1 J x - c_2 \alpha_{12} J_2 x - c_3 \alpha_{13} J_3 x,$$

$$\mathcal{J}_R(\pi_x)J_2 J x = c_2 J_2 J x + c_1 \alpha_{12} J_1 x - c_3 \alpha_{23} J_3 x,$$

$$\mathcal{J}_R(\pi_x)J_3 J x = c_3 J_3 J x + c_1 \alpha_{13} J_1 x + c_2 \alpha_{23} J_2 x.$$

Let $W(x) := \mathrm{Span}\{J_1 x, J_2 x, J_3 x, J_1 J x, J_2 J x, J_3 J x\} = \mathrm{Range}\{\mathcal{J}(\pi_x)\}$. If $\dim(W) = 6$, then the matrix associated to $\mathcal{J}_R(\pi_x)$ in W is

$$M = \begin{pmatrix} c_1 & 0 & 0 & 0 & c_1\alpha_{12} & c_1\alpha_{13} \\ 0 & c_2 & 0 & -c_2\alpha_{12} & 0 & c_2\alpha_{23} \\ 0 & 0 & c_3 & -c_3\alpha_{13} & -c_3\alpha_{23} & 0 \\ 0 & -c_1\alpha_{12} & -c_1\alpha_{13} & c_1 & 0 & 0 \\ c_2\alpha_{12} & 0 & -c_2\alpha_{23} & 0 & c_2 & 0 \\ c_3\alpha_{13} & c_3\alpha_{23} & 0 & 0 & 0 & c_3 \end{pmatrix}.$$

Computing the determinant yields

$$\det(\mathcal{J}(\pi_x)|_W) = (-1 + \alpha_{12}^2 + \alpha_{13}^2 + \alpha_{23}^2)^2 c_1^2 c_2^2 c_3^2.$$

Since $\dim\{V\} \geq 12$ and since $\mathcal{J}(\pi_x) = 0$ on $W^\perp$, the eigenvalue 0 has multiplicity at least 6. Thus by Theorem 5.2.1, the eigenvalue 0 has multiplicity at least $m - 4$. Consequently 0 is an eigenvalue of M and thus $\det(M) = 0$. This shows

$$\alpha_{12}^2 + \alpha_{13}^2 + \alpha_{23}^2 = 1.$$

As $\{J_1 J_2 x, J_1 J_3 x, J_2 J_3 x\}$ is an orthonormal set and as the coefficients $\{\alpha_{12}, \alpha_{13}, \alpha_{23}\}$ are the Fourier coefficients of a unit vector,

$$J x \in \mathrm{Span}\{J_1 J_2 x, J_1 J_3 x, J_2 J_3 x\} = \mathrm{Span}\{J_1^* x, J_2^* x, J_3^* x\}.$$

Consequently $W(x) = \mathrm{Span}\{J_1 x, J_2 x, J_3 x, J_1 J_2 J_3 x\}$ so $\dim(W) < 6$, which is a contradiction.

Suppose on the other hand that $\dim(W) \le 5$. We then have:

$$\dim\left\{\mathrm{Span}\{J_1 x, J_2 x, J_3 x\} \cap \mathrm{Span}\{J_1 J x, J_2 J x, J_3 J x\}\right\}$$
$$\ge 3 + 3 - 5 > 0.$$

This implies that we have a unit vector (a_1, a_2, a_3) so that

$$(a_1 J_1 + a_2 J_2 + a_3 J_3) J x \in \mathrm{Span}\{J_1 x, J_2 x, J_3 x\}.$$

Since $(a_1 J_1 + a_2 J_2 + a_3 J_3)^2 = -\,\mathrm{Id}$,

$$Jx \in (a_1 J_1 + a_2 J_2 + a_3 J_3)\,\mathrm{Span}\{J_1 x, J_2 x, J_3 x\}$$
$$= \mathrm{Span}\{x, J_1 J_2 x, J_1 J_3 x, J_2 J_3 x\}.$$

Since $Jx \perp x$, this implies

$$Jx \in \mathrm{Span}\{J_1^* x, J_2^* x, J_3^* x\}.$$

Since $m \ge 2\ell = 6$, Lemma 5.2.6 shows that $J \in \mathrm{Span}\{J_1^*, J_2^*, J_3^*\}$. Suppose $J_1 J_2 = \pm J_3$. Then $J \in \mathrm{Span}\{J_1, J_2, J_3\}$ and Lemma 5.5.2 may be used to reparametrize the family $\mathcal{F}$ so $J = J_1$. On the other hand, if $J_1 J_2 \ne \pm J_3$, then Lemma 5.5.3 may be used to reparametrize $\mathcal{F}$ so $J = J_2 J_3$. $\qquad\square$

We use Lemma 5.5.4 to complete the proof of Theorem 5.1.1 (4a). We reparametrize the family if necessary. If $J = J_1^*$, we have $J = J_2 J_3$ as desired. If $J = J_1$, then Lemma 5.5.2 (2) implies $J_1 J_2 = \pm J_3$ and again yields the desired conclusion. $\qquad\square$

5.5.3   *The tensor $A = c_0 A_{\langle\cdot,\cdot\rangle} + c_1 A_{J_1} + c_2 A_{J_2} + c_3 A_{J_3}$*

Assertion (4.b) of Theorem 5.1.1 will follow from the following result.

Lemma 5.5.5   *Let $\mathfrak{M} = (V, \langle\cdot,\cdot\rangle, J, A := c_0 A_{\langle\cdot,\cdot\rangle} + c_1 A_{J_1} + c_2 A_{J_2} + c_3 A_{J_3})$ be a complex Osserman 0-model which is given by a Clifford family of rank 3 where $c_0 \ne 0$, $c_1 \ne 0$, $c_2 \ne 0$, and $c_3 \ne 0$. Assume that $m \ge 16$. Then $J_1 J_2 = \pm J_3$ and there is a reparametrization of $\mathcal{F}$ so that $J = \tilde{J}_1$ and $A = c_0 A_{\langle\cdot,\cdot\rangle} + \tilde{c}_1 A_{\tilde{J}_1} + \tilde{c}_2 A_{\tilde{J}_2} + \tilde{c}_3 A_{\tilde{J}_3}$.*

Proof.   We change notation slightly and let

$$A = c_0 A_{\langle\cdot,\cdot\rangle} + \tfrac{1}{3}(c_1 A_{J_1} + c_2 A_{J_2} + c_3 A_{J_3}).$$

Let $V_\pm := \{x \in V : J_1 J_2 J_3 x = \pm x\}$. We wish to show

$$Jx \in \mathrm{Span}\{J_1 x, J_2 x, J_3 x\} \quad \text{for} \quad x \in V_\pm. \tag{5.5.c}$$

We suppose $x \in V_+$, the analysis is similar if $x \in V_-$. Set $\alpha_1 := \langle J_1 x, Jx \rangle$, $\alpha_2 := \langle J_2 x, Jx \rangle$, and $\alpha_3 := \langle J_3 x, Jx \rangle$. Then:

$$\mathcal{J}_R(\pi_x)x = c_0 x - c_1 \alpha_1 J_1 Jx - c_2 \alpha_2 J_2 Jx - c_3 \alpha_3 J_3 Jx,$$
$$\mathcal{J}_R(\pi_x)Jx = c_0 Jx + c_1 \alpha_1 J_1 x + c_2 \alpha_2 J_2 x + c_3 \alpha_3 J_3 x,$$
$$\mathcal{J}_R(\pi_x)J_1 x = -c_0 \alpha_1 Jx + (2c_0 + c_1)J_1 x + c_2 \alpha_3 J_2 Jx - c_3 \alpha_2 J_3 Jx,$$
$$\mathcal{J}_R(\pi_x)J_2 x = -c_0 \alpha_2 Jx + (2c_0 + c_2)J_2 x - c_1 \alpha_3 J_1 Jx + c_3 \alpha_1 J_3 Jx,$$
$$\mathcal{J}_R(\pi_x)J_3 x = -c_0 \alpha_3 Jx + (2c_0 + c_3)J_3 x + c_1 \alpha_2 J_1 Jx - c_2 \alpha_1 J_2 Jx,$$
$$\mathcal{J}_R(\pi_x)J_1 Jx = c_0 \alpha_1 x + (2c_0 + c_1)J_1 Jx - c_2 \alpha_3 J_2 x + c_3 \alpha_2 J_3 x,$$
$$\mathcal{J}_R(\pi_x)J_2 Jx = c_0 \alpha_2 x + (2c_0 + c_2)J_2 Jx + c_1 \alpha_3 J_1 x - c_3 \alpha_1 J_3 x,$$
$$\mathcal{J}_R(\pi_x)J_3 Jx = c_0 \alpha_3 x + (2c_0 + c_3)J_3 Jx - c_1 \alpha_2 J_1 x + c_2 \alpha_1 J_2 x.$$

We set $W := \{x, Jx, J_1 x, J_2 x, J_3 x, J_1 Jx, J_2 Jx, J_3 Jx\}$. We suppose $\dim(W) = 8$ and argue for a contradiction. Relative to this basis, the matrix of $\mathcal{J}(\pi_x)$ becomes:

$$M = \begin{pmatrix}
c_0 & 0 & 0 & 0 & 0 & c_0\alpha_1 & c_0\alpha_2 & c_0\alpha_3 \\
0 & c_0 & -c_0\alpha_1 & -c_0\alpha_2 & -c_0\alpha_3 & 0 & 0 & 0 \\
0 & c_1\alpha_1 & 2c_0 + c_1 & 0 & 0 & 0 & c_1\alpha_3 & -c_1\alpha_2 \\
0 & c_2\alpha_2 & 0 & 2c_0 + c_2 & 0 & -c_2\alpha_3 & 0 & c_2\alpha_1 \\
0 & c_3\alpha_3 & 0 & 0 & 2c_0 + c_3 & c_3\alpha_2 & -c_3\alpha_1 & 0 \\
-c_1\alpha_1 & 0 & 0 & -c_1\alpha_3 & c_1\alpha_2 & 2c_0 + c_1 & 0 & 0 \\
-c_2\alpha_2 & 0 & c_2\alpha_3 & 0 & -c_2\alpha_1 & 0 & 2c_0 + c_2 & 0 \\
-c_3\alpha_3 & 0 & -c_3\alpha_2 & c_3\alpha_1 & 0 & 0 & 0 & 2c_0 + c_3
\end{pmatrix}.$$

We then have

$$\det(M - 2c_0\,\mathrm{Id}) = c_0^2 c_1^2 c_2^2 c_3^2 (-1 + \alpha_1^2 + \alpha_2^2 + \alpha_3^2)^4.$$

We have $\dim(V) \geq 16$ and thus $\dim(W^\perp) \geq 6$. Thus 0 is an eigenvalue of multiplicity at least 6 and thus, by Theorem 5.2.1 (or, equivalently, by Corollary 5.2.1), zero is an eigenvalue of multiplicity at least $m - 4$ for the reduced complex Jacobi operator. Consequently 0 is an eigenvalue of $M - 2c_0\,\mathrm{Id}$ so $\det(M - 2c_0\,\mathrm{Id}) = 0$. Consequently

$$1 = \alpha_1^2 + \alpha_2^2 + \alpha_3^2 \quad \text{so} \quad Jx \in \mathrm{Span}\{J_1 x, J_2 x, J_3 x\}.$$

This shows that $W = \mathrm{Span}\{x, J_1 x, J_2 x, J_3 x\}$ so $\dim(W) = 4$ which is false.

We have shown that $\dim(W) \leq 7$. We set

$$U_1 := \operatorname{Span}\{x, J_1 x, J_2 x, J_3 x\},$$
$$U_2 := \operatorname{Span}\{Jx, J_1 Jx, J_2 Jx, J_3 Jx\}.$$

Then $\dim(U_1) = 4$ and $\dim(U_2) = 4$. Since

$$\dim(U_1) + \dim(U_2) > \dim(W),$$

$U_1 \cap U_2 \neq \{0\}$. Thus there exist constants a_i with $a_0^2 + a_1^2 + a_2^2 + a_3^2 = 1$ so

$$(a_0 + a_1 J_1 + a_2 J_2 + a_3 J_3)Jx \in \operatorname{Span}\{x, J_1 x, J_2 x, J_3 x\}.$$

Multiplying this relation by $(a_0 - a_1 J_1 - a_2 J_2 - a_3 J_3)$ yields

$$Jx \in \operatorname{Span}\{x, J_1 x, J_2 x, J_3 x\} \quad \text{if} \quad x \in V_\pm.$$

Since $Jx \perp x$, this establishes Eq. (5.5.c).

If $J_1 J_2 = \pm J_3$, then $V = V_+$ or $V = V_-$ so

$$Jx \in \operatorname{Span}\{J_1 x, J_2 x, J_3 x\} \quad \text{for all} \quad x \in V.$$

Lemma 5.2.6 then shows $J \in \operatorname{Span}\{J_1, J_2, J_3\}$ and the desired conclusion then follows from Lemma 5.5.2. If on the other hand $J_1 J_2 \neq \pm J_3$, then $\mathcal{N}$ is non-empty. Let $x = (x +_+ x_-)/\sqrt{2} \in \mathcal{N}$. Expand

$$Jx_\pm = a_1^\pm J_1 x_\pm + a_2^\pm J_2 x_\pm + a_3^\pm J_3 x_\pm.$$

Set $a_i := (a_i^+ + a_i^-)/2$ and $a_i^* := (a_i^+ - a_i^-)/2$. Since $J_i^* x_\pm = \pm J_i x_\pm$,

$$\sum_{i=1}^{3}(a_i J_i + a_i^* J_i^*)x = \tfrac{1}{\sqrt{2}}\sum_{i=1}^{3}\{(a_i + a_i^*)J_i x_+ + (a_i - a_i^*)J_i x_-\}$$
$$= \tfrac{1}{\sqrt{2}}(Jx_+ + Jx_-) = Jx.$$

Consequently

$$Jx \in \operatorname{Span}\{J_1 x, J_2 x, J_3 x, J_1^* x, J_2^* x, J_3^* x\} \quad \text{for all} \quad x \in \mathcal{N}.$$

Lemma 5.5.3 then shows $c_0 = 0$ which is false. $\qquad\square$

5.6 Tensors $A = c_1 A_{J_1} + ... + c_\ell A_{J_\ell}$ for $\ell \geq 4$

This section is devoted to the proof of Assertion (1) of Theorem 5.1.2. Let

$$\mathfrak{M} := (V, \langle \cdot, \cdot \rangle, J, A = c_1 A_{J_1} + ... + c_\ell A_{J_\ell})$$

be a complex Osserman 0-model on a vector space V of dimension m which is given by a Clifford family $\mathcal{F} := \{J_1, ..., J_\ell\}$. We shall suppose that $c_i \neq 0$ for all i and that $\ell \geq 4$ throughout this section. We begin by using Theorem 5.2.1 in an essential way to show:

Lemma 5.6.1 *Let $\mathfrak{M} := (V, \langle \cdot, \cdot \rangle, J, A = c_1 A_{J_1} + ... + c_\ell A_{J_\ell})$ be a complex Osserman 0-model which is given by a Clifford family of rank $\ell \geq 4$ where $c_i \neq 0$ for all i. If $\ell = 4$ or if $\ell = 5$, assume that $m \geq 2^\ell$. If $m \geq 6$, assume that $m \geq \ell(\ell - 1)$.*

(1) If $x \in S(V)$, then $\mathrm{Rank}\{\mathcal{J}_A(\pi_x)\} \leq 4$.
(2) If $x \in S(V)$, then $Jx \in \mathrm{Span}_{i \leq 4, i < j}\{J_i J_j x\}$.
(3) $\ell = 4$ or $\ell = 5$.

Proof. Since $c_0 = 0$, $\mathcal{J}(\pi_x) = \tilde{\mathcal{J}}(\pi_x)$. We note:

$$
\begin{aligned}
\ell = 4 &\quad \Rightarrow \quad m - 2\ell \geq 16 - 8 > 4, \\
\ell \geq 5 &\quad \Rightarrow \quad m - 2\ell \geq 32 - 10 > 4, \\
\ell \geq 6 &\quad \Rightarrow \quad m - 2\ell \geq \ell(\ell - 3) > 4.
\end{aligned}
$$

Thus by Corollary 5.2.1, 0 is an eigenvalue of multiplicity at least $m - 4$. This establishes Assertion (1).

Let $\alpha_i = \alpha_i(x) := \langle J_i x, J_1 J x \rangle$. Set

$$
\begin{aligned}
U(x) &:= \mathrm{Span}\{J_1 x, ..., J_\ell x, J_1 J x\}, \\
V(x) &:= \mathrm{Span}\{J_2 J x, ..., J_\ell J x\}, \\
W(x) &:= U(x) + V(x) \supset \mathrm{Range}\{\mathcal{J}_A(\pi_x)\}.
\end{aligned}
$$

Let ρ be projection on $W(x)/V(x)$. We may express:

$$\rho \mathcal{J}_A(\pi_x) J_i x = \rho\{3 c_i J_i x + 3 c_1 \alpha_i J_1 J x\}, \tag{5.6.a}$$

$$\rho \mathcal{J}_A(\pi_x) J_1 J x = \rho \left\{ \sum_i 3 c_i \alpha_i J_i x + 3 c_1 J J_1 x \right\}.$$

If $\dim\{U(x)\} \leq \ell$, then $J_1 J x \in \mathrm{Span}_i\{J_i x\}$ since $\{J_i x\}$ is an orthonormal set. Consequently, $Jx \in \mathrm{Span}_i\{J_1 J_i x\}$. As $\{x, J_1 J_2 x, ..., J_1 J_\ell x\}$ is an orthonormal set and as $Jx \perp x$, this implies $Jx \in \mathrm{Span}_{1 < i}\{J_1 J_i x\}$ which establishes Assertion (2) in this special case.

We therefore assume $\dim\{U(x)\} = \ell + 1$. We use Eq. (5.6.a) to see that $\rho \mathcal{J}_A(\pi_x) = \rho M$ on $U(x)$ where M is the following matrix:

$$M := 3 \begin{pmatrix} c_1 & 0 & 0 \, ... & 0 \, c_1\alpha_1 \\ 0 & c_2 & 0 \, ... & 0 \, c_1\alpha_2 \\ 0 & 0 & c_3 \, ... & 0 \, c_1\alpha_3 \\ ... & ... & ... \, ... & ... \\ 0 & 0 & 0 \, ... & c_\ell \, c_1\alpha_\ell \\ c_1\alpha_1 & c_2\alpha_2 & c_3\alpha_3 \, ... & c_\ell\alpha_\ell \quad c_1 \end{pmatrix}.$$

We have $\det(M) = 3^{\ell+1}c_1^2 c_2 ... c_\ell (1 - \alpha_1^2 - ... - \alpha_\ell^2)$. Since the α_i are the Fourier coefficients of $J_1 J x$ relative to the orthonormal basis $\{J_i x\}$ and since by hypothesis $J_1 J x \notin \mathrm{Span}_i\{J_i x\}$, $\alpha_1^2 + ... + \alpha_\ell^2 < 1$. This shows that M is invertible. Consequently

$$\dim\{\rho U(x)\} = \dim\{\rho M U(x)\} = \dim\{\rho \mathcal{J}_A(\pi_x) U(x)\}$$
$$\leq \mathrm{Rank}\{\mathcal{J}_A(\pi_x)\} \leq 4 \, .$$

We may now use the short exact sequence

$$0 \to V(x) \to W(x) \to W(x)/V(x) = \rho U(x) \to 0$$

to see that

$$\dim\{W(x)\} = \dim\{V(x)\} + \dim\{\rho U(x)\} \leq (\ell - 1) + 4 = \ell + 3 \, .$$

We can now estimate:

$$\dim\left\{ \mathrm{Span}\{J_1 x, J_2 x, J_3 x, J_4 x\} \cap \mathrm{Span}_i\{J_i J x\}\right\}$$
$$= 4 + \ell - \dim(W) \geq 4 + \ell - (\ell + 3) > 0 \, .$$

Thus we have unit vectors $\vec{a}, \vec{b}$ so

$$a_1 J_1 x + a_2 J_2 x + a_3 J_3 x + a_4 J_4 x = b_1 J_1 J x + ... + b_\ell J_\ell J x \, .$$

We invert this relation by multiplying by $(b_1 J_1 + ... + b_\ell J_\ell)$ to see

$$Jx \in \mathbb{R}x \oplus \mathrm{Span}_{i \leq 4, i \leq j}\{J_i J_j x\} \, .$$

Since this is an orthogonal direct sum and since $Jx \perp x$, we may take $i < j$ and establish Assertion (2) in this case as well.

Suppose that $\ell \geq 6$; we argue for a contradiction. As by assumption $m \geq \ell(\ell - 1)$, Lemma 5.2.7 (2) shows there exists x so $\{J_i J_j x\}_{i < j}$ forms a linearly independent set of $\frac{1}{2}\ell(\ell-1)$ vectors. We use Assertion (2) of Lemma 5.6.1 to expand

$$Jx = \sum_{i=1}^{6}\sum_{i<j} c_{ij}(x)J_iJ_jx;$$

the coefficients c_{ij} are uniquely determined. Since in fact, the sum may be restricted to $i \leq 4$, we have $c_{56}(x) = c_{5j}(x) = c_{6j}(x) = 0$ for $j \geq 7$. By permuting the roles of the indices $\{1,2,3,4,5,6\}$ we conclude that all the coefficients are 0 which is false. $\square$

The following Lemma is a crucial one; it will also be applied presently in Section 5.7; this is the crucial stage where we use the assumption $m \geq 32$ if $\ell = 5$ rather than just $m \geq 16$.

Lemma 5.6.2 *Let $\mathfrak{M} := (V, \langle\cdot,\cdot\rangle, J, A = c_1 A_{J_1} + ... + c_\ell A_{J_\ell})$ be a complex 0-model which is given by a Clifford family of rank $\ell = 4$ or $\ell = 5$ where $c_i \neq 0\ \forall i$ and where $m \geq 2^\ell$. Assume that A and J are compatible and that $Jx \in \mathrm{Span}_{i<j}\{J_iJ_jx\}$ for all $x \in V$. Then we can reparametrize $\mathcal{F}$ so $A = \sum_i \tilde{c}_i A_{\tilde{J}_i}$ and so $J = \tilde{J}_1\tilde{J}_2$.*

Proof. We have by assumption that $Jx \in \mathrm{Span}\{J_iJ_jx\}$ for all x. By Lemma 5.2.8 (2) and Lemma 5.2.9 (4), there exist $x,y \in S(V)$ so $\{J_iJ_jx, J_iJ_jy\}_{i<j}$ is a collection of $\ell(\ell-1)$ orthonormal vectors. Thus Lemma 5.2.5 (3) shows we may expand

$$J = \sum_{ij} a_{ij}J_iJ_j \quad \text{where} \quad a_{ij} = -a_{ji}.$$

We now apply an argument similar to that which was used to establish Lemma 5.2.4 (1). By Lemma 5.2.8 (1) and by Lemma 5.2.9 (2), we may choose x so $\langle J_1J_2J_3x, x\rangle = 1$ and so $\langle J_iJ_jJ_kx, x\rangle = 0$ if $\{i,j,k\}$ is not a permutation of $\{1,2,3\}$. We compute:

$$J\mathcal{A}(x,Jx)x = -J\mathcal{J}_A(x)Jx = -3J\sum_{ij}\sum_k a_{ij}c_k\langle J_iJ_jx, J_kx\rangle J_kx$$

$$= 6J(a_{23}c_1J_1x + a_{31}c_2J_2x + a_{12}c_3J_3x),$$

$$\mathcal{A}(x,Jx)Jx = \mathcal{J}_A(Jx)x = 3\sum_{ij}\sum_k a_{ij}c_k\langle x, J_kJ_iJ_jx\rangle J_kJx$$

$$= 6(a_{23}c_1J_1 + a_{31}c_2J_2 + a_{12}c_3J_3x)Jx.$$

Since J and A are compatible, $0 = \mathcal{A}(x, Jx)Jx - J\mathcal{A}(x, Jx)x$. Therefore:

$$0 = \{a_{23}c_1(J_1 J - J J_1) + a_{31}c_2(J_2 J - J J_2) + a_{12}c_3(J_3 J - J J_3)\}x.$$

We take the inner product with $J_1 x$ to see

$$0 = a_{12}a_{13}(c_2 - c_3).$$

We permute the indices to see that if $\{i, j, k\}$ are distinct, then

$$0 = a_{ij}a_{ik}(c_j - c_k). \tag{5.6.b}$$

Since J is an almost complex structure, $J^2 x = -x$. If $x \in S(V)$, then

$$-x = -\sum_{i<j} a_{ij}^2 x + 2\sum_{i<j}\sum_k a_{ik}a_{jk}J_i J_j x \tag{5.6.c}$$

$$+ 2\sum_{i<j<k<n}(a_{ij}a_{kn} + a_{in}a_{jk} - a_{ik}a_{jn})J_i J_j J_k J_n x.$$

We replace x by $J_1 x$ and then multiply the equation by J_1 to see

$$x = \sum_{i<j} a_{ij}^2 x + 2J_1\sum_{i<j}\sum_k a_{ik}a_{jk}J_i J_j J_1 x \tag{5.6.d}$$

$$+ 2J_1\sum_{i<j<k<n}(a_{ij}a_{kn} + a_{in}a_{jk} - a_{ik}a_{jn})J_i J_j J_k J_n J_1 x.$$

We add Eqs. (5.6.c) and (5.6.d). Because

$$J_1 J_i J_j J_1 = -J_i J_j, \quad \text{and}$$
$$J_1 J_i J_j J_k J_l J_1 = -J_i J_j J_k J_l \quad \text{for} \quad i > 1,$$

these terms cancel and after multiplying the result by J_1 we obtain

$$0 = \sum_{1<j}\sum_k a_{1k}a_{jk}J_j x \tag{5.6.e}$$

$$+ \sum_{1<j<k<l}(a_{1j}a_{kn} - a_{1n}a_{jk} + a_{1k}a_{jn})J_j J_k J_n x.$$

We apply Lemma 5.2.8 (1) and Lemma 5.2.9 (2) to see that there exists x so $\langle J_j J_k J_n x, x\rangle \neq 0$ only for $j = 2$, $k = 3$, $n = 4$. Taking the inner product of Eq. (5.6.e) with such an x shows

$$0 = a_{12}a_{34} + a_{14}a_{23} - a_{13}a_{24}.$$

We can then permute the indices to conclude that

$$0 = a_{ij}a_{kn} + a_{in}a_{jk} - a_{ik}a_{jn} \quad \text{for} \quad i < j < k < n. \tag{5.6.f}$$

Consequently

$$-x = \sum_{i<j} a_{ij}^2 x + 2 \sum_{i<j} \sum_{k} a_{ik}a_{jk} J_i J_j x.$$

Taking the inner product with x yields the identity $\sum_{i<j} a_{ij}^2 = 1$ so

$$0 = 2 \sum_{i<j} \sum_{k} a_{ik}a_{jk} J_i J_j x = 0.$$

By Lemmas 5.2.8 (2) and 5.2.9 (3), we can choose x so $\{J_i J_j x\}_{i<j}$ is an orthonormal set. Consequently:

$$\sum_{k} a_{ik}a_{jk} = 0 \quad \text{for} \quad i \neq j. \tag{5.6.g}$$

By permuting the indices, we may suppose that $a_{12} \neq 0$. We apply Eq. (5.6.b) to see that if $c_2 \neq c_3$, then $a_{13} = 0$. If, however, $c_2 = c_3$, then we can set

$$\begin{aligned}
\tilde{J}_2 &:= \cos\theta J_2 + \sin\theta J_3, & \tilde{J}_3 &:= -\sin\theta J_2 + \cos\theta J_3, \\
J_2 &= \cos\theta \tilde{J}_2 - \sin\theta \tilde{J}_3, & J_3 &= \sin\theta \tilde{J}_2 + \cos\theta \tilde{J}_3, \\
\tilde{a}_{12} &= \cos\theta a_{12} + \sin\theta a_{13}, & \tilde{a}_{13} &= -\sin\theta a_{12} + \cos\theta a_{13}.
\end{aligned}$$

Choose θ so $\tilde{a}_{13} = 0$. This yields $a_{13} = \sin\theta a_{12}/\cos\theta$. This then yields

$$\tilde{a}_{12} = \frac{a_{12}}{\cos\theta} \neq 0.$$

The arguments establishing Eqs. (5.6.b), (5.6.f), and (5.6.g) are unchanged by reparametrization; thus these equations continue to hold. We argue similarly to choose a reparametrization so $a_{1i} = 0$ for $i \geq 3$. We take $i = 1$ and $j \geq 3$ in Eq. (5.6.g) to see

$$0 = \sum_{k} a_{1k}a_{jk} = a_{12}a_{j2}$$

and hence $a_{j2} = 0$ for $j \geq 3$ as well. We now set $i = 1$, $j = 2$, and $3 \leq k < n$ in Eq. (5.6.f) to see

$$0 = a_{12}a_{kn} - a_{1n}a_{2k} + a_{1k}a_{2n} = a_{12}a_{kn}.$$

This shows that all the $a_{ij} = 0$ except a_{12}. $\qquad\square$

In light of Lemma 5.6.1 (1) and Lemma 5.6.2, we may complete the proof of Theorem 5.1.2 by showing:

Lemma 5.6.3   *Let* $\mathfrak{M} := (V, \langle \cdot, \cdot \rangle, J, A = c_1 A_{J_1} + ... + c_\ell A_{J_\ell})$ *be a 0-model which is given by a Clifford family of rank* $\ell = 4$ *or* $\ell = 5$ *where* $c_i \neq 0$ $\forall i$, *and where* $m \geq 2^\ell$. *Then* $\mathfrak{M}$ *is not complex Osserman.*

Proof.   Assume to the contrary that $\mathfrak{M}$ is complex Osserman. We apply Lemma 5.6.2 to assume without loss of generality that $J = J_1 J_2$. Suppose that $\ell = 4$. We may decompose $V = V_+ \oplus V_-$ where

$$V_\pm := \{x : J_1 J_2 J_3 x = \pm x\}.$$

Then V_+ and V_- are preserved by J_1, by J_2, and by J_3 while V_+ and V_- are intertwined by J_4. Let $x \in V_+$. Then

$$\{x, J_3 x, J_1 x, J_2 x, J_4 x, J_1 J_2 J_4 x\}$$

is an orthonormal set. Consequently

$$\mathcal{J}_R(\pi_x)y = \begin{cases} 3c_3 y & \text{if } y \in \text{Span}\{x, J_3 x\}, \\ (3c_1 + 3c_2)y & \text{if } y \in \text{Span}\{J_1 x, J_2 x\}, \\ 3c_4 y & \text{if } y \in \text{Span}\{J_4 x, J_1 J_2 J_4 x\}, \\ 0 & \text{if } y \perp \{x, J_3 x, J_1 x, J_2 x, J_4 x, J_1 J_2 J_4 x\}. \end{cases}$$

On the other hand, since $(J_1 J_2 J_3 J_4)^2 = \text{Id}$, we may choose z so $J_1 J_2 J_3 J_4 z = \pm z$. Then

$$\mathcal{J}_R(\pi_z)y = \begin{cases} 3(c_1 + c_2)y & \text{if } y \in \text{Span}\{J_1 x, J_2 x\}, \\ 3(c_3 + c_4)y & \text{if } y \in \text{Span}\{J_3 x, J_4 x\}, \\ 0 & \text{if } y \perp \{J_1 x, J_2 x, J_3 x, J_4 x\}. \end{cases}$$

As $\mathcal{J}(\pi_x)$ and $\mathcal{J}(\pi_z)$ have different eigenvalue structures, $\mathfrak{M}$ is not complex Osserman. This completes the proof if $\ell = 4$.

Suppose $\ell = 5$. By Lemma 5.2.9 (3), we may choose $x \in S(V)$ so that $\{J_i J_j x\}$ is an orthonormal set. These vectors are all orthogonal to x as well. Multiplying by J_1 then yields that

$$\{J_1 x, J_2 x, J_3 x, J_1 J_2 J_3 x, J_4 x, J_1 J_2 J_4 x, J_5 x, J_1 J_2 J_5 x\}$$

is an orthonormal set. We may then compute:

$$\mathcal{J}_R(\pi_x)y = \begin{cases} (3c_1 + 3c_2)y & \text{if } y \in \text{Span}\{J_1x, J_2x\}, \\ 3c_3y & \text{if } y \in \text{Span}\{J_3x, J_1J_2J_3x\}, \\ 3c_4y & \text{if } y \in \text{Span}\{J_4x, J_1J_2J_4x\}, \\ 3c_5y & \text{if } y \in \text{Span}\{J_5x, J_1J_2J_5x\}. \end{cases}$$

Since c_3, c_4, and c_5 are all non-zero, $\mathcal{J}_A(\pi_x)$ has rank at least 6. This contradicts Lemma 5.6.1 (1). $\qquad\square$

5.7 Tensors $A = c_0A_{\langle\cdot,\cdot\rangle} + c_1A_{J_1} + ... + c_\ell A_{J_\ell}$ for $\ell \geq 4$

This section is devoted to the proof of Assertion (2) of Theorem 5.1.2. Let

$$\mathfrak{M} := (V, \langle\cdot,\cdot\rangle, J, A = c_0A_{\langle\cdot,\cdot\rangle} + c_1A_{J_1} + ... + c_\ell A_{J_\ell})$$

be a complex Osserman 0-model on a vector space V of dimension m which is given by a Clifford family $\mathcal{F} := \{J_1, ..., J_\ell\}$. We shall suppose that $c_i \neq 0$ $\forall i$ and that $\ell \geq 4$. We begin our study by establishing an analog of Lemma 5.6.1 and reducing ourselves to considering the cases $\ell = 4$ and $\ell = 5$.

Lemma 5.7.1 *Let $\mathfrak{M} := (V, \langle\cdot,\cdot\rangle, J, A = c_0A_{\langle\cdot,\cdot\rangle} + c_1A_{J_1} + ... + c_\ell A_{J_\ell})$ be a complex Osserman 0 model which is given by a Clifford family of rank $\ell \geq 4$ where $c_i \neq 0$ $\forall i$. If $\ell = 4$, assume $m \geq 32$. If $5 \leq \ell \leq 7$, assume $m \geq 2^\ell$. If $\ell \geq 8$, assume $m \geq \ell(\ell - 1)$. Set $\tilde{\mathcal{J}}_x := \mathcal{J}(\pi_x) - 2c_0\,\text{Id}$.*

(1) $\dim \text{Rank}\{\tilde{\mathcal{J}}_x\} \leq 4$.
(2) If $x \in V$, then $Jx \in \text{Span}\{J_ix, J_jJ_kx\}_{j<k}$.
(3) Let $\ell \geq 6$. If $x \in V$, then $Jx \in \text{Span}\{J_iJ_jx\}_{i\leq 6, i<j}$.
(4) Either $\ell = 5$ and $J \notin \text{Span}\{J_iJ_j\}_{i<j}$ or $\ell = 4$.

Proof. Set $\tilde{\mathcal{J}}_x := \mathcal{J}(\pi_x) - 2c_0\,\text{id}$. Then:

$$\begin{aligned} \tilde{\mathcal{J}}_x y &= -c_0\langle y, x\rangle x - c_0\langle y, Jx\rangle Jx \\ &\quad + 3\sum_{i=1}^{\ell} c_i\{\langle y, J_ix\rangle J_ix + \langle y, J_iJx\rangle J_iJx\}. \end{aligned} \qquad (5.7.\text{a})$$

Thus $\dim\{\text{Range}\{\tilde{\mathcal{J}}_x\}\} \leq 2\ell + 2$ so 0 is an eigenvalue of multiplicity at least $m - 2\ell - 2$. We observe:

$$\begin{aligned} 4 \leq \ell \leq 7 &\quad\Rightarrow\quad m - 2\ell - 2 \geq 2^\ell - 2\ell - 2 > 4, \\ \ell \geq 8 &\quad\Rightarrow\quad m - 2\ell - 2 \geq \ell(\ell - 3) - 2 > 4. \end{aligned}$$

Corollary 5.2.1 yields 0 is an eigenvalue of multiplicity at least $m - 4$ for $\tilde{\mathcal{J}}_x$. Assertion (1) now follows.

To prove Assertion (2), we use Eq. (5.7.a) to compute:

$$\tilde{\mathcal{J}}_x x = -c_0 x + \sum_i 3c_i \langle x, J_i J x \rangle J_i J x,$$

$$\tilde{\mathcal{J}}_x J x = -c_0 J x + \sum_i 3c_i \langle J x, J_i x \rangle J_i x, \qquad (5.7.\text{b})$$

$$\tilde{\mathcal{J}}_x J_i x = 3c_i J_i x + \sum_j 3c_j \langle J_i x, J_j J x \rangle J_j J x - c_0 \langle J_i x, J x \rangle J x.$$

Let

$$M := \operatorname{diag}(-c_0, 3c_1, ..., 3c_\ell),$$
$$U(x) := \operatorname{Span}\{x, J_1 x, ..., J_\ell x\},$$
$$V(x) := \operatorname{Span}\{J x, J_1 J x, ..., J_\ell J x\},$$
$$W(x) := U(x) + V(x) \supset \operatorname{Range}\{\mathcal{J}(\pi_x)\}.$$

Let ρ be projection on $W(x)/V(x)$. Then $\rho \tilde{\mathcal{J}}_x = \rho M$ on U. Since M is invertible, we then have

$$\dim\{\rho U(x)\} = \dim\{\rho \tilde{\mathcal{J}}_x U(x)\} \leq 4,$$
$$\dim\{W(x)\} \leq 4 + \ell + 1,$$
$$\dim\{U(x) \cap V(x)\} \geq \ell + 1 + \ell + 1 - \ell - 5 = \ell - 3 > 0.$$

Thus there exists a non-trivial relationship

$$(a_0 + a_1 J_1 + ... + a_\ell J_\ell) J x = (b_0 + b_1 J_1 + ... + b_\ell J_\ell) x.$$

We invert this relationship to see $J x \in \operatorname{Span}_{j<k}\{x, J_i x, J_j J_k x\}$; eliminating the orthogonal summand $\operatorname{Span}\{x\}$ establishes Assertion (2) by showing:

$$J x \in \operatorname{Span}\{J_i x, J_j J_k x\}_{j<k}.$$

Suppose that $\ell \geq 6$. We can estimate similarly that:

$$\dim\left\{ \operatorname{Span}\{J_1 x, ..., J_6 x\} \cap \operatorname{Span}\{J_1 J x, ..., J_\ell J x\} \right\}$$
$$\geq 6 + \ell - \dim(W) \geq 6 + \ell - (5 + \ell) > 0.$$

Thus we have a non-trivial relationship from which Assertion (3) will follow after eliminating the orthogonal summand $\operatorname{Span}\{x\}$:

$$(a_1 J_1 + ... + a_\ell J_\ell) J x = (b_1 J_1 + ... + b_6 J_6) x.$$

Suppose $\ell \geq 8$; we argue for a contradiction. As $m \geq \ell(\ell - 1)$, Lemma 5.2.7 (2) shows there is $x \in V$ so that $\{J_i J_j x\}$ forms a linearly independent set with $\frac{1}{2}\ell(\ell - 1)$ elements. We use Assertion (3) to expand

$$Jx = \sum_{i=1}^{\ell-1} \sum_{j=i+1}^{\ell} c_{ij}(x) J_i J_j x;$$

the coefficients c_{ij} are uniquely determined. Since in fact, the sum may be restricted to $i \leq 6$ by Assertion (3), we have $c_{78}(x) = 0$. By permuting the indices, we conclude that all the coefficients are 0 which is false. Consequently,

$$4 \leq \ell \leq 7.$$

If $\ell = 6$, then $m \geq 64 > 60 = 12 \cdot 5$. On the other hand, if $\ell = 7$, then $m \geq 128 > 84 = 14 \cdot 6$. Thus $m \geq 2\ell(\ell - 1)$. By Lemma 5.7.1 (3),

$$Jx \in \mathrm{Span}_{i<j}\{J_i J_j x\}.$$

Thus by Lemma 5.2.7 (4),

$$J \in \mathrm{Span}\{J_i J_j\} \quad \text{if} \quad \ell = 6, 7.$$

Alternatively, if $\ell = 5$, simply assume that $J \in \mathrm{Span}_{i<j}\{J_i J_j\}$. Thus we may expand

$$J = \sum_{i=1}^{\ell-1} \sum_{j=i+1}^{\ell} a_{ij} J_i J_j.$$

By Lemma 5.2.9 and by Lemma 5.2.10, we may choose x so that

$$x \perp J_i J_j J_k x \quad \forall \quad i, j, k,$$
$$\{J_4 J_5 x\} \perp \mathrm{Span}\{J_i J_j x\}_{i<j, (i,j)\neq(4,5)}.$$

Fix such an x. Since $\langle x, J_i J x \rangle = 0$ and $\langle Jx, J_i x \rangle = 0$, Eq. (5.7.b) yields

$$\tilde{\mathcal{J}}_x x = -c_0 x, \qquad \tilde{\mathcal{J}}_x J x = -c_0 J x,$$
$$\tilde{\mathcal{J}}_x J_i x = 3c_i J_i x + \sum_{i=1}^{\ell} 3c_j \langle J_i x, J_j J x \rangle J_j J x.$$

Thus $Z(x) := \mathrm{Span}\{x, Jx\}$ is a 2-dimensional subspace which is invariant under $\tilde{\mathcal{J}}_x$. Consequently $\mathrm{Rank}\{\tilde{\mathcal{J}}_x|_{Z(x)^\perp}\} \leq 2$. We clear the previous notation. Define

$$U(x) := \mathrm{Span}\{J_1 x, ..., J_\ell x\},$$
$$V(x) := \mathrm{Span}\{J_1 J x, ..., J_\ell J x\},$$
$$W(x) := U(x) + V(x) \subset Z(x)^\perp .$$

Let ρ be the projection on $W(x)/V(x)$. We argue as above to see

$$\dim\{\rho U(x)\} = \dim\{\rho \tilde{J}_x\} \le 2 \quad \text{so} \quad \dim\{W(x)\} \le 2 + \ell .$$

Thus there exists a non-trivial relationship

$$(a_1 J_1 + a_2 J_2 + a_3 J_3)x = (b_1 J_1 + b_2 J_2 + ... + b_\ell J_\ell)J x .$$

This shows that

$$Jx \in \mathrm{Span}\{J_i J_j x\}_{i \le 3, i < j} .$$

Since $J_4 J_5 x \perp \mathrm{Span}_{i \le 3, i < j}\{J_i J_j\}x$, we may conclude that $a_{45} = 0$. We can permute the indices to see $a_{ij} = 0$ for all i, j which is impossible. This shows we have the two possibilities enumerated in Assertion (4). $\square$

We can now eliminate the case $\ell = 5$.

Lemma 5.7.2   *Let $\mathfrak{M} := (V, \langle \cdot, \cdot \rangle, J, A = c_0 A_{\langle \cdot, \cdot \rangle} + c_1 A_{J_1} + ... + c_5 A_{J_5})$ be a complex 0 model which is given by a Clifford family of rank $\ell = 5$ where $c_i \ne 0$ $\forall i$ and where $m \ge 32$. Then $\mathfrak{M}$ is not complex Osserman.*

Proof.   We suppose to the contrary that $\mathfrak{M}$ is complex Osserman and argue for a contradiction. Let

$$C := \{x \in V : Jx \in \mathrm{Span}\{J_i x\}\}$$
$$= \{x : Jx \wedge J_1 x \wedge ... \wedge J_5 x = 0\} . \tag{5.7.c}$$

Clearly C is closed. We wish to show C is nowhere dense. Suppose to the contrary that C contains a non-empty open subset of V. We argue for a contradiction. Since Eq. (5.7.c) is defined by a series of polynomial identities, this would imply that Eq. (5.7.c) holds identically and therefore that $Jx \in \mathrm{Span}\{J_i x\}$ for all $x \in V$. Since $m \ge 2\ell$, Lemma 5.2.6 (4) implies $J = \sum_i a_i J_i$. By Lemma 5.2.9 (3), we may choose x so $\{J_i J_j x\}_{i < j}$ is a linearly independent set. We then have

$$\tilde{\mathcal{J}}_x x = -c_0 Jx - \sum_i 3c_i a_i J_i Jx,$$

$$\tilde{\mathcal{J}}_x Jx = -c_0 Jx + \sum_i 3c_i a_i J_i x,$$

$$0 = \tilde{\mathcal{J}}_x x + J\tilde{\mathcal{J}}_x Jx = -\sum_i 3c_i a_i (J_i Jx - J J_i x)$$

$$= -\sum_{i \neq j} 6c_i a_i a_j J_i J_j x.$$

This leads to the relation $(c_i - c_j)a_i a_j = 0$. We choose the notation so $a_1 \neq 0$. Then if $a_j \neq 0$, $c_j = c_1$. Thus we may reparametrize the family in question so that $J = J_1$. By Lemma 5.2.9 (1), we may choose x so $x \perp J_i J_j J_k x$. It now follows from Eq. (5.7.b) that:

$$\tilde{\mathcal{J}}_x x = (3c_1 - c_0)x,$$
$$\tilde{\mathcal{J}}_x J_1 x = (3c_1 - c_0)x,$$
$$\tilde{\mathcal{J}}_x J_i x = 3c_i J_i x \quad \text{for} \quad i = 2, 3, 4, 5,$$
$$\tilde{\mathcal{J}}_x J_i J_1 x = 3c_i J_i J_1 x \quad \text{for} \quad i = 2, 3, 4, 5.$$

Because $J_{ijk} x \perp x$ for all i, j, k,

$$\{J_2 x, J_3 x, J_4 x, J_5 x, J_1 J_2 x, J_1 J_3 x, J_1 J_4 x, J_1 J_5 x\}$$

is an orthonormal set of 8 elements. Thus $\text{Rank}\{\tilde{\mathcal{J}}_x\} \geq 8$ which is false. This contradiction shows C is a closed nowhere dense set.

Let $x \in C^c$. Let $\alpha_i := \langle J_i x, Jx \rangle$ be the Fourier coefficients. Since $Jx \notin \text{Span}\{J_i x\}$, we sum the Fourier coefficients to see

$$\sum_{i=1}^{5} a_i^2 < 1.$$

Introduce the spaces

$$U(x) := \text{Span}\{J_1 x, ..., J_5 x, Jx\},$$
$$V(x) := \text{Span}\{J_1 Jx, ..., J_5 Jx\},$$
$$W(x) := U(x) + V(x).$$

Let ρ be projection from $W(x)$ to $W(x)/V(x)$. By Eq. (5.7.b)

$$\rho \tilde{\mathcal{J}}_x J_i x = \rho(3 c_i J_i x - c_0 \alpha_i J x),$$
$$\rho \tilde{\mathcal{J}}_x J x = \sum_i 3 c_i \alpha_i J_i x - c_0 J x.$$

Consider

$$M := \begin{pmatrix} 3c_1 & 0 & 0 & 0 & 0 & -c_0\alpha_1 \\ 0 & 3c_2 & 0 & 0 & 0 & -c_0\alpha_2 \\ 0 & 0 & 3c_3 & 0 & 0 & -c_0\alpha_3 \\ 0 & 0 & 0 & 3c_4 & 0 & -c_0\alpha_4 \\ 0 & 0 & 0 & 0 & 3c_5 & -c_0\alpha_5 \\ 3c_1\alpha_1 & 3c_2\alpha_2 & 3c_3\alpha_3 & 3c_4\alpha_4 & 3c_5\alpha_5 & -c_0 \end{pmatrix}.$$

Then

$$\det(M) = -c_0 c_1 c_2 c_3 c_4 c_5 (1 - \alpha_1^2 - \ldots - \alpha_5^2) \neq 0.$$

Thus we can invert this matrix to see

$$\dim\{\operatorname{Range}\rho\} = \dim\{\operatorname{Range}\rho\tilde{\mathcal{J}}_x\} \leq 4,$$
$$\dim\{W(x)\} = \dim\ker(\rho) + \dim\operatorname{Range}(\rho) \leq 5 + 4 = 9,$$
$$\dim\big\{\operatorname{Span}\{J_1 x, \ldots, J_5 x\} \cap \operatorname{Span}\{J_1 J x, \ldots, J_5 J x\}\big\} \geq 10 - 9 > 0.$$

Thus $Jx \in \operatorname{Span}\{J_i J_j x\}_{i<j}$ on an open dense subset of V. We now use in a crucial fashion the assumption that $m \geq 32$. Since C is a nowhere dense closed set, Lemma 5.2.5 (3) and Lemma 5.2.9 can be applied to see that $J \in \operatorname{Span}_{i<j}\{J_i J_j\}$. This contradicts Lemma 5.7.1 (4) and thereby completes the proof. $\qquad\square$

We conclude our study by considering the case $\ell = 4$.

Lemma 5.7.3   *Let $\mathfrak{M} := (V, \langle \cdot, \cdot \rangle, J, A = c_0 A_{\langle \cdot, \cdot \rangle} + c_1 A_{J_1} + \ldots + c_4 A_{J_4})$ be a complex 0 model which is given by a Clifford family of rank $\ell = 4$ where $c_i \neq 0 \; \forall i$ and where $m \geq 32$. Then $\mathfrak{M}$ is not complex Osserman.*

Proof.   Again, we suppose to the contrary that $\mathfrak{M}$ is complex Osserman. Since $m \geq 32$, Lemma 5.2.8 (4) shows there exist vectors $x, y \in V$ so that

$$\{J_i x, J_{jk} x, J_i y, J_{jk} y\}_{j<k}$$

forms a linearly independent set of 20 vectors. By Lemma 5.7.1 (2),

$$Jx \in \operatorname{Span}\{J_i x, J_{jk} x\}_{j<k} \quad \text{for all} \quad x \in V.$$

Consequently by Lemma 5.2.5 (3) there exist coefficients a_i and a_{jk} so that

$$J = \sum_{i=1}^{4} a_i J_i + \sum_{j=1}^{3} \sum_{k=j+1}^{4} a_{jk} J_{jk} \, .$$

Suppose first that $a_i \neq 0$ for some i. By permuting the indices if need be, we may assume without loss of generality that $a_1 \neq 0$. We shall argue for a contradiction. By Lemma 5.2.8 (3), we may choose x so $\{J_I x\}$ is an orthonormal set of 16 vectors. We set $a_{ji} = -a_{ij}$. We compute:

$$\tilde{\mathcal{J}}_x x = -c_0 x + \sum_{i=1}^{4} c_i \langle x, J_i J x \rangle J_i J x = -c_0 x - 3 \sum_{i=1}^{4} c_i a_i J_i J x,$$

$$\tilde{\mathcal{J}}_x J x = -c_0 J x + 3 \sum_{i=1}^{4} c_i a_i J_i x \, .$$

Since J and A are compatible, $0 = \tilde{\mathcal{J}}_x x + J \tilde{\mathcal{J}}_x J x$. Consequently:

$$0 = -3 \sum_{i,j=1}^{4} c_i a_i a_j (J_{ij} - J_{ji}) x \tag{5.7.d}$$

$$-3 \sum_{i=1}^{4} \sum_{j=1}^{3} \sum_{k=j+1}^{4} c_i a_i a_{jk} \{ J_{ijk} - J_{jki} \} x \, . \tag{5.7.e}$$

These relations decouple and the summation in Eq. (5.7.d) leads to the identity $(c_i - c_j) a_i a_j = 0$. We have assumed $a_1 \neq 0$. Thus if $a_j \neq 0$, we have $c_1 = c_j$. This permits us to reparametrize the family to assume

$$J = a_1 J_1 + \sum_{ij} a_{ij} J_{ij} \, .$$

The summation in Eq. (5.7.e) shows

$$\sum_{j<k} a_{jk} \{ J_{1jk} - J_{jk1} \} x = 0 \, .$$

This implies $a_{1j} = 0$ for $j = 2, 3, 4$. Consequently we have

$$J = a_1 J_1 + a_{23} J_{23} + a_{24} J_{24} + a_{34} J_{34} \, .$$

Squaring this relation and setting $J^2 x = -x$ then yields

$$2 a_1 a_{23} J_{123} x + 2 a_1 a_{24} J_{124} x + 2 a_1 a_{34} J_{134} x = 0 \, .$$

Thus $a_{23} = a_{24} = a_{34} = 0$. This shows $J = \pm J_1$. Consequently:

$$\tilde{\mathcal{J}}_x J_2 x = c_2 J_2 x, \qquad \tilde{\mathcal{J}}_x J_3 x = c_3 J_3 x, \qquad \tilde{\mathcal{J}}_x J_4 x = c_4 J_4 x,$$
$$\tilde{\mathcal{J}}_x J_2 J x = c_2 J_2 J x, \qquad \tilde{\mathcal{J}}_x J_3 J x = c_3 J_3 J x, \qquad \tilde{\mathcal{J}}_x J_4 J x = c_4 J_4 J x \,.$$

This implies $\mathrm{Rank}\{\tilde{\mathcal{J}}_x\} \geq 6$ which is false. This contradiction shows that

$$J = \sum_{i<j} a_{ij} J_i J_j \,.$$

Let $A_1 = c_1 A_{J_1} + c_2 A_{J_2} + c_3 A_{J_3} + c_4 A_{J_4}$. Since $c_0 A_{\langle \cdot, \cdot \rangle}$ is compatible with any almost complex structure J, A_1 and J are compatible. We can use Lemma 5.6.2 to reparametrize the family $\mathcal{F}$ so that

$$J = \tilde{J}_1 \tilde{J}_2 \quad \text{and} \quad A_1 = \tilde{c}_1 A_{\tilde{J}_1} + \tilde{c}_2 A_{\tilde{J}_2} + \tilde{c}_3 A_{\tilde{J}_3} + \tilde{c}_4 A_{\tilde{J}_4} \,.$$

Thus without loss of generality, we may assume that $J = J_1 J_2$. Again, we choose x so $\{J_I x\}$ is an orthogonal set of 16 vectors. We may then compute:

$$\tilde{\mathcal{J}}_x x = c_0 x, \qquad\qquad \tilde{\mathcal{J}}_x J_1 J_2 x = c_0 J_1 J_2 x,$$
$$\tilde{\mathcal{J}}_x J_3 x = 3 c_3 J_3 x, \qquad \tilde{\mathcal{J}}_x J_1 J_2 J_3 x = c_0 J_1 J_2 J_3 x,$$
$$\tilde{\mathcal{J}}_x J_4 x = 3 c_4 J_4 x, \qquad \tilde{\mathcal{J}}_x J_1 J_2 J_4 x = c_0 J_1 J_2 J_4 x \,.$$

This shows $\mathrm{Rank}\{\tilde{\mathcal{J}}_x\} \geq 6$ which is false. This final contradiction shows $\ell \neq 4$ and completes the proof of Theorem 5.1.2. $\qquad\qquad \square$

Chapter 6

Stanilov–Tsankov Theory

6.1 Introduction

In Chapter 6, we study commutativity properties of the Jacobi operator
and of the skew-symmetric curvature operator. M. Brozos-Vázquez is a
coauthor on this section and, except as will be noted via explicit references,
the material of this section is joint work with M. Brozos-Vázquez. It is a
pleasure to acknowledge the foundational contributions of Prof. Stanilov
who originally posed many of the questions being studied in this section. It
is also a pleasure to acknowledge the contributions of Prof. Tsankov who
proved the results which motivated much of our investigations. Professors
Nikčević and Videv also contributed to the development of the material of
this section.

Let $\mathfrak{M} = (V, \langle \cdot, \cdot \rangle, A)$ be a 0-model. Let A be the associated curvature
operator. The *Jacobi operator* $\mathcal{J}_x = \mathcal{J}(x)$ is given by:

$$\mathcal{J}_x z = \mathcal{J}(x)z := \mathcal{A}(z, x)x \,.$$

The Jacobi operator is quadratic in x and can be polarized to define a
bilinear operator $\mathcal{J}_{xy} = \mathcal{J}(x, y)$ by setting

$$\mathcal{J}_{xy} z = \mathcal{J}(x, y)z := \tfrac{1}{2}\{\mathcal{A}(z, x)y + \mathcal{A}(z, y)x\} \,.$$

This operator was first introduced and studied by Videv (1993). Note that

$$\mathcal{J}_x = \mathcal{J}_{xx}, \qquad \mathcal{J}_{xy}y = -\tfrac{1}{2}\mathcal{J}_y x,$$
$$\mathcal{J}_{\cos\theta x + \sin\theta y} = \cos^2\theta \mathcal{J}_x + 2\cos\theta\sin\theta\mathcal{J}_{xy} + \sin^2\theta\mathcal{J}_y \,. \tag{6.1.a}$$

If $\{e_1, e_2\}$ is an oriented orthonormal basis for an oriented non-degenerate

309

2-plane π, the *skew-symmetric curvature operator* $\mathcal{A}_\pi = \mathcal{A}(\pi)$ is given by:

$$\mathcal{A}_\pi z = \mathcal{A}(\pi)z := \mathcal{A}(e_1, e_2)z\,.$$

Let ρ be the Ricci operator. We begin by stating the basic definitions in the subject:

Definition 6.1.1   Let $\mathfrak{M} = (V, \langle \cdot, \cdot \rangle, A)$ be a 0-model.

(1) $\mathfrak{M}$ is *Jacobi Tsankov* if $\mathcal{J}_x \mathcal{J}_y = \mathcal{J}_y \mathcal{J}_x$ for all $x, y \in V$.
(2) $\mathfrak{M}$ is *orthogonally Jacobi Tsankov* if $\mathcal{J}_x \mathcal{J}_y = \mathcal{J}_y \mathcal{J}_x$ for all $x, y \in V$ with $x \perp y$.
(3) $\mathfrak{M}$ is *skew Tsankov* if $\mathcal{A}_\pi \mathcal{A}_\sigma = \mathcal{A}_\sigma \mathcal{A}_\pi$ for all non-degenerate oriented 2-planes π and σ.
(4) $\mathfrak{M}$ is *orthogonally skew Tsankov* if $\mathcal{A}_\pi \mathcal{A}_\sigma = \mathcal{A}_\sigma \mathcal{A}_\pi$ for all non-degenerate oriented 2-planes π and σ with $\pi \perp \sigma$.
(5) $\mathfrak{M}$ is *Stanilov–Tsankov* if $\mathcal{A}_{x_1 x_2} \mathcal{J}_{x_3} = \mathcal{J}_{x_3} \mathcal{A}_{x_1 x_2}$ for all $x_i \in V$.
(6) $\mathfrak{M}$ is *Jacobi Videv* if $\mathcal{J}_x \rho = \rho \mathcal{J}_x$ for all $x \in V$.
(7) $\mathfrak{M}$ is *skew Videv* if $\mathcal{R}_{xy} \rho = \rho \mathcal{R}_{xy}$ for all $x, y \in V$.

If $\mathcal{M} = (M, g)$ is a pseudo-Riemannian manifold, then $\mathcal{M}$ is said to have one of the properties discussed above if and only if the associated model $\mathfrak{M}(\mathcal{M}, P) = (T_P M, g_P, R_P)$ has this property for all points P of M.

Here is a brief outline to this introductory section. In Section 6.1.1, we discuss Jacobi Tsankov theory; this relates to the Jacobi operator. We sketch some of the results in this field which will be described in further detail in Sections 6.2, 6.3, and 6.4. Section 6.1.2 deals with the skew-symmetric curvature operator; we introduce some of the basic material in this area that will be treated more extensively in Section 6.4. Section 6.1.3 deals with Stanilov–Videv manifolds. We present the two main results in this area and refer to the literature for further details as we shall not pursue this matter further. Section 6.1.4 presents some foundational material relating to Jacobi Videv manifolds and models and serves as an introduction to the more complete discussion in Section 6.6.

6.1.1 *Jacobi Tsankov manifolds*

The notation "Jacobi Tsankov" is motivated by the seminal paper of Tsankov (2005) which studied hypersurfaces with these properties. Let L denote the second fundamental form of a hypersurface $\mathcal{M}$ in $\mathbb{R}^{m+1}$ and let $\{\lambda_1, ..., \lambda_n\}$ be the eigenvalues of the associated *shape operator* or *Weingarten operator*. Tsankov showed:

Theorem 6.1.1 [**Tsankov**] *A hypersurface in $\mathbb{R}^{m+1}$ is orthogonally Jacobi Tsankov if and only if either $\lambda_1 = ... = \lambda_n$ or $\lambda_1 = ... = \lambda_{n-1} = 0$.*

In Sections 6.2, 6.3, and 6.4, we shall present results of Brozos-Vázquez, Gilkey, and Nikčević (2006), of Brozos-Vázquez and Gilkey (2005), and of Brozos-Vázquez and Gilkey (2006). In Section 6.2, we study these questions in the Riemannian setting. Theorem 6.2.1 contains a complete classification of Riemannian Jacobi Tsankov 0-models and of Riemannian orthogonally Jacobi Tsankov 0-models. A corresponding classification in the geometric context is provided by Theorem 6.2.2.

In Section 6.3, we turn to the pseudo-Riemannian setting. Let $\mathfrak{M}$ be a pseudo-Riemannian Jacobi Tsankov 0-model. In Theorem 6.3.1, we show $\mathcal{J}_x^2 = 0$ for all $x \in V$. This implies $\mathfrak{M}$ is Osserman and shows $A = 0$ if $\mathfrak{M}$ is Lorentzian. In Theorem 6.3.2, we show there exist 0-models with $\mathcal{J}_x^2 = 0$ for all x which are not Jacobi Tsankov. In Theorem 6.3.3 we show that if $m \leq 13$ and if $\mathfrak{M}$ is Jacobi Tsankov, then $\mathcal{J}_x\mathcal{J}_y = 0$ for all $x, y \in V$. In Theorem 6.3.4, we classify all indecomposable 0-models satisfying the condition $\mathcal{J}_x\mathcal{J}_y = 0$ for all x, y.

In Section 6.4, we show the condition $\dim(V) \leq 13$ in Theorem 6.3.3 is sharp; there is a 14-dimensional 0-model of signature $(8, 6)$ which is Jacobi-Tsankov but which does not satisfy $\mathcal{J}_x\mathcal{J}_y = 0$ for all x, y. This model is geometrically realizable and we examine the properties of various geometrical realizations.

6.1.2 Skew Tsankov manifolds

One can raise similar questions for the skew-symmetric curvature operator. The following result of Tsankov (2005) is seminal in the subject.

Theorem 6.1.2 [**Tsankov**] *A hypersurface in $\mathbb{R}^{m+1}$ is orthogonally skew Tsankov if and only if $|\lambda_1| = ... = |\lambda_m|$, or $\lambda_1 = ... = \lambda_{m-1} = 0$ and $\lambda_m \neq 0$, or $\lambda_1 = ... = \lambda_{m-2} = 0$, and $\lambda_{m-1} \neq 0$ and $\lambda_n \neq 0$.*

We have additional examples provided by our previous discussion:

Theorem 6.1.3

(1) If $\mathcal{M}$ is the neutral signature generalized plane wave manifold of Definition 2.5.1, then $\mathcal{M}$ is skew Tsankov and Jacobi Tsankov.

(2) Let $\mathcal{M}$ be as in Definition 2.7.1. If $s = 2$, then $\mathcal{M}$ is skew Tsankov.

Proof.   Let $\mathcal{M}$ be as in Definition 2.5.1. Adopt the notation established in Section 2.5. By Lemma 2.5.1, $\mathcal{A}_{\xi_1\xi_2}\mathcal{A}_{\xi_3\xi_4} = 0$ for all $\xi_i \in TM$ and $\mathcal{J}_{\eta_1}\mathcal{J}_{\eta_2} = 0$ for all $\eta_i \in TM$.

Let $\mathcal{M}$ be as in Definition 2.7.1 with $s = 2$. Adopt the notation established in Section 2.7. The only non-zero quadratic curvatures are $\mathcal{A}_{U_iU_j}\mathcal{A}_{U_kU_l}$. Since $s = 2$, we need only consider $\mathcal{A}_{U_1U_2}$. It is now immediate that $\mathcal{M}$ is skew Tsankov. We remark that if $s = 3$, then $\mathcal{A}$ is not skew Tsankov since

$$\mathcal{A}_{U_1U_3}\mathcal{A}_{U_2U_3}U_2 = \mathcal{A}_{U_1U_3}(-T_3) = V_3,$$
$$\mathcal{A}_{U_2U_3}\mathcal{A}_{U_1U_3}U_2 = 0\,.$$

$\square$

In Section 6.5 we follow the treatment in Brozos-Vázquez and Gilkey (2006a). We shall present a complete classification of Riemannian skew Tsankov 0-models and we shall exhibit some irreducible 3-dimensional and 4-dimensional Riemannian skew Tsankov manifolds.

6.1.3   *Stanilov–Videv manifolds*

Let $\{e_i\}_{1\leq i\leq k}$ be a basis for a non-degenerate k-plane π. Recall that the *higher order Jacobi operator* $\mathcal{J}_\pi$ of Stanilov and Videv (1992) is defined by:

$$\mathcal{J}_\pi := \sum_{1\leq i,j\leq k} \xi^{ij}\mathcal{J}_{e_ie_j}$$

where $\xi_{ij} := \langle e_i, e_j\rangle$ and where ξ^{ij} is the inverse matrix. Note that if $\pi = V$ is the whole vector space, then $\rho = \mathcal{J}_V$ is the *Ricci operator*.

If $k = 2$, let $\mathcal{A}_\pi := \mathcal{A}_{e_1e_2}$ be the skew-symmetric curvature operator. Motivated by the discussion in Stanilov and Videv (2004), one says that $\mathfrak{M}$ is *Stanilov–Videv* if $\mathcal{A}_\pi\mathcal{J}_\pi = \mathcal{J}_\pi\mathcal{A}_\pi$ for all non-degenerate 2-planes π. One has the following result of Stanilov and Videv (2004), see also Videv (2005):

Theorem 6.1.4   *Let $\mathfrak{M}$ be a 4-dimensional Riemannian 0 model. Then $\mathfrak{M}$ is Einstein if and only if $\mathfrak{M}$ is Stanilov–Videv.*

There is also related work by Ivanova and Videv (2004) showing

Theorem 6.1.5   *Let $\mathfrak{M}$ be a 4-dimensional Riemannian 0-model. Then $\mathcal{J}_\pi\mathcal{A}_{\pi^\perp} = \mathcal{A}_{\pi^\perp}\mathcal{J}_\pi$ for all 2-planes π if and only if $\mathfrak{M}$ has constant sectional*

curvature.

6.1.4 *Jacobi Videv manifolds and 0-models*

We follow the discussion of Gilkey, Puffini, and Videv (2006) in Section 6.6. Let ρ be the Ricci operator of a 0-model $\mathfrak{M} = (V, \langle \cdot, \cdot \rangle, A)$. Recall that $\mathfrak{M}$ is said to be *Jacobi Videv* if $\rho \mathcal{J}(x) = \mathcal{J}(x)\rho$ for all $x \in V$. In Theorem 6.6.1, we give some equivalent characterizations of this property in terms of the higher order Jacobi operator. In Theorem 6.6.2, we establish some decomposition theorems; in particular we show that an indecomposable Riemannian model is Jacobi Videv if and only if it is Einstein; we present some examples showing this fails in the higher signature setting.

6.2   Riemannian Jacobi Tsankov Manifolds and 0-Models

In this section, we present work of Brozos-Vázquez and Gilkey (2005) that deals with the Riemannian context and postpone until Section 6.3 a discussion of the pseudo-Riemannian setting. We adopt the notational conventions established in Section 6.1. Recall that a 0-model $\mathfrak{M} = (V, \langle \cdot, \cdot \rangle, A)$ has *constant sectional curvature* if $A = cA_{\langle \cdot, \cdot \rangle}$ for some constant c where the associated curvature operator is

$$\mathcal{A}_{\langle \cdot, \cdot \rangle}(x, y)z := \langle y, z \rangle x - \langle x, z \rangle y\,.$$

Let Θ be a Hermitian almost complex structure on V; Θ exists, of course, if and only if m is even. Following Eq. (1.3.a), the curvature operator is defined by:

$$\mathcal{A}_{\Theta}(x, y)z := \langle \Theta y, z \rangle \Theta x - \langle \Theta x, z \rangle \Theta y - 2 \langle \Theta x, y \rangle \Theta z\,.$$

Let $W \subset V$. To simplify the notation, we let $S(W) := S(W, \langle \cdot, \cdot \rangle|_W)$ be the sphere of unit vectors in W. Let $x \in S(V)$. One then has

$$\mathcal{J}_{A_{\langle \cdot, \cdot \rangle}}(x)y = \begin{cases} y & \text{if } y \perp x, \\ 0 & \text{if } y \in \mathrm{Span}\{x\}, \end{cases} \tag{6.2.a}$$
$$\mathcal{J}_{A_{\Theta}}(x)y = 3\langle \Theta x, y \rangle \Theta x\,.$$

We have the following classification results:

Theorem 6.2.1   *Let* $\mathfrak{M} = (V, \langle \cdot, \cdot \rangle, A)$ *be a Riemannian 0-model.*

(1) $\mathfrak{M}$ *is Jacobi Tsankov if and only if* $A = 0$.

(2) $\mathfrak{M}$ is orthogonally Jacobi Tsankov if and only if either $\mathfrak{M}$ has constant sectional curvature or there is a Hermitian almost complex structure Θ so that A is a multiple of A_Θ.

This result has a corresponding geometric analog

Theorem 6.2.2   *Let $\mathcal{M}$ be a connected Riemannian manifold of dimension m.*

(1) $\mathcal{M}$ is Jacobi Tsankov if and only if $\mathcal{M}$ is flat.
(2) Let $m \geq 3$. $\mathcal{M}$ is orthogonally Jacobi Tsankov if and only if $\mathcal{M}$ has constant sectional curvature.

Here is a brief outline to Section 6.2. In Section 6.2.1, we shall establish Theorem 6.2.1 (1); Theorem 6.2.2 (1) then follows as an immediate consequence. In Section 6.2.2, we shall establish Theorem 6.2.1 (2) and in Section 6.2.3, we shall establish Theorem 6.2.2 (2).

6.2.1   *Riemannian Jacobi Tsankov 0-models*

Let $\mathfrak{M} = (V, \langle \cdot, \cdot \rangle, A)$ be a Riemannian Jacobi Tsankov 0-model. The Jacobi operators $\{\mathcal{J}_x\}_{x \in V}$ form a commuting family of self-adjoint operators. Such a family can be simultaneously diagonalized; there is an orthogonal direct sum decomposition:

$$V = \oplus_\lambda E_\lambda \quad \text{where} \quad \mathcal{J}_x \xi = \lambda(x)\xi \ \forall \ \xi \in E_\lambda, \ \forall \ x \in V .$$

Fix a unit vector $\eta_\lambda \in E_\lambda$. Then $\lambda(x) = \langle \mathcal{J}_x \eta_\lambda, \eta_\lambda \rangle$. Consequently the functions $x \to \lambda(x)$ are continuous functions of x.

Choose $\xi \in V$ and decompose $\xi = \sum_\lambda \xi_\lambda$ for $\xi_\lambda \in E_\lambda$. Let

$$\mathcal{O} := \{\xi \in V : \xi_\lambda \neq 0 \ \forall \ \lambda\} .$$

Then $\mathcal{O}$ is the complement of a finite number of hyperplanes and hence is a dense open subset of V. Let $\xi \in \mathcal{O}$. One then has:

$$0 = \mathcal{J}_\xi \xi = \sum_\lambda \lambda(\xi)\xi_\lambda .$$

Since the $\{\xi_\lambda\}$ are linearly independent, this implies $\lambda(\xi) = 0$ for all λ. As $\lambda(\cdot)$ vanishes on $\mathcal{O}$ which is an open dense subset of V, $\lambda(\cdot)$ vanishes identically. Thus $\mathcal{J}_x = 0$ for all $x \in V$. By Lemma 1.7.1, $A = 0$. This completes the proof of Theorem 6.2.1 (1). $\qquad\square$

We remark that another proof is given in Section 6.3.1.

6.2.2 *Riemannian orthogonally Jacobi Tsankov 0-models*

One implication of Theorem 6.2.1 (2) follows from:

Lemma 6.2.1 *Let $\mathfrak{M} = (V, \langle \cdot, \cdot \rangle, A)$ be a Riemannian 0-model. If there is a constant c so that either $A = cA_{\langle \cdot, \cdot \rangle}$ or $A = cA_\Theta$ for some Hermitian almost complex structure Θ on V, then $\mathfrak{M}$ is orthogonally Jacobi Tsankov.*

Proof. We apply Eq. (6.2.a). Let $x, y \in S(V)$ with $x \perp y$. Suppose $A = A_{\langle \cdot, \cdot \rangle}$. Then

$$\mathcal{J}_x \mathcal{J}_y z = \mathcal{J}_x \begin{cases} z & \text{if } z \perp y, \\ 0 & \text{if } z \in \mathrm{Span}\{y\}, \end{cases}$$

$$= \begin{cases} z & \text{if } z \perp x, y, \\ 0 & \text{if } z \in \mathrm{Span}\{x, y\}. \end{cases}$$

This is symmetric in the roles of x and y and thus $\mathcal{J}_x \mathcal{J}_y = \mathcal{J}_y \mathcal{J}_x$. If, on the other hand, $A = A_\Theta$, then $\langle \Theta x, \Theta y \rangle = \langle x, y \rangle = 0$ so

$$\mathcal{J}_x \mathcal{J}_y = 3 \mathcal{J}_x \{ \langle \Theta y, z \rangle \Theta y \} = 9 \langle \Theta y, z \rangle \langle \Theta y, \Theta x \rangle \Theta x = 0.$$

Again, this is symmetric in the roles of x and y so $\mathfrak{M}$ is orthogonally Jacobi Tsankov. $\qquad\square$

Before establishing the converse to Lemma 6.2.1 and completing the proof of Theorem 6.2.1 (2), we shall need a number of technical results. Let

$$r(x) := \mathrm{Rank}\{\mathcal{J}_x\}.$$

Since $\mathcal{J}_x x = 0$, $r(x) \leq m - 1$ for any $x \in V$.

Lemma 6.2.2 *Let $\mathfrak{M} = (V, \langle \cdot, \cdot \rangle, A)$ be a Riemannian orthogonally Jacobi Tsankov 0-model of dimension m. If there exists $x \in V$ with $r(x) = m - 1$, then A has constant sectional curvature.*

Proof. Let $\mathcal{O} := \{ x \in V : \mathrm{Rank}\{\mathcal{J}_x\} = m - 1 \}$. We suppose $\mathcal{O}$ is non-empty. We wish to show $\mathcal{O}$ is an open and dense subset of V. Let $x \in \mathcal{O}$. Let $\mathcal{B} := \{e_1, ..., e_{m-1}\}$ be an orthonormal basis for $x^\perp = \mathrm{Range}(\mathcal{J}_x)$. Let $\mathcal{J}_{ij}(z) := \langle \mathcal{J}_z e_i, e_j \rangle$. Set

$$p_\mathcal{B}(z) := \det(\mathcal{J}_{ij})(z).$$

Then $p_{\mathcal{B}}$ is polynomial in z and by hypothesis $p_{\mathcal{B}}(x) \neq 0$. Thus

$$\mathcal{O}_{\mathcal{B}} := \{z : p_{\mathcal{B}}(z) \neq 0\}$$

is a non-empty open dense subset of V which contains x. If $y \in \mathcal{O}_{\mathcal{B}}$, then $r(y) \geq m - 1$. Since $r(y) \leq m - 1$, $r(y) = m - 1$. This shows that $\mathcal{O}_{\mathcal{B}} \subset \mathcal{O}$ and $\mathcal{O}$ is a neighborhood of x; since x was arbitrary, $\mathcal{O}$ is open. Since $\mathcal{O}$ contains a dense subset, this shows, as desired, that $\mathcal{O}$ is an open dense subset of V.

Since $\mathfrak{M}$ has constant sectional curvature if $m = 2$, we suppose $m \geq 3$. Let $x \in \mathcal{O}$ and let $y \in x^{\perp}$. Then

$$\mathcal{J}_x \mathcal{J}_y x = \mathcal{J}_y \mathcal{J}_x x = 0 \quad \text{so} \quad \langle \mathcal{J}_y x, \mathcal{J}_x z \rangle = 0 \quad \text{for all} \quad z.$$

As $\mathrm{range}(\mathcal{J}_x) = x^{\perp}$, we have $\langle \mathcal{J}_y x, z \rangle = 0$ if $z \perp x$. Thus

$$A(x, y, y, z) = 0 \quad \text{if} \quad x \in \mathcal{O}, \ z \perp x, \ y \perp x. \tag{6.2.b}$$

Since $\mathcal{O}$ is dense, Eq. (6.2.b) holds for all $x \in V$. Thus if $\{e_i\}$ is an orthonormal basis for V and if $\{i, j, k\}$ are distinct indices,

$$A(e_i, e_j, e_j, e_k) = 0 \quad \text{for} \quad i, j, k \quad \text{distinct}.$$

Suppose that ℓ is a fourth distinct index; this is impossible, of course, if $m = 3$. Polarization yields

$$A(e_i, e_j, e_\ell, e_k) + A(e_i, e_\ell, e_j, e_k) = 0 \quad \text{for} \quad i, j, k, \ell \quad \text{distinct}.$$

The previous relation together with the first Bianchi identity and the other curvature identities of Eq. (1.2.g) show that:

$$\begin{aligned}
0 &= A(e_i, e_j, e_k, e_\ell) + A(e_i, e_k, e_\ell, e_j) + A(e_i, e_\ell, e_j, e_k) \\
&= A(e_i, e_j, e_k, e_\ell) - A(e_i, e_k, e_j, e_\ell) - A(e_i, e_j, e_\ell, e_k) \\
&= A(e_i, e_j, e_k, e_\ell) + A(e_i, e_j, e_k, e_\ell) + A(e_i, e_j, e_k, e_\ell) \\
&= 3A(e_i, e_j, e_k, e_\ell) \quad \text{for} \quad i, j, k, \ell \quad \text{distinct}.
\end{aligned}$$

Thus the only non-zero curvatures are $A(e_i, e_j, e_j, e_i) = c_{ij}$ for $i \neq j$. Consider the new basis

$$e_\nu(\theta) := \begin{cases} \cos\theta e_i + \sin\theta e_j & \text{if} \ \nu = i, \\ -\sin\theta e_i + \cos\theta e_j & \text{if} \ \nu = j, \\ e_\nu & \text{if} \ \nu \neq i, j. \end{cases}$$

If i, j, and k are distinct indices, then:

$$0 = A(e_i(\theta), e_k, e_k, e_j(\theta)) = \cos\theta \sin\theta \{-c_{ik} + c_{jk}\} \,.$$

It now follows that $c_{ik} = c_{jk}$ for i, j, k distinct and, consequently, A has constant sectional curvature. $\qquad\square$

In light of Lemma 6.2.2, we may suppose that $r(x) < m - 1$ for all x henceforth. Thus, in particular, given x, we can always choose y so $x \perp y$ and $\mathcal{J}_x y = 0$.

Lemma 6.2.3 *Let $\mathfrak{M} = (V, \langle \cdot, \cdot \rangle, A)$ be a Riemannian orthogonally Jacobi Tsankov 0-model of dimension m. Assume $r(x) < m - 1$ for all $x \in V$. Let $0 \neq x \in S(V)$. Choose $y \in S(V)$ so $y \perp x$ and so $\mathcal{J}_x y = 0$. Then:*

(1) $\mathcal{J}_y x = 0$ and $\mathcal{J}_x \mathcal{J}_y = 0$.

(2) $0 = \mathcal{J}_y^2 + \mathcal{J}_x^2 - 4\mathcal{J}_{xy}^2$, $\mathcal{J}_{xy} \mathcal{J}_x = \mathcal{J}_y \mathcal{J}_{xy}$, and $\mathcal{J}_x \mathcal{J}_{xy} = \mathcal{J}_{xy} \mathcal{J}_y$.

(3) Let $\{x, z_1, z_2\}$ be an orthonormal set. Suppose that $\mathcal{J}_x z_1 = \lambda_1 z_1$ and $\mathcal{J}_x z_2 = \lambda_2 z_2$ where $\lambda_1 \neq \lambda_2$. Then $\mathcal{J}_{z_1} z_2 = 0$.

(4) Let $\Xi = \mathrm{diag}(\lambda_1, ..., \lambda_r)$ where λ_i are the non-zero eigenvalues of $\mathcal{J}_x$, repeated according to multiplicity. We can choose an orthonormal basis for V so that

$$\mathcal{J}_x = \begin{pmatrix} \Xi & 0 & 0 \\ 0 & 0 & 0 \\ 0 & 0 & 0 \end{pmatrix}, \quad \mathcal{J}_y = \begin{pmatrix} 0 & 0 & 0 \\ 0 & \Xi & 0 \\ 0 & 0 & 0 \end{pmatrix}, \quad \mathcal{J}_{xy} = \frac{1}{2} \begin{pmatrix} 0 & \Xi & 0 \\ \Xi & 0 & 0 \\ 0 & 0 & 0 \end{pmatrix} \,.$$

Proof. Suppose that $\mathcal{J}_x y = 0$. By Eq. (6.1.a),

$$\{\mathcal{J}_{\cos\theta x + \sin\theta y} \mathcal{J}_{-\sin\theta x + \cos\theta y}\} y$$
$$= \mathcal{J}_{\cos\theta x + \sin\theta y} \{\sin^2\theta \mathcal{J}_x - 2\sin\theta\cos\theta \mathcal{J}_{xy} + \cos^2\theta \mathcal{J}_y\} y$$
$$= \{\cos^2\theta \mathcal{J}_x + 2\sin\theta\cos\theta \mathcal{J}_{xy} + \sin^2\theta \mathcal{J}_y\} \{\sin\theta\cos\theta \mathcal{J}_y x\}$$
$$= 2\sin^2\theta\cos^2\theta \mathcal{J}_{xy}\mathcal{J}_y x + \sin^3\theta\cos\theta \mathcal{J}_y\mathcal{J}_y x$$
$$= \{\mathcal{J}_{-\sin\theta x + \cos\theta y} \mathcal{J}_{\cos\theta x + \sin\theta y}\} y$$
$$= \mathcal{J}_{-\sin\theta x + \cos\theta y} \{\cos^2\theta \mathcal{J}_x + 2\cos\theta\sin\theta \mathcal{J}_{xy} + \sin^2\theta \mathcal{J}_y\} y$$
$$= \{\sin^2\theta \mathcal{J}_x - 2\cos\theta\sin\theta \mathcal{J}_{xy} + \cos^2\theta \mathcal{J}_y\} \{-\cos\theta\sin\theta \mathcal{J}_y x\}$$
$$= 2\cos^2\theta\sin^2\theta \mathcal{J}_{xy}\mathcal{J}_y x - \cos^3\theta\sin\theta \mathcal{J}_y\mathcal{J}_y x \,.$$

This equality for all θ shows $\mathcal{J}_y^2 x = 0$. Since $\mathcal{J}_y$ is self-adjoint and the metric is definite, $\mathcal{J}_y x = 0$.

We have $\mathcal{J}_x\mathcal{J}_y y = 0$ and $\mathcal{J}_x\mathcal{J}_y x = 0$. Let $z \perp \{x, y\}$. To complete the proof of Assertion (1), we must show $\mathcal{J}_x\mathcal{J}_y z = 0$. We compute:

$$
\begin{aligned}
0 &= \mathcal{J}_{\cos\theta x + \sin\theta z}\mathcal{J}_y x = \mathcal{J}_y\mathcal{J}_{\cos\theta x + \sin\theta z} x \\
&= \mathcal{J}_y\{\cos^2\theta\mathcal{J}_x + 2\cos\theta\sin\theta\mathcal{J}_{xz} + \sin^2\theta\mathcal{J}_z\}x \\
&= -\cos\theta\sin\theta\mathcal{J}_y\mathcal{J}_x z + \sin^2\theta\mathcal{J}_z\mathcal{J}_y x \\
&= -\cos\theta\sin\theta\mathcal{J}_y\mathcal{J}_x z \,.
\end{aligned}
$$

We now prove Assertion (2). Because $\mathcal{J}_{xy} x = -\frac{1}{2}\mathcal{J}_y x = 0$ and because $\mathcal{J}_{xy} y = -\frac{1}{2}\mathcal{J}_x y = 0$, we have:

$$
\begin{aligned}
&\mathcal{J}_{xy}\{-\sin\theta x + \cos\theta y\} = 0, \quad \text{so} \\
&\mathcal{J}_{\cos\theta x + \sin\theta y}\{-\sin\theta x + \cos\theta y\} = 0\,.
\end{aligned}
$$

Thus applying Assertion (1) to the pair $\{\cos\theta x + \sin\theta y,\, -\sin\theta x + \cos\theta y\}$ permits us to derive Assertion (2) from the following identity:

$$
\begin{aligned}
0 &= \mathcal{J}_{\cos\theta x + \sin\theta y}\mathcal{J}_{-\sin\theta x + \cos\theta y} \\
&= \{\cos^2\theta\mathcal{J}_x + 2\sin\theta\cos\theta\mathcal{J}_{xy} + \sin^2\theta\mathcal{J}_y\} \\
&\quad \cdot\{\sin^2\theta\mathcal{J}_x - 2\sin\theta\cos\theta\mathcal{J}_{xy} + \cos^2\theta\mathcal{J}_y\} \\
&= \cos^2\theta\sin^2\theta\{\mathcal{J}_y^2 + \mathcal{J}_x^2 - 4\mathcal{J}_{xy}^2\} + 2\sin^3\theta\cos\theta\{\mathcal{J}_{xy}\mathcal{J}_x - \mathcal{J}_y\mathcal{J}_{xy}\} \\
&\quad + 2\sin\theta\cos^3\theta\{\mathcal{J}_{xy}\mathcal{J}_y - \mathcal{J}_x\mathcal{J}_{xy}\}\,.
\end{aligned}
$$

Let $\{x, z_1, z_2\}$ be an orthonormal set with $\mathcal{J}_x z_i = \lambda_i z_i$ where $\lambda_1 \neq \lambda_2$. To prove Assertion (3), we compute

$$
\begin{aligned}
\mathcal{J}_x\mathcal{J}_{\cos\theta z_1 + \sin\theta z_2} z_1 &= \mathcal{J}_x\{2\cos\theta\sin\theta\mathcal{J}_{z_1 z_2} + \sin^2\theta\mathcal{J}_{z_2}\}z_1 \\
&= \mathcal{J}_x\{-\cos\theta\sin\theta\mathcal{J}_{z_1} z_2 + \sin^2\theta\mathcal{J}_{z_2} z_1\} \\
&= -\lambda_2\cos\theta\sin\theta\mathcal{J}_{z_1} z_2 + \lambda_1\sin^2\theta\mathcal{J}_{z_2} z_1 \\
&= \mathcal{J}_{\cos\theta z_1 + \sin\theta z_2}\mathcal{J}_x z_1 = \lambda_1\{2\cos\theta\sin\theta\mathcal{J}_{z_1 z_2} + \sin^2\theta\mathcal{J}_{z_2}\}z_1 \\
&= -\lambda_1\cos\theta\sin\theta\mathcal{J}_{z_1} z_2 + \lambda_1\sin^2\theta\mathcal{J}_{z_2} z_1\,.
\end{aligned}
$$

Assertion (3) now follows since we have

$$
\lambda_2\mathcal{J}_{z_1} z_2 = \lambda_1\mathcal{J}_{z_1} z_2\,.
$$

To prove Assertion (4), choose an orthonormal basis $\{e_1, ..., e_r\}$ for $\mathrm{Range}(\mathcal{J}_x)$ so

$$
\mathcal{J}_x e_i = \lambda_i e_i \quad \text{for} \quad \lambda_i \neq 0\,.
$$

We then have $\mathcal{J}_y e_i = 0$ and thus $4\mathcal{J}_{xy}^2 e_i = \lambda_i^2 e_i$. Define:

$$f_i := 2\lambda_i^{-1}\mathcal{J}_{xy} e_i.$$

The collection $\{f_1, ..., f_r\}$ is an orthonormal set since:

$$\langle f_i, f_j \rangle = 4\lambda_i^{-1}\lambda_j^{-1}\langle \mathcal{J}_{xy} e_i, \mathcal{J}_{xy} e_j \rangle = 4\lambda_i^{-1}\lambda_j^{-1}\langle \mathcal{J}_{xy}^2 e_i, e_j \rangle = \delta_{ij}.$$

Furthermore, $f_i \in \ker(\mathcal{J}_x) = \mathrm{Range}(\mathcal{J}_x)^\perp$ because:

$$\mathcal{J}_x f_i = 2\lambda_i^{-1}\mathcal{J}_x\mathcal{J}_{xy} e_i = 2\lambda_i^{-1}\mathcal{J}_{xy}\mathcal{J}_y e_i = 0.$$

This shows $\{e_1, ..., e_\ell, f_1, ..., f_\ell\}$ is an orthonormal set. We set

$$\Xi = \mathrm{diag}(\lambda_1, ..., \lambda_\ell).$$

Since $\mathcal{J}_y\mathcal{J}_{xy} = \mathcal{J}_{xy}\mathcal{J}_x$, we have $\mathcal{J}_y f_i = \lambda_i f_i$. Note that

$$\mathcal{J}_{xy} e_i = \tfrac{1}{2}\lambda_i f_i \quad \text{and} \quad \mathcal{J}_{xy} f_i = 2\lambda_i^{-1}\mathcal{J}_{xy}^2 e_i = \tfrac{1}{2}\lambda_i e_i.$$

On the subspace $\mathrm{Span}\{e_1, ..., e_\ell, f_1, ..., f_\ell\}$ one has that

$$\mathcal{J}_x = \begin{pmatrix} \Xi & 0 \\ 0 & 0 \end{pmatrix}, \quad \mathcal{J}_y = \begin{pmatrix} 0 & 0 \\ 0 & \Xi \end{pmatrix}, \quad \mathcal{J}_{xy} = \begin{pmatrix} 0 & \tfrac{1}{2}\Xi \\ \tfrac{1}{2}\Xi & 0 \end{pmatrix}.$$

On the other hand, clearly $\mathcal{J}_x = \mathcal{J}_y = 0$ on $\{\mathrm{Range}(\mathcal{J}_x) \oplus \mathrm{Range}(\mathcal{J}_y)\}^\perp$. If $\xi \in \{\mathrm{Range}(\mathcal{J}_x) \oplus \mathrm{Range}(\mathcal{J}_y)\}^\perp$, then $\mathcal{J}_x\xi = \mathcal{J}_y\xi = 0$ so Assertion (2) yields

$$\langle \mathcal{J}_{xy}\xi, \mathcal{J}_{xy}\xi \rangle = \langle \mathcal{J}_{xy}^2\xi, \xi \rangle = \tfrac{1}{4}\langle (\mathcal{J}_x^2 + \mathcal{J}_y^2)\xi, \xi \rangle = 0.$$

This shows $\mathcal{J}_{xy}\xi = 0$ as well and gives the desired decomposition. $\square$

We continue our study. Let

$$W(x) := \mathrm{Span}\{x\} \oplus \mathrm{Range}(\mathcal{J}_x).$$

Lemma 6.2.4 *Let $\mathfrak{M} = (V, \langle \cdot, \cdot \rangle, A)$ be a Riemannian orthogonally Jacobi Tsankov 0-model of dimension m. Assume that $r(x) < m - 1$ for all x. Let $x \in S(V)$. If $w \in S(W(x))$, then:*

(1) $\mathrm{Range}(\mathcal{J}_w) \subset W(x)$ and $\mathcal{J}_w$ vanishes on $W(x)^\perp$.
(2) $\mathcal{J}_w$ is similar to $\mathcal{J}_x$.
(3) $\mathcal{J}_x$ has at most one non-zero eigenvalue.

Proof.   Fix $w \in S(W(x))$. Expand $w = a_0 x + \sum a_i w_i$ where we have $\mathcal{J}_x w_i = \lambda_i w_i$ for $\lambda_i \neq 0$. Let $y \in W(x)^\perp \cap S(V)$. We apply Lemma 6.2.3. As $y \perp \mathrm{Range}(\mathcal{J}_x)$, $\mathcal{J}_x y = 0$ so $\mathcal{J}_y x = 0$. Furthermore since $\mathcal{J}_x y = 0$, since $\mathcal{J}_x w_i = \lambda_i w_i$, and since $\lambda_i \neq 0$, one has $\mathcal{J}_y w_i = 0$. Thus $\mathcal{J}_y w = 0$ and consequently $\mathcal{J}_w y = 0$ for all $y \in W(x)^\perp$. Thus

$$\mathrm{Range}(\mathcal{J}_w) \subset W(x) \quad \text{and} \quad \mathcal{J}_w = 0 \quad \text{on} \quad W(x)^\perp .$$

This proves Assertion (1). Furthermore $\mathcal{J}_x y = 0$ and $\mathcal{J}_w y = 0$ implies $\mathcal{J}_x$ is similar to $\mathcal{J}_y$ and $\mathcal{J}_w$ is similar to $\mathcal{J}_y$. This establishes Assertion (2).

To show that Assertion (3) is true, we apply Assertion (2) to see

$$\mathrm{Rank}(\mathcal{J}_w) = \mathrm{Rank}(\mathcal{J}_w|_{W(x)}) = \dim(W(x)) - 1 = r(x) .$$

Suppose $\mathcal{J}_x$ has two distinct non-zero eigenvalues $\lambda_i \neq \lambda_j$ for some $i < j$. Then $\mathcal{J}_{w_i} w_j = 0$. Since $\mathcal{J}_{w_i} w_i = 0$, we would have $\mathrm{Rank}\{\mathcal{J}_{w_i}\} \leq r - 1$ which is false. Thus $\mathcal{J}_x = \lambda\,\mathrm{id}$ on $\mathrm{Range}(\mathcal{J}_x)$. $\qquad\square$

Lemma 6.2.5   *Let $\mathfrak{M} = (V, \langle \cdot, \cdot \rangle, A)$ be a Riemannian orthogonally Jacobi Tsankov 0-model of dimension m where $A \neq 0$. Assume that $r(x) < m - 1$ for all x. Then $\mathfrak{M}$ is Osserman, $\mathcal{J}$ has only one non-zero eigenvalue λ on $S(V)$, and that λ has multiplicity 1.*

Proof.   Let $e_0 = x$ and $f_0 = y$. Let λ be the non-zero eigenvalue for $\mathcal{J}_x$ and let $r = r(x)$. Let the index i range from 1 through r and the index k range from 1 through $\dim(V) - 2r - 2$. By Lemma 6.2.3 there is an orthonormal basis $\{e_0, ..., e_r, f_0, ..., f_r, g_1, ..., g_\ell\}$ for V so that

$$
\begin{aligned}
\mathcal{J}_{e_0} e_0 &= 0, & \mathcal{J}_{f_0} e_0 &= 0, & \mathcal{J}_{e_0 f_0} e_0 &= 0, \\
\mathcal{J}_{e_0} f_0 &= 0, & \mathcal{J}_{f_0} f_0 &= 0, & \mathcal{J}_{e_0 f_0} f_0 &= 0, \\
\mathcal{J}_{e_0} e_i &= \lambda e_i, & \mathcal{J}_{f_0} e_i &= 0, & \mathcal{J}_{e_0 f_0} e_i &= \tfrac{1}{2}\lambda f_j & \text{for} \quad 1 \leq i \leq r, \\
\mathcal{J}_{e_0} f_i &= 0, & \mathcal{J}_{f_0} f_i &= \lambda f_i, & \mathcal{J}_{e_0 f_0} f_i &= \tfrac{1}{2}\lambda e_i & \text{for} \quad 1 \leq i \leq r, \\
\mathcal{J}_{e_0} g_k &= 0, & \mathcal{J}_{f_0} g_k &= 0, & \mathcal{J}_{e_0 f_0} g_k &= 0 & \text{for} \quad 1 \leq k \leq \ell.
\end{aligned}
$$

Suppose that $r(x) \geq 2$. Note that $\mathcal{J}_{e_1}$ preserves $\mathrm{Span}\{e_0, ..., e_r\} = W(e_0)$. Since λ is an eigenvalue of multiplicity r on $W(e_0)$, since $\mathcal{J}_{e_1}$ vanishes on $W(e_0)^\perp$, and since $\mathcal{J}_{e_1} e_1 = 0$,

$$\mathcal{J}_{e_1} e_2 = \lambda e_2, \quad \text{and} \quad \mathcal{J}_{e_1} f_2 = 0 .$$

Let $\xi = \frac{1}{\sqrt{2}}(e_0 + f_0)$; $\mathcal{J}_\xi = \frac{1}{2}\{\mathcal{J}_{e_0} + \mathcal{J}_{f_0} + 2\mathcal{J}_{e_0 f_0}\}$ so

$$\mathcal{J}_\xi \{e_2\} = \tfrac{1}{2}\lambda(e_2 + f_2) .$$

The following contradiction shows $r(x) \leq 1$:

$$\mathcal{J}_\xi \mathcal{J}_{e_1} f_2 = 0 \quad \text{and} \quad \mathcal{J}_{e_1} \mathcal{J}_\xi f_2 = \tfrac{1}{2}\lambda^2 e_2 \,.$$

Since $2 + 2r \leq \dim V$, one has that $m \geq 4$. Suppose that $m > 4$. Let $x, z \in S(V)$. Let

$$W(x, z) := \mathrm{Span}\{x, \mathrm{Range}(Jx), z, \mathrm{Range}(Jz)\} \,.$$

Since $\dim\{W(x, z)\} \leq 4 < \dim\{V\}$, we may choose $y \in S(V)$ so that $y \perp W(x, z)$. We then have $\mathcal{J}_x y = 0$ and $\mathcal{J}_z y = 0$. Thus $\mathcal{J}_x$ is similar to $\mathcal{J}_y$ and $\mathcal{J}_z$ is similar to $\mathcal{J}_y$. This shows $\mathcal{J}_x$ and $\mathcal{J}_z$ are similar and hence have the same eigenvalues as desired. This establishes the Lemma if $m \neq 4$.

Suppose that $m = 4$. Choose an orthonormal basis $\{e_1, e_2, e_3, e_4\}$ for V so that $\mathcal{J}_{e_1} e_3 = 0$ and $e_2 \in \mathrm{Range}\{\mathcal{J}_{e_1}\}$. We assume without loss of generality that $\lambda = 1$. Let $J_i := \mathcal{J}_{e_i}$ and let $J_{ij} := \mathcal{J}_{e_i e_j}$. By Lemmas 6.2.3 and 6.2.4, the non-zero components of J_i are:

$$J_1 e_2 = e_2, \quad J_2 e_1 = e_1, \quad J_3 e_4 = e_4, \quad J_4 e_3 = e_3 \,. \tag{6.2.c}$$

Since $J_1 e_2 = e_2$, Lemma 6.2.4 applies. We conclude that $J_{\cos\theta e_1 + \sin\theta e_2}$ preserves $\mathrm{Span}\{e_1, e_2\}$, has spectrum $\{0, 1\}$, and is zero on the orthogonal complement $\mathrm{Span}\{e_3, e_4\}$. Thus $J_{12} e_3 = J_{12} e_4 = 0$. By Eq. (6.1.a), we have $J_{ij} e_j = -\tfrac{1}{2} J_j e_i$. Consequently:

$$J_{12} e_1 = -\tfrac{1}{2} e_2, \quad J_{12} e_2 = -\tfrac{1}{2} e_1, \quad J_{12} e_3 = 0, \quad J_{12} e_4 = 0 \,.$$

This determines J_{12}; J_{34} is determined similarly. On the other hand, since $J_1 e_3 = 0$, we can apply Eq. (6.1.a) and Lemma 6.2.3 to see

$$J_{13} e_1 = 0, \quad J_{13} e_2 = \pm\tfrac{1}{2} e_4, \quad J_{13} e_3 = 0, \quad J_{13} e_4 = \pm\tfrac{1}{2} e_2 \,.$$

Similar arguments can be employed to study J_{14}, J_{23}, and J_{24}. This shows there exist constants $\varepsilon_{ij} = \pm\tfrac{1}{2}$ so the non-zero components of J_{ij} are:

$$\begin{aligned}
&J_{12} e_1 = -\tfrac{1}{2} e_2, \quad J_{12} e_2 = -\tfrac{1}{2} e_1, \quad J_{13} e_2 = \varepsilon_{13} e_4, \quad J_{13} e_4 = \varepsilon_{13} e_2, \\
&J_{14} e_2 = \varepsilon_{14} e_3, \quad J_{14} e_3 = \varepsilon_{14} e_2, \quad J_{23} e_1 = \varepsilon_{14} e_4, \quad J_{23} e_4 = \varepsilon_{14} e_1, \\
&J_{24} e_1 = \varepsilon_{24} e_3, \quad J_{24} e_3 = \varepsilon_{24} e_1, \quad J_{34} e_3 = -\tfrac{1}{2} e_4, \quad J_{34} e_4 = -\tfrac{1}{2} e_3 \,.
\end{aligned} \tag{6.2.d}$$

We consider the moving frame

$$\begin{aligned}
&e_1(\theta, \phi) := \cos\theta e_1 + \sin\theta e_2, \quad e_2(\theta, \phi) = -\sin\theta e_1 + \cos\theta e_2, \\
&e_3(\theta, \phi) := \cos\phi e_3 + \sin\phi e_4, \quad e_4(\theta, \phi) = -\sin\phi e_3 + \cos\phi e_4 \,.
\end{aligned}$$

We then have

$$\mathcal{J}_{e_1(\theta,\phi)}e_2(\theta,\phi) = e_2(\theta,\phi),$$
$$\mathcal{J}_{e_1(\theta,\phi)}e_3(\theta,\phi) = 0,$$
$$\mathcal{J}_{e_1(\theta,\phi)}e_4(\theta,\phi) = 0.$$

Similar calculations hold for $\mathcal{J}_{e_i(\theta,\phi)}$ for $i = 2, 3, 4$. Consequently Eqs. (6.2.c) and (6.2.d) continue to hold as functions of (θ, ϕ). Thus the signs $\varepsilon_{ij}(\theta, \phi)$ are constant. One has:

$$\varepsilon_{13} = \varepsilon_{13}(0,0), \qquad\qquad \varepsilon_{14} = -\varepsilon_{13}(0, \tfrac{1}{2}\pi) = -\varepsilon_{13},$$
$$\varepsilon_{23} = -\varepsilon_{13}(\tfrac{1}{2}\pi, 0) = -\varepsilon_{13}, \quad \varepsilon_{24} = \varepsilon_{13}(\tfrac{1}{2}\pi, \tfrac{1}{2}\pi) = \varepsilon_{13}.$$

By replacing e_2 by $-e_2$ if necessary, one can also assume that $\varepsilon_{13} = 1$. One then has $\varepsilon_{14} = -\tfrac{1}{2}$, $\varepsilon_{23} = -\tfrac{1}{2}$, and $\varepsilon_{24} = \tfrac{1}{2}$. This completely determines the structure of J_i and J_{ij} and thereby $\mathcal{J}$. By Lemma 1.7.1, this determines A. This shows that up to gauge equivalence there is only one orthogonally Jacobi Tsankov algebraic curvature tensor on V with a point $x \in S(V)$ so that $r(x) = 1$ and $\lambda = 1$.

Let Θ be any Hermitian almost complex structure on $\mathbb{R}^4$. Then $\tfrac{1}{3}R_\Theta$ defines a Riemannian orthogonally Jacobi Tsankov model with $r(x) = 1$ for any $x \in S(V)$ and with $\lambda = 1$. Thus A is isomorphic to $\tfrac{1}{3}A_\Theta$ which is Osserman. $\qquad\Box$

We have now established the necessary preliminaries to complete the proof of Theorem 6.2.1 (2). Suppose $\mathfrak{M} = (V, \langle\cdot,\cdot\rangle)$ is an orthogonally Jacobi Tsankov Riemannian 0-model. Suppose that $A \neq 0$. If $r(x) = m-1$ for any vector $x \in S(V)$, then A has constant sectional curvature by Lemma 6.2.2. On the other hand, if $r(x) < m - 1$ for every point $x \in S(V)$, then A is Osserman and $\mathrm{Rank}\{\mathcal{J}_x\} = 1$ for every $x \in S(V)$ by Lemma 6.2.5. Theorem 1.9.5 then shows $A = c_0 A_{\langle\cdot,\cdot\rangle} + c_1 A_\Theta$ for some Hermitian almost complex structure on V; necessarily $c_0 = 0$ since the rank is 1. $\qquad\Box$

6.2.3 *Riemannian Jacobi Tsankov manifolds*

Let $\mathcal{M} = (M, g)$ be a connected Riemannian manifold. We shall apply Theorem 6.2.1 to prove Theorem 6.2.2. Suppose first that $\mathcal{M}$ is Jacobi Tsankov. Then $R = 0$ so $\mathcal{M}$ is flat.

Suppose $\mathcal{M}$ is orthogonally Jacobi Tsankov and $m \geq 3$. Let $\mathcal{O}$ be open subset of points $P \in M$ so that there exists a unit tangent vector $x(P)$ with $r(x(P)) = m - 1$. Then $g|_\mathcal{O}$ has constant sectional curvature. Thus

$R_P = c_P R_{\langle\cdot,\cdot\rangle}$ on $\mathcal{O}$. Since $m \geq 3$, the multiple c_P is locally constant. Thus $R = cA_{\langle\cdot,\cdot\rangle}$ on the closure of $\mathcal{O}$. Thus $\mathcal{O}$ is an open and closed subset of M and hence all of M.

Thus if $\mathcal{M}$ does not have constant sectional curvature, $r(x) < m - 1$ for all $x \in S(M, g)$. Hence $r(x) = 1$ for all $x \in S(\mathcal{M})$. Furthermore m is even and there is an almost complex structure $\Theta(Q)$ defined on T_Q for every $Q \in M$ so that $R = \lambda A_\Theta$. In addition, the almost complex structure is uniquely determined up to sign. After a bit of technical fuss, one can see that Θ can be chosen to vary smoothly with Q, at least locally; global questions are irrelevant to our argument. The metric in question is Einstein and thus $\rho(x, x) = 3\lambda(x)$ is constant. Thus $\mathcal{M}$ is globally Osserman. This possibility is ruled out by Theorem 1.9.5; results of Tricerri and Vanhecke (1981) could also be used. The generalized complex space forms of Olszak (1989) play no role here. $\qquad\square$

6.3 Pseudo-Riemannian Jacobi Tsankov 0-Models

In this section, we present work of Brozos-Vázquez and Gilkey (2006). Let $\mathfrak{M} = (V, \langle\cdot,\cdot\rangle, A)$ be a 0-model. We set $\mathcal{J}_x := \mathcal{J}(x)$. Recall that $\mathfrak{M}$ is said to be *Jacobi Tsankov* if $\mathcal{J}_x\mathcal{J}_y = \mathcal{J}_y\mathcal{J}_x$ for all $x, y \in V$. We showed in Section 6.2 that any Riemannian Jacobi Tsankov 0-model is flat. Thus $A = 0$. In this section, we study the higher signature setting. The following result will be established in Section 6.3.1:

Theorem 6.3.1 *Let $\mathfrak{M} = (V, \langle\cdot,\cdot\rangle, A)$ be a Jacobi Tsankov 0-model. Then:*

(1) $\mathcal{J}_x^2 = 0$ for all $x \in V$.
(2) $\mathfrak{M}$ is Osserman.
(3) If $\mathfrak{M}$ is Riemannian or Lorentzian, then $A = 0$.

Theorem 6.3.1 has the following geometrical consequence:

Corollary 6.3.1 *Let M be a Jacobi Tsankov pseudo-Riemannian manifold of signature (p, q). Then M is nilpotent Osserman. If $p = 0$ or if $p = 1$, then M is flat.*

One might conjecture that the condition $\mathcal{J}_x^2 = 0$ for all $x \in V$ is sufficient to imply $\mathfrak{M}$ is Jacobi Tsankov. We show this is not the case in Section 6.3.2 by showing

Theorem 6.3.2   *There exists a 0-model $\mathfrak{M} = (V, \langle \cdot, \cdot \rangle, A)$ which is not Jacobi Tsankov but which has $\mathcal{J}_x^2 = 0$ for all $x \in V$.*

If $\mathcal{J}_x \mathcal{J}_y = 0$ for all $x, y \in V$, then necessarily $\mathfrak{M}$ is Jacobi Tsankov. In Section 6.3.3, we will show that the converse holds in low dimensions:

Theorem 6.3.3   *Let $\mathfrak{M} = (V, \langle \cdot, \cdot \rangle, A)$ be a Jacobi Tsankov 0-model of dimension m. If $m \leq 13$, then $\mathcal{J}_x \mathcal{J}_y = 0$ for all $x, y \in V$.*

The following classification theorem will be established in Section 6.3.4:

Theorem 6.3.4   *Let $\mathfrak{M} = (V, \langle \cdot, \cdot \rangle, A)$ be a 0-model. The following statements are equivalent:*

(1) $\mathfrak{M}$ is indecomposable and $\mathcal{J}_x \mathcal{J}_y = 0$ for all $x, y \in V$.
(2) $\mathfrak{M}$ is indecomposable and $A_{x_1 x_2} A_{x_3 x_4} = 0$ for all $x_i \in V$.
(3) We can decompose $V = W \oplus \bar{W}$ and $A = A_W \oplus 0$ where (W, A_W) is an irreducible weak 0-model and where W and $\bar{W}$ are totally isotropic subspaces of V.

In Section 6.4, we will show the assumption $m \leq 13$ in Theorem 6.3.3 is sharp by studying the geometry of a model of signature $(8, 6)$ which is Jacobi Tsankov, which is indecomposable, but which does not have the form given in Theorem 6.3.4.

6.3.1   *Jacobi Tsankov 0-models*

This section is devoted to the proof of Theorem 6.3.1. We adopt the notation of Section 6.1 to define $\mathcal{J}_x$ and $\mathcal{J}_{xy}$. A central role in our discussion will be played by Eq. (6.1.a). Let $\mathfrak{M} = (V, \langle \cdot, \cdot \rangle, A)$ be a Jacobi Tsankov 0-model. Polarizing the identity $\mathcal{J}_x \mathcal{J}_y = \mathcal{J}_y \mathcal{J}_x$ yields:

$$\mathcal{J}_{x_1 x_2} \mathcal{J}_{x_3 x_4} = \mathcal{J}_{x_3 x_4} \mathcal{J}_{x_1 x_2} \quad \text{for all} \quad x_1, x_2, x_3, x_4 \in V \,.$$

As $\mathcal{J}_x x = 0$, Eq. (6.1.a) yields Assertion (1) of Theorem 6.3.1 since

$$0 = \mathcal{J}_{xy} \mathcal{J}_{xx} x = \mathcal{J}_{xx} \mathcal{J}_{xy} x = -\tfrac{1}{2} \mathcal{J}_x \mathcal{J}_x y \,.$$

Since the Jacobi operator is nilpotent, $\{0\}$ is the only eigenvalue of $\mathcal{J}$. This shows that A is Osserman. If $p = 0$, then $\mathcal{J}_x$ is diagonalizable. Consequently, $\mathcal{J}_x^2 = 0$ implies $\mathcal{J}_x = 0$ for all x. It now follows $A = 0$. If $p = 1$, then A is Osserman implies A has constant sectional curvature c by Theorem 1.9.7. Since $\mathcal{J}_x^2 = 0$, $c = 0$ which again implies $A = 0$. This completes the proof of Theorem 6.3.1. $\square$

In fact, it is possible to work in a slightly more general setting. Following Bokan (1990), one says that C is a *generalized curvature operator* if it has the symmetries of the curvature operator which is defined by a torsion free connection. This means that one imposes the symmetries:

$$\mathcal{C}(x,y)z = -\mathcal{C}(y,x)z,$$
$$\mathcal{C}(x,y)z + \mathcal{C}(y,z)x + \mathcal{C}(z,x)y = 0.$$

This is a slightly more general class than the set of affine curvature tensors described in Definition 1.3.3. The proof given above then generalizes immediately to yield:

Corollary 6.3.2 *If C is a generalized curvature operator on V which is Jacobi Tsankov, then $\mathcal{J}_C$ is Osserman and $\mathcal{J}_C(x)^2 = 0$ for all $x \in V$.*

6.3.2 *Non Jacobi Tsankov 0-models with $\mathcal{J}_x^2 = 0 \ \forall \ x$*

We showed in Theorem 6.3.1 that if $\mathfrak{M} = (V, \langle \cdot, \cdot \rangle, A)$ is Jacobi Tsankov, then $\mathcal{J}_x^2 = 0$ for all $x \in V$. In this section, we exhibit a 0-model such that $\mathcal{J}_x^2 = 0$ for all $x \in V$ but which is not Jacobi Tsankov. This will complete the proof of Theorem 6.3.2.

Let ϕ be a skew-symmetric endomorphism of V. Following Eq. (1.3.a), define an algebraic curvature tensor A_ϕ and associated Jacobi operator $\mathcal{J}_\phi$:

$$A_\phi(x,y,z,w) := \langle \phi y, z \rangle \langle \phi x, w \rangle - \langle \phi x, z \rangle \langle \phi y, w \rangle - 2\langle \phi x, y \rangle \langle \phi z, w \rangle,$$
$$\mathcal{J}_\phi(x)y = 3\langle y, \phi x \rangle \phi x.$$

Let $\mathbb{R}^{(p,q)}$ denote Euclidean space with a metric of signature (p,q). Theorem 6.3.2 is a consequence of the following result:

Lemma 6.3.1 *There exist skew-symmetric endomorphisms $\{\Phi, \Psi\}$ of $\mathbb{R}^{(\ell,\ell)}$ for some ℓ so that $\Phi^2 = \Psi^2 = 0$, so that $\Phi\Psi + \Psi\Phi = 0$, and so that $\Phi\Psi \neq 0$. Set $\mathfrak{M} := (\mathbb{R}^{(\ell,\ell)}, \langle \cdot, \cdot \rangle, A)$ where $A = \frac{1}{3}\{A_\Phi + A_\Psi\}$. Then $\mathcal{J}_x^2 = 0$ for all x and $\mathfrak{M}$ is not Jacobi Tsankov.*

Proof. By Lemma 4.4.1, there is a collection of skew-symmetric endomorphisms of $\mathbb{R}^{(\ell,\ell)}$ so that

$$\phi_1^2 = \phi_2^2 = \mathrm{id}, \quad \phi_3^2 = \phi_4^2 = -\,\mathrm{id}, \quad \phi_i\phi_j + \phi_j\phi_i = 0 \text{ for } i \neq j.$$

Lemma 1.4.5 of Gilkey (2002) shows one may take $\ell = 4$. One may then set $\Phi = \phi_1 + \phi_3$ and set $\Psi = \phi_2 + \phi_4$ to construct skew-adjoint endomorphisms so

$$\Phi^2 = \Psi^2 = 0, \quad \Phi\Psi + \Psi\Phi = 0.$$

Suppose that $\Phi\Psi = 0$. We argue for a contradiction. We conjugate the identity

$$(\phi_1 + \phi_3)(\phi_2 + \phi_4) = 0$$

by ϕ_1 yields

$$0 = (-\phi_1 + \phi_3)(\phi_2 + \phi_4).$$

Adding these two equations yields $\phi_3(\phi_2 + \phi_4) = 0$. Multiplying by ϕ_3 implies $\phi_2 + \phi_4 = 0$. Conjugating this identity by ϕ_2 yields $\phi_2 - \phi_4 = 0$ and thus $\phi_2 = 0$. This is not possible. Thus $\Phi\Psi \neq 0$. One computes:

$$\mathcal{J}_x y = \langle y, \Phi x\rangle \Phi x + \langle y, \Psi x\rangle \Psi x,$$
$$\mathcal{J}_u \mathcal{J}_v y = \langle y, \Phi v\rangle \langle \Phi v, \Phi u\rangle \Phi u + \langle y, \Phi v\rangle \langle \Phi v, \Psi u\rangle \Psi u$$
$$+ \langle y, \Psi v\rangle \langle \Psi v, \Phi u\rangle \Phi u + \langle y, \Psi v\rangle \langle \Psi v, \Psi u\rangle \Psi u$$
$$= \langle y, \Phi v\rangle \langle \Phi v, \Psi u\rangle \Psi u + \langle y, \Psi v\rangle \langle \Psi v, \Phi u\rangle \Phi u.$$

Since $\langle \Phi x, \Psi x\rangle = -\langle \Psi\Phi x, x\rangle = \langle \Phi\Psi x, x\rangle = -\langle \Psi x, \Phi x\rangle$, we have, as desired,

$$\mathcal{J}_x \mathcal{J}_x = 0.$$

Choose u so $\Psi\Phi u \neq 0$. Set $y = \Phi u$. Choose v so $\langle \Phi u, \Psi v\rangle \neq 0$. Then:

$$\mathcal{J}_u \mathcal{J}_v y = \langle \Phi u, \Phi v\rangle \langle \Phi v, \Psi u\rangle \Psi u + \langle \Phi u, \Psi v\rangle \langle \Psi v, \Phi u\rangle \Phi u$$
$$= \langle \Phi u, \Psi v\rangle^2 \Phi u \neq 0,$$
$$\mathcal{J}_v \mathcal{J}_u y = \langle \Phi u, \Phi u\rangle \langle \Phi u, \Psi v\rangle \Psi v + \langle \Phi u, \Psi u\rangle \langle \Psi u, \Phi v\rangle \Phi v$$
$$= 0.$$

Then $\mathcal{J}_u \mathcal{J}_v y \neq 0$ while $\mathcal{J}_v \mathcal{J}_u y = 0$. Consequently $\mathfrak{M}$ is not a Jacobi Tsankov 0-model. $\qquad\square$

6.3.3   *0-models with* $\mathcal{J}_x \mathcal{J}_y = 0 \; \forall \; x, y \in V$

Theorem 6.3.3 will follow from the following result which also plays a central role in motivating the example to be studied presently in Section 6.3.4:

Lemma 6.3.2   *Let* $\mathfrak{M} := (V, \langle \cdot, \cdot\rangle, A)$ *be a Jacobi Tsankov 0-model. Suppose that there exist* $x, y \in V$ *so that* $\mathcal{J}_x \mathcal{J}_y \neq 0$.

(1) There exists $w \in V$ so $\langle \mathcal{J}_x \mathcal{J}_y w, w \rangle = \langle \mathcal{J}_y \mathcal{J}_w x, x \rangle = \langle \mathcal{J}_w \mathcal{J}_x y, y \rangle \neq 0$.

(2) $S := \{e_1, ..., e_5, f_1, ..., f_5, g_1, ..., g_4\}$ is a linearly independent set where

$$e_1 := w, \quad e_2 := \mathcal{J}_x \mathcal{J}_y w, \quad e_3 := \mathcal{J}_x w, \quad e_4 := \mathcal{J}_y w, \quad e_5 := \mathcal{J}_{xy} w,$$
$$f_1 := x, \quad f_2 := \mathcal{J}_y \mathcal{J}_w x, \quad f_3 := \mathcal{J}_y x, \quad f_4 := \mathcal{J}_w x, \quad f_5 := \mathcal{J}_{yw} x,$$
$$g_1 := y, \quad g_2 := \mathcal{J}_w \mathcal{J}_x y, \quad g_3 := \mathcal{J}_w y, \quad g_4 := \mathcal{J}_x y, \quad g_5 := \mathcal{J}_{wx} y.$$

(3) $e_5 + f_5 + g_5 = 0$.

(4) $\dim(V) \geq 14$.

Proof. Choose w and f so $\langle \mathcal{J}_x \mathcal{J}_y w, f \rangle \neq 0$. Set $w(\varepsilon) := w + \varepsilon f$. Then

$$p(\varepsilon) := \langle w(\varepsilon), \mathcal{J}_x \mathcal{J}_y w(\varepsilon) \rangle$$
$$= \langle w, \mathcal{J}_x \mathcal{J}_y w \rangle + 2\varepsilon \langle \mathcal{J}_x \mathcal{J}_y w, f \rangle + \varepsilon^2 \langle \mathcal{J}_x \mathcal{J}_y f, f \rangle.$$

As $\langle \mathcal{J}_x \mathcal{J}_y w, f \rangle \neq 0$, $p(\varepsilon)$ is a non-trivial polynomial in ε. Thus it is non-zero for a suitable choice of ε. Thus after replacing w by $w(\varepsilon)$ for suitably chosen ε, we see that there is $w \in V$ with $\langle w, \mathcal{J}_x \mathcal{J}_y w \rangle \neq 0$. Applying Eq. (6.1.a) yields:

$$\langle \mathcal{J}_y \mathcal{J}_w x, x \rangle = -2\langle \mathcal{J}_y \mathcal{J}_{wx} w, x \rangle = -2\langle \mathcal{J}_y w, \mathcal{J}_{wx} x \rangle$$
$$= \langle \mathcal{J}_y w, \mathcal{J}_x w \rangle = \langle \mathcal{J}_x \mathcal{J}_y w, w \rangle.$$

Similarly, $\langle \mathcal{J}_w \mathcal{J}_x y, y \rangle = \langle \mathcal{J}_x \mathcal{J}_y w, w \rangle$ and Assertion (1) follows.

Because $\mathcal{J}_{x+\varepsilon y} \mathcal{J}_{x+\varepsilon y} = 0$ for every $\varepsilon \in \mathbb{R}$ and because $\mathfrak{M}$ is Jacobi Tsankov, we have the following relations:

$$\mathcal{J}_x^2 = 0, \qquad \mathcal{J}_y^2 = 0, \qquad \mathcal{J}_x \mathcal{J}_y = \mathcal{J}_y \mathcal{J}_x,$$
$$\mathcal{J}_x \mathcal{J}_{xy} = \mathcal{J}_{xy} \mathcal{J}_x = 0, \quad \mathcal{J}_y \mathcal{J}_{xy} = \mathcal{J}_{xy} \mathcal{J}_y = 0, \quad \mathcal{J}_{xy}^2 = -\tfrac{1}{2} \mathcal{J}_x \mathcal{J}_y.$$

We have $\mathcal{J}_w \mathcal{J}_y x \neq 0$ and $\mathcal{J}_w \mathcal{J}_x y \neq 0$ by Assertion (1). To prove Assertion (2), suppose there is a non-trivial dependence relation among the elements of the set S:

$$0 = \sum_{i=1}^{5} \{a_i e_i + b_i f_i + c_i g_i\}$$
$$= a_1 w + a_2 \mathcal{J}_x \mathcal{J}_y w + a_3 \mathcal{J}_x w + a_4 \mathcal{J}_y w + a_5 \mathcal{J}_{xy} w \qquad \text{(6.3.a)}$$
$$+ b_1 x + b_2 \mathcal{J}_y \mathcal{J}_w x + b_3 \mathcal{J}_y x + b_4 \mathcal{J}_w x + b_5 \mathcal{J}_{yw} x$$
$$+ c_1 y + c_2 \mathcal{J}_w \mathcal{J}_x y + c_3 \mathcal{J}_w y + c_4 \mathcal{J}_x y + c_5 \mathcal{J}_{wx} y,$$

where, since $g_5 \notin S$, we suppose $c_5 = 0$.

We can apply $\mathcal{J}_x\mathcal{J}_y$ to Eq. (6.3.a) to see $a_1\mathcal{J}_x\mathcal{J}_yw = 0$. Since, by Assertion (1), $\mathcal{J}_x\mathcal{J}_yw \neq 0$, $a_1 = 0$. Similarly $b_1 = c_1 = 0$. If we now apply $\mathcal{J}_x$ to Eq. (6.3.a), we see

$$a_4\mathcal{J}_x\mathcal{J}_yw + c_3\mathcal{J}_x\mathcal{J}_wy = 0 \quad \text{so}$$
$$0 = \langle a_4\mathcal{J}_x\mathcal{J}_yw + c_3\mathcal{J}_x\mathcal{J}_wy, w\rangle = a_4\langle \mathcal{J}_x\mathcal{J}_yw, w\rangle \,.$$

By Assertion (1), $a_4 = 0$. Similarly, $a_3 = b_3 = b_4 = c_3 = c_4 = 0$. Thus Eq. (6.3.a) simplifies to become

$$0 = a_2\mathcal{J}_x\mathcal{J}_yw + a_5\mathcal{J}_{xy}w + b_2\mathcal{J}_y\mathcal{J}_wx + b_5\mathcal{J}_{yw}x + c_2\mathcal{J}_w\mathcal{J}_xy + c_5\mathcal{J}_{wx}y\,.$$

Applying $\mathcal{J}_{xy}$ then yields

$$\begin{aligned}
0 &= a_5\mathcal{J}_{xy}^2w + b_5\mathcal{J}_{xy}\mathcal{J}_{yw}x + c_5\mathcal{J}_{xy}\mathcal{J}_{wx}y \\
&= (a_5\mathcal{J}_{xy}^2 + \tfrac{1}{4}(b_5 + c_5)\mathcal{J}_x\mathcal{J}_y)w \\
&= (a_5 - \tfrac{1}{2}(b_5 + c_5))\mathcal{J}_{xy}^2w\,.
\end{aligned}$$

This shows $a_5 = \tfrac{1}{2}(b_5 + c_5)$; since a_5, b_5, and c_5 play symmetric roles, we obtain $a_5 = b_5 = c_5$. Since $c_5 = 0$, we have $a_5 = b_5 = 0$. Taking the inner product with x, y, and w then yields, respectively $b_2 = 0$, $c_2 = 0$, and $a_2 = 0$, which completes the proof of Assertion (2).

To prove Assertion (3), we use the curvature symmetries to compute:

$$\begin{aligned}
e_5 + f_5 + g_5 &= \mathcal{J}_{xy}w + \mathcal{J}_{yw}x + \mathcal{J}_{wx}y \\
&= \tfrac{1}{2}\{\mathcal{A}_{wx}y + \mathcal{A}_{wy}x + \mathcal{A}_{xy}w + \mathcal{A}_{xw}y + \mathcal{A}_{yw}x + \mathcal{A}_{yx}w\} \\
&= 0\,.
\end{aligned}$$

Assertion (4) is immediate from Assertion (2). $\qquad\qquad\square$

6.3.4   *0-models with* $\mathcal{A}_{xy}\mathcal{A}_{zw} = 0 \; \forall \; x, y, z, w \in V$

In this section, we prove Theorem 6.3.4. The following Lemma shows that Assertion (3) implies Assertion (2) in Theorem 6.3.4.

Lemma 6.3.3   *Let $\mathfrak{M}$ be as in Theorem 6.3.4 (3). Then $\mathfrak{M}$ is indecomposable and $\mathcal{A}_{x_1x_2}\mathcal{A}_{x_3x_4} = 0$ for all $x_i \in V$.*

Proof.   Let $\mathfrak{M} = (V, \langle\cdot,\cdot\rangle, A)$ be as in Theorem 6.3.4 (3). This means that $V = W \oplus \bar{W}$, that $A = A_W \oplus 0$, that W and $\bar{W}$ are totally isotropic, and that (W, A_W) is irreducible. Suppose there is a non-trivial decomposition

$$\mathfrak{M} = \mathfrak{M}_1 \oplus \mathfrak{M}_2 \quad \text{where} \quad \mathfrak{M}_i = (V_i, \langle\cdot,\cdot\rangle_i, A_i)\,.$$

This would then induce a non-trivial decomposition of (W, A_W). Since (W, A_W) is assumed indecomposable, either $W \subset V_1$ or $W \subset V_2$; we suppose without loss of generality that $W \subset V_1$. As W and $\bar{W}$ are totally isotropic and as $V = W \oplus \bar{W}$, we have $W^\perp = W$. Because $V_2 \perp W$, we may conclude that $V_2 \subset W$. Since $V_2 \cap V_1 = \{0\}$, this implies $V_2 = \{0\}$ which is false. Consequently $\mathfrak{M}$ is indecomposable.

Choose a basis $\{e_i\}$ for W and choose a basis $\{\bar{e}_i\}$ for $\bar{W}$ so the only non-zero components of the inner product are $\langle e_i, \bar{e}_i \rangle = 1$. The only non-zero components of $\mathcal{A}$ are

$$\mathcal{A}(e_i, e_j)e_k = \sum_l A_W(e_i, e_j, e_k, e_l)\bar{e}_l \,.$$

This shows $\mathcal{A}_{x_1 x_2}\mathcal{A}_{x_3 x_4} = 0$. $\qquad\square$

We now show Assertion (2) implies Assertion (1) in Theorem 6.3.4.

Lemma 6.3.4 *Let* $\mathfrak{M} := (V, \langle \cdot, \cdot \rangle, A)$. *If* $\mathcal{A}_{x_1 x_2}\mathcal{A}_{x_3 x_4} = 0$ *for all* $x_i \in V$, *then* $\mathcal{J}_x \mathcal{J}_y = 0$ *for all* $x, y \in V$.

Proof. Suppose $\mathcal{A}_{x_1 x_2}\mathcal{A}_{x_3 x_4} = 0$ for all $x_i \in V$. One then may compute:

$$0 = -\langle \mathcal{A}_{x_1 x_2}\mathcal{A}_{x_3 x_4}x_4, x_2 \rangle = \langle \mathcal{A}_{x_3 x_4}x_4, \mathcal{A}_{x_1 x_2}x_2 \rangle$$
$$= \langle \mathcal{J}_{x_4}x_3, \mathcal{J}_{x_2}x_1 \rangle = \langle \mathcal{J}_{x_2}\mathcal{J}_{x_4}x_3, x_1 \rangle \,.$$

This shows $\mathcal{J}_{x_2}\mathcal{J}_{x_4} = 0$ for all $x_2, x_4 \in V$. $\qquad\square$

Before completing the proof of Theorem 6.3.4, we must establish a technical result.

Lemma 6.3.5 *Let* $\mathfrak{M} := (V, \langle \cdot, \cdot \rangle, A)$. *Suppose that* $\mathcal{J}_x y = 0$ *for all* $x \in V$. *Then* $A(x_1, x_2, x_3, y) = 0$ *for all* $x_i \in V$.

Proof. We compute:

$$A(x_1, x_2, x_3, y) + A(x_1, x_3, x_2, y) = 2\langle \mathcal{J}_{x_2 x_3}x_1, y \rangle$$
$$= 2\langle x_1, \mathcal{J}_{x_2 x_3}y \rangle = 0 \,.$$

Consequently $A(x_1, x_2, x_3, y) = -A(x_1, x_3, x_2, y)$ for all $x_i \in V$. Thus

$$0 = A(x_1, x_2, x_3, y) + A(x_2, x_3, x_1, y) + A(x_3, x_1, x_2, y)$$
$$= A(x_1, x_2, x_3, y) - A(x_2, x_1, x_3, y) - A(x_1, x_3, x_2, y)$$
$$= A(x_1, x_2, x_3, y) + A(x_1, x_2, x_3, y) + A(x_1, x_2, x_3, y)$$
$$= 3A(x_1, x_2, x_3, y) \,. \qquad\square$$

We complete our discussion by showing that Assertion (1) implies Assertion (3) in Theorem 6.3.4. Suppose that $\mathfrak{M}$ is indecomposable and that $\mathcal{J}_x\mathcal{J}_y = 0$ for all $x, y \in V$. Set

$$W^* := \mathrm{Span}_{v_1, v_2 \in V}\{\mathcal{J}_{v_1} v_2\},$$
$$U := \{v \in V : \mathcal{J}_{v_1} v = 0 \;\forall v_1 \in V\}.$$

Then by assumption, $W^* \subset U$. Furthermore, by Lemma 6.3.5, $A(v_1, v_2, v_3, v_4) = 0$ if any of the $v_i \in U$. Choose a complementary subspace W_1 so that $V = U \oplus W_1$.

If $w^* \in W^*$, then $w^* = \sum_j \mathcal{J}_{x_j} y_j$. Thus if $u \in U$,

$$\langle w^*, u \rangle = \Big\langle \sum_j \mathcal{J}_{x_j} y_j, u \Big\rangle = \sum_j \langle y_j, \mathcal{J}_{x_j} u \rangle = 0. \qquad (6.3.\text{b})$$

As $\langle \cdot, \cdot \rangle$ is non-degenerate, there must exist $\tilde{w} \in W_1$ so $\langle \tilde{w}, w^* \rangle \neq 0$. Fix a basis $\{w_1^*, ..., w_k^*\}$ for W^*. The argument given above shows we can find corresponding elements $\{\tilde{w}_1, ..., \tilde{w}_k\}$ in W_1 so

$$\langle \tilde{w}_i, w_j^* \rangle = \delta_{ij}.$$

If $\{\tilde{w}_1, ..., \tilde{w}_k\}$ does not span W_1, choose $0 \neq \tilde{w} \in W_1$ so that $\tilde{w} \perp W^*$. Since $\tilde{w} \notin U$, there exists y so that $\mathcal{J}_y \tilde{w} \neq 0$. Choose $z \in V$ so

$$0 \neq \langle \mathcal{J}_y \tilde{w}, z \rangle = \langle \tilde{w}, \mathcal{J}_y z \rangle.$$

This contradicts the fact that $\tilde{w} \perp W^*$. Thus $\{\tilde{w}_1, ..., \tilde{w}_k\}$ is a basis for W_1. We set $w_i := \tilde{w}_i - \frac{1}{2}\sum_j \langle \tilde{w}_i, \tilde{w}_j \rangle w_j^*$. We then have

$$W := \mathrm{Span}\{w_i\}, \quad W^* = \mathrm{Span}\{w_i^*\}, \quad V = W \oplus U,$$
$$\langle w_i, w_j \rangle = 0, \qquad \langle w_i^*, w_j^* \rangle = 0, \qquad \langle w_i, w_j^* \rangle = \delta_{ij}.$$

Let $\{w_1^*, ..., w_k^*, \tilde{u}_1, ..., \tilde{u}_l\}$ be a basis for U. Set

$$u_i := \tilde{u}_i - \sum_j \langle w_j, \tilde{u}_i \rangle w_j^*.$$

By Eq. (6.3.b), $\langle w_i^*, \tilde{u}_j \rangle = 0$. Consequently $\langle u_i, w_j \rangle = \langle u_i, w_j^* \rangle = 0$ for $1 \leq i \leq l$ and $1 \leq j \leq k$. Let $T := \mathrm{Span}\{u_i\}$. Then:

$$(V, \langle \cdot, \cdot \rangle, A) = (W \oplus W^*, \langle \cdot, \cdot \rangle|_{W \oplus W^*}, A|_W \oplus 0) \;\oplus\; (T, \langle \cdot, \cdot \rangle|_T, 0).$$

Since $(V, \langle \cdot, \cdot \rangle, A)$ is indecomposable, $T = \{0\}$. Since W and W^* are totally isotropic, the Lemma follows. $\qquad\qquad\square$

6.4 A Jacobi Tsankov 0-Model with $\mathcal{J}_x\mathcal{J}_y \neq 0$ for some x, y

The condition $\dim\{V\} \leq 13$ in Theorem 6.3.2 is sharp. In this section, we present work of Brozos-Vázquez, Gilkey, and Nikčević (2006) showing that there is a counter example if $m = 14$. The study of this tensor in the algebraic context and then subsequently in the geometric context will form the focus of this section. We begin by specifying the 0-model of interest:

Definition 6.4.1 Let $\{\alpha_i, \alpha_i^*, \beta_{i,1}, \beta_{i,2}, \beta_{4,1}, \beta_{4,2}\}$ be a basis for $\mathbb{R}^{14}$ where we shall let the index i range from 1 through 3. Let $\mathfrak{M}_{14} := (\mathbb{R}^{14}, \langle\cdot,\cdot\rangle, A)$ be the 0-model where the non-zero components of $\langle\cdot,\cdot\rangle$ and of A are given, up to the usual symmetries, by:

$$\begin{aligned}
&\langle\alpha_i, \alpha_i^*\rangle = \langle\beta_{i,1}, \beta_{i,2}\rangle = 1, \\
&\langle\beta_{4,1}, \beta_{4,1}\rangle = \langle\beta_{4,2}, \beta_{4,2}\rangle = -\tfrac{1}{2}, \quad \langle\beta_{4,1}, \beta_{4,2}\rangle = \tfrac{1}{4}, \\
&A(\alpha_2, \alpha_1, \alpha_1, \beta_{2,1}) = A(\alpha_3, \alpha_1, \alpha_1, \beta_{3,1}) = 1, \\
&A(\alpha_3, \alpha_2, \alpha_2, \beta_{3,2}) = A(\alpha_1, \alpha_2, \alpha_2, \beta_{1,2}) = 1, \\
&A(\alpha_1, \alpha_3, \alpha_3, \beta_{1,1}) = A(\alpha_2, \alpha_3, \alpha_3, \beta_{2,2}) = 1, \\
&A(\alpha_1, \alpha_2, \alpha_3, \beta_{4,1}) = A(\alpha_1, \alpha_3, \alpha_2, \beta_{4,1}) = -\tfrac{1}{2}, \\
&A(\alpha_2, \alpha_3, \alpha_1, \beta_{4,2}) = A(\alpha_2, \alpha_1, \alpha_3, \beta_{4,2}) = -\tfrac{1}{2}.
\end{aligned} \qquad (6.4.\text{a})$$

One may then check by inspection that the $\mathbb{Z}_2$ symmetries of Definition 1.3.1 are satisfied as is the first Bianchi identity. Thus A is an algebraic curvature tensor; the metric $\langle\cdot,\cdot\rangle$ has signature $(8, 6)$.

Let $\mathrm{Sl}_\pm(3)$ be the group of all 3×3 matrices of determinant ± 1 and let $\mathcal{G}(\mathfrak{M}_{14})$ be the group of isomorphisms of $\mathfrak{M}_{14}$. We will establish the following result in Section 6.4.1 that describes the basic properties of the model $\mathfrak{M}_{14}$.

Theorem 6.4.1 *Let $\mathfrak{M}_{14}$ be the 0-model of Definition 6.4.1.*

(1) $\mathfrak{M}_{14}$ is Jacobi Tsankov.

(2) There exist $x_i \in V$ so $A_{x_1x_2}A_{x_3x_4} \neq A_{x_3x_4}A_{x_1x_2}$. Thus $\mathfrak{M}_{14}$ is not skew Tsankov. Furthermore, there exist $x, y \in V$ so $\mathcal{J}_x\mathcal{J}_y \neq 0$.

(3) There is a short exact sequence $1 \to \mathbb{R}^{21} \to \mathcal{G}(\mathfrak{M}_{14}) \to \mathrm{Sl}_\pm(3) \to 1$.

(4) One has $A_{x_1x_2}\mathcal{J}_{x_3} = \mathcal{J}_{x_3}A_{x_1x_2}$ for all $x_i \in V$. Thus, in particular, $\mathfrak{M}_{14}$ is Stanilov–Tsankov.

In Section 6.4.2, we show $\mathfrak{M}_{14}$ is geometrically realizable by considering the following family of examples:

Definition 6.4.2   Let $\{x_i, x_i^*, y_{i,1}, y_{i,2}, y_{4,1}, y_{4,2}\}$ be coordinates on $\mathbb{R}^{14}$ where the index i ranges from 1 through 3. Suppose given a collection of functions $\Phi := \{\phi_{i,1}, \phi_{i,2}\} \in C^\infty(\mathbb{R})$ with $\phi_{i,1}' \phi_{i,2}' = 1$. Let $\mathcal{M}_\Phi := (\mathbb{R}^{14}, g_\Phi)$ where the non-zero components of g_Φ are, up to the usual $\mathbb{Z}_2$ symmetry, given by:

$$g_\Phi(\partial_{x_i}, \partial_{x_i^*}) = g_\Phi(\partial_{y_{i,1}}, \partial_{y_{i,2}}) = 1,$$
$$g_\Phi(\partial_{y_{4,1}}, \partial_{y_{4,1}}) = g_\Phi(\partial_{y_{4,2}}, \partial_{y_{4,2}}) = -\tfrac{1}{2}, \quad g_\Phi(\partial_{y_{4,1}}, \partial_{y_{4,2}}) = \tfrac{1}{4},$$
$$g_\Phi(\partial_{x_1}, \partial_{x_1}) = -2\phi_{2,1}(x_2)y_{2,1} - 2\phi_{3,1}(x_3)y_{3,1},$$
$$g_\Phi(\partial_{x_2}, \partial_{x_2}) = -2\phi_{3,2}(x_3)y_{3,2} - 2\phi_{1,2}(x_1)y_{1,2},$$
$$g_\Phi(\partial_{x_3}, \partial_{x_3}) = -2\phi_{1,1}(x_1)y_{1,1} - 2\phi_{2,2}(x_2)y_{2,2},$$
$$g_\Phi(\partial_{x_1}, \partial_{x_3}) = x_2 y_{4,2}, \quad g_\Phi(\partial_{x_2}, \partial_{x_3}) = x_1 y_{4,1}.$$

Theorem 6.4.2   *Let $\mathcal{M}_\Phi := (\mathbb{R}^{14}, g_\Phi)$ be as in Definition 6.4.2. Then $\mathcal{M}_\Phi$ is a generalized plane wave manifold which has 0-model $\mathfrak{M}_{14}$.*

If we specialize the construction, we can say a bit more. We will establish the following result in Section 6.4.3 by constructing isometry invariants:

Theorem 6.4.3   *In Definition 6.4.2, set $\phi_{2,1}(x_2) = \phi_{2,2}(x_2) = x_2$ and set $\phi_{3,1}(x_3) = \phi_{3,2}(x_3) = x_3$. Let $\{\phi_{1,1}, \phi_{1,2}\}$ be real analytic with $\phi_{1,1}' \phi_{1,2}' = 1$ and with $\phi_{1,j}'' \neq 0$. Then*

(1) $\Xi := \{1 - \phi_{1,1}' \phi_{1,1}''' (\phi_{1,1}'')^{-2}\}^2$ is a local isometry invariant of $\mathcal{M}_\Phi$.
(2) If $\phi_{1,1}'(x_1) \neq b e^{c x_1}$, then Ξ is not locally constant and hence $\mathcal{M}_\Phi$ is not locally homogeneous.

There are symmetric spaces which have model $\mathfrak{M}_{14}$.

Definition 6.4.3   Let $\{x_i, x_i^*, y_{i,1}, y_{i,2}, y_{4,1}, y_{4,2}\}$ for $1 \leq i \leq 3$ be coordinates on $\mathbb{R}^{14}$. Let $A := \{a_{i,j}\}$ be a collection of real constants. Let $\mathcal{M}_A := (\mathbb{R}^{14}, g_A)$ where the non-zero components of g_A are given, up to the usual $\mathbb{Z}_2$ symmetry, by:

$$g_A(\partial_{x_i}, \partial_{x_i^*}) = g_A(\partial_{y_{i,1}}, \partial_{y_{i,2}}) = 1,$$
$$g_A(\partial_{y_{4,1}}, \partial_{y_{4,1}}) = g_A(\partial_{y_{4,2}}, \partial_{y_{4,2}}) = -\tfrac{1}{2}, \quad g_A(\partial_{y_{4,1}}, \partial_{y_{4,2}}) = \tfrac{1}{4},$$
$$g_A(\partial_{x_1}, \partial_{x_1}) = -2a_{2,1}x_2 y_{2,1} - 2a_{3,1}x_3 y_{3,1},$$
$$g_A(\partial_{x_2}, \partial_{x_2}) = -2a_{3,2}x_3 y_{3,2} - 2a_{1,2}x_1 y_{1,2},$$
$$g_A(\partial_{x_3}, \partial_{x_3}) = -2a_{1,1}x_1 y_{1,1} - 2a_{2,2}x_2 y_{2,2},$$
$$g_A(\partial_{x_1}, \partial_{x_2}) = 2(1 - a_{2,1})x_1 y_{2,1} + 2(1 - a_{1,2})x_2 y_{1,2},$$
$$g_A(\partial_{x_2}, \partial_{x_3}) = x_1 y_{4,1} + 2(1 - a_{3,2})x_2 y_{3,2} + 2(1 - a_{2,2})x_3 y_{2,2},$$

$$g_A(\partial_{x_1}, \partial_{x_3}) = x_2 y_{4,2} + 2(1 - a_{3,1})x_1 y_{3,1} + 2(1 - a_{1,1})x_3 y_{1,1}.$$

We will establish the following result in Section 6.4.4:

Theorem 6.4.4 *Let $\mathcal{M}_A$ be described by Definition 6.4.3. Then $\mathcal{M}_A$ is a generalized plane wave manifold with 0-model $\mathfrak{M}_{14}$. Furthermore $\mathcal{M}_A$ is locally symmetric if and only if*

$$a_{1,1} + a_{2,2} + a_{3,1}a_{3,2} = 2,$$
$$3a_{2,1} + 3a_{3,1} + 3a_{1,2}a_{1,1} = 4, \text{ and}$$
$$3a_{1,2} + 3a_{3,2} + 3a_{2,1}a_{2,2} = 4.$$

6.4.1 *The model $\mathfrak{M}_{14}$*

We study the algebraic properties of the model $\mathfrak{M}_{14}$ which was introduced in Definition 6.4.1 to establish Theorem 6.4.1. We establish the following notational conventions. The following spaces are invariantly defined:

$$V_{\beta,\alpha^*} := \mathrm{Span}_{\xi_i \in \mathbb{R}^{14}}\{\mathcal{J}_{\xi_1}\xi_2\} = \mathrm{Span}_{1 \le i \le 3, 1 \le j \le 2}\{\beta_{i,j}, \beta_{4,j}, \alpha_i^*\},$$
$$V_{\alpha^*} := \mathrm{Span}_{\xi_i \in \mathbb{R}^{14}}\{\mathcal{J}_{\xi_1}\mathcal{J}_{\xi_2}\xi_3\} = \mathrm{Span}_{1 \le i \le 3}\{\alpha_i^*\} . \tag{6.4.b}$$

Define

$$\beta_{4,1}^* := -\tfrac{8}{3}\beta_{4,1} - \tfrac{4}{3}\beta_{4,2}, \quad \beta_{4,2}^* := -\tfrac{4}{3}\beta_{4,1} - \tfrac{8}{3}\beta_{4,2} .$$

One then has that

$$\langle \beta_{4,i}^*, \beta_{4,j} \rangle = \delta_{ij} .$$

Proof of Theorem 6.4.1 (1). If $\xi \in \mathbb{R}^{14}$, then

$$\mathcal{J}_\xi \alpha_i \subset V_{\beta,\alpha^*}, \quad \mathcal{J}_\xi \beta_{ij} \subset V_{\alpha^*}, \quad \text{and} \quad \mathcal{J}_\xi \alpha_i^* = 0 .$$

Thus to show $\mathcal{J}_x \mathcal{J}_y = \mathcal{J}_y \mathcal{J}_x$ for all x, y, it suffices to show

$$\mathcal{J}_x \mathcal{J}_y \alpha_i = \mathcal{J}_y \mathcal{J}_x \alpha_i$$

for all x, y, i. Since $\mathcal{J}_x \mathcal{J}_y \alpha_i \in V_{\alpha^*}$, this can be done by establishing:

$$\langle \mathcal{J}_x \alpha_i, \mathcal{J}_y \alpha_j \rangle = \langle \mathcal{J}_y \alpha_i, \mathcal{J}_x \alpha_j \rangle$$

for all x, y, i, j. Since $\mathcal{J}_{x_1 x_2}\alpha_i \in V_{\alpha^*}$ if either x_1 or $x_2 \in V_{\beta,\alpha^*}$, we may take $x_1 = \alpha_i$ and $x_2 = \alpha_j$. Let $\mathcal{J}_{ijk} := \mathcal{J}_{\alpha_i \alpha_j}\alpha_k$. We must show:

$$\langle \mathcal{J}_{i_1 i_2 i_3}, \mathcal{J}_{j_1 j_2 j_3} \rangle = \langle \mathcal{J}_{i_1 i_2 j_3}, \mathcal{J}_{j_1 j_2 i_3} \rangle \quad \forall i_1 i_2 i_3 j_1 j_2 j_3 .$$

The non-zero components of $\mathcal{J}_{ijk} = \mathcal{J}_{jik}$ are:

$$\mathcal{J}_{112} = \beta_{2,2}, \qquad \mathcal{J}_{113} = \beta_{3,2}, \qquad \mathcal{J}_{221} = \beta_{1,1},$$

$$\mathcal{J}_{223} = \beta_{3,1}, \qquad \mathcal{J}_{331} = \beta_{1,2}, \qquad \mathcal{J}_{332} = \beta_{2,1},$$

$$\mathcal{J}_{121} = -\tfrac{1}{2}\beta_{2,2}, \qquad \mathcal{J}_{122} = -\tfrac{1}{2}\beta_{1,1}, \qquad \mathcal{J}_{131} = -\tfrac{1}{2}\beta_{3,2},$$

$$\mathcal{J}_{133} = -\tfrac{1}{2}\beta_{1,2}, \qquad \mathcal{J}_{232} = -\tfrac{1}{2}\beta_{3,1}, \qquad \mathcal{J}_{233} = -\tfrac{1}{2}\beta_{2,1},$$

$$\mathcal{J}_{132} = \tfrac{1}{4}\beta_{4,1}^{*} - \tfrac{1}{2}\beta_{4,2}^{*} = \beta_{4,2},$$

$$\mathcal{J}_{231} = -\tfrac{1}{2}\beta_{4,1}^{*} + \tfrac{1}{4}\beta_{4,2}^{*} = \beta_{4,1},$$

$$\mathcal{J}_{123} = \tfrac{1}{4}\beta_{4,1}^{*} + \tfrac{1}{4}\beta_{4,2}^{*} = -\beta_{4,1} - \beta_{4,2}.$$

The non-zero inner products are given by:

$$\langle \mathcal{J}_{112}, \mathcal{J}_{332}\rangle = 1, \qquad \langle \mathcal{J}_{112}, \mathcal{J}_{233}\rangle = -\tfrac{1}{2}, \qquad \langle \mathcal{J}_{121}, \mathcal{J}_{332}\rangle = -\tfrac{1}{2},$$

$$\langle \mathcal{J}_{121}, \mathcal{J}_{233}\rangle = \tfrac{1}{4}, \qquad \langle \mathcal{J}_{113}, \mathcal{J}_{223}\rangle = 1, \qquad \langle \mathcal{J}_{113}, \mathcal{J}_{232}\rangle = -\tfrac{1}{2},$$

$$\langle \mathcal{J}_{131}, \mathcal{J}_{223}\rangle = -\tfrac{1}{2}, \qquad \langle \mathcal{J}_{232}, \mathcal{J}_{131}\rangle = \tfrac{1}{4}, \qquad \langle \mathcal{J}_{221}, \mathcal{J}_{331}\rangle = 1,$$

$$\langle \mathcal{J}_{221}, \mathcal{J}_{133}\rangle = -\tfrac{1}{2}, \qquad \langle \mathcal{J}_{122}, \mathcal{J}_{331}\rangle = -\tfrac{1}{2}, \qquad \langle \mathcal{J}_{122}, \mathcal{J}_{133}\rangle = \tfrac{1}{4},$$

$$\langle \mathcal{J}_{123}, \mathcal{J}_{123}\rangle = \star, \qquad \langle \mathcal{J}_{123}, \mathcal{J}_{132}\rangle = \tfrac{1}{4}, \qquad \langle \mathcal{J}_{123}, \mathcal{J}_{231}\rangle = \tfrac{1}{4},$$

$$\langle \mathcal{J}_{132}, \mathcal{J}_{132}\rangle = \star, \qquad \langle \mathcal{J}_{132}, \mathcal{J}_{231}\rangle = \tfrac{1}{4}, \qquad \langle \mathcal{J}_{231}, \mathcal{J}_{231}\rangle = \star.$$

The desired symmetries are now immediate:

$$\langle \mathcal{J}_{112}, \mathcal{J}_{233}\rangle = -\tfrac{1}{2} = \langle \mathcal{J}_{113}, \mathcal{J}_{232}\rangle, \qquad \langle \mathcal{J}_{123}, \mathcal{J}_{132}\rangle = \tfrac{1}{4} = \langle \mathcal{J}_{122}, \mathcal{J}_{133}\rangle,$$

$$\langle \mathcal{J}_{121}, \mathcal{J}_{332}\rangle = -\tfrac{1}{2} = \langle \mathcal{J}_{122}, \mathcal{J}_{331}\rangle, \qquad \langle \mathcal{J}_{123}, \mathcal{J}_{231}\rangle = \tfrac{1}{4} = \langle \mathcal{J}_{121}, \mathcal{J}_{233}\rangle,$$

$$\langle \mathcal{J}_{131}, \mathcal{J}_{223}\rangle = -\tfrac{1}{2} = \langle \mathcal{J}_{133}, \mathcal{J}_{221}\rangle, \qquad \langle \mathcal{J}_{132}, \mathcal{J}_{231}\rangle = \tfrac{1}{4} = \langle \mathcal{J}_{131}, \mathcal{J}_{232}\rangle.$$

This completes the proof of Theorem 6.4.1 (1). $\square$

Proof of Theorem 6.4.1 (2). It is immediate from the definition that

$$\mathcal{J}_{\alpha_3}\mathcal{J}_{\alpha_2}\alpha_1 = \mathcal{J}_{\alpha_3}\beta_{1,1} = \alpha_1^{*}$$

so there exist $x, y \in V$ so $\mathcal{J}_x\mathcal{J}_y \neq 0$. Let $\mathcal{A}_{ij} := \mathcal{A}(\alpha_i, \alpha_j)$. One shows $\mathfrak{M}_{14}$ is not skew Tsankov by computing:

$$\mathcal{A}_{12}\mathcal{A}_{13}\alpha_3 = \mathcal{A}_{12}\beta_{1,2} = -\alpha_2^{*},$$

$$\mathcal{A}_{13}\mathcal{A}_{12}\alpha_3 = -\tfrac{1}{2}\mathcal{A}_{13}\{\beta_{4,1}^{*} - \beta_{4,2}^{*}\} = \mathcal{A}_{13}\{\tfrac{2}{3}\beta_{4,1} - \tfrac{2}{3}\beta_{4,2}\} = \tfrac{1}{3}\alpha_2^{*}.$$

This establishes Theorem 6.4.1 (2). $\square$

Proof of Theorem 6.4.1 (3). Let $\mathcal{G} = \mathcal{G}(\mathfrak{M}_{14})$ be the group of symmetries of the model $\mathfrak{M}_{14}$. Note that the spaces V_{β,α^*} and V_{α^*} defined in Eq. (6.4.b) are preserved by $\mathcal{G}$. Consequently one has that

$$TV_{\alpha^*} \subset V_{\alpha^*} \quad \text{and} \quad TV_{\beta,\alpha^*} \subset V_{\beta,\alpha^*} \quad \text{if} \quad T \in \mathcal{G}. \tag{6.4.c}$$

Let $\tau : \mathcal{G} \to \mathrm{Gl}(3)$ be the restriction of T to $V_{\alpha^*} = \mathbb{R}^3$. We will prove Theorem 6.4.1 (3) by showing:

$$\mathrm{Sl}_{\pm}(3) = \tau(\mathcal{G}) \quad \text{and} \quad \ker(\tau) = \mathbb{R}^{21} \,.$$

We argue as follows to show $\mathrm{Sl}_{\pm}(3) \subset \tau(\mathcal{G})$. Let $\beta_{4,3} := -\beta_{4,1} - \beta_{4,2}$. Define a linear map T of $\mathbb{R}^{14}$ which interchanges the first 2 coordinates of $\mathbb{R}^3$ by setting

$$T : \alpha_1 \leftrightarrow \alpha_2, \quad T : \alpha_3 \leftrightarrow \alpha_3, \quad T : \alpha_1^* \leftrightarrow \alpha_2^*, \quad T : \alpha_3^* \leftrightarrow \alpha_3^*,$$
$$T : \beta_{1,1} \leftrightarrow \beta_{2,2}, T : \beta_{1,2} \leftrightarrow \beta_{2,1}, T : \beta_{3,1} \leftrightarrow \beta_{3,2}, T : \beta_{4,1} \leftrightarrow \beta_{4,2} \,.$$

It is then immediate by inspection that T preserves the relations of Definition 6.4.1 and hence $T \in \mathcal{G}$. One may define a map $T \in \mathcal{G}$ which interchanges the first and third coordinates by setting:

$$T : \alpha_1 \leftrightarrow \alpha_3, \quad T : \alpha_2 \leftrightarrow \alpha_2, \quad T : \alpha_1^* \leftrightarrow \alpha_3^*, \quad T : \alpha_2^* \leftrightarrow \alpha_2^*,$$
$$T : \beta_{1,1} \leftrightarrow \beta_{3,1}, \quad T : \beta_{1,2} \leftrightarrow \beta_{3,2}, \quad T : \beta_{2,1} \leftrightarrow \beta_{2,2}, \quad T : \beta_{4,1} \leftrightarrow \beta_{4,3},$$
$$T : \beta_{4,2} \leftrightarrow \beta_{4,2} \,.$$

To define a map $T \in \mathcal{G}$ which induces a rotation in the first two coordinates, we set

$$T_\theta : \alpha_1 \to \cos\theta\,\alpha_1 + \sin\theta\,\alpha_2,$$
$$T_\theta : \alpha_2 \to -\sin\theta\,\alpha_1 + \cos\theta\,\alpha_2,$$
$$T_\theta : \alpha_1^* \to \cos\theta\,\alpha_1^* + \sin\theta\,\alpha_2^*,$$
$$T_\theta : \alpha_2^* \to -\sin\theta\,\alpha_1^* + \cos\theta\,\alpha_2^*,$$
$$T_\theta : \alpha_3 \to \alpha_3, \quad T_\theta : \alpha_3^* \to \alpha_3^*,$$
$$T_\theta : \beta_{1,1} \to \cos\theta\,\beta_{1,1} + \sin\theta\,\beta_{2,2},$$
$$T_\theta : \beta_{1,2} \to \cos\theta\,\beta_{1,2} + \sin\theta\,\beta_{2,1},$$
$$T_\theta : \beta_{2,1} \to -\sin\theta\,\beta_{1,2} + \cos\theta\,\beta_{2,1},$$
$$T_\theta : \beta_{2,2} \to -\sin\theta\,\beta_{1,1} + \cos\theta\,\beta_{2,2},$$
$$T_\theta : \beta_{3,1} \to \sin^2\theta\,\beta_{3,2} - 2\sin\theta\cos\theta\,\beta_{4,3} + \cos^2\theta\,\beta_{3,1},$$
$$T_\theta : \beta_{3,2} \to \cos^2\theta\,\beta_{3,2} + 2\cos\theta\sin\theta\,\beta_{4,3} + \sin^2\theta\,\beta_{3,1},$$
$$T_\theta : \beta_{4,1} \to \tfrac{1}{2}\sin\theta\cos\theta\,\beta_{3,2} - \tfrac{1}{2}\sin\theta\cos\theta\,\beta_{3,1} - \sin^2\theta\,\beta_{4,2} + \cos^2\theta\,\beta_{4,1},$$
$$T_\theta : \beta_{4,2} \to \tfrac{1}{2}\sin\theta\cos\theta\,\beta_{3,2} - \tfrac{1}{2}\sin\theta\cos\theta\,\beta_{3,1} + \cos^2\theta\,\beta_{4,2} - \sin^2\theta\,\beta_{4,1}.$$

Finally, we show that the dilatations of determinant 1 belong to $\mathrm{Range}\{\tau\}$. Suppose $a_1 a_2 a_3 = 1$. We may define $T \in \mathcal{G}$ by setting:

$$T\alpha_1 = a_1\alpha_1, \quad T\alpha_2 = a_2\alpha_2, \quad T\alpha_3 = a_3\alpha_3, \quad T\alpha_1^* = \tfrac{1}{a_1}\alpha_1^*,$$

$$T\alpha_2^* = \tfrac{1}{a_2}\alpha_2^*, \quad T\alpha_3^* = \tfrac{1}{a_3}\alpha_3^*, \quad T\beta_{1,1} = \tfrac{a_2}{a_3}\beta_{1,1}, \ T\beta_{1,2} = \tfrac{a_3}{a_2}\beta_{1,2},$$

$$T\beta_{2,1} = \tfrac{a_3}{a_1}\beta_{2,1}, \ T\beta_{2,2} = \tfrac{a_1}{a_3}\beta_{2,2}, \ T\beta_{3,1} = \tfrac{a_2}{a_1}\beta_{3,1}, \ T\beta_{3,2} = \tfrac{a_1}{a_2}\beta_{3,2},$$

$$T\beta_{4,1} = \beta_{4,1}, \quad T\beta_{4,2} = \beta_{4,2} \,.$$

Since these elements acting on V_{α^*} generate $\mathrm{Sl}_\pm(3)$,

$$\mathrm{Sl}_\pm(3) \subset \tau(\mathcal{G}) \,.$$

Conversely, let $T \in \mathcal{G}$. We must show $\tau(T) \in \mathrm{Sl}_\pm(3)$. Since one has that $\mathrm{Sl}_\pm(3) \subset \mathrm{Range}(\tau)$, there exists $S \in \mathcal{G}$ so that $\tau(TS)$ is diagonal. Thus without loss of generality, we may assume $\tau(T)$ is diagonal and hence:

$$T\alpha_i = a_i\alpha_i + \sum_\nu b_i^\nu \beta_\nu + \sum_j c_i^j \alpha_j^*,$$

$$T\beta_\nu = b_\nu \beta_\nu + \sum_i d_\nu^i \alpha_i^*,$$

$$T\alpha_i^* = a_i^{-1}\alpha_i^* \,.$$

We have the relations

$$-\tfrac{1}{2} = A(T\alpha_1, T\alpha_2, T\alpha_3, T\beta_{4,1}) = -\tfrac{1}{2}a_1 a_2 a_3 b_{4,1},$$

$$-\tfrac{1}{2} = \langle T\beta_{4,1}, T\beta_{4,1}\rangle = -\tfrac{1}{2}b_{4,1}b_{4,1} \,.$$

These relations show that $b_{4,1}^2 = 1$ and thus $a_1 a_2 a_3 = \pm 1$. Consequently, $\mathrm{Range}(\tau) = \mathrm{Sl}_\pm(3)$.

We complete the proof of Theorem 6.4.1 (3) by studying $T \in \ker(\tau)$. One has

$$T\alpha_i = \alpha_i + \sum_\nu b_i^\nu \beta_\nu + \sum_j c_i^j \alpha_j^*,$$

$$T\beta_\nu = \beta_\nu + \sum_i d_\nu^i \alpha_i^*,$$

$$T\alpha_i^* = \alpha_i^* \,.$$

Using the relations $A(\alpha_i, \alpha_j, \alpha_k, \alpha_l) = 0$ then leads to the following 6 linear equations the coefficients b_i^ν must satisfy:

$$0 = A(T\alpha_2, T\alpha_1, T\alpha_1, T\alpha_2)$$
$$= 2A(b_2^{2,1}\beta_{2,1}, \alpha_1, \alpha_1, \alpha_2) + 2A(b_1^{1,2}\beta_{1,2}, \alpha_2, \alpha_2, \alpha_1) = 2b_2^{2,1} + 2b_1^{1,2},$$
$$0 = A(T\alpha_3, T\alpha_1, T\alpha_1, T\alpha_3)$$
$$= 2A(b_3^{3,1}\beta_{3,1}, \alpha_1, \alpha_1, \alpha_3) + 2A(b_1^{1,1}\beta_{1,1}, \alpha_3, \alpha_3, \alpha_1) = 2b_3^{3,1} + 2b_1^{1,1},$$
$$0 = A(T\alpha_3, T\alpha_2, T\alpha_2, T\alpha_3)$$
$$= 2A(b_3^{3,2}\beta_{3,2}, \alpha_2, \alpha_2, \alpha_3) + 2A(b_2^{2,2}\beta_{2,2}, \alpha_3, \alpha_3, \alpha_2) = 2b_3^{3,2} + 2b_2^{2,2},$$
$$0 = A(T\alpha_2, T\alpha_1, T\alpha_1, T\alpha_3)$$
$$= A(b_2^{3,1}\beta_{3,1}, \alpha_1, \alpha_1, \alpha_3) + A(\alpha_2, \alpha_1, \alpha_1, b_3^{2,1}\beta_{2,1})$$
$$+ A(\alpha_2, b_1^{4,1}\beta_{4,1} + b_1^{4,2}\beta_{4,2}, \alpha_1, \alpha_3) + A(\alpha_2, \alpha_1, b_1^{4,1}\beta_{4,1} + b_1^{4,2}\beta_{4,2}, \alpha_3)$$
$$= b_2^{3,1} + b_3^{2,1} - \tfrac{1}{2}b_1^{4,1} - \tfrac{1}{2}b_1^{4,1} + \tfrac{1}{2}b_1^{4,2},$$
$$0 = A(T\alpha_1, T\alpha_2, T\alpha_2, T\alpha_3)$$
$$= A(b_1^{3,2}\beta_{3,2}, \alpha_2, \alpha_2, \alpha_3) + A(\alpha_1, \alpha_2, \alpha_2, b_3^{1,2}\beta_{1,2})$$
$$+ A(\alpha_1, b_2^{4,1}\beta_{4,1} + b_2^{4,2}\beta_{4,2}, \alpha_2, \alpha_3) + A(\alpha_1, \alpha_2, b_2^{4,1}\beta_{4,1} + b_2^{4,2}\beta_{4,2}, \alpha_3)$$
$$= b_1^{3,2} + b_3^{1,2} - \tfrac{1}{2}b_2^{4,2} + \tfrac{1}{2}b_2^{4,1} - \tfrac{1}{2}b_2^{4,2},$$
$$0 = A(T\alpha_1, T\alpha_3, T\alpha_3, T\alpha_2)$$
$$= A(b_1^{2,2}\beta_{2,2}, \alpha_3, \alpha_3, \alpha_2) + A(\alpha_1, \alpha_3, \alpha_3, b_2^{1,1}\beta_{1,1})$$
$$+ A(\alpha_1, b_3^{4,1}\beta_{4,1} + b_3^{4,2}\beta_{4,2}, \alpha_3, \alpha_2) + A(\alpha_1, \alpha_3, b_3^{4,1}\beta_{4,1} + b_3^{4,2}\beta_{4,2}, \alpha_2)$$
$$= b_1^{2,2} + b_2^{1,1} + \tfrac{1}{2}b_3^{4,2} + \tfrac{1}{2}b_3^{4,1}.$$

These equations are linearly independent so there are 18 degrees of freedom in choosing the b's. Once the b's are known, the coefficients d_ν^i are determined;

$$0 = \langle T\alpha_i, T\beta_\nu \rangle = d_\nu^i + \sum_\mu \langle \beta_\nu, \beta_\mu \rangle b_i^\mu .$$

The relation $\langle T\alpha_i, T\alpha_j \rangle = \delta_{ij}$ implies $c_i^j + c_j^i = 0$; this creates an additional 3 degrees of freedom. Thus $\ker(\tau)$ is isomorphic to the additive group $\mathbb{R}^{21}$. This completes the proof of Theorem 6.4.1 (3). $\qquad\square$

Proof of Theorem 6.4.1 (4). Let $\xi_i \in V$. We wish to show

$$\mathcal{A}_{\xi_1\xi_2}\mathcal{J}_{\xi_3} = \mathcal{J}_{\xi_3}\mathcal{A}_{\xi_1\xi_2} \quad \text{for all} \quad \xi_i \in V .$$

Since $\mathcal{A}_{\xi_1\xi_2}\mathcal{J}_{\xi_3} = \mathcal{J}_{\xi_3}\mathcal{A}_{\xi_1\xi_2} = 0$ if any of the $\xi_i \in V_{\beta,\alpha^*}$, we may work modulo V_{β,α^*} and suppose that $\xi_i \in \mathrm{Span}\{\alpha_i\}$. Since $\mathcal{A}_{\xi_1\xi_2} = 0$ if the ξ_i are linearly dependent, we suppose ξ_1 and ξ_2 are linearly independent.

There are 2 cases to be considered. We first suppose $\xi_3 \in \mathrm{Span}\{\xi_1, \xi_2\}$. The argument given above shows that a subgroup of $\mathcal{G}$ isomorphic to $\mathrm{Sl}_{\pm}(3)$ acts on $\mathrm{Span}\{\alpha_i\}$. After reparametrizing by this action, we may suppose

$$\mathrm{Span}\{\xi_1, \xi_2\} = \mathrm{Span}\{\alpha_1, \alpha_2\} \quad \text{and} \quad \xi_3 = \alpha_1 \, .$$

Furthermore, because $\mathcal{A}_{\xi_1 \xi_2} = c\mathcal{A}_{\alpha_1 \alpha_2}$, we may also assume $\xi_1 = \alpha_1$ and $\xi_2 = \alpha_2$. We set $\mathcal{A}_{ij} := \mathcal{A}_{\alpha_i \alpha_j}$ and $\mathcal{J}_k := \mathcal{J}_{\alpha_k}$. The desired result is obtained by computing:

$$
\begin{aligned}
\mathcal{A}_{12}\mathcal{J}_1\alpha_1 &= 0, & \mathcal{J}_1\mathcal{A}_{12}\alpha_1 &= -\mathcal{J}_1\beta_{2,2} = 0, \\
\mathcal{A}_{12}\mathcal{J}_1\alpha_2 &= \mathcal{A}_{12}\beta_{2,2} = 0, & \mathcal{J}_1\mathcal{A}_{12}\alpha_2 &= \mathcal{J}_1\beta_{1,1} = 0, \\
\mathcal{A}_{12}\mathcal{J}_1\alpha_3 &= \mathcal{A}_{12}\beta_{3,2} = 0, & \mathcal{J}_1\mathcal{A}_{12}\alpha_3 &= \tfrac{1}{2}\mathcal{J}_1(-\beta^*_{4,1} + \beta^*_{4,2}) = 0 \, .
\end{aligned}
$$

On the other hand, if $\{\xi_1, \xi_2, \xi_3\}$ are linearly independent, we can apply a symmetry in $\mathcal{G}$ and rescale to assume $\xi_i = \alpha_i$. We compute:

$$
\begin{aligned}
\mathcal{A}_{12}\mathcal{J}_3\alpha_1 &= \mathcal{A}_{12}\beta_{1,2} = -\alpha^*_2, & \mathcal{J}_3\mathcal{A}_{12}\alpha_1 &= -\mathcal{J}_3\beta_{2,2} = -\alpha^*_2, \\
\mathcal{A}_{12}\mathcal{J}_3\alpha_2 &= \mathcal{A}_{12}\beta_{2,1} = \alpha^*_1, & \mathcal{J}_3\mathcal{A}_{12}\alpha_2 &= \mathcal{J}_3\beta_{1,1} = \alpha^*_1, \\
\mathcal{A}_{12}\mathcal{J}_3\alpha_3 &= 0, & \mathcal{J}_3\mathcal{A}_{12}\alpha_3 &= \tfrac{1}{2}\mathcal{J}_3(-\beta^*_{4,1} + \beta^*_{4,2}) = 0 \, .
\end{aligned}
$$

Theorem 6.4.1 follows. $\qquad\qquad\qquad\qquad\qquad\qquad\qquad\qquad\qquad\square$

6.4.2   *A geometric realization of $\mathfrak{M}_{14}$*

The following is an ansatz which constructs generalized plane wave manifolds; this ansatz is an extension of Definition 2.9.1. The crucial fact is that certain variables appear linearly as warping functions.

Definition 6.4.4   Let indices i, j range from 1 through a and indices μ, ν range from 1 through b. Let $\{x_i, x_i^*, y_\mu\}$ be coordinates on $\mathbb{R}^{2a+b}$. We suppose given a non-degenerate symmetric matrix $C_{\mu\nu}$ and smooth functions $\psi_{ij\mu} = \psi_{ij\mu}(\vec{x})$ with $\psi_{ij\mu} = \psi_{ji\mu}$. Let $\mathcal{M}_{C,\psi} := (\mathbb{R}^{2a+b}, g_{C,\psi})$, where

$$
\begin{aligned}
g_{C,\psi}(\partial_{x_i}, \partial_{x_j}) &= 2\sum_k y_\mu \psi_{ij\mu}, \\
g_{C,\psi}(\partial_{x_i}, \partial_{x_i^*}) &= 1, \\
g_{C,\psi}(\partial_{y_\mu}, \partial_{y_\nu}) &= C_{\mu\nu} \, .
\end{aligned}
$$

Lemma 6.4.1   *Let $\mathcal{M}_{C,\psi} = (\mathbb{R}^{2a+b}, g_{C,\psi})$ be as in Definition 6.4.4. Then $\mathcal{M}_{C,\psi}$ is a generalized plane wave manifold and the possibly non-zero com-*

ponents of the curvature tensor are, up to the usual $\mathbb{Z}_2$ symmetries,

$$R(\partial_{x_i}, \partial_{x_j}, \partial_{x_k}, \partial_{y_\nu}) = -\partial_{x_i}\psi_{jk\nu} + \partial_{x_j}\psi_{ik\nu},$$

$$R(\partial_{x_i}, \partial_{x_j}, \partial_{x_k}, \partial_{x_l}) = \sum_{\nu\mu} C^{\nu\mu}\{\psi_{ik\mu}\psi_{jl\nu} - \psi_{il\mu}\psi_{jk\nu}\}$$

$$+ \sum_\nu y_\nu \{\partial_{x_i}\partial_{x_k}\psi_{jl\nu} + \partial_{x_j}\partial_{x_l}\psi_{ik\nu} - \partial_{x_i}\partial_{x_l}\psi_{jk\nu} - \partial_{x_j}\partial_{x_k}\psi_{il\nu}\}.$$

Proof. The non-zero Christoffel symbols of the first kind are given by:

$$g(\nabla_{\partial_{x_i}}\partial_{x_j}, \partial_{x_k}) = \sum_{\mu=1}^{b}\{\partial_{x_i}\psi_{jk\mu} + \partial_{x_j}\psi_{ik\mu} - \partial_{x_k}\psi_{ij\mu}\}y_\mu,$$

$$g(\nabla_{\partial_{x_i}}\partial_{x_j}, \partial_{y_\nu}) = -\psi_{ij\nu},$$

$$g(\nabla_{\partial_{x_i}}\partial_{y_\nu}, \partial_{x_k}) = g(\nabla_{\partial_{y_\nu}}\partial_{x_i}, \partial_{x_k}) = \psi_{ik\nu},$$

and the non-zero Christoffel symbols of the second kind are given by:

$$\nabla_{\partial_{x_i}}\partial_{x_j} = \sum_{\mu=1}^{b} y_\mu\{\partial_{x_i}\psi_{jk\mu} + \partial_{x_j}\psi_{ik\mu} - \partial_{x_k}\psi_{ij\mu}\}\partial_{x_k^*}$$

$$- \sum_{\mu=1}^{b}\sum_{\nu=1}^{b} C^{\nu\mu}\psi_{ij\nu}\partial_{y_\mu},$$

$$\nabla_{\partial_{x_i}}\partial_{y_\nu} = \nabla_{\partial_{y_\nu}}\partial_{x_i} = \sum_{k=1}^{a} \psi_{ik\nu}\partial_{x_k^*}.$$

This shows that $\mathcal{M}$ is a generalized plane wave manifold; furthermore, the curvature can now be computed. $\square$

Proof of Theorem 6.4.2. We use Lemma 6.4.1 to see $\mathcal{M}_\Phi$ is a generalized plane wave manifold and that the possibly non-zero components of the curvature tensor defined by the metric of Definition 6.4.2 are:

$$R(\partial_{x_{i_1}}, \partial_{x_{i_2}}, \partial_{x_{i_3}}, \partial_{x_{i_4}}) = \star,$$

$$R(\partial_{x_1}, \partial_{x_2}, \partial_{y_{2,1}}, \partial_{x_1}) = \partial_{x_2}\phi_{2,1}, \quad R(\partial_{x_1}, \partial_{x_3}, \partial_{y_{3,1}}, \partial_{x_1}) = \partial_{x_3}\phi_{3,1},$$

$$R(\partial_{x_2}, \partial_{x_3}, \partial_{y_{3,2}}, \partial_{x_2}) = \partial_{x_3}\phi_{3,2}, \quad R(\partial_{x_2}, \partial_{x_1}, \partial_{y_{1,2}}, \partial_{x_2}) = \partial_{x_1}\phi_{1,2},$$

$$R(\partial_{x_3}, \partial_{x_1}, \partial_{y_{1,1}}, \partial_{x_3}) = \partial_{x_1}\phi_{1,1}, \quad R(\partial_{x_3}, \partial_{x_2}, \partial_{y_{2,2}}, \partial_{x_3}) = \partial_{x_2}\phi_{2,2},$$

$$R(\partial_{x_2}, \partial_{x_1}, \partial_{y_{4,1}}, \partial_{x_3}) = R(\partial_{x_3}, \partial_{x_1}, \partial_{y_{4,1}}, \partial_{x_2}) = -\tfrac{1}{2},$$

$$R(\partial_{x_1}, \partial_{x_2}, \partial_{y_{4,2}}, \partial_{x_3}) = R(\partial_{x_3}, \partial_{x_2}, \partial_{y_{4,2}}, \partial_{x_1}) = -\tfrac{1}{2}.$$

We introduce the following basis as a first step in showing $\mathcal{M}_{14}$ is a 0-model for $\mathcal{M}_{14}$. Let the index i range from 1 to 3 and the index j run from 1 to 2. Set:

$$\bar{\alpha}_i := \partial_{x_i}, \qquad \alpha_i^* := \partial_{x_i^*},$$
$$\bar{\beta}_{4,j} := \partial_{y_{4,j}}, \quad \bar{\beta}_{i,j} := \{\phi_{i,j}'\}^{-1}\partial_{y_{i,j}}. \tag{6.4.d}$$

Since $\phi_{i,1}' \cdot \phi_{i,2}' = 1$, the relations of Eq. (6.4.a) are satisfied. However, we still have the following potentially non-zero terms to deal with:

$$g(\bar{\alpha}_i, \bar{\alpha}_j) = \star \quad \text{and} \quad R(\bar{\alpha}_i, \bar{\alpha}_j, \bar{\alpha}_k, \bar{\alpha}_l) = \star.$$

To deal with the extra curvature terms, we introduce a modified basis setting:

$$\tilde{\alpha}_1 := \bar{\alpha}_1 + R(\bar{\alpha}_1, \bar{\alpha}_2, \bar{\alpha}_3, \bar{\alpha}_1)\bar{\beta}_{4,1} - \tfrac{1}{2}R(\bar{\alpha}_1, \bar{\alpha}_2, \bar{\alpha}_2, \bar{\alpha}_1)\bar{\beta}_{1,2},$$
$$\tilde{\alpha}_2 := \bar{\alpha}_2 + R(\bar{\alpha}_2, \bar{\alpha}_1, \bar{\alpha}_3, \bar{\alpha}_2)\bar{\beta}_{4,2} - \tfrac{1}{2}R(\bar{\alpha}_2, \bar{\alpha}_3, \bar{\alpha}_3, \bar{\alpha}_2)\bar{\beta}_{2,2},$$
$$\tilde{\alpha}_3 := \bar{\alpha}_3 - 2R(\bar{\alpha}_3, \bar{\alpha}_1, \bar{\alpha}_2, \bar{\alpha}_3)\bar{\beta}_{4,1} - \tfrac{1}{2}R(\bar{\alpha}_1, \bar{\alpha}_3, \bar{\alpha}_3, \bar{\alpha}_1)\bar{\beta}_{3,1},$$
$$\beta_{1,1} := \bar{\beta}_{1,1} + \tfrac{1}{2}R(\bar{\alpha}_1, \bar{\alpha}_2, \bar{\alpha}_2, \bar{\alpha}_1)\alpha_1^*, \qquad \beta_{1,2} := \bar{\beta}_{1,2},$$
$$\beta_{2,1} := \bar{\beta}_{2,1} + \tfrac{1}{2}R(\bar{\alpha}_2, \bar{\alpha}_3, \bar{\alpha}_3, \bar{\alpha}_2)\alpha_2^*, \qquad \beta_{2,2} := \bar{\beta}_{2,2},$$
$$\beta_{3,2} := \bar{\beta}_{3,2} + \tfrac{1}{2}R(\bar{\alpha}_1, \bar{\alpha}_3, \bar{\alpha}_3, \bar{\alpha}_1)\alpha_3^*, \qquad \beta_{3,1} := \bar{\beta}_{3,1}, \tag{6.4.e}$$
$$\beta_{4,1} := \bar{\beta}_{4,1} + \tfrac{1}{2}R(\bar{\alpha}_1, \bar{\alpha}_2, \bar{\alpha}_3, \bar{\alpha}_1)\alpha_1^* - \tfrac{1}{4}R(\bar{\alpha}_2, \bar{\alpha}_1, \bar{\alpha}_3, \bar{\alpha}_2)\alpha_2^*$$
$$\qquad\qquad - R(\bar{\alpha}_3, \bar{\alpha}_1, \bar{\alpha}_2, \bar{\alpha}_3)\alpha_3^*,$$
$$\beta_{4,2} := \bar{\beta}_{4,2} - \tfrac{1}{4}R(\bar{\alpha}_1, \bar{\alpha}_2, \bar{\alpha}_3, \bar{\alpha}_1)\alpha_1^* + \tfrac{1}{2}R(\bar{\alpha}_2, \bar{\alpha}_1, \bar{\alpha}_3, \bar{\alpha}_2)\alpha_2^*$$
$$\qquad\qquad + \tfrac{1}{2}R(\bar{\alpha}_3, \bar{\alpha}_1, \bar{\alpha}_2, \bar{\alpha}_3)\alpha_3^*.$$

All the normalizations of Eq. (6.4.a) are satisfied except for the unwanted metric terms $g(\tilde{\alpha}_i, \tilde{\alpha}_j)$. To eliminate these terms and to exhibit a basis with the required normalizations, we set:

$$\alpha_i := \tilde{\alpha}_i - \tfrac{1}{2}\sum_{j=1}^{3} g(\tilde{\alpha}_i, \tilde{\alpha}_j)\alpha_j^*. \qquad \square \tag{6.4.f}$$

6.4.3 *Isometry invariants*

We now turn to the task of constructing invariants.

Lemma 6.4.2 *Adopt the assumptions of Theorem 6.4.3. Let $\{\alpha_i\}$ be defined by Eq. (6.4.f), let $\{\beta_\nu\}$ be defined by Eq. (6.4.e), and let $\{\alpha_i^*\}$ be defined by Eq. (6.4.d). Set $\phi_1 := \phi_{1,1}'$ and $\phi_2 := \phi_{1,2}'$.*

(1) $\nabla R(v_1, v_2, v_3, v_4; v_5) = 0$ *if at least one of the* $v_i \in V_{\alpha^*}$.

(2) $\nabla R(v_1, v_2, v_3, v_4; v_5) = 0$ *if at least two of the* $v_i \in V_{\beta,\alpha^*}$.

(3) $\nabla^k R(\alpha_1, \alpha_2, \alpha_2, \beta_{1,2}; \alpha_1, ..., \alpha_1) = \phi_2^{-1} \phi_2^{(k)}$.

(4) $\nabla^k R(\alpha_1, \alpha_3, \alpha_3, \beta_{1,1}; \alpha_1, ..., \alpha_1) = \phi_1^{-1} \phi_1^{(k)}$.

(5) $\nabla R(\alpha_i, \alpha_j, \alpha_k, \beta_\nu; \alpha_{l_1}, ..., \alpha_{l_k}) = 0$ *in cases other than those given in* *(3) and (4) up to the usual* $\mathbb{Z}_2$ *symmetry in the first 2 entries.*

Proof. Let v_i be coordinate vector fields. To prove Assertion (1), we suppose some $v_i \in V_{\alpha^*}$. We use the second Bianchi identity and the other curvature symmetries to assume without loss of generality that $v_1 \in V_{\alpha^*}$. Since $\nabla_{v_5} v_1 = 0$ and since $R(v_1, \cdot, \cdot, \cdot) = 0$, Assertion (1) follows. The proof of Assertion (2) is similar and uses the fact that $R(\cdot, \cdot, \cdot, \cdot) = 0$ if 2-entries belong to V_{β,α^*}. The proof of the remaining assertions is similar and uses the particular form of the warping functions $\phi_{i,j}$; the factor of $\phi_{1,j}^{-1}$ arising from the normalization in Eq. (6.4.d). $\qquad\square$

Definition 6.4.5 We say that a basis $\tilde{\mathcal{B}} := \{\tilde{\alpha}_i, \tilde{\beta}_\nu, \tilde{\alpha}_i^*\}$ is *0-normalized* if the normalizations of Eq. (6.4.a) are satisfied and *1-normalized* if it is 0-normalized and if additionally

$$\nabla R(\tilde{\alpha}_1, \tilde{\alpha}_3, \tilde{\alpha}_3, \tilde{\beta}_{1,1}; \tilde{\alpha}_1) = -\nabla R(\tilde{\alpha}_3, \tilde{\alpha}_1, \tilde{\alpha}_3, \tilde{\beta}_{1,1}; \tilde{\alpha}_1) \neq 0,$$
$$\nabla R(\tilde{\alpha}_1, \tilde{\alpha}_2, \tilde{\alpha}_2, \tilde{\beta}_{1,2}; \tilde{\alpha}_1) = -\nabla R(\tilde{\alpha}_2, \tilde{\alpha}_1, \tilde{\alpha}_2, \tilde{\beta}_{1,2}; \tilde{\alpha}_1) \neq 0,$$
$$\nabla R(\tilde{\alpha}_i, \tilde{\alpha}_j, \tilde{\alpha}_k, \tilde{\beta}_\nu; \tilde{\alpha}_l) = 0 \quad \text{otherwise}.$$

Lemma 6.4.3 *Adopt the assumptions of Theorem 6.4.3. Then:*

(1) There exists a 1-normalized basis.

(2) If $\tilde{\mathcal{B}}$ is a 1-normalized basis, then there exist constants a_i so $a_1 a_2 a_3 = \varepsilon$ for $\varepsilon = \pm 1$ and so that exactly one of the following conditions holds:

(a) $\tilde{\alpha}_1 = a_1 \alpha_1$, $\tilde{\alpha}_2 = a_2 \alpha_2$, $\tilde{\alpha}_3 = a_3 \alpha_3$, $\tilde{\beta}_{1,1} = \varepsilon \frac{a_2}{a_3} \beta_{1,1}$, $\tilde{\beta}_{1,2} = \varepsilon \frac{a_3}{a_2} \beta_{1,2}$.

(b) $\tilde{\alpha}_1 = a_1 \alpha_1$, $\tilde{\alpha}_2 = a_3 \alpha_3$, $\tilde{\alpha}_3 = a_2 \alpha_2$, $\tilde{\beta}_{1,1} = \varepsilon \frac{a_3}{a_2} \beta_{1,2}$, $\tilde{\beta}_{1,2} = \varepsilon \frac{a_2}{a_3} \beta_{1,1}$.

Proof. We use Eq. (6.4.d), Eq. (6.4.e), and Eq. (6.4.f) to construct a 0-normalized basis and then apply Lemma 6.4.2 to see that this basis is 1-normalized. On the other hand, if $\tilde{\mathcal{B}}$ is a 1-normalized basis, we may expand:

$$\tilde{\alpha}_1 = a_{11} \alpha_1 + a_{12} \alpha_2 + a_{13} \alpha_3 + ...,$$
$$\tilde{\alpha}_2 = a_{21} \alpha_1 + a_{22} \alpha_2 + a_{23} \alpha_3 + ..., \quad \tilde{\beta}_{1,2} = b_{21} \beta_{1,1} + b_{22} \beta_{1,2} + ...,$$
$$\tilde{\alpha}_3 = a_{31} \alpha_1 + a_{32} \alpha_2 + a_{33} \alpha_3 + ..., \quad \tilde{\beta}_{1,1} = b_{11} \beta_{1,1} + b_{12} \beta_{1,2} + ...$$

Because

$$0 \neq \nabla R(\tilde{\alpha}_1, \tilde{\alpha}_2, \tilde{\alpha}_2, \tilde{\beta}_{1,2}; \tilde{\alpha}_1)$$
$$= a_{11} \left\{ (a_{11}a_{22} - a_{12}a_{21})a_{22}b_{22})\phi_2^{-1}\phi_2' \right.$$
$$\left. + (a_{11}a_{33} - a_{13}a_{31})a_{33}b_{21})\phi_1^{-1}\phi_1' \right\},$$

we have $a_{11} \neq 0$. Because

$$0 = \nabla R(\tilde{\alpha}_1, \tilde{\alpha}_2, \tilde{\alpha}_2, \tilde{\beta}_{1,2}; \tilde{\alpha}_2) = \tfrac{a_{21}}{a_{11}} \nabla R(\tilde{\alpha}_1, \tilde{\alpha}_2, \tilde{\alpha}_2, \tilde{\beta}_{1,2}; \tilde{\alpha}_1),$$

we have $a_{21} = 0$; similarly $a_{31} = 0$. Since $\mathrm{Span}\{\alpha_i\} = \mathrm{Span}\{\tilde{\alpha}_i\} \bmod V_{\beta,\alpha*}$,

$$a_{22}a_{33} - a_{23}a_{32} \neq 0.$$

By hypothesis $R(\tilde{\alpha}_1, \tilde{\alpha}_2, \tilde{\alpha}_3, \beta; \tilde{\alpha}_1) = 0$ for any β which belongs to $\mathrm{Span}\{\tilde{\beta}_\nu, \tilde{\alpha}_i^*\} = V_{\beta,\alpha*}$ so

$$0 = R(\tilde{\alpha}_1, \tilde{\alpha}_2, \tilde{\alpha}_3, \beta_{1,2}; \tilde{\alpha}_1) = a_{11}^2 a_{22}a_{32}\phi_2^{-1}\phi_2',$$
$$0 = R(\tilde{\alpha}_1, \tilde{\alpha}_2, \tilde{\alpha}_3, \beta_{1,1}; \tilde{\alpha}_1) = a_{11}^2 a_{23}a_{33}\phi_1^{-1}\phi_1'.$$

Suppose that $a_{22} \neq 0$. Since $a_{11}^2 a_{22}a_{32} = 0$ and since $a_{11} \neq 0$, we conclude $a_{32} = 0$. Because $a_{22}a_{33} - a_{23}a_{32} \neq 0$, $a_{33} \neq 0$. As $a_{11}^2 a_{23}a_{33} = 0$, we also have $a_{23} = 0$. Since the basis is also 0-normalized,

$$\mathrm{diag}(a_{11}^{-1}, a_{22}^{-1}, a_{33}^{-1}) \in \mathrm{Sl}_\pm(3)$$

from the discussion in Section 6.4.1. Thus

$$\varepsilon := a_{11}a_{22}a_{33} = \pm 1, \quad b_{11} = \varepsilon \tfrac{a_{33}}{a_{22}}, \quad b_{22} = \varepsilon \tfrac{a_{22}}{a_{33}}.$$

These are the relations of Assertion (2a). The argument is similar if $a_{32} \neq 0$; we simply reverse the roles of $\tilde{\alpha}_2$ and $\tilde{\alpha}_3$ to establish the relations of Assertion (2b). $\qquad\square$

Proof of Theorem 6.4.3. Let

$$\Xi(\mathcal{B}) :=$$
$$\frac{1}{4}\left\{ \frac{\nabla^2 R(\tilde{\alpha}_1, \tilde{\alpha}_2, \tilde{\alpha}_2, \tilde{\beta}_{1,2}; \tilde{\alpha}_1, \tilde{\alpha}_1)}{\{\nabla R(\tilde{\alpha}_1, \tilde{\alpha}_2, \tilde{\alpha}_2, \tilde{\beta}_{1,2}; \tilde{\alpha}_1)\}^2} - \frac{\nabla^2 R(\tilde{\alpha}_1, \tilde{\alpha}_3, \tilde{\alpha}_3, \tilde{\beta}_{1,1}; \tilde{\alpha}_1, \tilde{\alpha}_1)}{\{\nabla R(\tilde{\alpha}_1, \tilde{\alpha}_3, \tilde{\alpha}_3, \tilde{\beta}_{1,1}; \tilde{\alpha}_1)\}^2} \right\}^2.$$

We apply Lemma 6.4.3. Suppose the conditions of Assertion (2a) hold. Then:

$$\nabla R(\tilde{\alpha}_1, \tilde{\alpha}_2, \tilde{\alpha}_2, \tilde{\beta}_{1,2}; \tilde{\alpha}_1) = a_1 \phi_2^{-1} \phi_2',$$
$$\nabla^2 R(\tilde{\alpha}_1, \tilde{\alpha}_2, \tilde{\alpha}_2, \tilde{\beta}_{1,2}; \tilde{\alpha}_1, \tilde{\alpha}_1) = a_1^2 \phi_2^{-1} \phi_2'',$$
$$\nabla R(\tilde{\alpha}_1, \tilde{\alpha}_3, \tilde{\alpha}_3, \tilde{\beta}_{1,1}; \tilde{\alpha}_1) = a_1 \phi_1^{-1} \phi_1',$$
$$\nabla^2 R(\tilde{\alpha}_1, \tilde{\alpha}_3, \tilde{\alpha}_3, \tilde{\beta}_{1,1}; \tilde{\alpha}_1, \tilde{\alpha}_1) = a_1^2 \phi_1^{-1} \phi_1'' .$$

Consequently one has that:

$$\Xi(\mathcal{B}) = \frac{1}{4} \left\{ \frac{\phi_2 \phi_2''}{\phi_2' \phi_2'} - \frac{\phi_1 \phi_1''}{\phi_1' \phi_1'} \right\}^2 .$$

The roles of ϕ_1 and ϕ_2 are reversed if Assertion (2b) holds. It now follows that Ξ is a local isometry invariant. Since

$$\phi_2 = \phi_1^{-1}, \quad \phi_2' = -\phi_1^{-2} \phi_1', \quad \phi_2'' = 2\phi_1^{-3} \phi_1' \phi_1' - \phi_1^{-2} \phi_1''$$

we may establish Assertion (1) of Theorem 6.4.3 by computing

$$\frac{\phi_2 \phi_2''}{\phi_2' \phi_2'} = \frac{\phi_1^{-1}(2\phi_1^{-3} \phi_1' \phi_1' - \phi_1^{-2} \phi_1'')}{\phi_1^{-4} \phi_1' \phi_1'} = 2 - \frac{\phi_1 \phi_1''}{\phi_1' \phi_1'} \quad \text{so}$$
$$\Xi = \frac{1}{4} \left\{ 2 - 2\frac{\phi_1 \phi_1''}{\phi_1' \phi_1'} \right\}^2 .$$

If $\mathcal{M}_\Phi$ is locally homogeneous, then Ξ must be constant. Conversely, if Ξ is constant, then $\phi_1 \phi_1'' = k \phi_1' \phi_1'$ for some $k \in \mathbb{R}$. Lemma 1.5.5 applies. The solutions to this ordinary differential equation take the form $\phi_1(t) = a(t+b)^c$ if $k \neq 1$ and $\phi_1(t) = a e^{bt}$ if $k = 1$ for suitably chosen constants a and b and for $c = c(k)$. The first family is ruled out as ϕ_1 and ϕ_1' must be invertible for all t. Thus $\phi_1(t)$ is a pure exponential; Assertion (2) of Theorem 6.4.3 follows. $\qquad\square$

6.4.4   *A symmetric space with model $\mathfrak{M}_{14}$*

We give the proof of Theorem 6.4.4 as follows. Let $\mathcal{M}_A$ be as described by Definition 6.4.3. By Lemma 6.4.1 one has that:

$$R(\partial_{x_2}, \partial_{x_1}, \partial_{x_1}, \partial_{y_{2,1}}) = R(\partial_{x_3}, \partial_{x_1}, \partial_{x_1}, \partial_{y_{3,1}}) = 1,$$
$$R(\partial_{x_3}, \partial_{x_2}, \partial_{x_2}, \partial_{y_{3,2}}) = R(\partial_{x_1}, \partial_{x_2}, \partial_{x_2}, \partial_{y_{1,2}}) = 1,$$
$$R(\partial_{x_1}, \partial_{x_3}, \partial_{x_3}, \partial_{y_{1,1}}) = R(\partial_{x_2}, \partial_{x_3}, \partial_{x_3}, \partial_{y_{2,2}}) = 1,$$
$$R(\partial_{x_1}, \partial_{x_2}, \partial_{x_3}, \partial_{y_{4,1}}) = R(\partial_{x_1}, \partial_{x_3}, \partial_{x_2}, \partial_{y_{4,1}}) = -\tfrac{1}{2},$$

$$R(\partial_{x_2}, \partial_{x_3}, \partial_{x_1}, \partial_{y_{4,2}}) = R(\partial_{x_2}, \partial_{x_1}, \partial_{x_3}, \partial_{y_{4,2}}) = -\tfrac{1}{2}.$$

The same argument constructing a 0-normalized basis which was given in the proof of Theorem 6.4.1 can then be used to construct a 0-normalized basis in this setting and establish that $\mathcal{M}_A$ has 0-model $\mathfrak{M}_{14}$.

We can also apply Lemma 6.4.1 to see:

$$R(\partial_{x_1}, \partial_{x_2}, \partial_{x_2}, \partial_{x_1}) = -a_{3,1}a_{3,2}x_3^2,$$
$$R(\partial_{x_1}, \partial_{x_3}, \partial_{x_3}, \partial_{x_1}) = -\tfrac{1}{3}(2 + 3a_{2,1}a_{2,2})x_2^2,$$
$$R(\partial_{x_3}, \partial_{x_2}, \partial_{x_2}, \partial_{x_3}) = -\tfrac{1}{3}(2 + 3a_{1,1}a_{1,2})x_1^2,$$
$$R(\partial_{x_2}, \partial_{x_1}, \partial_{x_1}, \partial_{x_3}) = (1 - a_{1,1} - a_{1,2} + a_{1,1}a_{1,2} + a_{2,1}$$
$$-a_{2,1}a_{2,2} + a_{3,1} - a_{3,1}a_{3,2})x_2x_3,$$
$$R(\partial_{x_1}, \partial_{x_2}, \partial_{x_2}, \partial_{x_3}) = (1 + a_{1,2} - a_{2,1} - a_{1,1}a_{1,2} - a_{2,2}$$
$$+a_{2,1}a_{2,2} + a_{3,2} - a_{3,1}a_{3,2})x_1x_3,$$
$$R(\partial_{x_1}, \partial_{x_3}, \partial_{x_3}, \partial_{x_2}) = (\tfrac{2}{3} + a_{1,1} - a_{1,1}a_{1,2} + a_{2,2}$$
$$-a_{2,1}a_{2,2} - a_{3,1} - a_{3,2} + a_{3,1}a_{3,2})x_1x_2.$$

The Christoffel symbols describing $\nabla_{\partial_{x_i}}\partial_{x_j}$ are given by:

$$\nabla_{\partial_{x_1}}\partial_{x_1} = (2 - a_{2,1})y_{2,1}\partial_{x_2^*} + (2 - a_{3,1})y_{3,1}\partial_{x_3^*} + a_{2,1}x_2\partial_{y_{2,2}}$$
$$+a_{3,1}x_3\partial_{y_{3,2}},$$
$$\nabla_{\partial_{x_2}}\partial_{x_2} = (2 - a_{1,2})y_{1,2}\partial_{x_1^*} + (2 - a_{3,2})y_{3,2}\partial_{x_3^*} + a_{1,2}x_1\partial_{y_{1,1}}$$
$$+a_{3,2}x_3\partial_{y_{3,1}},$$
$$\nabla_{\partial_{x_3}}\partial_{x_3} = (2 - a_{1,1})y_{1,1}\partial_{x_1^*} + (2 - a_{2,2})y_{2,2}\partial_{x_2^*} + a_{2,2}x_2\partial_{y_{2,1}}$$
$$+a_{1,1}x_1\partial_{y_{1,2}},$$
$$\nabla_{\partial_{x_1}}\partial_{x_2} = -a_{2,1}y_{2,1}\partial_{x_1^*} - a_{1,2}y_{1,2}\partial_{x_2^*} + \tfrac{y_{4,1}+y_{4,2}}{2}\partial_{x_3^*}$$
$$+(a_{1,2} - 1)x_2\partial_{y_{1,1}} + (a_{2,1} - 1)x_1\partial_{y_{2,2}},$$
$$\nabla_{\partial_{x_1}}\partial_{x_3} = -a_{3,1}y_{3,1}\partial_{x_1^*} + \tfrac{y_{4,1}-y_{4,2}}{2}\partial_{x_2^*} - a_{1,1}y_{1,1}\partial_{x_3^*}$$
$$+(a_{1,1} - 1)x_3\partial_{y_{1,2}} + (a_{3,1} - 1)x_1\partial_{y_{3,2}} + \tfrac{2x_2}{3}\partial_{y_{4,1}} + \tfrac{4x_2}{3}\partial_{y_{4,2}},$$
$$\nabla_{\partial_{x_2}}\partial_{x_3} = \tfrac{-y_{4,1}+y_{4,2}}{2}\partial_{x_1^*} - a_{3,2}y_{3,2}\partial_{x_2^*} - a_{2,2}y_{2,2}\partial_{x_3^*}$$
$$+(a_{2,2} - 1)x_3\partial_{y_{2,1}} + (a_{3,2} - 1)x_2\partial_{y_{3,1}} + \tfrac{4x_1}{3}\partial_{y_{4,1}} + \tfrac{2x_1}{3}\partial_{y_{4,2}}.$$

It is now easy to show that the non-zero components of ∇R are:

$$\nabla R(\partial_{x_1}, \partial_{x_2}, \partial_{x_2}, \partial_{x_1}; \partial_{x_3}) = -2(-2 + a_{1,1} + a_{2,2} + a_{3,1}a_{3,2})x_3,$$
$$\nabla R(\partial_{x_1}, \partial_{x_3}, \partial_{x_3}, \partial_{x_1}; \partial_{x_2}) = -\tfrac{2}{3}(-4 + 3a_{1,2} + 3a_{3,2} + 3a_{2,1}a_{2,2})x_2,$$
$$\nabla R(\partial_{x_2}, \partial_{x_3}, \partial_{x_3}, \partial_{x_2}; \partial_{x_1}) = -\tfrac{2}{3}(-4 + 3a_{2,1} + 3a_{3,1} + 3a_{1,1}a_{1,2})x_1,$$
$$\nabla R(\partial_{x_2}, \partial_{x_1}, \partial_{x_1}, \partial_{x_3}; \partial_{x_2}) = (2 - a_{1,1} - a_{1,2} + a_{2,1} - a_{2,2}$$
$$+a_{3,1} - a_{3,2} + a_{1,1}a_{1,2} - a_{2,1}a_{2,2} - a_{3,1}a_{3,2})x_3,$$

$$\nabla R(\partial_{x_2}, \partial_{x_1}, \partial_{x_1}, \partial_{x_3}; \partial_{x_3}) = (2 - a_{1,1} - a_{1,2} + a_{2,1} - a_{2,2}$$
$$+ a_{3,1} - a_{3,2} + a_{1,1}a_{1,2} - a_{2,1}a_{2,2} - a_{3,1}a_{3,2})x_2,$$

$$\nabla R(\partial_{x_1}, \partial_{x_2}, \partial_{x_2}, \partial_{x_3}; \partial_{x_1}) = (2 - a_{1,1} + a_{1,2} - a_{2,1} - a_{2,2}$$
$$- a_{3,1} + a_{3,2} - a_{1,1}a_{1,2} + a_{2,1}a_{2,2} - a_{3,1}a_{3,2})x_3,$$

$$\nabla R(\partial_{x_1}, \partial_{x_2}, \partial_{x_2}, \partial_{x_3}; \partial_{x_3}) = (2 - a_{1,1} + a_{1,2} - a_{2,1} - a_{2,2}$$
$$- a_{3,1} + a_{3,2} - a_{1,1}a_{1,2} + a_{2,1}a_{2,2} - a_{3,1}a_{3,2})x_1,$$

$$\nabla R(\partial_{x_1}, \partial_{x_3}, \partial_{x_3}, \partial_{x_2}; \partial_{x_1}) = (\tfrac{2}{3} + a_{1,1} - a_{1,2} - a_{2,1} + a_{2,2}$$
$$- a_{3,1} - a_{3,2} - a_{1,1}a_{1,2} - a_{2,1}a_{2,2} + a_{3,1}a_{3,2})x_2,$$

$$\nabla R(\partial_{x_1}, \partial_{x_3}, \partial_{x_3}, \partial_{x_2}; \partial_{x_2}) = (\tfrac{2}{3} + a_{1,1} - a_{1,2} - a_{2,1} + a_{2,2}$$
$$- a_{3,1} - a_{3,2} - a_{1,1}a_{1,2} - a_{2,1}a_{2,2} + a_{3,1}a_{3,2})x_1.$$

We set $\nabla R = 0$ to obtain the desired equations of Theorem 6.4.4; the first 3 equations generate the last 6. $\qquad\square$

6.5 Riemannian Skew Tsankov Models and Manifolds

Recall that a 0-model $\mathfrak{M}$ is said to be *skew Tsankov* if

$$\mathcal{A}_{xy}\mathcal{A}_{uv} = \mathcal{A}_{uv}\mathcal{A}_{xy} \ \forall \ x, y, u, v \in V.$$

A pseudo-Riemannian manifold $\mathcal{M}$ is said to be skew Tsankov if the 0-model $\mathfrak{M}(\mathcal{M}, P) := (T_P M, g_P, R_P)$ is skew Tsankov for every $P \in M$. We shall follow the discussion in Brozos-Vázquez and Gilkey (2006a) to give a complete classification of Riemannian skew Tsankov 0-models and to exhibit some irreducible Riemannian skew Tsankov manifolds of dimension 3 and of dimension 4.

Consider the following family of algebraic curvature tensors:

Definition 6.5.1 Let $\langle \cdot, \cdot \rangle$ be a positive definite inner product on a finite dimensional real vector space V. Let $\{\Psi_1, ..., \Psi_k\}$ be a collection of self-adjoint linear maps of V which satisfy

$$\Psi_i \Psi_j = \delta_{ij} \Psi_i \quad \text{and} \quad \text{Rank}\{\Psi_i\} = 2.$$

Let $\pi_i := \text{Range}\{\Psi_i\}$; Ψ_i is orthogonal projection on π and $\pi_i \perp \pi_j$ for $i \neq j$. Following the discussion in Section 1.3.2, we introduce the associated canonical algebraic curvature tensors and algebraic curvature operators by setting:

$$A_{\Psi_i}(x,y,z,w) := \langle \Psi_i x, w\rangle \langle \Psi_i y, z\rangle - \langle \Psi_i y, w\rangle \langle \Psi_i x, z\rangle,$$
$$\mathcal{A}_{\Psi_i}(x,y)z := \langle \Psi_i z, y\rangle \Psi_i x - \langle \Psi_i z, x\rangle \Psi_i y\,.$$

Let a_i be real constants and let

$$A := a_1 A_{\Psi_1} + \ldots + a_k A_{\Psi_k}\,.$$

Let $\pi_0 := (\pi_1 \oplus \ldots \oplus \pi_k)^\perp$ and let $\langle \cdot, \cdot \rangle_i := \langle \cdot, \cdot\rangle|_{\pi_i}$. Let

$$\mathfrak{M} := (V, \langle \cdot, \cdot\rangle, A), \quad \mathfrak{M}_0 := (\pi_0, \langle \cdot, \cdot\rangle_0, 0), \quad \mathfrak{M}_i := (\pi_i, \langle \cdot, \cdot\rangle_i, a_i A_{\Psi_i})\,.$$

We then have $\mathfrak{M} = \mathfrak{M}_0 \oplus \mathfrak{M}_1 \oplus \ldots \oplus \mathfrak{M}_k$.

The following result will be established in Section 6.5.1:

Theorem 6.5.1   *Let $\mathfrak{M}$ be a Riemannian 0-model. Then $\mathfrak{M}$ is skew Tsankov if and only if A has the form given in Definition 6.5.1. Such a 0-model $\mathfrak{M}$ is indecomposable if and only if $\dim(V) = 2$ and $k = 1$.*

This result gives a complete algebraic classification of Riemannian skew Tsankov 0-models. Despite the fact that the algebraic classification is complete in the Riemannian setting, the geometric classification is incomplete. Since the Cartesian product of skew Tsankov models or manifolds is again skew Tsankov, it is natural to look for indecomposable examples. In Section 6.5.2, we establish the following result which exhibits irreducible 3-dimensional skew Tsankov manifolds arising as a warped product of an interval with a Riemann surface:

Theorem 6.5.2   *Let $\mathcal{N} := (N, g_N)$ be a Riemann surface which does not have constant sectional curvature $+1$. Let $\mathcal{M} := (N \times (0, \infty), g_M)$ where $g_M := dx_3^2 + x_3^2 g_N$. Then $\mathcal{M}$ is an irreducible skew Tsankov manifold with scalar curvature $\tau_M = x_3^{-2}\{\tau_N - 2\}$.*

Theorem 6.5.2 immediately specializes to yield the following special case which is closely related to the original construction of Tsankov.

Corollary 6.5.1   *Let f be an isometric embedding of a Riemann surface N in the round sphere $S^3 \subset \mathbb{R}^4$. Let $M := N \times (0, \infty)$. Let $F(x,t) := tf(x)$ define an embedding of M in $\mathbb{R}^4$ and let g_M be the induced metric. Then $\mathcal{M}$ is skew Tsankov.*

In Section 6.5.3, we construct irreducible 4-dimensional skew Tsankov manifolds by taking a warped product of two flat 2-dimensional manifolds. Let (x_1, x_2, x_3, x_4) be the usual coordinates on $\mathbb{R}^4$. Let $\partial_i := \partial_{x_i}$ and let

$$\mathcal{O} := \{(x_1, x_2, x_3, x_4) \in \mathbb{R}^4 : x_3 > 0, x_4 > 0\}.$$

Theorem 6.5.3 *For $\beta > 0$, let $\mathcal{M}_\beta := (\mathcal{O}, g_\beta)$ where*

$$g_\beta(\partial_1, \partial_1) = x_3^2, \quad g_\beta(\partial_2, \partial_2) = (x_3 + \beta x_4)^2,$$
$$g_\beta(\partial_3, \partial_3) = 1, \quad g_\beta(\partial_4, \partial_4) = 1.$$

Then $\mathcal{M}_\beta$ is an indecomposable skew Tsankov manifold with scalar curvature $\tau_\beta = -2x_3^{-1}(x_3 + \beta x_4)^{-1}$. Furthermore, $\mathcal{M}_{\beta_1}$ is not isometric to $\mathcal{M}_{\beta_2}$ for $\beta_1 \neq \beta_2$.

We remark that the curves $t \to (x, t)$ in the manifolds of Theorem 6.5.2 and the curves $t \to (x_1, x_2, t, x_4)$ in the manifolds of Theorem 6.5.3 are unit speed geodesics along which the scalar curvature blows up as $t \to 0^+$. Thus these manifolds are geodesically incomplete and may not be embedded as an open subset of a geodesically complete manifold.

We now study manifolds where $\mathcal{R}$ has rank 2; this corresponds to taking $k = 1$ in Theorem 6.5.1. The geometric structures are quite rigid; Theorems 6.5.2 and 6.5.3 provide examples of these structures. In Section 6.5.4, we will prove:

Theorem 6.5.4 *Let $\mathcal{M}$ be a Riemannian skew Tsankov manifold. Assume $\mathrm{Rank}\{\mathcal{R}\} = 2$. Let $\mathcal{E} := \mathrm{Range}\{\mathcal{R}\}$ and let $\mathcal{F} := \mathcal{E}^\perp$. Then:*

(1) The distribution $\mathcal{F}$ is integrable.
(2) Let X be a leaf of the foliation defined by $\mathcal{F}$.

(a) X is totally geodesic and flat.
(b) The normal distribution $\mathcal{E}$ is parallel along X.

Note that in Theorem 6.5.2 one has $\mathcal{F} = \mathrm{Span}\{\partial_{x_3}\}$ and that in Theorem 6.5.3 one has $\mathcal{F} = \mathrm{Span}\{\partial_{x_3}, \partial_{x_4}\}$. These are flat. In these examples, $\mathcal{E}$ is integrable as well and $\mathcal{M}$ is a warped product of $\mathcal{E}$ over $\mathcal{F}$. We do not know if this is always the case.

6.5.1 *Riemannian skew Tsankov models*

This section is devoted to the proof of Theorem 6.5.1. Suppose that

$$A = a_1 A_{\Psi_1} + \ldots + a_k A_{\Psi_k}$$

is as in Definition 6.5.1. Let $\{e_i^1, e_i^2\}$ be an orthonormal basis for the 2-dimensional subspaces $\pi_i := \text{Range}\{\Psi_i\}$; one has $\pi_i \perp \pi_j$ for $i \neq j$. Define $\pi_0 := (\pi_1 \oplus ... \oplus \pi_k)^\perp$. This yields an orthogonal direct sum decomposition

$$V = \pi_0 \oplus ... \oplus \pi_k \,.$$

Decompose

$$x = \sum_i (x_i^1 e_i^1 + x_i^2 e_i^2) + x_0 \quad \text{and} \quad y = \sum_i (y_i^1 e_i^1 + y_i^2 e_i^2) + y_0$$

for $x_0, y_0 \in \pi_0$. Set

$$\varepsilon_i(x, y) := x_i^1 y_i^2 - x_i^2 y_i^1 = \langle x, e_i^1 \rangle \langle y, e_i^2 \rangle - \langle x, e_i^2 \rangle \langle y, e_i^1 \rangle \,.$$

One may then express:

$$\mathcal{A}_i(x, y)\xi = \begin{cases} -a_i \varepsilon_i(x, y) e_i^2 & \text{if } \xi = e_i^1, \\ a_i \varepsilon_i(x, y) e_i^1 & \text{if } \xi = e_i^2, \\ 0 & \text{if } \xi \in \pi_i^\perp \,. \end{cases}$$

Consequently

$$\mathcal{A}_{xy} \mathcal{A}_{\bar{x}\bar{y}} \xi = \begin{cases} -a_i^2 \varepsilon_i(x, y) \varepsilon_i(\bar{x}, \bar{y}) \xi & \text{if } \xi \in \pi_i, i > 0, \\ 0 & \text{if } \xi \in \pi_0 \,. \end{cases}$$

Since (x, y) and $(\bar{x}, \bar{y})$ play symmetric roles, $\mathfrak{M}$ is skew Tsankov. One may let $\tilde{A}_0 := 0$ and, similarly, one may let $\tilde{A}_i := A_i|_{\pi_i}$. Then:

$$V = \pi_0 \oplus \pi_1 \oplus ... \oplus \pi_k \quad \text{and} \quad \tilde{A} = \tilde{A}_0 \oplus \tilde{A}_1 \oplus ... \oplus \tilde{A}_k \,.$$

Thus $\mathfrak{M}$ is indecomposable if and only if $\dim(V) = 2$ and $k = 1$.

Conversely, suppose that $\mathfrak{M}$ is skew Tsankov. We simultaneously skew-diagonalize the collection $\{\mathcal{A}_{xy}\}_{x,y \in V}$ of commuting skew-adjoint linear operators. Let $\{e_i^1, e_i^2\}$ be an orthonormal basis for 2-dimensional subspaces π_i where $\pi_i \perp \pi_j$ for $i \neq j$. Let $\pi_0 = (\pi_1 \oplus ... \oplus \pi_k)^\perp$. For $1 \leq i \leq k$, let $\varepsilon_i(x, y)$ be the associated skew-eigenfunctions so that

$$\mathcal{A}_{xy}\xi = \begin{cases} -\varepsilon_i(x, y) e_i^2 & \text{if } \xi = e_i^1, \\ \varepsilon_i(x, y) e_i^1 & \text{if } \xi = e_i^2, \\ 0 & \text{if } \xi \in \pi_0 \,. \end{cases}$$

Let $\{f_1, ..., f_l\}$ be an orthonormal basis for π_0. Then

$$\mathcal{B} := \{e_1^1, e_1^2, ..., e_k^1, e_k^2, f_1, ..., f_l\}$$

is an orthonormal basis for V. The only non-zero entries in the curvature tensor relative to this base are

$$A(\cdot,\cdot,e_i^1,e_i^2) = -A(\cdot,\cdot,e_i^2,e_i^1)\,.$$

Interchanging the first 2 entries with the last 2 entries shows the only non-zero curvatures are

$$A(e_i^1,e_i^2,e_j^1,e_j^2)\,.$$

On the other hand, if $i \neq j$, we can use the first Bianchi identity to express

$$A(e_i^1,e_i^2,e_j^1,e_j^2) = -A(e_i^1,e_j^1,e_j^2,e_i^2) - A(e_i^1,e_j^2,e_i^2,e_j^1) = 0$$

and thus the only non-zero curvatures are $a_i := A(e_i^1,e_i^2,e_i^2,e_i^1)$. Setting Ψ_i to be orthogonal projection on $\mathrm{Span}\{e_i^1,e_i^2\}$ then permits us to express

$$A = \sum_i a_i A_{\Psi_i}$$

and complete the proof of Theorem 6.5.1. $\qquad\qquad\square$

6.5.2 *3-dimensional skew Tsankov manifolds*

Any 2-dimensional Riemannian manifold is necessarily skew Tsankov. What is perhaps somewhat surprising is that there are irreducible examples of higher dimension. In this section, we discuss the warped product construction of Theorem 6.5.2. Choose isothermal coordinates to express

$$ds_N^2 = e^{2\alpha}(dx_1^2 + dx_2^2),$$

at least locally. Let $\alpha_i := \partial_i(\alpha)$ and $\alpha_{ij} := \partial_i\partial_j(\alpha)$. We take

$$g(\partial_1,\partial_1) = g(\partial_2,\partial_2) = x_3^2 e^{2\alpha}, \quad g(\partial_3,\partial_3) = 1\,.$$

The non-zero Christoffel symbols of the first kind must have at least one repeated index different from 3:

$$
\begin{aligned}
&\Gamma_{111} = \alpha_1 x_3^2 e^{2\alpha}, && \Gamma_{112} = -\alpha_2 x_3^2 e^{2\alpha}, && \Gamma_{113} = -x_3 e^{2\alpha}, \\
&\Gamma_{121} = \Gamma_{211} = \alpha_2 x_3^2 e^{2\alpha}, && \Gamma_{131} = \Gamma_{311} = x_3 e^{2\alpha}, && \\
&\Gamma_{221} = -\alpha_1 x_3^2 e^{2\alpha}, && \Gamma_{222} = \alpha_2 x_3^2 e^{2\alpha}, && \Gamma_{223} = -x_3 e^{2\alpha}, \\
&\Gamma_{122} = \Gamma_{212} = \alpha_1 x_3^2 e^{2\alpha}, && \Gamma_{322} = \Gamma_{232} = x_3 e^{2\alpha}\,.
\end{aligned}
$$

Since the metric is diagonal, we can raise indices to see:

$$\nabla_{\partial_1}\partial_1 = \alpha_1\partial_1 - \alpha_2\partial_2 - x_3 e^{2\alpha}\partial_3,$$
$$\nabla_{\partial_1}\partial_2 = \nabla_{\partial_2}\partial_1 = \alpha_2\partial_1 + \alpha_1\partial_2,$$
$$\nabla_{\partial_1}\partial_3 = \nabla_{\partial_3}\partial_1 = x_3^{-1}\partial_1,$$
$$\nabla_{\partial_2}\partial_2 = -\alpha_1\partial_1 + \alpha_2\partial_2 - x_3 e^{2\alpha}\partial_3,$$
$$\nabla_{\partial_2}\partial_3 = \nabla_{\partial_3}\partial_2 = x_3^{-1}\partial_2\,.$$

We can now compute the curvatures. We begin by computing:

$$
\begin{aligned}
\mathcal{R}(\partial_1,\partial_2)\partial_1 &= (\nabla_{\partial_1}\nabla_{\partial_2} - \nabla_{\partial_2}\nabla_{\partial_1})\partial_1 \\
&= \nabla_{\partial_1}(\alpha_2\partial_1 + \alpha_1\partial_2) - \nabla_{\partial_2}(\alpha_1\partial_1 - \alpha_2\partial_2 - x_3 e^{2\alpha}\partial_3) \\
&= \alpha_{12}\partial_1 + \alpha_{11}\partial_2 + \alpha_2(\alpha_1\partial_1 - \alpha_2\partial_2 - x_3 e^{2\alpha}\partial_3) \\
&\quad + \alpha_1(\alpha_2\partial_1 + \alpha_1\partial_2) - \alpha_{12}\partial_1 + \alpha_{22}\partial_2 + 2x_3\alpha_2 e^{2\alpha}\partial_3 \\
&\quad - \alpha_1(\alpha_2\partial_1 + \alpha_1\partial_2) + \alpha_2(-\alpha_1\partial_1 + \alpha_2\partial_2 - x_3 e^{2\alpha}\partial_3) \\
&\quad + x_3 e^{2\alpha} x_3^{-1}\partial_2 \\
&= (\alpha_{11} + \alpha_{22} + e^{2\alpha})\partial_2\,.
\end{aligned}
$$

Similarly $\mathcal{R}(\partial_1,\partial_2)\partial_2 = -(\alpha_{11} + \alpha_{22} + e^{2\alpha})\partial_1$. Consequently:

$$R(\partial_1,\partial_2,\partial_2,\partial_1) = -x_3^2 e^{2\alpha}(\alpha_{11} + \alpha_{22} + e^{2\alpha}),$$
$$R(\partial_1,\partial_2,\partial_2,\partial_3) = 0,\quad R(\partial_1,\partial_2,\partial_1,\partial_3) = 0\,.$$

We also compute:

$$
\begin{aligned}
\mathcal{R}(\partial_1,\partial_3)\partial_3 &= (\nabla_{\partial_1}\nabla_{\partial_3} - \nabla_{\partial_3}\nabla_{\partial_1})\partial_3 \\
&= -\nabla_{\partial_3}(x_3^{-1}\partial_1) = -(-x_3^{-2}\partial_1 + x_3^{-1}x_3^{-1}\partial_1) = 0\,.
\end{aligned}
$$

Similar computations yield $\mathcal{R}(\partial_2,\partial_3)\partial_3 = 0$. This shows

$$R(\partial_1,\partial_3,\partial_3,\partial_1) = R(\partial_1,\partial_3,\partial_3,\partial_2) = R(\partial_2,\partial_3,\partial_3,\partial_2) = 0\,.$$

Thus the only curvature is $R(\partial_1,\partial_2,\partial_2,\partial_1)$ so, by Theorem 6.5.1, $\mathcal{M}$ is skew Tsankov. One also has:

$$
\begin{aligned}
\tau_{\mathcal{M}} &= 2g(\partial_1,\partial_1)^{-1} g(\partial_2,\partial_2)^{-1} R(\partial_1,\partial_2,\partial_2,\partial_1) \\
&= x_3^{-2}\{-2e^{-2\alpha}(\alpha_{11} + \alpha_{22}) - 2\}\,.
\end{aligned}
$$

An analogous computation on $\mathcal{N}$ yields:

$$\Gamma_{111} = \alpha_1 x_3^2 e^{2\alpha},\qquad \Gamma_{112} = -\alpha_2 x_3^2 e^{2\alpha},\qquad \Gamma_{121} = \Gamma_{211} = \alpha_2 x_3^2 e^{2\alpha},$$
$$\Gamma_{221} = -\alpha_1 x_3^2 e^{2\alpha},\qquad \Gamma_{222} = \alpha_2 x_3^2 e^{2\alpha},\qquad \Gamma_{122} = \Gamma_{212} = \alpha_1 x_3^2 e^{2\alpha}.$$

Thus the Christoffel symbols of the second kind are given by:

$$\nabla_{\partial_1}\partial_1 = \alpha_1\partial_1 - \alpha_2\partial_2,$$
$$\nabla_{\partial_1}\partial_2 = \nabla_{\partial_2}\partial_1 = \alpha_2\partial_1 + \alpha_1\partial_2,$$
$$\nabla_{\partial_2}\partial_2 = -\alpha_1\partial_1 + \alpha_2\partial_2,$$

and the curvature is given by

$$
\begin{aligned}
\mathcal{R}(\partial_1,\partial_2)\partial_1 &= (\nabla_{\partial_1}\nabla_{\partial_2} - \nabla_{\partial_2}\nabla_{\partial_1})\partial_1 \\
&= \nabla_{\partial_1}(\alpha_2\partial_1 + \alpha_1\partial_2) - \nabla_{\partial_2}(\alpha_1\partial_1 - \alpha_2\partial_2) \\
&= \alpha_{12}\partial_1 + \alpha_{11}\partial_2 + \alpha_2(\alpha_1\partial_1 - \alpha_2\partial_2) + \alpha_1(\alpha_2\partial_1 + \alpha_1\partial_2) \\
&\quad - \alpha_{12}\partial_1 + \alpha_{22}\partial_2 - \alpha_1(\alpha_2\partial_1 + \alpha_1\partial_2) + \alpha_2(-\alpha_1\partial_1 + \alpha_2\partial_2) \\
&= (\alpha_{11} + \alpha_{22})\partial_2\,.
\end{aligned}
$$

Theorem 6.5.2 now follows; one can see that $\mathcal{M}$ is indecomposable since

$$\text{Range}\{\mathcal{R}\} = \text{Span}\{\partial_1,\partial_2\}$$

and since $\tau_{\mathcal{M}}$ exhibits non-trivial dependence on x_3. This establishes Theorem 6.5.2. $\qquad\square$

6.5.3 *Irreducible 4-dimensional skew Tsankov manifolds*

The metric takes the form:

$$g(\partial_1,\partial_1) = x_3^2, \quad g(\partial_2,\partial_2) = (x_3 + \beta x_4)^2, \quad g(\partial_3,\partial_3) = g(\partial_4,\partial_4) = 1\,.$$

The non-zero Christoffel symbols are therefore:

$$
\begin{array}{ll}
\Gamma_{113} = -x_3, & \Gamma_{131} = \Gamma_{311} = x_3, \\
\Gamma_{223} = -(x_3 + \beta x_4), & \Gamma_{232} = \Gamma_{322} = x_3 + \beta x_4, \\
\Gamma_{224} = -\beta(x_3 + \beta x_4), & \Gamma_{242} = \Gamma_{422} = \beta(x_3 + \beta x_4)\,.
\end{array}
$$

Since the metric is diagonal, we may raise indices to compute:

$$
\begin{aligned}
\nabla_{\partial_1}\partial_1 &= -x_3\partial_3, \\
\nabla_{\partial_1}\partial_3 &= \nabla_{\partial_3}x_1 = x_3^{-1}\partial_1, \\
\nabla_{\partial_2}\partial_2 &= -(x_3 + \beta x_4)\partial_3 - \beta(x_3 + \beta x_4)\partial_4, \\
\nabla_{\partial_2}\partial_3 &= \nabla_{\partial_3}\partial_2 = (x_3 + \beta x_4)^{-1}\partial_2, \\
\nabla_{\partial_2}\partial_4 &= \nabla_{\partial_4}\partial_2 = \beta(x_3 + \beta x_4)^{-1}\partial_2\,.
\end{aligned}
$$

The curvature operator can now be studied:

$$\mathcal{R}(\partial_1,\partial_2)\partial_1 = -\nabla_{\partial_2}\nabla_{\partial_1}\partial_1 = \nabla_{\partial_2}\{x_3\partial_3\} = x_3(x_3+\beta x_4)^{-1}\partial_2,$$

$$\mathcal{R}(\partial_1,\partial_2)\partial_2 = \nabla_{\partial_1}\nabla_{\partial_2}\partial_2 = -\nabla_{\partial_1}\{(x_3+\beta x_4)\partial_3 + \beta_4(x_3+\beta x_4)\partial_4\}$$
$$= x_3^{-1}(x_3+\beta x_4)\partial_1,$$

$$\mathcal{R}(\partial_1,\partial_2)\partial_3 = \nabla_{\partial_1}\{(x_3+\beta x_4)^{-1}\partial_2\} - \nabla_{\partial_2}(x_3^{-1}\partial_1) = 0,$$

$$\mathcal{R}(\partial_1,\partial_2)\partial_4 = \nabla_{\partial_1}\{\beta(x_3+\beta x_4)^{-1}\partial_2\} = 0,$$

$$\mathcal{R}(\partial_1,\partial_3)\partial_1 = \nabla_{\partial_1}\{x_3^{-1}\partial_1\} - \nabla_{\partial_3}\{-x_3\partial_3\} = -\partial_3 + \partial_3 = 0,$$

$$\mathcal{R}(\partial_1,\partial_3)\partial_2 = \nabla_{\partial_1}\{(x_3+\beta x_4)^{-1}\partial_2\} = 0,$$

$$\mathcal{R}(\partial_1,\partial_3)\partial_3 = -\nabla_{\partial_3}\{x_3^{-1}\partial_1\} = (-x_3^{-2}\partial_1 + x_3^{-1}x_3^{-1}\partial_1) = 0,$$

$$\mathcal{R}(\partial_1,\partial_3)\partial_4 = 0,$$

$$\mathcal{R}(\partial_1,\partial_4)\partial_1 = -\nabla_{\partial_4}\{-x_3\partial_3\} = 0,$$

$$\mathcal{R}(\partial_1,\partial_4)\partial_2 = \nabla_{\partial_1}\{\beta(x_3+\beta x_4)^{-1}\partial_2\} = 0,$$

$$\mathcal{R}(\partial_1,\partial_4)\partial_3 = -\nabla_{\partial_4}\{x_3^{-1}\partial_1\} = 0,$$

$$\mathcal{R}(\partial_1,\partial_4)\partial_4 = 0.$$

The curvature symmetries can now be used to reduce the number of remaining computations that must be performed by eliminating the index 1. We have

$$\mathcal{R}(\partial_2,\partial_3)\partial_2 = \nabla_{\partial_2}\{(x_3+\beta x_4)^{-1}\partial_2\}$$
$$-\nabla_{\partial_3}\{-(x_3+\beta x_4)\partial_3 - \beta(x_3+\beta x_4)\partial_4\}$$
$$= (x_3+\beta x_4)^{-1}\{-(x_3+\beta x_4)\partial_3 - \beta(x_3+\beta x_4)\partial_4\}+\partial_3+\beta\partial_4 = 0,$$

$$\mathcal{R}(\partial_2,\partial_3)\partial_3 = -\nabla_{\partial_3}\{(x_3+\beta x_4)^{-1}\partial_2\}$$
$$= -\{-(x_3+\beta x_4)^{-2} + (x_3+\beta x_4)^{-1}(x_3+\beta x_4)^{-1}\}\partial_2 = 0,$$

$$\mathcal{R}(\partial_2,\partial_3)\partial_4 = -\nabla_{\partial_3}\{\beta(x_3+\beta x_4)^{-1}\partial_2\}$$
$$= -\{-\beta(x_3+\beta x_4)^{-2} + \beta(x_3+\beta x_4)^{-1}(x_3+\beta x_4)^{-1}\}\partial_2 = 0,$$

$$\mathcal{R}(\partial_2,\partial_4)\partial_2 = \nabla_2\{\beta(x_3+\beta x_4)^{-1}\partial_2\}$$
$$+\nabla_{\partial_4}\{(x_3+\beta x_4)\partial_3 + \beta(x_3+\beta x_4)\partial_4\}$$
$$= -\beta\partial_3 - \beta^2\partial_4 + \beta\partial_3 + \beta^2\partial_4 = 0,$$

$$\mathcal{R}(\partial_2,\partial_4)\partial_3 = -\nabla_{\partial_4}\{(x_3+\beta x_4)^{-1}\partial_2$$
$$= -\{-\beta(x_3+\beta x_4)^{-2} + (x_3+\beta x_4)^{-1}\beta(x_3+\beta x_4)^{-1}\}\partial_2 = 0,$$

$$\mathcal{R}(\partial_2,\partial_4)\partial_4 = -\nabla_{\partial_4}\{\beta(x_3+\beta x_4)^{-1}\partial_2\}$$
$$= -\{-\beta^2(x_3+\beta x_4)^{-2} + \beta(x_3+\beta x_4)^{-1}\beta(x_3+\beta x_4)^{-1}\}\partial_2 = 0.$$

Thus we may eliminate the index 2. The only remaining curvatures are

$$\mathcal{R}(\partial_3,\partial_4)\partial_3 = \mathcal{R}(\partial_3,\partial_4)\partial_4 = 0.$$

This shows that the only non-zero curvature is

$$\mathcal{R}(\partial_1, \partial_2, \partial_2, \partial_1) = -x_3(x_3 + \beta x_4)$$

and hence $\mathcal{M}$ is skew Tsankov by Theorem 6.5.1.

The calculations performed above show that the scalar curvature is:

$$\tau = -2x_3^{-1}(x_3 + \beta x_4)^{-1}.$$

Let

$$\mathcal{E} := \mathrm{Range}\{\mathcal{R}\} = \mathrm{Span}\{\partial_1, \partial_2\},$$
$$\mathcal{F} := \mathcal{E}^{\perp} = \mathrm{Span}\{\partial_3, \partial_4\}.$$

These spaces are invariantly defined. Let $H := \nabla^2\{-\ln\tau\}|_{\mathcal{F}}$. Since $\nabla_{\partial_a}\partial_b = 0$ for $a = 3, 4$ and $b = 3, 4$,

$$H(\partial_3, \partial_3) = \partial_3^2\{-\ln\tau\} = x_3^{-2} + (x_3 + \beta x_4)^{-2},$$
$$H(\partial_3, \partial_4) = H(\partial_4, \partial_3) = \partial_3\partial_4\{-\ln\tau\} = \beta(x + \beta x_4)^{-2},$$
$$H(\partial_4, \partial_4) = \partial_4^2\{-\ln\tau\} = \beta^2(x_3 + \beta x_4)^{-2},$$
$$\det(H) = \beta^2 x_3^{-2}(x_3 + \beta x_4)^{-2} = \tfrac{1}{4}\beta\tau^2.$$

This shows that β is an isometry invariant of $\mathcal{M}_\beta$. Furthermore since $H|_{\mathcal{F}}$ has rank 2, $\mathcal{M}$ is irreducible. The remaining assertions of Theorem 6.5.3 now follow. $\qquad\square$

6.5.4 *Flats in a Riemannian skew Tsankov manifold*

Let $\mathcal{M}$ be a Riemannian skew Tsankov manifold. We apply Theorem 6.5.1 to classify the 1-model. We suppose $k = 1$ and hence $R_P = a(P)R_{\Psi(P)}$ where Ψ is a self-adjoint map of rank 2 with $\Psi(P)^2 = \Psi(P)$. After a minor bit of technical fuss, one can show that $\Psi(P)$ and $a(P)$ can be chosen to vary smoothly with P, at least locally. We begin the proof of Theorem 6.5.4 with

Lemma 6.5.1 *Let $\mathcal{M}$ be skew Tsankov with curvature tensor $R = aR_\Psi$. Let $\{e_i\}$ be a local orthonormal frame field where $\{e_1, e_2\}$ is a frame for $\mathrm{Range}\{\Psi\}$. Let $\Psi_i := \nabla_{e_i}\Psi$ and let $a_i := e_i(a_i)$. Then*

(1) If $\mu \geq 3$, $\Psi_i e_\mu \in \mathrm{Span}\{e_1, e_2\}$. If $\mu \leq 2$, $\Psi_i e_\mu \in \mathrm{Span}\{e_3, ..., e_m\}$.
(2) If $\mu, \nu \geq 3$, then $\Psi_\mu e_\nu = \Psi_\nu e_\mu$.
(3) If $\nu \leq 2$ and $\mu \geq 3$, then $\Psi_\mu e_\nu = 0$.
(4) If $\mu \geq 3$, then $a_\mu = a\langle\Psi_1 e_\mu, e_1\rangle + a\langle\Psi_2 e_\mu, e_2\rangle$.

(5) If $\mu \geq 3$, then $\Psi_\mu = 0$.

Proof. Since $\Psi^2 = \Psi$, we have $\Psi_i \Psi + \Psi \Psi_i = \Psi_i$. Assertion (1) follows from:

$$\langle \Psi_i e_j, e_k \rangle = \langle \Psi \Psi_i e_j, e_k \rangle + \langle \Psi_i \Psi e_j, e_k \rangle = 0 \quad \text{if} \quad j, k \geq 3,$$
$$\langle \Psi_i e_j, e_k \rangle = \langle \Psi \Psi_i e_j, e_k \rangle + \langle \Psi_i \Psi e_j, e_k \rangle = 2\langle \Psi_i e_j, e_k \rangle \quad \text{if} \quad j, k \leq 2.$$

We compute:

$$\nabla_{e_i} R(e_j, e_k) e_\ell = a_i \langle \Psi e_k, e_\ell \rangle \Psi e_j + a \langle \Psi_i e_k, e_\ell \rangle \Psi e_j + a \langle \Psi e_k, e_\ell \rangle \Psi_i e_j$$
$$- a_i \langle \Psi e_j, e_\ell \rangle \Psi e_k - a \langle \Psi_i e_j, e_\ell \rangle \Psi e_k - a \langle \Psi e_j, e_\ell \rangle \Psi_i e_k.$$

The first Bianchi identity yields no information giving only a trivial identity where all the terms cancel. However, the second Bianchi identity yields non-trivial information:

$$0 = \nabla_{e_i} R(e_j, e_k) e_\ell + \nabla_{e_j} R(e_k, e_i) e_\ell + \nabla_{e_k} R(e_i, e_j) e_\ell$$
$$= a_i \langle \Psi e_k, e_\ell \rangle \Psi e_j + a \langle \Psi_i e_k, e_\ell \rangle \Psi e_j + a \langle \Psi e_k, e_\ell \rangle \Psi_i e_j$$
$$- a_i \langle \Psi e_j, e_\ell \rangle \Psi e_k - a \langle \Psi_i e_j, e_\ell \rangle \Psi e_k - a \langle \Psi e_j, e_\ell \rangle \Psi_i e_k$$
$$+ a_j \langle \Psi e_i, e_\ell \rangle \Psi e_k + a \langle \Psi_j e_i, e_\ell \rangle \Psi e_k + a \langle \Psi e_i, e_\ell \rangle \Psi_j e_k$$
$$- a_j \langle \Psi e_k, e_\ell \rangle \Psi e_i - a \langle \Psi_j e_k, e_\ell \rangle \Psi e_i - a \langle \Psi e_k, e_\ell \rangle \Psi_j e_i$$
$$+ a_k \langle \Psi e_j, e_\ell \rangle \Psi e_i + a \langle \Psi_k e_j, e_\ell \rangle \Psi e_i + a \langle \Psi e_j, e_\ell \rangle \Psi_k e_i$$
$$- a_k \langle \Psi e_i, e_\ell \rangle \Psi e_j - a \langle \Psi_k e_i, e_\ell \rangle \Psi e_j - a \langle \Psi e_i, e_\ell \rangle \Psi_k e_j.$$

We set $i = \ell = 1$ and assume $2 < j < k$.

$$0 = a_1 \langle \Psi e_k, e_1 \rangle \Psi e_j + a \langle \Psi_1 e_k, e_1 \rangle \Psi e_j + a \langle \Psi e_k, e_1 \rangle \Psi_1 e_j$$
$$- a_1 \langle \Psi e_j, e_1 \rangle \Psi e_k - a \langle \Psi_1 e_j, e_1 \rangle \Psi e_k - a \langle \Psi e_j, e_1 \rangle \Psi_1 e_k$$
$$+ a_j \langle \Psi e_1, e_1 \rangle \Psi e_k + a \langle \Psi_j e_1, e_1 \rangle \Psi e_k + a \langle \Psi e_1, e_1 \rangle \Psi_j e_k$$
$$- a_j \langle \Psi e_k, e_1 \rangle \Psi e_1 - a \langle \Psi_j e_k, e_1 \rangle \Psi e_1 - a \langle \Psi e_k, e_1 \rangle \Psi_j e_1$$
$$+ a_k \langle \Psi e_j, e_1 \rangle \Psi e_1 + a \langle \Psi_k e_j, e_1 \rangle \Psi e_1 + a \langle \Psi e_j, e_1 \rangle \Psi_k e_1$$
$$- a_k \langle \Psi e_1, e_1 \rangle \Psi e_j - a \langle \Psi_k e_1, e_1 \rangle \Psi e_j - a \langle \Psi e_1, e_1 \rangle \Psi_k e_j$$
$$= a \Psi_j e_k - a \langle \Psi_j e_k, e_1 \rangle e_1 + a \langle \Psi_k e_j, e_1 \rangle e_1 - a \Psi_k e_j$$
$$= \langle a \Psi_j e_k, e_1 \rangle e_1 + \langle a \Psi_j e_k, e_2 \rangle e_2 - a \langle \Psi_j e_k, e_1 \rangle e_1$$
$$+ a \langle \Psi_k e_j, e_1 \rangle e_1 - a \langle \Psi_k e_j, e_1 \rangle e_1 - a \langle \Psi_k e_j, e_2 \rangle e_2$$
$$= a \langle \Psi_k e_j - \Psi_j e_k, e_2 \rangle e_2.$$

A similar argument shows $\langle \Psi_k e_j - \Psi_j e_k, e_1 \rangle e_1 = 0$. Assertion (2) now follows from Assertion (1).

We set $i = \ell = 1$ and assume $2 = j < k$.

$$0 = a_1 \langle \Psi e_k, e_1 \rangle \Psi e_2 + a \langle \Psi_1 e_k, e_1 \rangle \Psi e_2 + a \langle \Psi e_k, e_1 \rangle \Psi_1 e_2$$
$$- a_1 \langle \Psi e_2, e_1 \rangle \Psi e_k - a \langle \Psi_1 e_2, e_1 \rangle \Psi e_k - a \langle \Psi e_2, e_1 \rangle \Psi_1 e_k$$

$$+a_2\langle\Psi e_1,e_1\rangle\Psi e_k + a\langle\Psi_2 e_1,e_1\rangle\Psi e_k + a\langle\Psi e_1,e_1\rangle\Psi_2 e_k$$
$$-a_2\langle\Psi e_k,e_1\rangle\Psi e_1 - a\langle\Psi_2 e_k,e_1\rangle\Psi e_1 - a\langle\Psi e_k,e_1\rangle\Psi_2 e_1$$
$$+a_k\langle\Psi e_2,e_1\rangle\Psi e_1 + a\langle\Psi_k e_2,e_1\rangle\Psi e_1 + a\langle\Psi e_2,e_1\rangle\Psi_k e_1$$
$$-a_k\langle\Psi e_1,e_1\rangle\Psi e_2 - a\langle\Psi_k e_1,e_1\rangle\Psi e_2 - a\langle\Psi e_1,e_1\rangle\Psi_k e_2$$
$$= a\langle\Psi_1 e_k,e_1)e_2 + a\Psi_2 e_k - a\langle\Psi_2 e_k,e_1)e_1 - a_k e_2 - a\Psi_k e_2$$
$$= a\langle\Psi_1 e_k,e_1)e_2 + a\langle\Psi_2 e_k,e_2)e_2 - a_k e_2 - a\Psi_k e_2.$$

Since $\Psi_k e_2 \in \mathrm{Span}\{e_3,...,e_m\}$, $\Psi_k e_2 = 0$; Assertions (3) and (4) now follow.

If we were to set $i=1, \ell=2$ and assume $3 \le j,k$, no new information is obtained. Similarly, if we were to set $\ell=1$ and let $3 \le i < j < k$, no new information would be obtained. We set $i=1$, $j=2$, $2 < k$, $2 < \ell$, to see:

$$0 = a_1\langle\Psi e_k,e_\ell\rangle\Psi e_2 + a\langle\Psi_1 e_k,e_\ell\rangle\Psi e_2 + a\langle\Psi e_k,e_\ell\rangle\Psi_1 e_2$$
$$-a_1\langle\Psi e_2,e_\ell\rangle\Psi e_k - a\langle\Psi_1 e_2,e_\ell\rangle\Psi e_k - a\langle\Psi e_2,e_\ell\rangle\Psi_1 e_k$$
$$+a_2\langle\Psi e_1,e_\ell\rangle\Psi e_k + a\langle\Psi_2 e_1,e_\ell\rangle\Psi e_k + a\langle\Psi e_1,e_\ell\rangle\Psi_2 e_k$$
$$-a_2\langle\Psi e_k,e_\ell\rangle\Psi e_1 - a\langle\Psi_2 e_k,e_\ell\rangle\Psi e_1 - a\langle\Psi e_k,e_\ell\rangle\Psi_2 e_1$$
$$+a_k\langle\Psi e_2,e_\ell\rangle\Psi e_1 + a\langle\Psi_k e_2,e_\ell\rangle\Psi e_1 + a\langle\Psi e_2,e_\ell\rangle\Psi_k e_1$$
$$-a_k\langle\Psi e_1,e_\ell\rangle\Psi e_2 - a\langle\Psi_k e_1,e_\ell\rangle\Psi e_2 - a\langle\Psi e_1,e_\ell\rangle\Psi_k e_2$$
$$= a\langle\Psi_k e_2,e_\ell)e_1 - a\langle\Psi_k e_1,e_\ell)e_2.$$

It now follows that $\langle\Psi_k e_2,e_\ell\rangle = 0$. Since $\Psi_k e_2 \in \mathrm{Span}\{e_3,...,e_m\}$ this means that $\Psi_k e_2 = 0$. Similarly $\Psi_k e_1 = 0$. Thus $\Psi_k = 0$ on $\mathrm{Span}\{e_1,e_2\}$. On the other hand $\langle\Psi_k e_\ell,e_1\rangle = \langle e_\ell,\Psi_k e_1\rangle = 0$ and thus $\Psi_k = 0$ on $\mathrm{Span}\{e_3,...,e_m\}$ as well. Assertion (5) now follows. Were we to set $i=1$, $3 \le j < k$, $3 \le \ell$, no additional information would be obtained. $\qquad\square$

We can now give the proof of Theorem 6.5.4. Let $\mathcal{E} := \mathrm{Span}\{e_1,e_2\}$ and let $\mathcal{F} := \mathrm{Span}\{e_3,...,e_m\}$. Then $\Psi e = e$ if $e \in \mathcal{E}$ while $\Psi f = 0$ if $f \in \mathcal{F}$. Let $j,k \ge 3$. We compute:

$$g([e_j,e_k],e_1) = g(\nabla_{e_j}e_k - \nabla_{e_k}e_j,e_1)$$
$$= g(\nabla_{e_j}e_k - \nabla_{e_k}e_j,\Psi e_1)$$
$$= g(\Psi\nabla_{e_j}e_k - \Psi\nabla_{e_k}e_j,e_1)$$
$$= g(\nabla_{e_j}\Psi e_k - \nabla_{e_k}\Psi e_j - \Psi_j e_k + \Psi_k e_j,e_1)$$
$$= 0.$$

Thus $[e_j,e_k]$ belongs to $\mathcal{F}$ as well. Assertion (1) of Theorem 6.5.4 now follows.

Let $\{f_1,f_2\}$ be vector fields tangent to a leaf X. We can extend $\{f_1,f_2\}$ to be vector fields on M which remain tangent to the leaves. Thus $\Psi_{;f_i} = 0$.

If e is a vector field which is perpendicular to the leaf, then

$$g(\nabla_{f_1} f_2, e) = g(\nabla_{f_1} f_2, \Psi e) = g(\Psi \nabla_{f_1} f_2, e)$$
$$= g(\nabla_{f_1} \Psi f_2 - \Psi_{;f_1} f_2, e) = 0.$$

Thus the second fundamental form of X in M vanishes; equivalently, the Levi-Civita connection of M restricts to the Levi-Civita connection on X. The restriction of the curvature of M to X is the curvature of X; Assertion (2a) now follows. Assertion (2b) follows similarly since

$$g(\nabla_{f_1} e, f_2) = g(e, \nabla_{f_1} f_2) = 0.$$

This completes the proof of Theorem 6.5.4.

6.6   Jacobi Videv Models and Manifolds

Let $\langle \cdot, \cdot \rangle$ be an inner product of signature (p, q) on V. Recall that $\mathrm{Gr}_{r,s}(V, \langle \cdot, \cdot \rangle)$ is the Grassmannian of all non-degenerate linear subspaces of V which have signature (r, s); the pair (r, s) is said to be *admissible* if and only if $\mathrm{Gr}_{r,s}(V, \langle \cdot, \cdot \rangle)$ is non-empty and does not consist of a single point or, equivalently, if one has the inequalities:

$$0 \leq r \leq p, \quad 0 \leq s \leq q, \quad 1 \leq r + s \leq m - 1.$$

Let $\mathcal{J}(\pi)$ be the *higher order Jacobi operator* associated to a non-degenerate plane $\pi \in \mathrm{Gr}_{r,s}(V, \langle \cdot, \cdot \rangle)$. We shall prove the following result in Section 6.6.1.

Theorem 6.6.1   *Let $\mathfrak{M} = (V, \langle \cdot, \cdot \rangle, A)$ be a 0-model. The following assertions are equivalent; if any is satisfied, then we shall say that $\mathfrak{M}$ is a* Jacobi Videv *0-model.*

(1) There exists (r_0, s_0) admissible so that $\mathcal{J}(\pi)\mathcal{J}(\pi^\perp) = \mathcal{J}(\pi^\perp)\mathcal{J}(\pi)$ for all π in $\mathrm{Gr}_{r_0,s_0}(V, \langle \cdot, \cdot \rangle)$.
(2) There exists (r_0, s_0) admissible so that $\mathcal{J}(\pi)\rho = \rho\mathcal{J}(\pi)$ for all π in $\mathrm{Gr}_{r_0,s_0}(V, \langle \cdot, \cdot \rangle)$.
(3) $\mathcal{J}(\pi)\mathcal{J}(\pi^\perp) = \mathcal{J}(\pi^\perp)\mathcal{J}(\pi)$ for every non-degenerate subspace π.
(4) $\mathcal{J}(\pi)\rho = \rho\mathcal{J}(\pi)$ for every non-degenerate subspace π.

It follows from Theorem 6.6.1 that the direct sum of Jacobi Videv 0-models is again such a model. By Theorem 6.6.1, any Einstein 0-model is Jacobi Videv. More generally, the direct sum of Jacobi Videv models is again Jacobi Videv. One says that a 0-model is *pseudo-Einstein* either if the Ricci operator ρ has only one real eigenvalue λ or if the Ricci operator ρ has two complex eigenvalues λ_1, λ_2 with $\bar{\lambda}_1 = \lambda_2$. This does not imply that ρ is diagonalizable in the higher signature setting and hence $\mathfrak{M}$ need not be Einstein. We shall establish the following decomposition result in Section 6.6.2:

Theorem 6.6.2 *Let $\mathfrak{M} = (V, \langle \cdot, \cdot \rangle, A)$ be a 0-model of arbitrary signature.*

(1) If $\mathfrak{M}$ is Jacobi Videv, then we may decompose $\mathfrak{M}$ as the direct sum of pseudo-Einstein 0-models.

(2) If $\mathfrak{M}$ is Riemannian and indecomposable, then $\mathfrak{M}$ is Jacobi Videv if and only if $\mathfrak{M}$ is Einstein.

We note that Theorem 6.6.2 fails on the geometric level. The manifolds described by Tsankov in Theorem 6.1.2 or, more generally, those detailed in Section 6.5 are irreducible Riemannian skew Tsankov manifolds which are Jacobi Videv but which are not Einstein. At each point of the manifold, the 0-model decomposes as the direct sum of Einstein models. However the 1-model is irreducible.

There are indecomposable pseudo-Riemannian Jacobi Videv manifolds which are not Ricci flat but where $\rho^2 = 0$; thus not every indecomposable Jacobi Videv manifold is Einstein. There are indecomposable pseudo-Riemannian manifolds where $\rho^2 \neq 0$ but $\rho^3 = 0$ which are not Jacobi Videv; thus not every pseudo-Einstein manifold is Jacobi Videv. Finally, there are pseudo-Riemannian Jacobi Videv models where $\rho^2 = -\,\mathrm{Id}$; thus 0 is not the critical eigenvalue. We refer to Gilkey and Nikčević (2006e), which is work in progress, for further details.

6.6.1 *Equivalent properties characterizing Jacobi Videv models*

We shall follow the discussion in Gilkey, Puffini, and Videv (2006). Let $\mathfrak{M} = (V, \langle \cdot, \cdot \rangle, A)$ be a 0-model. If $\{v_1, ..., v_k\}$ is a basis for a non-degenerate k-plane π, recall that

$$\mathcal{J}(\pi) := \sum_{i=1}^{k} \sum_{j=1}^{k} \xi^{ij} \mathcal{J}(v_i, v_j). \tag{6.6.a}$$

Here $\xi_{ij} := \langle v_i, v_j \rangle$ and ξ^{ij} is the inverse matrix. Let ρ be the Ricci operator. Since $\mathcal{J}_\pi + \mathcal{J}_{\pi^\perp} = \rho$, one may conclude that

$$
\begin{aligned}
\mathcal{J}_\pi &\mathcal{J}_{\pi^\perp} - \mathcal{J}_{\pi^\perp} \mathcal{J}_\pi \\
&= \mathcal{J}_\pi \{ \mathcal{J}_\pi + \mathcal{J}_{\pi^\perp} \} - \{ \mathcal{J}_\pi + \mathcal{J}_{\pi^\perp} \} \mathcal{J}_\pi \\
&= \mathcal{J}_\pi \rho - \rho \mathcal{J}_\pi \, .
\end{aligned}
$$

This establishes the equivalence of Assertions (1) and (2) and of Assertions (3) and (4) in Theorem 6.6.1. It is immediate that Assertion (4) implies Assertion (2). Assume there exists (r_0, s_0) admissible so that

$$
\mathcal{J}(\pi)\mathcal{J}(\pi^\perp) = \mathcal{J}(\pi^\perp)\mathcal{J}(\pi) \; \forall \; \pi \in \mathrm{Gr}_{r_0, s_0}(V, \langle \cdot, \cdot \rangle) \, .
$$

Let $1 \le \kappa := r_0 + s_0 < m := \dim(V)$. Let $\{e_1, ..., e_\kappa, e_{\kappa+1}, ..., e_m\}$ be an orthonormal basis for V where $\{e_1, ..., e_\kappa\}$ spans a non-degenerate plane π of signature (r_0, s_0). Let $\varepsilon_i := \langle e_i, e_i \rangle$. Then

$$
\mathcal{J}(\pi) := \sum_{i=1}^{\kappa} \varepsilon_i \mathcal{J}(e_i) \, .
$$

We distinguish two cases. Suppose first that $\varepsilon_1 = \varepsilon_{\kappa+1}$. Set

$$
e_1(\theta) := \cos(\theta)e_1 + \sin(\theta)e_{\kappa+1} \, .
$$

Then $\{e_1(\theta), e_2, ..., e_\kappa\}$ is an orthonormal basis for a non-degenerate plane $\pi(\theta)$ of signature (r_0, s_0). One has for any θ that:

$$
\begin{aligned}
0 = [\rho, \mathcal{J}(\pi(\theta)) - \mathcal{J}(\pi)] &= 0 \\
= [\rho, (\cos^2 \theta &- 1)\mathcal{J}(e_1) \\
&+ 2\sin\theta\cos\theta\mathcal{J}(e_1, e_{\kappa+1}) + \sin^2\theta\mathcal{J}(e_{\kappa+1})] \, .
\end{aligned}
$$

This identity for all θ implies

$$
[\rho, \mathcal{J}(e_1) - \mathcal{J}(e_{\kappa+1})] = 0 \quad \text{if} \quad \varepsilon_1 = \varepsilon_{\kappa+1} \, .
$$

Suppose next that $\varepsilon_1 = -\varepsilon_{\kappa+1}$. Set $e_1(\theta) := \cosh(\theta)e_1 + \sinh(\theta)e_{\kappa+1}$. A similar computation, after paying attention to the signs involved, yields:

$$
\begin{aligned}
0 = [\rho, (\cosh^2 \theta - 1)\mathcal{J}(e_1) &- 2\sinh\theta\cosh\theta\mathcal{J}(e_1, e_{\kappa+1}) \\
&+ \sinh^2\theta\mathcal{J}(e_{\kappa+1})] \quad \forall \quad \theta \, .
\end{aligned}
$$

This yields the identity

$$
0 = [\rho, \mathcal{J}(e_1) + \mathcal{J}(e_{\kappa+1})] \, .
$$

We combine these two calculations to see that for all $1 \leq i, j \leq m$ one has:

$$\varepsilon_i[\rho, \mathcal{J}(e_i)] = \varepsilon_j[\rho, \mathcal{J}(e_j)] . \tag{6.6.b}$$

We use Eq. (6.6.b) to see that

$$0 = [\rho, \mathcal{J}(\pi)] = \sum_{i=1}^{\kappa} \varepsilon_i[\rho, \mathcal{J}(e_i)] = \kappa \varepsilon_1[\rho, \mathcal{J}(e_1)]$$

and thus $[\rho, \mathcal{J}(e_1)] = 0$. This shows that $[\rho, \mathcal{J}(v)] = 0$ for every unit spacelike vector if $s_0 > 0$ and for every unit timelike vector if $r_0 > 0$. We can rescale to conclude $[\rho, \mathcal{J}(v)] = 0$ on a non-empty open subset of V and hence, as this is a polynomial identity, conclude $[\rho, \mathcal{J}(v)] = 0$ for all $v \in V$. It then follows from Eq. (6.6.a) that $[\rho, \mathcal{J}(\pi)] = 0$ for every non-degenerate k-plane π. $\qquad\square$

6.6.2 *Decomposing Jacobi Videv models*

We now establish Theorem 6.6.2. If $\mathfrak{M}$ is Einstein, then $\rho = c\,\mathrm{Id}$ is a scalar multiple of the identity. Consequently Condition (4) holds in Theorem 6.6.1. Conversely, suppose that $\mathfrak{M}$ is an indecomposable Riemannian 0-model such that $\mathcal{J}_x \rho = \rho \mathcal{J}_x$ for all $x \in V$. Decompose $V = \oplus_i V_i$ into the eigenspaces of the Ricci operator corresponding to distinct eigenvalues λ_i. Then $\mathcal{J}_x$ preserves each eigenspace. We suppose that $\mathfrak{M}$ is not Einstein and argue for a contradiction.

Choose $x_i \in S(V_{\lambda_i})$. Polarization yields $\mathcal{J}_{x_2 x_3}$ preserves each eigenspace. Suppose that $\lambda_1 \neq \lambda_4$. One then has $\langle \mathcal{J}(x_2, x_3)x_1, x_4 \rangle = 0$. Consequently

$$A(x_1, x_2, x_3, x_4) = -A(x_1, x_3, x_2, x_4) \quad \text{if} \quad \lambda_1 \neq \lambda_4 . \tag{6.6.c}$$

Suppose that $\lambda_1 \neq \lambda_4$ and that $\lambda_2 \neq \lambda_4$. We use Eq. (6.6.c), the first Bianchi identity, and the other curvature identities to see:

$$A(x_1, x_2, x_3, x_4) = -A(x_2, x_3, x_1, x_4) - A(x_3, x_1, x_2, x_4)$$
$$= A(x_2, x_1, x_3, x_4) + A(x_1, x_3, x_2, x_4)$$
$$= -A(x_1, x_2, x_3, x_4) - A(x_1, x_2, x_3, x_4) .$$

This shows

$$A(x_1, x_2, x_3, x_4) = 0 \quad \text{if} \quad \lambda_1 \neq \lambda_4 \text{ and } \lambda_2 \neq \lambda_4 . \tag{6.6.d}$$

Suppose that $A(x_1, x_2, x_3, x_4) \neq 0$ and that $\lambda_1 \neq \lambda_4$. By Eq. (6.6.d), $\lambda_2 = \lambda_4$ and, similarly, $\lambda_3 = \lambda_1$. Applying Eq. (6.6.c) then yields

$$A(x_1, x_2, x_3, x_4) = -A(x_1, x_3, x_2, x_4).$$

A final use of Eq. (6.6.d) then shows $A(x_1, x_3, x_2, x_4) = 0$ as $\lambda_1 = \lambda_3 \neq \lambda_4$. Consequently

$$A(x_1, x_2, x_3, x_4) \neq 0 \quad \text{implies} \quad \lambda_1 = \lambda_2 = \lambda_3 = \lambda_4. \qquad (6.6.e)$$

Let $A_i = A|_{V_i}$. By Eq. (6.6.e), $A = \oplus A_i$. If ρ has more than one eigenvalue, this gives a non-trivial decomposition of $\mathfrak{M}$ which contradicts the assumption that $\mathfrak{M}$ is indecomposable. This completes the proof of Theorem 6.6.2 (2). A similar argument using the Jordan normal form decomposition rather than the eigenvector decomposition then suffices to establish Theorem 6.6.2 (1). $\qquad \qquad \square$

Bibliography

Adams, J., 'Vector fields on spheres', *Ann. of Math.* **75** (1962), 603–632.

Adkins, W., and Weintraub, S., *Algebra - an approach via module theory*, Graduate Texts in Mathematics **136**, Sprinter-Verlag, New York (1992).

Alekseevskiĭ, D. V., 'Riemannian manifolds with exceptional holonomy groups', *Funct. Anal. Appl.* **2** (1968), 97–105.

Arvanitoyeorgos, A., 'An introduction to Lie groups and the geometry of homogeneous spaces', *Student Mathematical Library* **22**, Amer. Math. Soc. (2003).

Atiyah, M.F., Bott, R., and Patodi, V.K., 'On the heat equation and the index theorem', *Invent. Math.* **13** (1973), 279–330; Errata **28** (1975), 277–280.

Atiyah, M. F., Bott, R., and Shapiro, A., 'Clifford Modules', *Topology* **3** suppl. 1 (1964), 3–38.

Belger, M. and Kowalski, O., 'Riemannian metrics with the prescribed curvature tensor and all its covariant derivatives at one point', *Math. Nachr.* **168** (1994), 209–225.

Belger, M., and Stanilov, G., 'About the Riemannian geometry of some curvature operators', *Annuaire Univ. Sofia* **1** (1995), in press.

Berger, M., 'Riemannian geometry during the second half of the twentieth century', *Jahresber. Deutsch. Math.-Verein.* **100** (1998), 45–208.

Berndt, J., and Samiou, E., 'Rank rigidity, cones, and curvature-homogeneous Hadamard manifolds', *Osaka J. Math.* **39** (2002), 383–394.

Besse, A. L., *Manifolds all of whose Geodesics are Closed*, Ergebnisse der Mathematik und ihrer Grenzgebiete, Springer, Berlin, 1978.

Blažić, N., 'Natural curvature operators of bounded spectrum', Differential Geom. Appl. **24** (2006), to appear.

Blažić, N., Bokan, N., and Gilkey, P., 'A Note on Osserman Lorentzian manifolds', *Bull. London Math. Soc.* **29** (1997), 227–230.

Blažić, N., Bokan, N., Gilkey, P., and Rakić, Z., 'Pseudo-Riemannian Osserman manifolds', *Balkan J. Geom. Appl.* **2** (1997), 1–12.

Blažić, N., Bokan, N., and Rakić, Z., 'The first order PDE system for type III Osserman manifolds', *Publ. Inst. Math. (Beograd) (N.S.)* **62** (1997), 113–119.

Blažić, N., Bokan, N., and Rakić, Z., 'Nondiagonalizable timelike (spacelike) Osserman $(2,2)$ manifolds', *Saitama Math. J.* **16** (1998), 15–22.

Blažić, N., Bokan, N., and Rakić, Z., 'A note on the Osserman conjecture and isotropic covariant derivative of curvature', *Proc. Amer. Math. Soc.* **128** (2000), 245–253.

Blažić, N., Bokan, N., and Rakić, Z., 'Osserman pseudo-Riemannian manifolds of signature $(2,2)$', *Aust. Math. Soc.* **71** (2001), 367–395.

Blažić, N., and Gilkey, P., 'Conformally Osserman manifolds and conformally complex space forms', *Int. J. Geom. Methods Mod. Phys.* **1** (2004), 97–106.

Blažić, N., and Gilkey, P., 'Conformally Osserman manifolds and self-duality in Riemannian geometry', math.DG/0504498.

Blažić, N., Gilkey, P., Nikčević, S., and Simon, U., 'The spectral geometry of the Weyl conformal tensor', *Banach Center Publ.* **69** (2005), 195–203.

Blažić, N., Gilkey, P., Nikčević, S., Simon, U., 'Algebraic theory of affine curvature tensors', preprint.

Blažić, N., and Vukmirović, S., 'Examples of Self-dual, Einstein metrics of $(2,2)$-signature', *Math. Scand.* **94** (2003), 1–12.

Boeckx, E., Kowalski, O., and Vanhecke, L., *Riemannian Manifolds of Conullity Two*, World Scientific (1996).

Boeckx, E., and Vanhecke, L., 'Curvature homogeneous unit tangent sphere bundles', *Publ. Math. Debrecen* **53** (1998), 389–413.

Bokan, N., 'On the complete decomposition of curvature tensors of Riemannian manifolds with symmetric connection', *Rend. Circ. Mat. Palermo* **XXIX** (1990), 331–380.

Bonome, A., Castro, R., García-Río, E., 'Four-Dimensional Generalized Osserman Manifolds', *Classical Quantum Gravity* **18** (2001), 4813–4822.

Bonome, A., Castro, R., García-Río, E., Hervella, L., and Vázquez-Lorenzo, R., 'Nonsymmetric Osserman indefinite Kähler manifolds', *Proc. Amer. Math. Soc.* **126** (1998), 2763–2769.

Borel, A., 'Sur la cohomologie des espaces fibrés principaux et des spaces homogènes de groupes de Lie compactes', *Ann. of Math.* **57** (1953), 115–207.

Bott, R., and Loring, T., *Differential Forms in Algebraic Topology*, Graduate Text in Mathematics, Springer-Verlag, Berlin, (1982).

Brozos-Vázquez, M., García-Río, E., and Vázquez-Lorenzo, R., 'Conformally Osserman four-dimensional manifolds whose conformal Jacobi operators have complex eigenvalues', *Proc. Royal Society A* **462** (2006), 1425–1441.

Brozos-Vázquez, M., García-Río, E., and Gilkey, P., 'Relating the curvature tensor and the complex Jacobi operator of an almost Hermitian manifold', preprint.

Brozos-Vázquez, M., and Gilkey, P., 'Manifolds with Commuting Jacobi Operators', math.DG/0507554.

Brozos-Vázquez, M., and Gilkey, P., 'Pseudo-Riemannian manifolds with Commuting Jacobi Operators', math.DG/0608707.

Brozos-Vázquez, M., and Gilkey, P., 'The global geometry of Riemannian manifolds with commuting curvature operators', math.DG/0609500.

Brozos-Vázquez, M., Gilkey, P., and Nikčević, S., 'Jacobi–Tsankov manifolds which are not 2-step nilpotent', math.DG/0609565.

Bryant, R., 'Pseudo-Riemannian metrics with parallel spinor fields and vanishing Ricci tensor', *Global Analysis and Harmonic Analysis (Marseille-Luminy, 1999)*, Sémin. Congr. **4**, Soc. Math. France, Paris 2000, 53–94.

Bueken, P., 'On curvature homogeneous three-dimensional Lorentzian manifolds', *J. Geom. Phys.* **22** (1997), 349–362.

Bueken, P., 'Three-dimensional Lorentzian manifolds with constant principal Ricci curvatures $\rho_1 = \rho_2 \neq \rho_3$', *J. Math. Phys.* **38** (1997), 1000–1013.

Bueken, P., and Djoric, M., 'Three-dimensional Lorentz metrics and curvature homogeneity of order one', *Ann. Global Anal. Geom.* **18** (2000), 85–103.

Bueken, P., Gillard, J., and Vanhecke, L., 'Mean and scalar curvature homogeneous Riemannian manifolds', *Gen. Math.* **5** (1997), 67–83.

Bueken, P., and Vanhecke, L., 'Examples of curvature homogeneous Lorentz metrics', *Classical Quantum Gravity* **14** (1997), L93–L96.

Bueken, P., and Vanhecke, L., 'Three- and four-dimensional Einstein-like manifolds and homogeneity', *Geom. Dedicata* **75** (1999), 123–136.

Cahen, M., Leroy, J., Parker, M., Tricerri, F., and Vanhecke, L., 'Lorentz manifolds modelled on a Lorentz symmetric space', *J. Geom. Phys.* **7** (1990), 571–581.

Calvaruso, G., Marinosci, R. A., and Perrone, D., 'Three-dimensional curvature homogeneous hypersurfaces', *Arch. Math. (Brno)* **36** (2000), 269–278.

Carpenter, P., Gray, A., and Willmore, T. J., 'The curvature of Einstein symmetric spaces', *Quart. J. Math. Oxford* **33** (1982), 45–64.

Chaichi, M., García-Río, E., and Matsushita, Y., 'Curvature properties of four-dimensional Walker metrics', *Classical Quantum Gravity* **22** (2005), 559–577.

Chaichi, M., García-Río, E., and Vázquez-Abal, M. E., 'Three dimensional Lorentz manifolds admitting a parallel null vector field', *J. Phys. A* **38** (2005), 841–850.

Chi, Q. S., 'A curvature characterization of certain locally rank one symmetric spaces', *J. Differential Geom.* **28** (1988), 187–202.

Chi, Q. S., 'Quaternionic Kähler manifolds and a characterization of two-point homogeneous spaces', *Illinois J. Math.* **35** (1991), 408–418.

Chi, Q. S., 'Curvature characterization and classification of rank-one symmetric spaces', *Pacific J. Math.* **150** (1991), 31–42.

Damek, E., and Ricci, F., 'A class of nonsymmetric harmonic Riemannian spaces', *Bull. Amer. Math. Soc.* **27** (1992), 139–142.

Derdzinski, A., 'Einstein metrics in dimension four', *Handbook of Differential Geometry. Vol. I* Edited by Dillen and Verstraelen, North-Holland, Amsterdam, (2000), 419–707.

Derdzinski, A., 'Curvature-homogeneous indefinite Einstein metrics in dimension four: the diagonalizable case', *Contemp. Math.* **337**, Amer. Math. Soc., Providence, RI (2003), 21–38.

Díaz-Ramos, J. C., and García-Río, E., 'A note on the structure of algebraic curvature tensors', *Linear Algebra Appl.* **382** (2004), 271–277.

Díaz-Ramos, J. C., García-Río, E., and Vázquez-Lorenzo, R., 'New examples of Osserman metrics with nondiagonalizable Jacobi operators', *Differential Geom. Appl.* **24** (2006), 433–442.

Díaz-Ramos, J. C., Fiedler, B., García-Río, E., and Gilkey, P., 'The structure of algebraic covariant derivative curvature tensors', *Int. J. Geom. Methods Mod. Phys.* **6** (2004), 711–720.

Djorić, M., and Vanhecke, L., 'Almost Hermitian geometry, geodesic spheres and symmetries', *Math. J. Okayama Univ.* **32** (1990), 187–206.

Dotti, I., and Druetta, M., 'Negatively curved homogeneous Osserman spaces', *Differential Geom. Appl.* **11** (1999), 163–178.

Dotti, I., and Druetta, M., 'Osserman p-spaces of Iwasawa type', *Differential geometry and Applications (Brno, 1998)*, Masaryk Univ., Brno, 1999, 61–72.

Dotti, I., and Druetta, M., 'Damek–Ricci spaces satisfying the Osserman p-condition', *Rev. Un. Mat. Argentina* **41** (2000), 27–32.

Dunn, C., *Curvature Homogeneous Pseudo-Riemannian Manifolds* (Ph. D. Thesis), Univ. Oregon (2006).

Dunn, C., and Gilkey, P., 'Curvature homogeneous pseudo-Riemannian manifolds which are not locally homogeneous', *Complex, Contact and Symmetric Manifolds*, Progr. Math. **234**, Birkhäuser Boston, Boston, MA, (2005), 145–152.

Dunn, C., Gilkey, P., and Nikčević, S., 'Curvature homogeneous signature $(2,2)$ manifolds', *Differential Geometry and its Applications, Proc. Conf. Prague, August 30-September 3, 2004 Charles University, Prague (Czech Republic)*, editors Bures, Kowalski, Krupka, Slovak, MATFYZPRESS (2005), 29–44.

Eisenhart, L., *Non-Riemannian geometry*, Amer. Math. Soc. Colloquium Publications **8**, 5th printing (1964), Providence RI.

Falcitelli, M., Farinola, A., and Salamon, S., 'Almost-Hermitian geometry', *Differential Geom. Appl.* **4** (1994), 259-282.

Ferus, D., Karcher, H., Münzner, H., 'Cliffordalgebren und neue isoparametrische Hyperflächen', *Math. Z.* **177** (1981), 479–502.

Fiedler, B., 'On the symmetry classes of the first covariant derivatives of tensor fields', *Sém. Lothar. Combin.* **49** (2002/04), Art. **B49f**, 22 pp. (electronic).

Fiedler, B., 'Determination of the structure of algebraic curvature tensors by means of Young symmetrizers', *Seminaire Lotharingien de Combinatoire* **B48d** (2003a). 20 pp. Electronically published: http://www.mat.univie.ac.at/~slc/; see also math.CO/0212278.

Fiedler, B., 'Generators of algebraic covariant derivative curvature tensors and Young symmetrizers', to appear in *Progress in Computer Science Research*, ed. F. Columbus, Nova Science Publishers, Inc., see also math.CO/0310020.

Fielder, B., and Gilkey, P., 'Nilpotent Szabó, Osserman and Ivanov–Petrova pseudo-Riemannian manifolds', *Contemp. Math.* **337** (2003), 53–64.

Friedrich, Th., Kath, I., Moroianu, A., and Semmelmann, U., 'On Nearly Parallel G2 Structures', *J. Geom. Phys.* **23** (1997), 259–286.

Gadea, P. M., and Masqué, J. M., 'Classification of non-flat para-Kählerian space forms', *Houston J. Math.* **21** (1995), 89–94.

García-Rió, E., and Kupeli, D., '4-Dimensional Osserman Lorentzian Manifolds', *New Developments in Differential Geometry*, Proceedings of the Colloquium on Differential Geometry, Debrecen, Hungary (1996), Kluwer Academic Publishers **350**, 201–211.

García-Río, E., Kupeli, D., and Vázquez-Abal, M. E., 'On a problem of Osserman in Lorentzian geometry', *Differential Geom. Appl.* **7** (1997), 85–100.

García-Río, Kupeli, D., E., Vázquez-Abal, M. E., and Vázquez-Lorenzo, R., 'Affine Osserman connections and their Riemann extensions', *Differential Geom. Appl.* **11** (1999), 145–153.

García-Río, E., Kupeli, D., and Vázquez-Lorenzo, R., *Osserman Manifolds in Semi-Riemannian Geometry*, Lecture Notes in Mathematics **1777**, Springer-Verlag, Berlin, 2002.

García-Río, E., Rakić, Z., and Vázquez-Abal, M. E., 'Four-dimensional indefinite Kaehler Osserman manifolds', J. Math. Phys. **46** (2005), 11pp.

García-Río, E., Vázquez-Abal, M. E., and Vázquez-Lorenzo, R., 'Nonsymmetric Osserman pseudo Riemannian manifolds', *Proc. Amer. Math. Soc.* **126** (1998), 2771–2778.

Gilkey, P., 'Manifolds whose higher odd order curvature operators have constant eigenvalues at the basepoint', *J. Geom. Anal.* **2** (1992), 151–156.

Gilkey, P., 'Manifolds whose curvature operator has constant eigenvalues at the basepoint', *J. Geom. Anal.* **4** (1994), 155–158.

Gilkey, P., *Invariance Theory, the Heat Equation, and the Atiyah-Singer Index Theorem (2nd ed)*, CRC Press (1995).

Gilkey, P., 'Generalized Osserman Manifolds', *Abh. Math. Sem. Univ. Hamburg* **68** (1998), 125–127.

Gilkey, P., 'Riemannian manifolds whose skew-symmetric curvature operator has constant eigenvalues II', *Differential Geometry and Applications*, eds. Kolar, Kowalski, Krupka and Slovák, Publ. Massaryk University, Brno, Czech Republic, ISBN 80-210-2097-0 (1999), 73–87.

Gilkey, P., 'Relating algebraic properties of the curvature tensor to geometry', *Novi Sad J. Math* **29** (1999), 109–119.

Gilkey, P., 'Bundles over projective spaces and algebraic curvature tensors', *J. Geom* **71** (2001), 54–67.

Gilkey, P., 'Algebraic curvature tensors which are p-Osserman', *Differential Geom. Appl.* **14** (2001), 297–311.

Gilkey, P., *Geometric Properties of Natural Operators Defined by the Riemann Curvature Tensor*, World Scientific (2001).

Gilkey, P., and Ivanova, R., 'Complex IP pseudo-Riemannian algebraic curvature tensors', *Banach Center Publ.* **57**, Polish Acad. Sci., Warsaw (2002), 195–202.

Gilkey, P., and Ivanova, R., 'Geometric consequences of some algebraic properties of the curvature tensor', *Bull. Mathematiques Soc. Math. Roumanie* **93** (2000), 255–265.

Gilkey, P., and Ivanova, R., 'The geometry of the skew-symmetric curvature operator in the complex setting', *Contemp. Math.* **288** (2001), 325–333.

Gilkey, P., and Ivanova, R., 'The Jordan normal form of Osserman algebraic curvature tensors', *Results Math.* **40** (2001), 192–204.

Gilkey, P., and Ivanova, R., 'The Jordan normal form of higher order Osserman algebraic curvature tensors', *Comment. Math. Univ. Carolinae* **43** (2002), 231–242.

Gilkey, P., and Ivanova, R., 'Spacelike Jordan Osserman algebraic curvature tensors in the higher signature setting', *Differential Geometry, Valencia 2001*, World Scientific (2002), 179–186.

Gilkey, P., Ivanova, R., and Stavrov, I., 'Jordan Szabó algebraic covariant derivative curvature tensors', *Contemp. Math.* **337** (2003), 65–76.

Gilkey, P., Ivanova, R., and Zhang, T., 'Higher order Jordan Osserman, pseudo-Riemannian manifolds', *Classical Quantum Gravity* **19** (2002), 4543-4551.

Gilkey, P., Ivanova, R., and Zhang, T., 'Szabo Osserman IP pseudo-Riemannian manifolds', *Publ. Math. Debrecen* **62** (2003), 387–401.

Gilkey, P., Leahy, J.V., and Sadofsky, H., 'Riemannian manifolds whose skew-symmetric curvature operator has constant eigenvalues', *Indiana Univ. Math. J.* **48** (1999), 615–634.

Gilkey, P., and Nikčević, S., 'Curvature homogeneous spacelike Jordan Osserman pseudo-Riemannian manifolds', *Classical Quantum Gravity* **21** (2004), 497–507.

Gilkey, P., and Nikčević, S., 'Nilpotent spacelike Jordan Osserman pseudo-Riemannian manifolds', *Rend. Circ. Mat. Palermo Suppl.* **72** (2004), 99–105.

Gilkey, P., and Nikčević, S., 'Complete curvature homogeneous pseudo-Riemannian manifolds', *Classical Quantum Gravity* **21** (2004), 3755–3770.

Gilkey, P., and Nikčević, S., 'Complete k-curvature homogeneous pseudo-Riemannian manifolds', *Ann. Global Anal. Geom.* **27** (2005), 87–100.

Gilkey, P., and Nikčević, S., 'Generalized plane wave manifolds', *Kragujevac J. Math.* **28** (2005), 113–138.

Gilkey, P., and Nikčević, S., 'Complete k-curvature homogeneous pseudo-Riemannian manifolds 0-modeled on an irreducible symmetric space', *Topics in Almost Hermitian Geometry and the Related Fields*, World Scientific (2005), 87–104.

Gilkey, P., and Nikčević, S., 'Isometry groups of k-curvature homogeneous pseudo-Riemannian manifolds', to appear *Rend. Circ. Mat. Palermo, (Proceedings of the 25th Winter School Geometry and Physics, Srni 2005)*, math.DG/0505598.

Gilkey, P., and Nikčević, S., 'Affine curvature homogeneous 3-dimensional Lorentz Manifolds', *Int. J. Geom. Methods Mod. Phys.* **2** (2005), 737–749.

Gilkey, P., and Nikčević, S., 'Pseudo-Riemannian Jacobi–Videv Manifolds', work in progress.

Gilkey, P., Nikčević, S., and Videv, V., 'Manifolds which are Ivanov–Petrova or k-Stanilov', *J. Geom.* **80** (2004), 82–94.

Gilkey, P., Puffini, E., and Videv, V., 'Puffini–Videv models and manifolds', math.DG/0605464.

Gilkey, P., and Semmelmann, U., 'Spinors, self-duality, and IP algebraic curvature tensors of rank 4', *Proceedings of the Symposium on Contemp. Math. devoted to the 125 Anniversary of Faculty of Mathematics in Belgrade*, Published by the Faculty of Mathematics of the University of Belgrade (2000), ISBN 86-7589-014-1, Ed. Neda Bokan, 1–12.

Gilkey, P., Stanilov, G., and Videv, V., 'Pseudo Riemannian Manifolds whose Generalized Jacobi Operator has Constant Characteristic Polynomial', *J. Geom.* **62** (1998), 144–153.

Gilkey, P., Swann, A., and Vanhecke, L., 'Isoparametric geodesic spheres and a conjecture of Osserman regarding the Jacobi Operator', *Quart. J. Math. Oxford* **46** (1995), 299–320.

Gilkey, P., and Stavrov, I., 'Curvature tensors whose Jacobi or Szabó operator is nilpotent on null vectors', *Bull. London Math. Soc.* **34** (2002), 650–658.

Gilkey, P., and Zhang, T., 'Algebraic curvature tensors whose skew-symmetric curvature operator has constant rank 2', *Period. Math. Hungar.* **44** (2002), 7–26.

Gilkey, P., and Zhang, T., 'Algebraic curvature tensors for indefinite metrics whose skew-symmetric curvature tensor has constant Jordan normal form', *Houston Math J.* **28** (2002), 311–328.

Glover, H., Homer, W., and Stong, R., 'Splitting the tangent bundle of projective space', *Indiana Univ. Math. J.* **31** (1982), 161–166.

Gray, A., 'The volume of a small geodesic ball of a Riemannian manifold', *Michigan Math. J.* **20** (1973), 329–344.

Gray, A., 'Classification des variétés approximativement kählériennes de courbure sectionelle holomorphe constante', *C. R. Acad. Sci. Paris* **279** (1974), 797–800.

Gray, A., 'Curvature identities for Hermitian and almost Hermitian manifolds', *Tôhoku Math. J.* **28** (1976), 601-612.

Gromov, M., *Partial differential relations*, Ergeb. Math. Grenzgeb 3. Folge, Band **9**, Springer-Verlag (1986).

Hahn, J., *Homogene Hyperflächen in der pseudoremannschen Geometrie*, Bonner Mathematische Schriften **172**, Universität Bonn Mathematische Institute, Bonn (1986).

Helgason, S., *Differential Geometry, Lie Groups and Symmetric Spaces*, Academic Press (New York), 1978.

Ivanov, S., and Petrova, I., 'Riemannian manifolds in which certain curvature operator has constant eigenvalues along each circle', *Ann. Global Anal. Geom.* **15** (1997), 157–171.

Ivanov, S., and Petrova, I., 'Riemannian manifold in which the skew-symmetric curvature operator has pointwise constant eigenvalues', *Geom. Dedicata* **70** (1998), 269–282.

Ivanova, M., and Videv, V., 'Four-dimensional Riemannian manifolds with commuting Stanilov curvature operators', *Mathematics and Education in Mathematics (Bulgarian)*, Union of Bulgarian Math., Sofia, (2004), 184–188.

Ivanova, R., 'About a classification of some Einstein manifolds', *Mathematics and Education in Mathematics, Proc.* **XXIII** *Spring Conference of the Union of Bulgarian Mathematicians, Stara Zagora*, (1994), 175–182.

Ivanova, R., 'Necessary and sufficient conditions for the point-wise constancy of a curvature operator's characteristical coefficient', *Ann. de'l Universite de Sofia, Fac. de Math. et Inf.* **88** (1994), 17–25.

Ivanova, R., 'Point-wise constancy of the skew-symmetric curvature operator's characteristical coefficients', *Tensor, N. S.,* **57** (1996), 97–104.

Ivanova, R., 'On a property of the 4-dimensional Einstein manifolds', *Proceedings of the 4th International Congress of Geometry (Thessaloniki)* (1996), 198–207.

Ivanova, R., 'Two systems of curvature conditions for some 4-dimensional Riemannian manifolds', *God. Univ. Arkhit. Stroit. Geod. Sofiya Svit k II Mat. Mekh.* **39** (1996/97), 33–44.

Ivanova, R., 'An orthogonal tangential group of transformations of a 4-dimensional Riemannian manifold', *Math. Balkanica* **11** (1997), 53–63.

Ivanova, R., '4-dimensional Riemannian manifolds characterized by a skew-symmetric curvature operator', *Tensor, N. S.* **60** (1998), 293–302.

Ivanova, R., 'Generalization of a property of some 4 dimensional Einstein manifolds', *Tensor, N.S.* **60** (1998), 303–308.

Ivanova, R., and Stanilov, G., 'A skew-symmetric curvature operator in Riemannian geometry', *Symposia Gaussiana (1994), Conf. A: Mathematics, Eds. Behara, Fritsch and Lintz* (1995), 391–395.

Karoubi, M., *K-theory. An introduction*, Grundlehren der Mathematischen Wissenschaften, Band **226**, Springer-Verlag, Berlin (1978).

Kath, I., *Killing Spinors on Pseudo-Riemannian Manifolds, Habilitation*, Humboldt Universität zu Berlin (2000).

Klebaner, F., Sudbury, A., and Watterson, G. A., 'The volumes of simplices, or, find the penguin', *J. Austral. Math. Soc. Ser. A* **47** (1989), 263–268.

Kobayashi, S., and Nomizu, K., *Foundations of Differential Geometry*, Interscience Publishers (1963).

Kofinas, G., 'General brane cosmology with ^4R term in $(A)dS_5$ or Minkowski bulk', *J. High Energy Phys.* **2001** #8 Paper 34, 21pp.

Koutras, A., and McIntosh, C., 'A metric with no symmetries or invariants', *Classical Quantum Gravity* **13** (1996), L47–L49.

Kowalski, O., 'On curvature homogeneous spaces', *Proceedings of the Workshop on Recent Topics in Differential Geometry*, Santiago de Compostela, Spain, July 16–19, 1997. Santiago de Compostela: Universidade de Santiago de Compostela. Publ. Dep. Geom. Topologá, Univ. Santiago Compostela **89** (1998), 193–205.

Kowalski, O., Opozda, B., and Vlášek, Z., 'Curvature homogeneity of affine connections on two-dimensional manifolds', *Colloq. Math.* **81** (1999), 123–139.

Kowalski, O., Opozda, B., and Vlášek, Z., 'A classification of locally homogeneous affine connections with skew-symmetric Ricci tensor on 2 dimensional manifolds', *Monatsh. Math.* **130** (2000), 109–125.

Kowalski, O., Opozda, B., and Vlášek, Z., 'A classification of locally homogeneous connections on 2-dimensional manifolds via group-theoretical approach', *Cent. Eur. J. Math.* **2** (2004), 87–102.

Kowalski, O., and Prüfer, F., 'Curvature tensors in dimension four which do not belong to any curvature homogeneous space', *Arch. Math. (Brno)* **30** (1994), 45–57.

Kowalski, O., Sekizawa, M., and Vlášek, Z., 'Can tangent sphere bundles over Riemannian manifolds have strictly positive sectional curvature?', *Contemp. Math.* **288**, Amer. Math. Soc., Providence, RI, 2001.

Kowalski, O., Tricerri, F., and Vanhecke, L., 'New examples of non-homogeneous Riemannian manifolds whose curvature tensor is that of a Riemannian symmetric space', *C. R. Acad. Sci.* Paris, Sér. I **311** (1990), 355–360.

Kowalski, O., Tricerri, F., and Vanhecke, L., 'Curvature homogeneous Riemannian manifolds', *J. Math. Pures Appl.* **71** (1992a), 471–501.

Kowalski, O., Tricerri, F., and Vanhecke, L., 'Curvature homogeneous spaces with a solvable Lie group as homogeneous model', *J. Math. Soc. Japan* **44** (1992b), 461–484.

Kowalski, O., and Vlášek, Z., 'On special Riemannian 3-manifolds with distinct constant Ricci eigenvalues', *Math. Bohem.* **124** (1999), 45–66.

Kowalski, O., and Vlášek, Z., 'Classification of Riemannian 3-manifolds with distinct constant principal Ricci curvatures', *Bull. Belg. Math. Soc. Simon Stevin* **5** (1998), 59–68.

Kowalski, O., and Vlášek, Z., 'On 3-dimensional Riemannian manifolds with prescribed Ricci eigenvalues', *Progr. Math.* **234** (2005), 187–208.

Li, Y., and Ouyang, C., 'Curvature homogeneous real hypersurfaces immersed in the complex projective space', *J. Math. Wuhan Univ.* **15** (1995), 405–408.

Marchiafava, S., 'Variétés riemanniennes dont le tenseur de courbure est celui d'un espace symétrique de rang un', *C. R. Acad. Sci. Paris* **295** (1982), 463–466.

Milnor, J., and Stasheff, J., *Characteristic Classes*, Annals of Math. Studies, Princeton University Press (1974).

Nash, J., 'The embedding problem for Riemannian manifolds', *Ann. of Math.* **63** (1956), 20–63.

Nicolodi, L., and Tricerri, F., 'On two theorems of I. M. Singer about homogeneous spaces', *Ann. Global Anal. Geom.* **8** (1990), 193–200.

Nikolayevsky, Y., 'Osserman manifolds and Clifford Structures', *Houston J. Math* **29** (2003), 59–75.

Nikolayevsky, Y., 'Two theorems on Osserman manifolds', *Differential Geom. Appl.* **18** (2003b), 239–253.

Nikolayevsky, Y., 'Osserman manifolds of dimension 8', *Manuscr. Math.* **115** (2004), 31–53.

Nikolayevsky, Y., 'Osserman Conjecture in dimension $n \neq 8, 16$', *Mat. Annalen* **331** (2005), 505–522.

Nikolayevsky, Y., *On Osserman manifolds of dimension 16*, (preprint).

Nikolayevsky, Y., 'Riemannian manifolds whose curvature operator $R(X, Y)$ has constant eigenvalues', *Bull. Austral. Math. Soc.* **70** (2004), 301–319.

Newlander, A., Nirenberg, L., 'Complex analytic coordinates in almost complex manifolds' *Ann. of Math.* **65** (1957), 391–404.

Olszak, Z., 'On the existence of generalized complex space forms' *Israel J. Math.* **65** (1989), 214–218.

Opozda, B., 'An intrinsic characterization of developable surfaces', *Results Math.* **27** (1995), 97–104.

Opozda, B., 'On curvature homogeneous and locally homogeneous affine connections', *Proc. Amer. Math. Soc.* **124** (1996), 1889–1893.

Opozda, B., 'Affine versions of Singer's theorem on locally homogeneous spaces', *Ann. Global Anal. Geom.* **15** (1997), 187–199.

Oproiu, V., 'Harmonic maps between tangent bundles', *Rend. Sem. Mat. Univ. Politec. Torino* **47** (1989), 47–55.

Osserman, R., 'Curvature in the eighties', *Amer. Math. Monthly* **97** (1990), 731–756.

Overduin, J. M., and Wesson, P. S., 'Kaluza-Klein gravity', *Phys. Rep.* **283** (1997), 303–378.

Patrangenaru, V., 'Constant gravitational fields and redshift of light', *J. Geom. Phys.* **26** (1998), 227–246.

Patrangenaru, V., 'Three-dimensional metrics with a spherical homogeneous model', *J. Math. Phys.* **39** (1998), 1189–1198.

Patrangenaru, V., 'Lorentz manifolds with the three largest degrees of symmetry', *Geom. Dedicata* **102** (2003), 251–33.

Petrova, I., and Stanilov, G., 'A generalized Jacobi operator on the 4 dimensional Riemannian geometry', *Annuaire de l'Univ. Sofia 'St. Kl. Ochridski'* **85** (1991), 55–64.

Pinkall, U., Schwenk-Schellschmidt, A., and Simon, U., 'Geometric methods for solving Codazzi and Monge-Ampère equations', *Math. Annalen* **298** (1994), 89–100.

Podesta, F., and Spiro, A., 'Four-dimensional Einstein-like manifolds and curvature homogeneity', *Geom. Dedicata* **54** (1995), 225–243.

Podesta, F., and Spiro, A., 'Introduzione ai Gruppi di Trasformazioni', *Volume of the Preprint Series of the Mathematics Department* V. Volterra of the University of Ancona, Via delle Brecce Bianche, Ancona, ITALY (1996).

Prüfer, F., Tricerri, F., and Vanhecke, L., 'Curvature invariants, differential operators and local homogeneity', *Trans. Amer. Math. Soc.* **348** (1996), 4643–4652.

Pravda, V., Pravdová, A., Coley, A., and Milson, R., 'All spacetimes with vanishing curvature invariants', *Classical Quantum Gravity* **19** (2002), 6213–6236.

Rakić, Z., 'An example of rank two symmetric Osserman space', *Bull. Austral. Math. Soc.* **56** (1997), 517–521.

Rakić, Z., 'On duality principle in Osserman manifolds', *Linear Algebra Appl.* **296** (1999), 183–189.

Ryan, P. J., 'Homogeneity and some curvature conditions for hypersurfaces', *Tôhoku Math. J.* **21** (1969), 363–388.

Sahni, V., and Shtanov, Y., 'Bouncing braneworlds', *Phys. Lett.* B **557** (2003), 1–6.

Sekigawa, K., 'On some 3-dimensional curvature homogeneous spaces', *Tensor (N.S.)* **31** (1977), 87–97.

Sekigawa, K., 'Notes on some curvature homogeneous spaces', *Tensor (N.S.)* **29** (1975), 255–258.

Sekigawa, K., Suga, H., and Vanhecke, L., 'Four-dimensional curvature homogeneous spaces', *Commentat. Math. Univ. Carol.* **33** (1992), 261–268.

Sekigawa, K., Suga, H., and Vanhecke, L., 'Curvature homogeneity for four-dimensional manifolds', *J. Korean Math. Soc.* **32** (1995), 93–101.

Singer, I. M., 'Infinitesimally homogeneous spaces', *Commun. Pure Appl. Math.* **13** (1960), 685–697.

Stanilov, G., 'On the geometry of the Jacobi operators on 4-dimensional Riemannian manifolds', *Tensor (N.S.)* **51** (1992), 9–15.

Stanilov, G., 'Curvature operators based on the skew-symmetric curvature operator and their place in the Differential Geometry', preprint (2000).

Stanilov, G., 'Higher order skew-symmetric and symmetric curvature operators', *C. R. Acad. Bulgare Sci.* **57** (2004), 9–12.

Stanilov, G., and Videv, V., 'On a generalization of the Jacobi operator in the Riemannian geometry', *Annuaire Univ. Sofia Fac. Math. Inform.* **86** (1992), 27–34.

Stanilov, G., and Videv, V., 'On Osserman conjecture by characteristical coefficients', *Algebras, Groups and Geometries* **12** (1995), 157–163.

Stanilov, G., and Videv, V., 'Four dimensional pointwise Osserman manifolds', *Abh. Math. Sem. Univ. Hamburg* **68** (1998), 1–6.

Stanilov, G., and Videv, V., 'On the commuting of curvature operators', *Mathematics and Education in Mathematics*, Proceedings of the Thirty Third Spring Conference of the Union of Bulgarian Mathematicians Borovtes, April 1-4, 2004, Union of Bulgarian Mathematicians, Sofia (2004), 176–179.

Stavrov, I., *Spectral Geometry of the Riemann Curvature Tensor* (Ph. D. Thesis), Univ. Oregon (2003).

Stavrov, I., 'Locally Isotropic pseudo-Riemannian Manifolds', math.DG/0409189.

Stavrov, I., 'Vector bundles over Grassmannians and the skew-symmetric curvature operator', math.DG/0407284.

Stong, R., 'Splitting the universal bundles over Grassmannians', *Algebraic and Differential Topology - Global Differential Geometry. On the Occasion of the 90th Anniversary of M. Morse's Birth*, Teubner-Texte Math. **70** (1984), 275–287.

Szabó, Z. I., 'A simple topological proof for the symmetry of 2 point homogeneous spaces', *Invent. Math.* **106** (1991), 61–64.

Takagi, H., 'On curvature homogeneity of Riemannian manifolds', *Tôhoku Math. J.* **26** (1974), 581–585.

Tomassini, A., 'Curvature homogeneous metrics on principal fibre bundles', *Ann. Mat. Pura Appl.* **172** (1997), 287–295.

Tricerri, F., 'Riemannian manifolds with the same curvature as a homogeneous space, and a conjecture of Gromov', *Riv. Mat. Univ. Parma* **14** (1988), 91–104.

Tricerri, F., and Vanhecke, L., 'Curvature tensors on almost Hermitian manifolds', *Trans. Amer. Math. Soc.* **267** (1981), 365–398.

Tricerri, F., and Vanhecke, L., 'Variétés riemanniennes dont le tenseur de courbure est celui d'un espace symétrique riemannien irréductible', *C. R. Acad. Sci. Paris, Sér. I* **302** (1986), 233–235.

Tricerri, F., and Vanhecke, L., 'Curvature homogeneous Riemannian manifolds', *Ann. Sci. Ecole Norm. Sup.* **22** (1989), 535–554.

Tsankov, Y., 'A characterization of n-dimensional hypersurface in $\mathbb{R}^{n+1}$ with commuting curvature operators', *Banach Center Publ.* **69** (2005), 205–209.

Tsukada, K., 'Curvature homogeneous hypersurfaces immersed in a real space form', *Tôhoku Math. J.* **40** (1988), 221–244.

Tsukada, K., 'Curvature homogeneous spaces whose curvature tensors have large symmetries', *Comment. Math. Univ. Carolin.* **43** (2002), 283–297.

Tsukada, K., and Videv, V., 'A Riemannian pointwise Stanilov manifolds of type (n,k)', *Abstracts of 4th International Conference on Geometry and Applications*, Varna, 1999, 62–63.

Vanhecke, L., 'Curvature homogeneity and related problems', *Proceedings of the Workshop on Recent Topics in Differential Geometry* (Puerto de la Cruz, 1990), 103–122, Informes **32**, Univ. La Laguna, La Laguna, 1991.

Vanhecke, L., and Willmore, T. J., 'Interaction of tubes and spheres', *Math. Ann.* **263** (1983), 31–42.

Verdiani, L., 'Curvature homogeneous metrics of cohomogeneity one on vector spaces', *Riv. Mat. Univ. Parma* **6** (1997), 179–200.

Videv, V., 'A characteristic of the real space forms by a linear operator', Plovdiv University [Paisii Hilendarski], Bulgaria, Scientific Works-Mathematics, **30** (1993), 5–8.

Videv, V., 'Characterization of the four dimensional Einstein Riemannian manifolds using curvature operators', (preprint).

Videv, V., Stoeva, M., Stajkov, Y., 'An application of the skew-symmetric Stanilov curvature operator in the classical ecology', *Mathematics and Education in Mathematics (Bulgarian) (Sunny Beach, 2003)*, Union of Bulgarian Mathematicians, Sofia, (2003), 208–212.

Walker, A. G., 'Canonical form for a Riemannian space with a parallel field of null planes', *Quart. J. Math. Oxford* **2** (1950), 69–79.

Weyl, H., *The Classical Groups*, Princeton University Press, Princeton, 1946.

Wolf, J., 'Homogeneity and bounded isometries in manifolds of negative curvature', *Ill. J. Math.* **8** (1964), 14–18.

Wolf, J., *Spaces of Constant Curvature*, 5th ed., Publish or Perish Inc. (1984).

Yamato, K., 'Algebraic Riemann manifolds', *Nagoya Math. J.* **115** (1989), 87–104.

Zhang, T., *Manifolds with Indefinite Metrics whose Skew-symmetric Curvature Operator has Constant Eigenvalues* (Ph. D. Thesis), Univ. Oregon, (2000).

Zhang, T., 'Applications of algebraic topology in bounding the rank of the skew-symmetric curvature operator', *Topology Appl.* **124** (2002), 9–24.

Index